Agricultural Crop Genetics

Agricultural Crop Genetics

Editor: Harvey Parker

R CALLISTO
REFERENCE

www.callistoreference.com

Callisto Reference,
118-35 Queens Blvd., Suite 400,
Forest Hills, NY 11375, USA

Visit us on the World Wide Web at:
www.callistoreference.com

ISBN: 978-1-63239-802-4 (Hardback)

The publisher's policy is to use permanent paper from mills that operate a sustainable forestry policy. Furthermore, the publisher ensures that the text paper and cover boards used have met acceptable environmental accreditation standards.

Trademark Notice: Registered trademark of products or corporate names are used only for explanation and identification without intent to infringe.

Printed in the United States of America.

Cataloging-in-publication Data

Agricultural crop genetics / edited by Harvey Parker.
p. cm.
Includes bibliographical references and index.
ISBN 978-1-63239-802-4
1. Crops--Genetics. 2. Plant genetics. 3. Crops--Nutrition--Genetic aspects. 4. Agriculture. I. Parker, Harvey.
SB106.G46 A37 2017
631.523 3--dc23

Table of Contents

Preface...IX

Chapter 1 **Effects of Land Management Strategies on the Dispersal Pattern of a Beneficial Arthropod**..1
Chiara Marchi, Liselotte Wesley Andersen, Volker Loeschcke

Chapter 2 **Genetic Diversity and Demographic History of *Cajanus spp.* Illustrated from Genome-Wide SNPs** 8
Rachit K. Saxena, Eric von Wettberg, Hari D. Upadhyaya, Vanessa Sanchez, Serah Songok, Kulbhushan Saxena, Paul Kimurto, Rajeev K. Varshney

Chapter 3 **Genetic Diversity and Ecological Niche Modelling of Wild Barley: Refugia, Large-Scale Post-LGM Range Expansion and Limited Mid-Future Climate Threats?** 17
Joanne Russell, Maarten van Zonneveld, Ian K. Dawson, Allan Booth, Robbie Waugh, Brian Steffenson

Chapter 4 **Exploring Germplasm Diversity to Understand the Domestication Process in *Cicer spp.* using SNP and DArT Markers**..27
Manish Roorkiwal, Eric J. von Wettberg, Hari D. Upadhyaya, Emily Warschefsky, Abhishek Rathore, Rajeev K. Varshney

Chapter 5 **Mapping Genetic Diversity of Cherimoya (*Annona cherimola* Mill.): Application of Spatial Analysis for Conservation and use of Plant Genetic Resources** 37
Maarten van Zonneveld, Xavier Scheldeman, Pilar Escribano, María A. Viruel, Patrick Van Damme, Willman Garcia, César Tapia, José Romero, Manuel Sigueñas, JoséI. Hormaza

Chapter 6 **Impact of Transgenic Wheat with *wheat yellow mosaic virus* Resistance on Microbial Community Diversity and Enzyme Activity in Rhizosphere Soil** 51
Jirong Wu, Mingzheng Yu, Jianhong Xu, Juan Du, Fang Ji, Fei Dong, Xinhai Li, Jianrong Shi

Chapter 7 **Single-Nucleotide Polymorphism Markers from De-Novo Assembly of the Pomegranate Transcriptome Reveal Germplasm Genetic Diversity** 62
Ron Ophir, Amir Sherman, Mor Rubinstein, Ravit Eshed, Michal Sharabi Schwager, Rotem Harel-Beja, Irit Bar-Ya' akov, Doron Holland

Chapter 8 **Unraveling the Complex Trait of Harvest Index with Association Mapping in Rice (*Oryza sativa* L.)** 74
Xiaobai Li, Wengui Yan, Hesham Agrama, Limeng Jia, Aaron Jackson, Karen Moldenhauer, Kathleen Yeater, Anna McClung, Dianxing Wu

Chapter 9   **Geographical Gradient of the *eIF4E* Alleles Conferring Resistance to Potyviruses
in Pea (*Pisum*) Germplasm** 84
Eva Konečná, Dana Šafářová, Milan Navrátil, Pavel Hanáček, Clarice Coyne,
Andrew Flavell, Margarita Vishnyakova, Mike Ambrose, Robert Redden,
Petr Smýkal

Chapter 10   **Allele Distributions at Hybrid Incompatibility Loci Facilitate the Potential for
Gene Flow between Cultivated and Weedy Rice in the US** 95
Stephanie M. Craig, Michael Reagon, Lauren E. Resnick, Ana L. Caicedo

Chapter 11   **Variation in Broccoli Cultivar Phytochemical Content under Organic and
Conventional Management Systems: Implications in Breeding for Nutrition** 106
Erica N. C. Renaud, Edith T. Lammerts van Bueren, James R. Myers,
Maria João Paulo, Fred A. van Eeuwijk, Ning Zhu, John A. Juvik

Chapter 12   **Identification of Suitable Reference Genes for Gene Expression Normalization
in qRT-PCR Analysis in Watermelon** 122
Qiusheng Kong, Jingxian Yuan, Lingyun Gao, Shuang Zhao, Wei Jiang,
Yuan Huang, Zhilong Bie

Chapter 13   **Whole-Genome Quantitative Trait Locus Mapping Reveals Major Role
of Epistasis on Yield of Rice** 133
Anhui Huang, Shizhong Xu, Xiaodong Cai

Chapter 14   **Australian Wild Rice Reveals Pre-Domestication Origin of Polymorphism
Deserts in Rice Genome** 144
Gopala Krishnan S, Daniel L. E. Waters, Robert J. Henry

Chapter 15   **Population Genetic Structure of the Cotton Bollworm *Helicoverpa armigera*
(Hübner) (Lepidoptera: Noctuidae) in India as Inferred from EPIC-PCR
DNA Markers** 150
Gajanan Tryambak Behere, Wee Tek Tay, Derek Alan Russell, Keshav Raj Kranthi,
Philip Batterham

Chapter 16   **High-Throughput Sequencing and Mutagenesis to Accelerate the Domestication
of Microlaena stipoides as a New Food Crop** 164
Frances M. Shapter, Michael Cross, Gary Ablett, Sylvia Malory, Ian H. Chivers,
Graham J. King, Robert J. Henry

Chapter 17   **Diversifying Selection on Flavanone 3-Hydroxylase and Isoflavone Synthase
Genes in Cultivated Soybean and its Wild Progenitors** 173
Hao Cheng, Jiao Wang, Shanshan Chu, Hong Lang Yan, Deyue Yu

Chapter 18   **Influence of Ethnolinguistic Diversity on the Sorghum Genetic Patterns
in Subsistence Farming Systems in Eastern Kenya** 182
Vanesse Labeyrie, Monique Deu, Adeline Barnaud, Caroline Calatayud,
Marylène Buiron, Peterson Wambugu2, Stéphanie Manel, Jean-Christophe
Glaszmann, Christian Leclerc

Chapter 19   **Massive Sorghum Collection Genotyped with SSR Markers to Enhance use
of Global Genetic Resources** 196
Claire Billot, Punna Ramu, Sophie Bouchet, Jacques Chantereau, Monique Deu,
Laetitia Gardes, Jean-Louis Noyer, Jean-François Rami, Ronan Rivallan, Yu Li,
Ping Lu, Tianyu Wang, Rolf T. Folkertsma, Elizabeth Arnaud, Hari D. Upadhyaya,
Jean- Christophe Glaszmann, C. Thomas Hash

Chapter 20   **Managing Potato Biodiversity to Cope with Frost Risk in the High Andes: A Modeling Perspective** 212
Bruno Condori, Robert J. Hijmans, Jean Francois Ledent, Roberto Quiroz

Chapter 21   **Crop Diversity for Yield Increase**.. 223
Chengyun Li, Xiahong He, Shusheng Zhu, Huiping Zhou, Yunyue Wang, Yan Li,
Jing Yang, Jinxiang Fan, Jincheng Yang, Guibin Wang, Yunfu Long, Jiayou Xu,
Yongsheng Tang, Gaohui Zhao, Jianrong Yang, Lin Liu, Yan Sun, Yong Xie,
Haining Wang, Youyong Zhu

Chapter 22   **Nucleotide Polymorphisms and Haplotype Diversity of *RTCS* Gene in China Elite Maize Inbred Lines**... 229
Enying Zhang, Zefeng Yang, Yifan Wang, Yunyun Hu, Xiyun Song, Chenwu Xu

Chapter 23   **Selection Strategies for the Development of Maize Introgression Populations**.................................... 235
Eva Herzog, Karen Christin Falke, Thomas Presterl, Daniela Scheuermann,
Milena Ouzunova, Matthias Frisch

Permissions

List of Contributors

Index

Preface

Agricultural crop genetics is the study of crop heredity and inherited characteristics among agricultural crop. There is an impressive history of crop modification in the past three decades and this book on agricultural crop genetics deals with advanced practices in this area. Recent studies in plant biology have focused on disease-resistant crops, high-yield varieties and conditions of tissue culture. Inter-disciplinary research in fields such as plant genetics, and evolutionary genetics has also contributed much to this field. Some of the diverse topics covered in this book address the varied branches that fall under this category. This brings forth some of the most innovative concepts and elucidates the unexplored aspects of crop genetics. For all readers who all are interested in the field of crop genetics, the case studies included in this book will serve as an excellent guide to develop comprehensive understanding.

The researches compiled throughout the book are authentic and of high quality, combining several disciplines and from very diverse regions from around the world. Drawing on the contributions of many researchers from diverse countries, the book's objective is to provide the readers with the latest achievements in the area of research. This book will surely be a source of knowledge to all interested and researching the field.

In the end, I would like to express my deep sense of gratitude to all the authors for meeting the set deadlines in completing and submitting their research chapters. I would also like to thank the publisher for the support offered to us throughout the course of the book. Finally, I extend my sincere thanks to my family for being a constant source of inspiration and encouragement.

Editor

Effects of Land Management Strategies on the Dispersal Pattern of a Beneficial Arthropod

Chiara Marchi[1]*, Liselotte Wesley Andersen[2], Volker Loeschcke[1]

1 Department of Bioscience, Aarhus University, Aarhus, Denmark, **2** Department of Bioscience, Aarhus University, Rønde, Denmark

Abstract

Several arthropods are known to be highly beneficial to agricultural production. Consequently it is of great relevance to study the importance of land management and land composition for the conservation of beneficial aphid-predator arthropod species in agricultural areas. Therefore our study focusing on the beneficial arthropod *Bembidion lampros* had two main purposes: I) identifying the physical barriers to the species' dispersal in the agricultural landscape, and II) assessing the effect of different land management strategies (i.e. use of pesticides and intensiveness) on the dispersal patterns. The study was conducted using genetic analysis (microsatellite markers) applied to samples from two agricultural areas (in Denmark) with different agricultural intensity. Land management effects on dispersal patterns were investigated with particular focus on: physical barriers, use of pesticide and intensity of cultivation. The results showed that *Bembidion lampros* disperse preferably through hedges rather than fields, which act as physical barriers to gene flow. Moreover the results support the hypothesis that organic fields act as reservoirs for the re-colonization of conventional fields, but only when cultivation intensity is low. These results show the importance of non-cultivated areas and of low intensity organic managed areas within the agricultural landscape as corridors for dispersal (also for a species typically found within fields). Hence, the hypothesis that pesticide use cannot be used as the sole predictor of agriculture's effect on wild species is supported as land structure and agricultural intensity can be just as important.

Editor: Anna-Liisa Laine, University of Helsinki, Finland

Funding: Funded by the Ministry of Food, Agriculture and Fisheries under the Finance and Appropriation Act, Sections 24.33.02.10. Part of the Ph.D. project is funded by AGSoS (Aarhus Graduate School of Science). The funders had no role in study design, data collection and analysis, decision to publish, or preparation of the manuscript.

Competing Interests: The authors have declared that no competing interests exist.

* E-mail: chiarapuspa@gmail.com

Introduction

A number of arthropods, typically living in agricultural areas, have been shown to be highly beneficial to agricultural production, because they act as predators on pest species [1]. These beneficial species are of major importance especially in relation to organic farming, and their abundance and dispersal pattern can be expected to be influenced by land management.

Bembidion lampros, a small polyphagus arthropod (of the Carabidae family) has been shown to be efficient as aphid pest controller in agricultural areas [1–3] and can also be used as indicator for insect abundance and diversity [4]. Effects of landscape structure and land use on *B. lampros* have been widely investigated, but focusing only on the species abundance rather than on the drivers of dispersal in the agricultural landscape.

The typical *B. lampros* life cycle is marked by an overwintering stage where individuals reside in hedges at the field margins and a dispersal stage in early spring where they move into the field [5]. The timing of dispersal is determined by air temperature and occurs in Denmark usually between early March and early April [5]. After reproducing in the field the individuals return in early autumn to the hedges to overwinter [6,7]. *B. lampros* is univoltine [6] and short-lived but it lives mostly more than a year. It is one of the early spring-moving species [5], thus it has probably a relatively larger effect on aphid pest control compared to later-moving species [3]. While its life cycle has been thoroughly

studied, the dispersal routes within the agricultural land are largely unknown. In particular it is not known whether *B. lampros* cross hedgerows, tend to return to the same hedgerow after reproduction or easily cross ploughed fields. Such knowledge however is crucial for understanding the population dynamics of this highly beneficial species [2]. Since this information is difficult to get through mark-recapture studies (given the very high population number and the small size of this species), the use of molecular genetic markers may help to elucidate this part of its ecology. Knowing which factors act as barrier to dispersal can be used to predict the probability of an area to receive immigrants and consequently be re-colonized in case of a sudden drop in abundance (e.g. after pesticides spraying). This knowledge could allow the implementation of management strategies that prevent populations going extinct or suffer severe bottlenecks (which could then lead to local extinction).

Hedgerow types, soil type, presence of pesticides and number of mechanical soil treatments have a clear effect on the abundance of *B. lampros* [8–10]. In particular many commonly used insecticides are shown to be detrimental to this species, both under laboratory and field conditions [11,12]. Even pesticides legally used in organic fields in many European countries (but not in Denmark) have been proven to be harmful [9]. The use of pesticides in the field has led to a reduction in *B. lampros* abundance persisting up to four weeks after spraying; the following return to abundance levels preceding the spraying was ascribed to an effect of re-colonization from untreated nearby areas [13]. This might suggest that fields grown organically can act as genetic reservoirs for the re-colonization of conventional fields.

Several studies have shown the importance of cultivation intensity for the abundance and mobility of *B. lampros*, in particular of low yield cultivation with high weed abundance [14,15]. Therefore the effect of agricultural intensiveness was included in our study by comparing two areas with very different agricultural intensity. The organic fields in the two areas also differed in time since conversion from conventional fields.

The main objectives of our study were :

i) to analyze the population structure and gene flow pattern in the agricultural landscape and thereby identifying possibly dispersal barriers (with particular attention to fields, hedgerows and roads).

ii) to test the hypothesis that organically grown fields act as genetic reservoirs for conventional grown fields and the hypothesis that areas with low intensity agricultural land use have a higher genetic diversity, effective population size and comparatively higher gene flow.

Materials and Methods

Sampling design

We selected two agricultural areas with different intensity of cultivation composed of conventional and organic fields separated by tree-covered hedgerows. Sampling was performed for two consecutive years in early autumn (when individuals return to the hedgerows to overwinter) at sampling sites along the hedgerows.

Individuals were sampled along the hedgerows from two agricultural areas in Denmark, Kalø 56.363° North 9.700° East, and Bjerringbro 56.290° Nord 10.500° East. The two areas represented an intensive (Bjerringbro) and an extensive (Kalø) type of management: Bjerringbro area had a higher average amount of cultivated areas (66%) compared to the national level (61%), while the opposite was true for the Kalø area where only 52% of land is used for agriculture (Beate Strandberg, pers. comm.). In Kalø, the fields have been organically managed for many years (>15 years), longer than in Bjerringbro, and the weed cover in the fields in Kalø was also much higher (Beate Strandberg pers. comm.). The sampling scheme was designed to obtain samples from hedgerows dividing two organic or two conventional fields as well as from hedgerows dividing organic and conventional fields (on both sides). Regarding the pesticide regime the organic fields were never treated with pesticides/herbicides while these were routinely applied to conventional fields. Therefore conventional vs. organic classification defines pesticide regime. Each hedgerow presented four sampling sites 50 meters apart, two on both sites just opposite each other with a width of one meter. A total of 1140 samples were hand-picked during the two seasons (see Table 1 and Fig. 1 for details). Before DNA-extraction all individuals were identified to species morphologically.

DNA extraction and PCR details

Samples were frozen dry upon collection and stored at −20°C. DNA was extracted using a standard CTAB buffer and proteinase-K procedure [16,17]. Two multiplex PCR runs were used to amplify 15 microsatellite markers developed specifically for this species (Marchi *et al.* unpublished) using Qiagen Multiplex enzyme following the manufactories instructions. First multiplex: BL8, BL11, BL19, BL22, BL28, BL28b, BL30, BL34; second multiplex: BL2, BL5, BL6, BL12, BL14, BL24, BL27). Thermal profile, for both multiplex runs, was: 95°C for 15 min; then 35 cycles of 30 s at 94°C, 45 s at 57°C and 60 s at 72°C, with a final extension at 60°C for 30 min. PCR products were analysed using an ABI 3730 automated sequencer and typed using GENEMAPPER version 4.1 (Applied Biosystems).

Genetic diversity and effective population size analysis

Genetic diversity for each sampling site and each year was estimated as allelic richness with FSTAT [18] and expected heterozygosity with GENALEX [19]. Tests for goodness of fit to Hardy-Weinberg expectations and linkage equilibrium were performed in FSTAT. Moreover effective population size (N_e) was estimated with NEESTIMATOR [20] for each sampling site for both years using "heterozygote excess method" [21] and Waples' [22] moments based approach. The differences in genetic diversity and N_e among conventional and organic sampling sites were tested using a Student t-test. The same tests were repeated, grouping the samples with regard to neighboring field's management as well (sampling site in conventional field with neighboring organic field versus sampling site in conventional field with neighboring conventional field and so on for all pairs).

Genetic structure analysis

Population structure was evaluated using a Bayesian based cluster analysis implemented in GENELAND R package [23]. The analyses were performed for Kalø and Bjerringbro data using geographic coordinates as prior and k(max) = 10 (based on five separate runs each with 20,000/100,000 burn-in/sampling iterations).

The population structure was also estimated using pairwise multilocus F_{ST} between sampling sites using ARLEQUIN v. 3.5 [24].

Gene flow analysis

The effect of environmental variables on the pattern of gene flow between sampling sites were assessed using BIMr [25]. BIMr estimates recent gene flow to find the best explanatory factors for the recovered pattern among the given environmental distance matrices using a linear model fit with the gene flow matrix acting as dependent variable (a null model, including none of the specified factors, is always included in the analysis). The variables included in the analysis were: distance, hedgerow of belonging, field of belonging (hedgerow or field nearest to the sampling site), management type (conventional or organic) and road side (only for Bjerringbro). The inclusion of distance between sampling sites was used to account for possible effects of spatial autocorrelation. BIMr analyses were run ten times for each area (using 1,000,000(burn in)/1,000,000(sampling) runs) and the gene flow values were obtained from the best run (which was selected on the basis of $D_{assignment}$ as suggested by Faubet et al. [26]; when more than one run presented the same $D_{assignment}$, posterior probability and the visual analyses of likelihood plot for each run were used to determine the best run). BIMr results were analyzed also on the basis of the alpha value recovered for each factor (or combination of factor): "the sign and the magnitude of the alpha's tell us about the direction and the strength of the environmental factors" [25].

Assignment test

Isolation of the clusters detected by GENELAND was checked using a series of assignment tests conducted in GENECLASS [27]. We performed four self-assignment tests using data from 2008 and 2009, separately, for each area. Then two assignment tests were performed assigning 2009 individuals using 2008 individuals as reference. For all six tests the individuals were divided according to

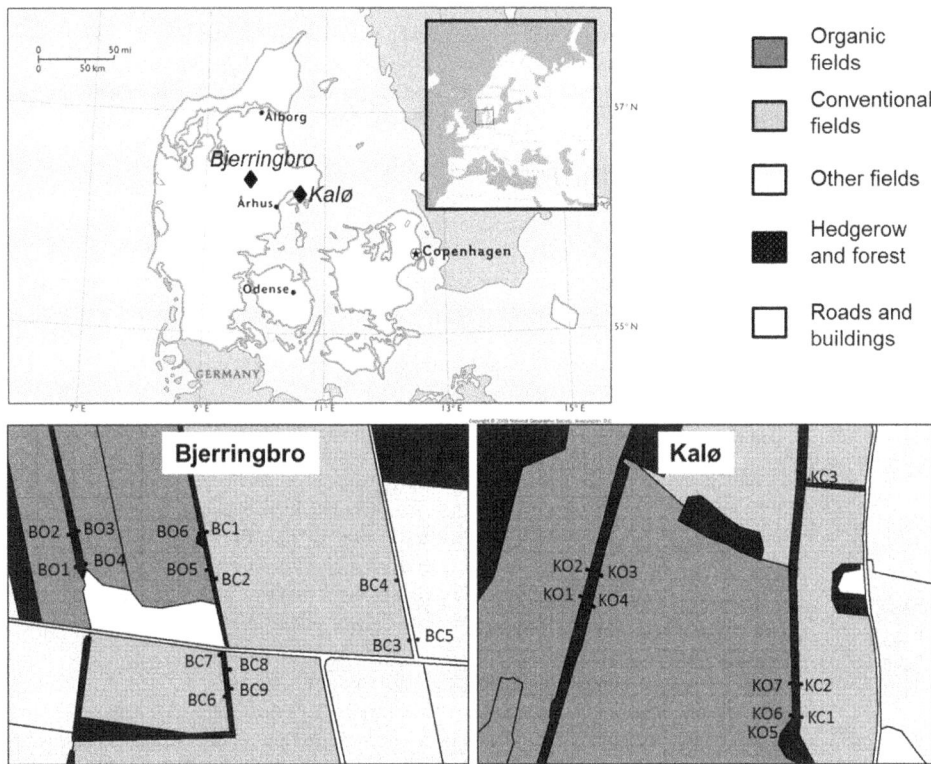

Figure 1. Map of sampling sites. The top map shows the location of the two areas used in this study. The maps at the bottom show the location and name of each sampling site for both areas (B = Bjerringbro, K = Kalø) and management types of the field (C for conventional or O for organic) and ordinal number.

the grouping recovered by the genetic structure analysis in GENELAND. All tests were performed using Rannala and Mountain [28] method; 10,000 simulations with Paetkau et al. [29] method and a type I error rate $P<0.01$.

Results

Genetic diversity and effective population size analysis

No significant deviation from Hardy-Weinberg- or linkage equilibrium was found. The genetic diversity indexes were very similar across sampling sites with different management and across years. The Kalø area showed a significantly higher allelic richness compared to Bjerringbro area (Table S1). Effective population size did not show any difference across area, year or management type (Tables S1 and S2).

Genetic structure analysis

GENELAND analysis performed on samples from both years detected five clusters in Kalø and eight in Bjerringbro. In both cases no migrants were identified and samples from 2008 clustered together with the corresponding ones from 2009. All the sampling sites that did not constitute a cluster of their own were grouped accordingly to the nearest hedgerow (Fig. 2).

The pairwise F_{ST} analyses showed higher F_{ST} values in Bjerringbro than in Kalø; despite the presence of significantly differentiated pairs, no isolated (group of) sampling sites were found (Fig. 3). However, in Kalø there were two more distinct groups: KO1-KO2-KO3-KO4 and KO7-KC1-KC2-KC3 (Fig. 3) which partly agreed with the results from GENELAND (Fig. 2) as the sampling sites in the two groups belonged to the same hedgerow (confirming the effect of fields as genetic barriers).

Gene flow analysis

BIMr analyses of Bjerringbro data converged in nine out of ten cases. "Hedgerow of belonging" was the most important factor in three out of nine runs. Further, in two single runs "road side" and "field of belonging", respectively, were the most important explanatory factors. In other three runs "distance" (two runs) and "distance plus road side" (one run) were most important. When two factors were combined, "road side" and "field of belonging" were the least important ones (lowest absolute value of alpha). One run found the best fit to be the null model. The highest posterior probability and highest alpha were given to "hedgerow of belonging" alone. For the best run, the most important factor was "hedgerow of belonging" and the gene flow was higher for pairs of sampling sites belonging to the same hedgerow (see alpha values in Table 2).

All ten runs regarding Kalø samples converged. In nine out of ten cases "management type" was the most important factor alone (four runs), together with "hedgerow of belonging" (4 runs) or together with" field of belonging" (1 run). In cases where two factors were present, "management type" was the most important one (highest absolute number of alpha). The best run found "management type" alone to be the most important factor (with a posterior probability of 0.582, see Table 3) and, as the correlation coefficient (alpha) was negative, the migration was lowest from conventional to organic sampling sites and highest from organic to conventional ones (see Materials and Methods for details).

In all cases, in both areas, belonging to the same hedgerow was found to increase gene flow (see alpha values for hedgerow in Tables 2 and 3).

Table 1. Sampling details.

	2008	2009	Total	Management
Bjerringbro				
BO1	28	0	28	Organic
BO2	28	0	28	Organic
BO3	24	17	41	Organic
BO4	29	27	56	Organic
BO5	34	28	62	Organic
BO6	20	22	42	Organic
BC1	30	29	59	Conventional
BC2	20	30	50	Conventional
BC3	23	9	32	Conventional
BC4	5	29	34	Conventional
BC5	0	25	25	Conventional
BC6	15	17	32	Conventional
BC7	9	7	16	Conventional
BC8	27	28	55	Conventional
BC9	26	26	52	Conventional
Total	318	294	612	
Kalø				
KO1	18	30	48	Organic
KO2	27	28	55	Organic
KO3	25	30	55	Organic
KO4	27	30	57	Organic
KO5	27	0	27	Organic
KO6	26	30	56	Organic
KO7	29	30	59	Organic
KC1	29	28	57	Conventional
KC2	30	30	60	Conventional
KC3	27	27	54	Conventional
Total	265	263	528	

Number of sampled individuals per sampling site, per year and in total (first to third column). Last column shows the type of management.

Assignment test

The self-assignment tests, performed on the samples from 2008 and 2009 for the two areas, were in good agreement with the population structure results as the percentage of mis-assigned individuals was generally low (Tables S3 and S4). The assignment test (2008 samples as reference for 2009) did not give clear cut results as most individuals could be assigned to more than one cluster. This was probably due to a lack of statistical power.

Discussion

The main objectives of this study were: I) to determine the gene flow and the population structure patterns of *B. lampros* in agricultural areas with respect to landscape features (particularly: fields, hedgerows and roads) and II) to determine the effect of different land management strategies (specifically, pesticide use and agricultural intensiveness) on such patterns.

Genetic diversity and effective population size analysis

Genetic diversity indexes and effective population sizes were similar across all sampling sites both within and across areas (with the exception of allelic richness which was much higher in Kalø). There was also no indication of an effect of management. The reason for this homogeneity might lie in the fact that the populations are not genetically isolated (as proven by the fact that the majority of pairs, in the F_{ST} analysis, did not differ significantly from each other).

Genetic structure analysis

The genetic structure analysis performed in GENELAND showed that fields act as barriers to gene flow. The GENELAND algorithm was chosen as it has been shown to be particularly good at finding genetic boundaries, especially when the population subdivision is recent [30]. Many sampling sites constituted a cluster of their own, while the ones that were grouped, always belonged to the same hedgerow. Management did not seem to affect the population structuring. It should also be noted that samples from the two separate years always clustered together showing that genetic structure is consistent through years.

The patterns recovered by the F_{ST} analysis were very different in the two areas (Fig. 4). The F_{ST} values were higher in Bjerringbro than in Kalø; however, there were very few significantly different pairs in Bjerringbro compared to Kalø. In Kalø the F_{ST} pattern is in good agreement with GENELAND results as sampling sites lying on different field's sides are more differentiated than the ones sharing the same hedgerow. This results suggest a higher gene flow (lower F_{ST} values), but an older isolation of sampling sites (more significantly different pairs) in Kalø compared to Bjerringbro. This difference might be due to a more consistent field structure in Kalø, where field management changed less over the years (Beate Strandberg, pers. comm.).

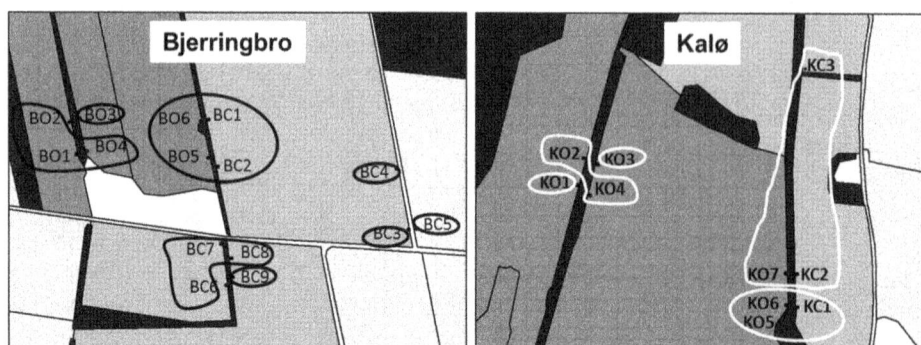

Figure 2. Genetic clusters recovered by GENELAND. The figure shows the genetic clusters for each area with each cluster being included in a solid line (black for Bjerringbro, white for Kalø).

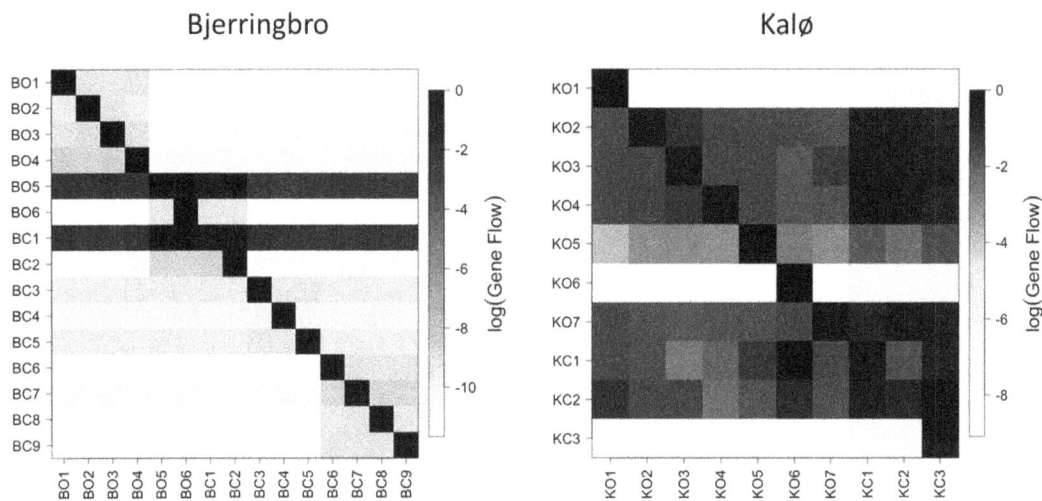

Figure 3. Pairwise F_{ST} values. The pairwise F_{ST} values for each pair of sampling sites are shown on a heatmap for both areas. Pairs that are significantly different, after sequential Bonferroni correction, are represented by a + sign on the heatmap.

Gene flow analysis

The analysis of gene flow confirmed the results of the F_{ST} analysis, showing a much higher gene flow in the Kalø area compared to Bjerringbro (Fig. 3). The barrier effect of fields was also confirmed and hedgerows were shown to be the main dispersal path in both areas (i.e. the highest gene flow occurred between sampling sites belonging to the same hedgerow, Fig. 4). A study by Eyre et al. [8] has shown the importance of vegetation cover and crop type on the activity of *B. lampros* and, despite not dealing directly with dispersal patterns, activity in the field was found not to be correlated with activity in the fields' margin. However, in the Kalø area there was a very clear effect of management on the strength and direction of gene flow. In particular, the highest gene flow went from organic towards conventional fields (Fig. 4 right, last three columns). The fact that this result was not found for Bjerringbro area suggests that the factor that makes a difference is the degree of intensity of the

management and that only a low intensity of cultivation allows organic fields to act as genetic reservoirs. The positive effect of low-intensity cultivation has been shown for many wild species such as: corn buntings [31] and farmland birds in general [32], European hare [33] and for a large number of arthropods [34]. Together with the previously mentioned studies, our work confirms, from a genetic point of view, the importance of low-intensity cultivations for the conservation of biodiversity. Moreover we show that, while organic fields can act as reservoirs for the re-colonization of conventional fields, the absence of pesticide use is not enough for a field to become a fauna-source for more managed areas. Therefore, our study discloses that (at least for this species) pesticide use and agricultural intensity are interdependent variables that should not be evaluated separately. It is however not possible to infer from these results the long-term effect of pesticide use and agricultural intensity due to the limited time-scale of the study (two years).

Table 2. Results of gene flow analyses (Bjerringbro).

Most significant factors	Number of runs (total of 9 runs)	Posterior probability of the model	Alpha of the corresponding model
Hedgerow	*3*	*0.399*	*1.83*
		0.182	0.986
		0.352	1.51
Hedgerow + Road Side	1	0.242	Hedgerow:1.18
			Road Side:−0.9
Hedgerow + Field	1	0.191	Hedgerow: 0.996
			Field:−0.71
Distance	2	0.271	−1.58
		0.166	0.407
Distance + Road Side	1	0.103	Distance:−0.746
			Road Side:−0.477
Null model	1	0.155	1.35

For each of the 9 runs performed with BIMR the most significant factor(s), their posterior probability and the alpha value for that factor(s) are reported. The results from the best run (see Materials and methods for details) are shown in italics.

Table 3. Results of gene flow analyses (Kalø).

Most significant factors	Number of runs (total of 10 runs)	Posterior probability of the model	Alpha of the corresponding model	
Management	4	*0.582*	*−1.58*	
		0.441	−1.23	
		0.368	−2.36	
		0.497	−2.06	
Hedgerow + Management	4		Hedgerow	Management
		0.490	0.497	−2.44
		0.310	0.571	−1.58
		0.359	0.625	−1.07
		0.267	0.732	−1.9
Field + Management	1	0.603	Field: −0.686	
			Management: −1.89	
Null model	1	0.300	0.339	

For each of the 10 runs performed with BIMR the most significant factor(s), their posterior probability and the alpha value for that factor(s) are reported. The results from the best run (see Materials and methods for details) are shown in italics.

Assignment tests

The results of self-assignment tests showed a good agreement with the genetic structure results as the number of mis-assigned individuals was generally low for each cluster. The results from Bjerringbro gene flow analyses showed that BO5 and BC1 are the most important source of emigrants; this is in accordance with the assignment tests from 2008 in which a lot of individuals from other clusters were assigned to the cluster containing these two populations. The results from Kalø for 2009 are also in agreement with the gene flow results as a lot of samples from other clusters were assigned to the clusters containing the conventional fields.

Conclusions

All the presented results suggest that tree-covered hedgerows are the preferred dispersal route used by *B. lampros*. On the other hand, fields represent a barrier to gene flow, suggesting a tendency for individuals to return to the same hedgerow. However, low-intensity organic fields can be used for dispersal, probably due to a lower death rate caused by mechanical treatments and a higher concentration of aphids. This finding confirms the hypothesis of Huusela-Veistola [13] which suggested that the increase in *B. lampros* abundance following a sharp decrease after spraying was due to immigration from neighboring organic fields. Going beyond this, our results also confirm the importance of intensity of cultivation, and in particular of a low yield cultivation with high weed abundance [14,15]. Future longer term and geographically broader studies might be able to determine with more accuracy the extent of the effect of each single variable.

Given the importance of *B. lampros* as a pest control agent, the results of this study can be important for the development of management strategies that maximize its beneficial effects [2]. Moreover, the fact that tree-covered hedgerows and low-intensity cultivation were found to be fundamental for the gene flow in agricultural areas strengthen the importance of including land-

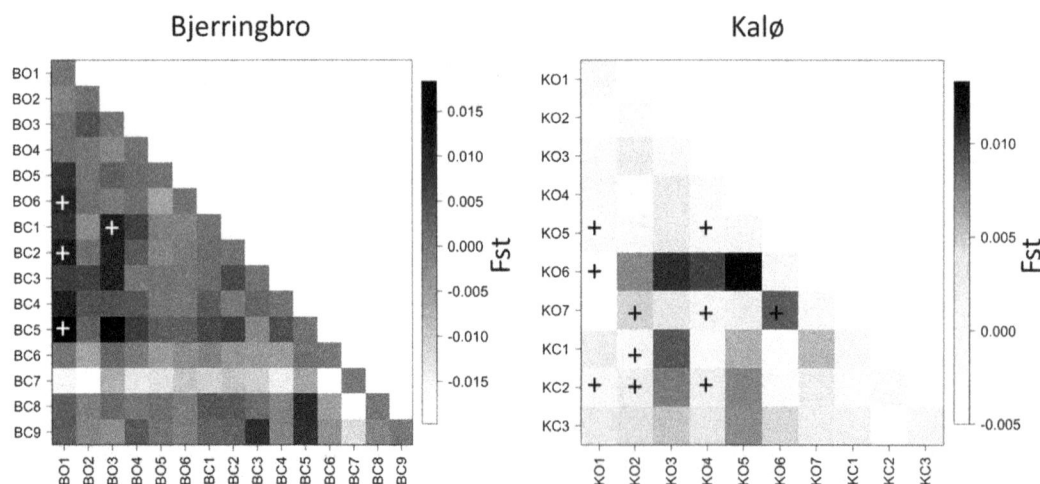

Figure 4. Gene flow values. The $\log_{10}$ of the gene flow values are shown on a heatmap for both areas. The populations of origin are shown on the vertical axis.

scape features and cultivation intensity as parameters for the evaluation of land management strategies.

Supporting Information

Table S1   Genetic diversity indexes for each sampling site and year. First column represent sampling site, (BO = Bjerringbro organic, BC = Bjerringbro conventional, KO = Kalø organic, KC = Kalø conventional) second column the year of sampling, third to fifth columns are the expected heterozigosity (H_e) allelic richness (Ar) and effective population size respectively (N_e) (see Materials and methods for details about their calculation).

Table S2   Effective population sizes for all sampling sites calculated with moment based method. BO = Bjerringbro organic, BC = Bjerringbro conventional, KO = Kalø organic, KC = Kalø conventional.

Table S3   Results of self-assignment tests for Bjerringbro. C1: BO1, BO2, BO4; C2: BO3; C3: BO5, BO6, BC1, BC2; C4: BC3; C5: BC4; C6: BC5; C7: BC6, BC7, BC8; C8: BC9.

Table S4   Results of self-assignment tests for Kalø. C1: KO; C2: KO2, BO4; C3: KO3; C4: KO5, KO6, KC1; C5: KO7, KC2, KC3.

Acknowledgments

We would like to thank all the people involved in the REFUGIA project for helpful discussions. Special thanks go to Zdenek Gavor, who performed the morphological identification of the beetle species.

Author Contributions

Conceived and designed the experiments: CM LWA VL. Performed the experiments: CM LWA. Analyzed the data: CM LWA VL. Contributed reagents/materials/analysis tools: CM LWA VL. Wrote the paper: CM LWA VL.

References

1. Edwards CA, Sunderland K, George K (1979) Studies on polyphagous predators of cereal aphids. Journal of Applied Ecology: 811–823.
2. Landis DA, Wratten SD, Gurr GM (2000) Habitat management to conserve natural enemies of arthropod pests in agriculture. Annual Review of Entomology 45: 175–201.
3. Chiverton PA (1986) Predator density manipulation and its effects on populations of Rhopalosiphumpadi (Horn.: Aphidiae) in spring barley. Annals of Applied Biology 109: 49–60.
4. Cameron KH, Leather SR (2012) How good are carabid beetles (Coleoptera, Carabidae) as indicators of invertebrate abundance and order richness? Biodiversity and Conservation 21: 763–779.
5. Petersen MK (1999) The timing of dispersal of the predatory beetles *Bembidion lampros* and *Tachyporus hypnorum* from hibernating sites into arable fields. Entomologia Experimentalis et Applicata 90: 221–224.
6. Wallin H (1989) Habitat selection, reproduction and survival of two small carabid species on arable land: a comparison between *Trechus secalis* and *Bembidion lampros*. Ecography 12: 193–200.
7. Pederson M, Pedersen L, Abildgaard K (1990) Annual and diurnal activity of some Tachyporus species (Coleoptera: Staphilinidae) in two spring barley fields and a hedge. Pedobiologia 34: 367–378.
8. Eyre MD, Labanowska-Bury D, Avayanos JG, White R, Leifert C (2009) Ground beetles (Coleoptera, Carabidae) in an intensively managed vegetable crop landscape in eastern England. Agriculture, Ecosystems & Environment 131: 340–346.
9. Jansen JP, Defrance T, Warnier AM (2010) Effects of organic-farming-compatible insecticides on four aphid natural enemy species. Pest management science 66: 650–656.
10. Purvis G, Fadl A (2002) The influence of cropping rotations and soil cultivation practice on the population ecology of carabids (Coleoptera: Carabidae) in arable land. Pedobiologia 46: 452–474.
11. Hassan S, Bigler F, Bogenschütz H, Boller E, Brun J, et al. (1991) Results of the fifth joint pesticide testing programme carried out by the IOBC/WPRS-Working Group "Pesticides and Beneficial Organisms". BioControl 36: 55–67.
12. Hassan S, Bigler F, Bogenschütz H, Boller E, Brun J, et al. (1988) Results of the fourth joint pesticide testing programme carried out by the IOBC/WPRS-Working Group "Pesticides and Beneficial Organisms". Journal of Applied Entomology 105: 321–329.
13. Huusela-Veistola E (1996) Effects of pesticide use and cultivation techniques on ground beetles (Col., Carabidae) in cereal fields. Annales Zoologici Fennici 33: 197–206.
14. Honek A, Jarosik V (2000) The role of crop density, seed and aphid presence in diversification of field communities of Carabidae (Coleoptera). European Journal of Entomology 97: 517–526.
15. Purvis G, Curry J (1984) The influence of weeds and farmyard manure on the activity of Carabidae and other ground-dwelling arthropods in a sugar beet crop. Journal of Applied Ecology: 271–283.
16. Milligan B, Hoelzel A (1992) Plant DNA isolation. Molecular genetic analysis of populations: a practical approach: 59–88.
17. Andersen LW, Fog K, Damgaard C (2004) Habitat fragmentation causes bottlenecks and inbreeding in the European tree frog (*Hyla arborea*). Proceedings of the Royal Society of London, Series B: Biological Sciences 271: 1293–1302.
18. Goudet J (1995) FSTAT (version 1.2): a computer program to calculate F-statistics. Journal of Heredity 86: 485.
19. Peakall R, Smouse PE (2006) GENALEX 6: genetic analysis in Excel. Population genetic software for teaching and research. Molecular Ecology Notes 6: 288–295.
20. Ovenden JR, Peel D, Street R, Courtney AJ, Hoyle SD, et al. (2007) The genetic effective and adult census size of an Australian population of tiger prawns *(Penaeus esculentus)*. Molecular Ecology 16: 127–138.
21. Pudovkin A, Zaykin D, Hedgecock D (1996) On the potential for estimating the effective number of breeders from heterozygote-excess in progeny. Genetics 144: 383–387.
22. Waples RS (1989) A generalized approach for estimating effective population size from temporal changes in allele frequency. Genetics 121: 379–391.
23. Guillot G, Mortier F, Estoup A (2005) GENELAND: a computer package for landscape genetics. Molecular Ecology Notes 5: 712–715.
24. Excoffier L, Lischer HEL (2010) Arlequin suite ver 3.5: a new series of programs to perform population genetics analyses under Linux and Windows. Molecular Ecology Resources 10: 564–567.
25. Faubet P, Gaggiotti OE (2008) A new Bayesian method to identify the environmental factors that influence recent migration. Genetics 178: 1491–1504.
26. Faubet P, Waples RS, Gaggiotti OE (2007) Evaluating the performance of a multilocus Bayesian method for the estimation of migration rates. Molecular Ecology 16: 1149–1166.
27. Piry S, Alapetite A, Cornuet JM, Paetkau D, Baudouin L, et al. (2004) GENECLASS2: a software for genetic assignment and first-generation migrant detection. Journal of Heredity 95: 536–539.
28. Rannala B, Mountain JL (1997) Detecting immigration by using multilocus genotypes. Proceedings of the National Academy of Sciences USA 94: 9197.
29. Paetkau D, Slade R, Burden M, Estoup A (2004) Genetic assignment methods for the direct, real-time estimation of migration rate: a simulation-based exploration of accuracy and power. Molecular Ecology 13: 55–65.
30. Safner T, Miller MP, McRae BH, Fortin MJ, Manel S (2011) Comparison of bayesian clustering and edge detection methods for inferring boundaries in landscape genetics. International Journal of Molecular Sciences 12: 865–889.
31. Brickle NW, Harper DGC, Aebischer NJ, Cockayne SH (2000) Effects of agricultural intensification on the breeding success of corn buntings *Miliaria calandra*. Journal of Applied Ecology 37: 742–755.
32. Smith HG, Danhardt J, Lindstrom A, Rundlof M (2010) Consequences of organic farming and landscape heterogeneity for species richness and abundance of farmland birds. Oecologia 162: 1071–1079.
33. Smith RK, Vaughan Jennings N, Harris S (2005) A quantitative analysis of the abundance and demography of European hares *Lepus europaeus* in relation to habitat type, intensity of agriculture and climate. Mammal Review 35: 1–24.
34. Hendrickx F, Maelfait J-P, Van Wingerden W, Schweiger O, Speelmans M, et al. (2007) How landscape structure, land-use intensity and habitat diversity affect components of total arthropod diversity in agricultural landscapes. Journal of Applied Ecology 44: 340–351.

Genetic Diversity and Demographic History of *Cajanus spp.* Illustrated from Genome-Wide SNPs

Rachit K. Saxena[1], Eric von Wettberg[2,3], Hari D. Upadhyaya[1], Vanessa Sanchez[4], Serah Songok[1,5], Kulbhushan Saxena[1], Paul Kimurto[5], Rajeev K. Varshney[1]*

1 International Crops Research Institute for the Semi-Arid Tropics (ICRISAT), Hyderabad, Andhra Pradesh, India, **2** Department of Biological Sciences, Florida International University, Miami, Florida, United States of America, **3** Fairchild Tropical Botanic Garden, Kushlan Institute for Tropical Science, Miami, Florida, United States of America, **4** Florida International University, Department of Earth and Environment, Miami, Florida, United States of America, **5** Egerton University, Egerton, Kenya

Abstract

Understanding genetic structure of *Cajanus* spp. is essential for achieving genetic improvement by quantitative trait loci (QTL) mapping or association studies and use of selected markers through genomic assisted breeding and genomic selection. After developing a comprehensive set of 1,616 single nucleotide polymorphism (SNPs) and their conversion into cost effective KASPar assays for pigeonpea (*Cajanus cajan*), we studied levels of genetic variability both within and between diverse set of *Cajanus* lines including 56 breeding lines, 21 landraces and 107 accessions from 18 wild species. These results revealed a high frequency of polymorphic SNPs and relatively high level of cross-species transferability. Indeed, 75.8% of successful SNP assays revealed polymorphism, and more than 95% of these assays could be successfully transferred to related wild species. To show regional patterns of variation, we used STRUCTURE and Analysis of Molecular Variance (AMOVA) to partition variance among hierarchical sets of landraces and wild species at either the continental scale or within India. STRUCTURE separated most of the domesticated germplasm from wild ecotypes, and separates Australian and Asian wild species as has been found previously. Among Indian regions and states within regions, we found 36% of the variation between regions, and 64% within landraces or wilds within states. The highest level of polymorphism in wild relatives and landraces was found in Madhya Pradesh and Andhra Pradesh provinces of India representing the centre of origin and domestication of pigeonpea respectively.

Editor: Manoj Prasad, National Institute of Plant Genome Research, India

Funding: We thank United States Agency for International Development (USAID)- India Mission and Department of Agriculture and Co-opeartion, Ministry of Agriculture, Government of India for financial support for the research work to RKV and support from Florida International University and Fairchild Tropical Botanic Garden to EvW. VS is thankful for financial support from USDA-NIFA-NNF 2011-38420-20053. The funders had no role in study design, data collection and analysis, decision to publish, or preparation of the manuscript. This work has been undertaken as part of the CGIAR Research Program on Grain Legumes. ICRISAT is a member of CGIAR Consortium.

Competing Interests: The authors have declared that no competing interests exist.

* E-mail: r.k.varshney@cgiar.org

Introduction

Understanding the germplasm diversity and relationships among breeding material is critical to crop improvement. Wild relatives of crops are crucial reservoirs of natural diversity, often possessing abiotic stress tolerance, disease resistance, and other characters that are absent or inadequate in breeding material. Natural selection, domestication and centuries long breeding practices for desirable traits have resulted in a loss of genetic diversity in most annual crop species [1–5] and this seems to be more severe in self-pollinated or partially out crossing species such as chickpea (*Cicer arietinum*) [6] and pigeonpea (*Cajanus cajan*) [7–9]. Wild relatives and landraces are the best source for increasing diversity in the breeding material as they can be crossed, albeit sometimes with some difficulty, into cultivated forms [2,10]. There are secondary and tertiary gene pools which can contribute to crop improvement, but may consist of several closely related species-complexes [11,12] and may require extensive work to cross into the cultivated gene pool. In many cases we know very little about the ecology and population biology of these taxa in their natural habitats, and species delineation may be rudimentary for most

crop wild relatives. Characterization of these resources is critical, as it can identify regions of diversity, and suggest areas where greater collections would be helpful.

Levels of genetic variation present in different wild relatives of a crop may vary due to different distributions and evolutionary histories. In species complexes related to crops, some clades may have colonized new areas relatively recently, such as since the last glaciation, and may have undergone colonization bottlenecks in that process [9,13,14]. These processes are poorly understood in most crop wild relatives, but may have a significant impact on the value of wild relatives for breeding programs. We can improve our understanding of the relationship of wild species to cultivated forms by localizing the region of domestication, even in cases where the wild progenitor is clear. If the wild progenitor varies spatially, the crop may most closely resemble the wild populations from a particular region, and may show evidence of multiple regions of domestication [15]. However, the signal of regional contribution to domesticated material depends on the scale of sampling and the pace and intensity of domestication [16,17]. Spatial variation in wild relatives also may serve as a bridge for introgression, allowing more distant relatives to be crossed into an intermediate that is compatible with the cultivated form. Finally, variation in wild relatives may also give us insight into locally adaptive variants in wild species that can be harnessed to provide local adaptation to a crop [18]. Archaeological evidence, high diversity of wild species and cultural usage have supported India as the domestication centre of pigeonpea [19,20]. This evidence is further supported by recent molecular studies that are providing insights in to pigeonpea domestication [9].

Cultivated pigeonpea suffers from low levels of genetic diversity [21] and existing genetic diversity in wild relatives has received relative little attention or limited systematic use [22]. In order to broaden the genetic diversity in the cultivated gene pool, it is imperative to understand the genetic diversity present in wild relatives in a systematic manner with the genome wide markers. In the past a number of marker systems such as random amplified polymorphic DNA (RAPD) [23], diversity array technology markers (DArT) [24] and simple sequence repeats (SSRs) [21] have been used for detecting genetic diversity in the cultivated gene pool and limited number of wild relatives. Single nucleotide polymorphisms (SNPs) are now markers of choice for various genome wide analysis due to their higher levels of polymorphism, accuracy and automated genotyping methods [25]. A number of high-throughput SNP genotyping platforms are available for the community to make SNP genotyping cost-effective such as BeadXpress and GoldenGate assays from Illumina Inc. Many of these platforms have been developed and used in several crop species like barley [26], wheat [27], maize [28] oilseed rape [29], soybean [30], cowpea [31] and pea [32]. Such platforms, however, not found cost-effective when a variable number of SNPs are required for a number of applications in the same species with a variable size of genotypes. In such cases, **C**ompetitive **A**llele **S**pecific **P**CR (KASPar) assay from KBiosciences (www. kbioscience.co.uk) seems to be an effective marker assay. Because of the importance of KASPar assays in SNP genotyping more samples with a few SNPs, they have been developed in wheat [33], common bean [34], chickpea [35], pigeonpea [8] and recently in peanut [36].

This study reports the genetic diversity and insights in to *Cajanaus* origin using a broad panel of 184 genotypes representing 18 *Cajanus* species across the primary (77), secondary (69) and tertiary gene pools (38), as well as cultivated germplasm from three continents (Figure 1) representing a range of forms from landraces to elite breeding materials using 1,616 SNP markers through KASPar genotyping platform.

Methods

Germplasm and DNA isolation

A total of 184 accessions representing 18 *Cajanus* species were selected from >13,000 *Cajanus* accessions deposited in GeneBank and parental lines of mapping populations (Table S1). Total DNA was isolated from two to three young leaves following a standard DNA isolation protocol [37]. The DNA quantity for each sample was assessed on 0.8% agarose gel.

Single nucleotide polymorphism and KASPar genotyping

SNPs were identified by using next generation sequencing (Illumina GA IIx) technology on 12 parental genotypes of mapping populations [8]. In brief a total of 128.9 million, 36 bp short single end reads were generated from these genotypes. Subsequently SNPs were identified by aligning of sequence reads generated from each of the counter genotypes against the reference assembly, i.e pigeonpea transcriptome assembly that was developed by Kudapa et al. [38]. High quality SNPs were

selected for **C**ompetitive **A**llele **S**pecific **P**CR (KASPar) assay from KBiosciences assay and pigeonpea specific assays were developed as described in Saxena et al. [8].

Data analysis

To assess genetic diversity within groups formed on the basis of biological status (passport data) and geographical origin, we used Genalex 6.3 [39] to estimate observed heterozygosity (Ho), expected heterozygosity (He), fixation index (Fst), and % polymorphism. We subdivided the germplasm several ways: as primary, secondary and tertiary gene pools; as wild species, landraces, and breeding lines, and geographically by continent, country, and within India, by region and state. Based on these categories, we hierarchically analyzed variation with an Analysis of Molecular Variance (AMOVA), implemented in Genalex 6.3. We assessed spatial variation in the groups of germplasm by calculating spatial autocorrelation, implemented in Genalex 6.3. In a complementary analysis, SNPs having mapping positions were used to assess gene diversity according to 11 linkage groups [8] in wild species, landraces, breeding lines and across the germplasm by using PowerMarker software (http://statgen.ncsu. edu/powermarker/). The polymorphism information content or PIC values for developed makers across 184 accessions were calculated by using PowerMarker software (http://statgen.ncsu. edu/powermarker/).

As our analysis of the germplasm depends on the accuracy of the passport data, we verified the groupings by STRUCTURE analyses [40]. We did this with the primary, secondary, and tertiary gene pools, and with the Indian landraces and wilds in two separate STRUCTURE analyses. For both sets of analyses, we ran STRUCTURE on our full dataset of 1,616 SNPs without mapping information, using an admixture model and the default settings. We used Structure harvester [41] and the Evanno method [42] to determine the most likely number of populations (k) present in a sample. To cluster the genetic variation, we also performed a principal component analysis in Genalex 6.3 [39]. Pairwise relatedness was calculated as genetic distance with Genalex 6.3 [39]. The matrix of genetic distances was used to create a neighbour-joining tree with Mega 5.05[43].

Results

SNP marker polymorphism

A total of 1,616 SNPs were used for polymorphism screening on 184 *Cajanus* accessions representing cultivated *C. cajan* (77 accessions) and its wild relatives (107 accessions) (Table S1). The wild accessions represent 18 wild relative species taxonomically placed in gene pool II (GP II) and gene pool III (GP III). The cultivated accessions include elite cultivars and landraces. All the sampled accessions in this study representing widespread geographical regions, ranging from Africa, Asia, Latin America and Australia (Figure 1). From entire set of SNPs we used, 1,615 and 1,504 could be amplified in GP II and GP III respectively (Table 1). A total of 1,226 markers from the set of 1,616 markers were found to be polymorphic across 184 *Cajanus* accessions (Table S2). The polymorphic information content (PIC) for the 1,226 markers ranged from 0.02 to 0.50, with an average of 0.16 for all examined accessions (Figure 2). In the case of cultivated accessions 210 markers were found polymorphic, whereas 1,016 SNPs were polymorphic among the wild accessions.

Genetic diversity in *Cajanus*

SNP genotyping data obtained for all polymorphic markers on 184 accessions were used to assess the genetic diversity harboured

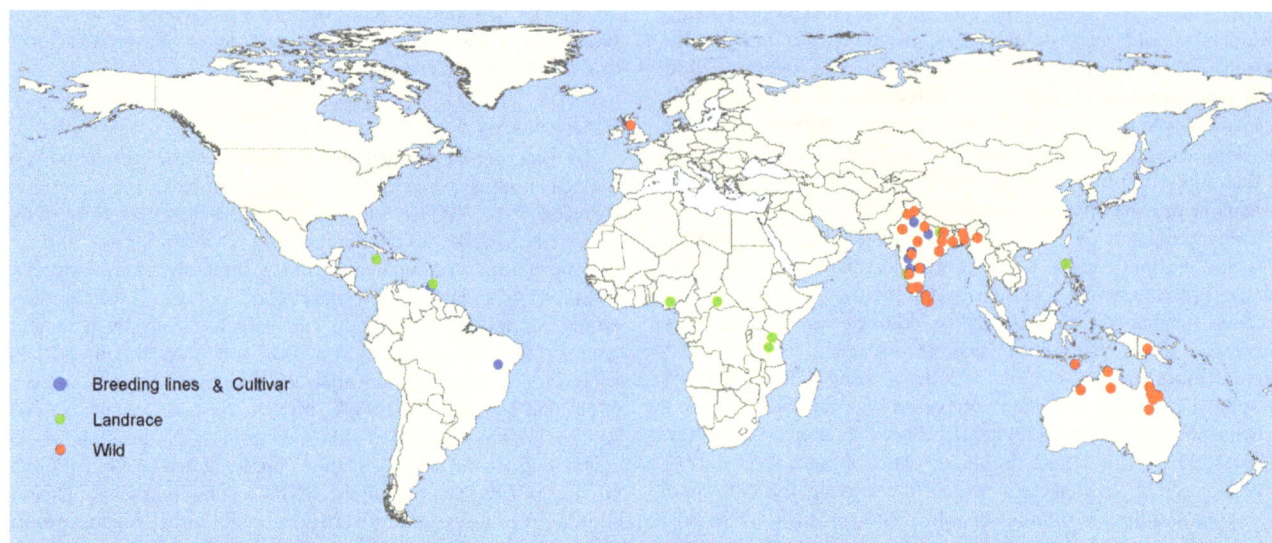

Figure 1. Geographical distribution of the collection sites for cultivated and wild *Cajanus* accessions.

within the germplasm. The average gene diversity across the 56 breeding lines was lowest (0.01) followed by 21 landraces (0.02). In wild relatives, 69 accessions from GP II have a higher (0.26) gene diversity as compare to 38 accessions from GP III (0.2) (Table 1). By using the SNP genotyping data, gene diversity, as measured by expected heterozygosity (He), ranged from 0.022 in GP I to 0.214 in GP II (Table 2). In the case of breeding lines, landraces and wild relatives expected heterozygosity (He) was estimated as 0.02, 0.027 and 0.2 respectively (Table S3). To estimate the gene diversity at the level of linkage groups (LGs) across the breeding lines, landraces and wild relatives, 875 mapped markers were used. Across 184 accessions the average gene diversity of these mapped markers was 0.35, whereas it was highest in wild relatives (0.26) followed by landraces (0.02) and breeding lines (0.01) (Table 3). While comparing average gene diversity of the mapped markers on the individual LGs, all the LGs showed loss of gene diversity during the course of domestication (wild relatives to landraces) and selection (landraces to breeding lines) (Figure S1). Interestingly, average gene diversity in CcLG06 was the most differentiated among the wild relatives (0.263), landraces (0.003) and breeding lines (0.00) (Table 3).

Relatedness of cultivated and wild species

Because breeders often use a limited range of material, assessing the relatedness of cultivars in germplasm collections can assist with selecting distantly related lines for breeding programs. For this purpose, we present pairwise relatedness through neighbour-joining trees based on pairwise genetic distances (Figure 3). All 184 accessions were classified into three main clusters: cluster 'I', cluster 'II' and cluster 'III'. Cluster 'I' contained 18 cultivated accessions; cluster 'II' contained 20 cultivated accessions while the remaining 146 cultivated and wild accessions were grouped in cluster 'III'. Under each of the main clusters, accessions were grouped further into sub-clusters. It is interesting to note that cluster 'I' and cluster 'II' were made up solely of cultivated accessions, whereas, in cluster 'III' cultivated accessions were grouped together with the wild relatives. For instance, 13 breeding lines and 5 landraces were grouped in cluster 'I' and 18 breeding lines and 2 landraces were grouped in cluster 'II'. In the case of cluster 'III' 107 wild accessions representing 18 wild relative species were grouped together with the 25 breeding lines and 14 landraces. Accessions from 18 wild relative species were found scattered and no clear grouping could be detected in cluster 'III'. In order to check the effect of possible cross pollination on varietal maintenance, SNP genotyping data was also used to detect the heterogeneity present in two leading varieties (ICPL 87119 or

Table 1. SNP marker polymorphism status across cultivated and wild *Cajanus* accessions.

	Cultivated (77)		Wild (107)	
	Breeding lines (56)	Landraces (21)	Gene pool II (69)	Gene pool III (38)
No. of markers used	1616	1616	1616	1616
No. of markers amplified	1616	1616	1615	1504
No. of polymorphic markers	134	210	1181	722
Average PIC value of polymorphic markers	0.19	0.17	0.24	0.24
Average gene diversity of polymorphic markers	0.24	0.2	0.29	0.3
Average diversity across	0.01	0.02	0.26	0.2

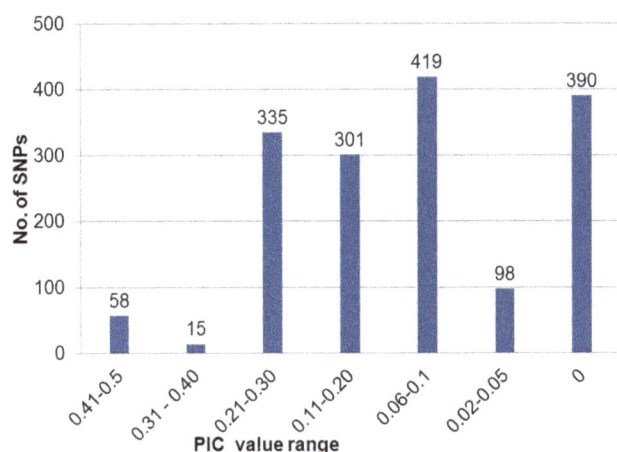

Figure 2. Polymorphism information content (PIC) value range of 1,616 PKAM screened over 184 *Cajanus* accessions.

ASHA and ICP 8863 or Maruthi). It was anticipated that there could be variation from plant to plant at the genome level, and hence samples were collected from two different sources (ICRISAT-Patancheru and UAS-Bangalore). No significant differences were identified and both samples from this variety grouped in close proximity in cluster 'III'.

We used STRUCTURE to assess the clustering of cultivated and wild genotypes. STRUCTURE divided the wild and cultivated accessions into two groups, representing cultivated and wild gene pools. Several wild lines did show evidence of admixture with cultivated material. To further assess relationships among accessions we separated the accessions into gene pools. When the germplasm in the primary, secondary, and tertiary genepools was analyzed with STRUCTURE, the three gene pools were classified into just two groups, with the primary gene pool distinguished from the secondary and tertiary gene pools (Figure S2a). We also conducted a Principal Coordinate Analysis (PCoA) to distinguish among the primary, secondary, and tertiary gene pools. Accessions representing GP I clustered in a tight group, whereas accessions from GP II and GP III were scattered about. We found substantial overlap among the gene pools. The first two discriminant axes accounted for 76% and 10% of the genetic variation, respectively (Figure S2b).

Regional patterns of variation

In order to find the regional patterns of variation, landraces and wild accessions were classified by their continent, country and province of origin. At the continental scale, accessions were grouped as Meso America, South Asia, sub-Saharan Africa and Australia-Oceania. The highest per cent polymorphism was identified within landraces (79.76%) and wild relatives (96.60%) present in South Asia. Variation measured by expected heterozygosity (0.48 in wilds and 0.38 in landraces) was highest in South Asia (Table S4). Analysis of Molecular Variance (AMOVA) was used to partition variance among hierarchical sets of landraces and wild species. At the continental scale 69% of the variation segregated between landraces and wilds, and 31% within continents, with no variation among continents (Figure S3).

To further asses the regional diversity at the country scale, accessions were grouped as India, Tanzania, Myanmar, Sri Lanka, Australia and Papua New Guinea. The highest level of polymorphism was observed within wild relatives (96.47%) and landraces (76.49%) present in India (Table S5). Similarly expected heterozygosity was found to be highest in wild relatives (0.48) and landraces (0.38) originating in India. These results verify the previous postulations of India being the centre of origin and primary domestication centre [9,19,20]. Genetic polymorphism was highest in wild and landrace groups of Indian origin, although surprising amounts of landrace variation were present in some of the landrace material from Meso America and sub-Saharan Africa as well. Further attempts were made to narrow down and mark the centre of origin and domestication within India; accessions from India were grouped according to province (Table S6). Genetic polymorphism within wild relatives were found to be highest in Andhra Pradesh (93.50%) followed by Madhya Pradesh (92.45%) as compare to other provinces in India. We also found the highest polymorphism in Andhra Pradesh (75.43%) followed by Madhya Pradesh (75.31%). The remainder of the South Indian landraces had greater diversity than landraces from other regions of India. Among Indian regions and province within regions, we found 36% of the variation between regions, and 64% within landraces or wilds within provinces, with no variation among provinces (Figure 4). A further principal coordinate analysis of the Indian landrace and wild material did not cluster genotypes by region or wild/landrace (Figure 4). To investigate genetic relationships among accessions and to search for evidence of genetic admixture between landraces and wild accessions, we performed a further STRUCTURE analysis on material from different provinces of

Table 2. Diversity in three different gene pools (GP) of pigeonpea germplasm.

GP	Sample size		N	Na	Ne	I	Ho	He	UHe	F	%P
GP I	77	Mean	76.277	1.154	1.037	0.036	0.01	0.022	0.022	0.679	15.41%
		SE	0.042	0.009	0.004	0.003	0.002	0.002	0.002	0.013	
GP II	69	Mean	43.39	1.730	1.342	0.333	0.013	0.214	0.217	0.928	73.08%
		SE	0.639	0.011	0.008	0.006	0.001	0.004	0.004	0.005	
GP III	38	Mean	22	1.377	1.146	0.206	0.006	0.133	0.136	0.935	44.68%
		SE	0.396	0.015	0.011	0.006	0.001	0.004	0.004	0.005	

Na = No. of Different Alleles, Ne = No. of Effective Alleles = 1 / (Sum pi^2), I = Shannon's Information Index = −1* Sum (pi * Ln (pi)), Ho = Observed Heterozygosity = No. of Hets / N, He = Expected Heterozygosity = 1 - Sum pi^2, UHe = Unbiased Expected Heterozygosity = (2N / (2N-1)) * He, F = Fixation Index = (He − Ho) / He = 1 − (Ho / He) (Where pi is the frequency of the ith allele for the population & Sum pi^2 is the sum of the squared population allele frequencies), %P = percent of loci polymorphic.

Table 3. Gene diversity across breeding lines, landraces and wild relatives estimated by the 875 mapped PKAM.

Linkage group	Gene diversity			
	Breeding lines	Landraces	Wild	Across
CcLG01	0.007	0.012	0.253	0.320
CcLG02	0.007	0.019	0.357	0.357
CcLG03	0.006	0.020	0.243	0.351
CcLG04	0.019	0.027	0.252	0.359
CcLG05	0.009	0.021	0.253	0.299
CcLG06	0.000	0.003	0.263	0.368
CcLG07	0.022	0.032	0.243	0.370
CcLG08	0.009	0.018	0.243	0.350
CcLG09	0.017	0.034	0.291	0.388
CcLG10	0.017	0.028	0.277	0.347
CcLG11	0.021	0.029	0.271	0.376
Average	0.012	0.022	0.268	0.353

India. At a K of 2, the wild species and landraces from different provinces consistently shared partial genetic composition (Figure 4). Landraces from Madhya Pradesh, Bihar, Orissa and Andhra Pradesh clearly separated from their wild ancestors. The genetic composition of wild relatives from different provinces had shown admixture in few accessions which were potentially the progenitor of these landraces. This shared genetic composition is not unexpected as domesticated *C. cajan* is derived from the wild accessions from India.

Several studies have shown that the highest heterozygosity is present in accessions from centre of origin [44]. The maximum expected heterozygosity found in wild relatives was 0.49 within the accessions from Madhya Pradesh and 0.47 in Andhra Pradesh (Table S6). It is important to mention here that size of the analysed samples was highly variable and low. As Madhya Pradesh was represented by only two accessions from landraces and two wild relative species (three accessions from *C. cajanifolius* and one accession from *C. scarabaeoides*) and Andhra Pradesh had five accessions from landraces and 10 accessions from wild relatives representing five species (*C.albicans, C.cajanifolius, C.crassus, C.scarabaeoides* and *C.sericeus*). However, based on current sampling, the higher heterozygosity is consistent with Madhya Pradesh being the centre of origin of pigeonpea. Expected heterozygosity in landraces was similar (0.37) in both the states (Table S6). Here it might be a function of sampling size used for the current study.

Discussion

This study reports the patterns of variation in cultivated pigeonpea and its wild relatives using SNP markers. Polymorphism survey of sampled *Cajanus* accessions indicated that cultivated pigeonpea is missing significant genetic diversity that was found in wild relatives. The wild relatives of pigeonpea remain the most critical source for increasing the available variation for pigeonpea breeding [45], even if their use has been limited due to a combination of poor agronomic traits, incomplete characterization, and limited collections.

Utility of KASPar assays for germplasm charterization

A number of marker systems have been developed for pigeonpea such as random amplified polymorphic DNA (RAPD) [23], amplified fragment length polymorphism (AFLP) [46], diversity array technology markers (DArT) [47], single feature polymorphism (SFP) [48] and simple sequence repeats (SSRs) [21]. Recently SNPs markers have also been developed and converted to cost effective genotyping platforms such as KASPar (PKAM [8]: **P**igeonpea **K**aspar **A**ssay **M**arkers) and BeadXpress assays [49]. KASPar assays provide flexibility in terms of number of SNPs used for genotyping. This feature provides upper edge to KASPar assays as compared to other SNP genotyping assays such as BeadXpress and Infinium assays. KASPar assays have been used for linkage mapping and parental polymorphism estimation [8], however these assays have not been used for large scale germplasm characterization in pigeonpea. KASPar assays have been found suitable for diversity estimation in common bean [34], chickpea [35] and peanut [36]. In the present study 75.86% PKAMs were found polymorphic while screening on 184 *Cajanus* accessions representing elite breeding lines, landraces and wild relatives, which is fractionally short from parental polymorphism identified in 24 pigeonpea genotypes (77.4%) [8] and peanut (80%) [36] and higher than chickpea (66.8%) [35]. PKAM categorization of germplasm agrees with the previous analysis of extent of diversity present in cultivated pool and wild relatives of pigeonpea conducted with AFLP [46] and DArT [24] markers. In terms of sub-divisions of *Cajanus* accessions, PKAM allowed the identification of two separate clusters corresponding to cultivated pigeonpea and one cluster corresponding to both wild relatives and cultivated pigeonpea. No clear groupings were identified in terms of genepools, however in cluster 'III', GP I accessions showed sub-grouping. GP II and GP III accessions were scattered in the cluster 'III'. Nevertheless, the *Cajanifolius* wild genotypes were closer to the cultivated pigeonpea than other wild species as revealed in previous marker based studies [9,50].

Variation across linkage groups

Great strides have been made in both sequencing the pigeonpea genome [51] and in placing a range of markers from SSRs to ESTs onto the linkage groups [21,38,47]. This study has assisted in the next step in providing information on sampled loci across the pigeonpea genome harboring high diversity. These sites may harbor unique features, from loci under different forms of natural selection to locations of inversions as discovered in case of chickpea by re-sequencing of cultivated and wild accessions [6]. Genotyping data suggested major loss of diversity across the pigeonpea genome during the course of domestication and further by modern breeding. These findings indicate that the cultivated pigeonpea has a narrowed genetic reservoir and possibly a reduced capacity to respond to future needs. Therefore, new methods must be applied to reintroduce adaptive diversity lost through domestication and breeding. This study emphasizes the need for support and planning for on-going, new, or novel efforts to maintain genetic diversity using wild relatives. Future crop production challenges will include new or more virulent diseases, environmental changes, degradation of agricultural land, etc., necessitating alternatives. Therefore, a diverse genetic reservoir in crop production remains as crucial as ever.

Insights into domestication

This study used high-throughput SNP genotyping for investigating the genetic diversity in cultivated pigeonpea and its wild relatives towards understanding the domestication and centre of origin. These analysis have provided better understanding about

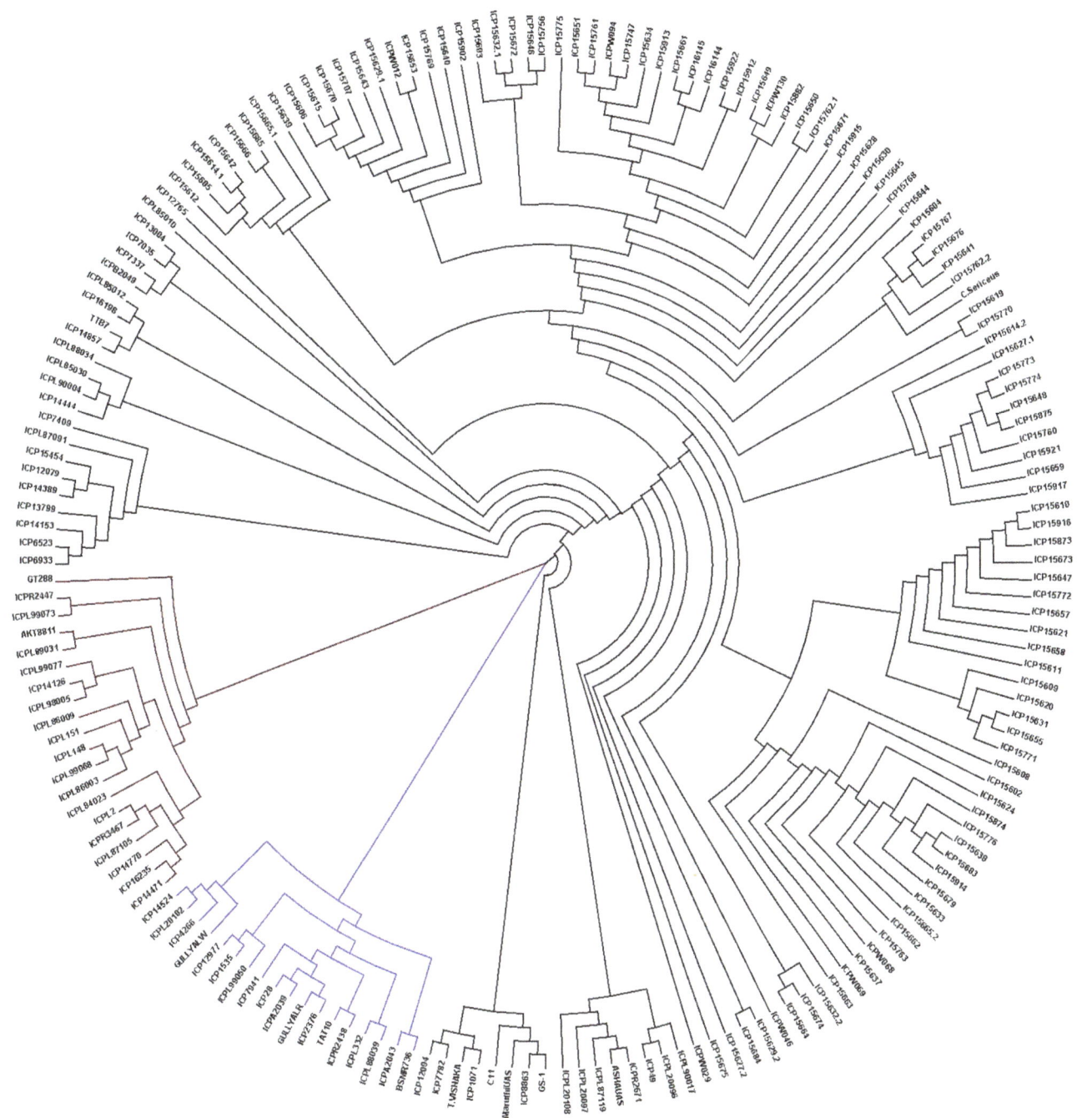

Figure 3. Neighbor-joinging tree of pairwise relatedness among 184 accessions.

the genetic diversity present in *Cajanus* as compared to previous studies [8,9,24,49]. This study was in congruence with some of the previous findings based on Archaeological [19,20] and molecular evidence [9] supported India as the domestication centre of pigeonpea. These results also assigned *C. cajanifolius* as the closest wild relative of cultivated pigeonpea and most likely progenitor species. Based on genetic diversity and heterozygosity, in the present study Madhya Pradesh (central province in India) has been designated as centre of origin of pigeonpea, however, almost similar levels of diversity were found in both wild relatives and landraces in the two Indian states namely Andhra Pradesh and Madhya Pradesh. Andhra Pradesh and Madhya Pradesh have

been designated as centre of domestication and centre of origin respectively in past [19,20]. However, our sample sizes were restricted by the size of existing collections of wild relatives and primitive landraces, and were insufficient to have complete confidence in Andhra Pradesh being the centre of domestication or diversification. Even if Andhra Pradesh or a nearby state is the centre of domestication, likely other regions, such as the more topologically and edaphically diverse Western Ghats region of India were also important areas of diversification of wild *Cajanus* species. And the relatively open breeding system of cultivated *C. cajan* makes it distinctly possible that pollen from wild relatives has entered the cultivated gene pool across areas of cultivation in

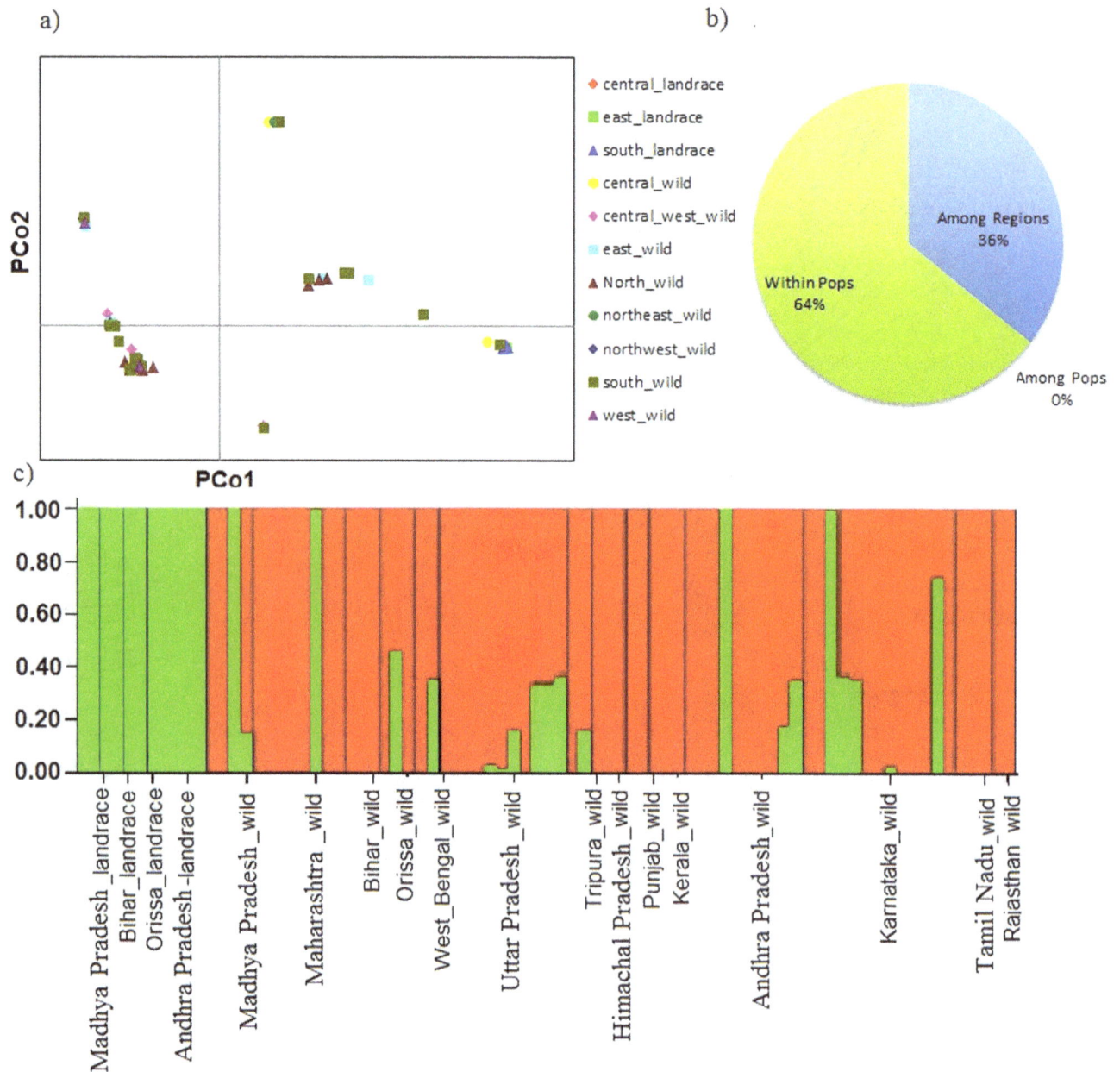

Figure 4. Population analysis of *Cajanus* **accessions present in Indian regions and provinces** *a)* Principal coordinates analysis of domesticated pigeonpea and wild relatives in 11 defined zones *b)* Analysis of molecular variance (AMOVA) in 11 defined zones *c)* Structure results across gene pools at the province scale

South Asia that overlap with the ranges of closely related wild species such as *C. cajanifolius*. Intra-specific patterns of variation in the wild relatives may be substantial. For traits such as flowering time that varies latitudinal, diverse range-wide collections of wild relatives would be particularly useful for introgressing desirable flowering time variation into cultivated pigeonpea. This could be particularly desirable to adapt it to new regions, or expand the range of seasons in which fresh pigeonpeas are available for markets where the fresh pigeonpeas are in demand.

Needs for more germplasm collection?

To increase genetic diversity of pigeonpea breeding material, new diversity from wild relatives will be extremely useful. Although we find substantial variation in existing collections, we are

certainly under sampling diversity within wild *Cajanus* species. Existing collections are inadequate for in-depth analysis of genetic variation between different *Cajanus* species. In particular, we expect to find substantial variation within species along climatic gradients across India. We advocate for systematic sampling from Madhya Pradesh and Andhra Pradesh to locate the exact geographical location of origin and first domestication event. Sampling from other potential regions would be beneficial to understand the movement of pigeonpea from its origin, and patterns of ongoing hybridization with wild relatives. This would also be helpful in assessing the outcrossing limits of pigeonpea, and allow a determining of isolation distances required for pigeonpea hybrid seed production.

Supporting Information

Figure S1 Estimated genome wide (CcLG01 to CcLG11) gene diversity using 875 mapped loci. "X" axis represents the length of each linkage group (CcLG) in cM and "Y" axis represents the value of gene diversity.

Figure S2 Population analysis of gene pools of *Cajanus* *a)* Structure results across gene pools. Groups 1, 2, and 3 represent the primary, secondary, and tertiary gene pools *b)* Principal coordinates analysis of domesticated pigeonpea and wild relatives. Red diamonds, primary gene pool; green squares, secondary gene pool; dark blue triangles, tertiary gene pool.

Figure S3 Analysis of molecular variance (AMOVA) at the continent scale.

Table S1 Details on 184 *Cajanus* accessions used for diversity and population analysis.

Table S2 Genotyping data generated using 1,616 PKAM on 184 *Cajanus* accessions.

Table S3 Diversity in breeding lines, landraces and wild relatives.

Table S4 Diversity in landraces and wild relatives at the continent scale.

Table S5 Diversity in landraces and wild relatives at the country scale.

Table S6 Diversity in landraces and wild relatives at the province scale with in India.

Acknowledgments

We thank Doug Cook, Mulualem Kassa and R VermaPenmetsa for helpful discussions.

Author Contributions

Conceived and designed the experiments: RKV. Performed the experiments: RKS EvW VS SS PK. Analyzed the data: EvW RKS RKV. Contributed reagents/materials/analysis tools: RKV HDU KBS. Wrote the paper: RKS EvW VS RKV.

References

1. Hawkes JG (1991) The importance of genetic resources in plant breeding. Biological journal of the linnean society 43: 3–10.
2. Tanksley SD, McCouch SR (1997) Seed banks and molecular maps: unlocking genetic potential from the wild. Science 277: 1063–1066.
3. Buckler ES, Thornsberry JM, Kresovich S (2001) Molecular diversity, structure and domestication of grasses. Genetical research 77: 213–218.
4. Gepts P (2004) Who owns biodiversity, and how should the owners be compensated? Plant physiology 134: 1295–1307.
5. Gross BL, Olsen KM (2010) Genetic perspectives on crop domestication, Trends in plant science 15:529–537.
6. Varshney RK, Song C, Saxena RK, Azam S, Sheng Y, et al. (2013) Draft genome sequence of chickpea (*Cicer arietinum*) provides a resource for trait improvement. Nature biotechnology 31: 240–246.
7. Saxena RK, Prathima C, Saxena KB, Hoisington DA, Singh NK, et al. (2010). Novel SSR markers for polymorphism detection in pigeonpea (*Cajanus* spp.). Plant breeding 129: 142–148.
8. Saxena RK, VarmaPenmetsa R, Upadhyaya HD, Kumar A, Carrasquilla-Garcia N, et al. (2012) Large-scale development of cost-effective single-nucleotide polymorphism marker assays for genetic mapping in pigeonpea and comparative mapping in legumes. DNA research 19: 449–461.
9. Kassa MT, VarmaPenmetsa R, Carrasquila-Garcia N, Sarma BK, Datta S, et al. (2012) Genetic patterns of domestication in pigeonpea (*Cajanus cajan* (L) Millsp) and wild *Cajanus* relatives. PloS one 7: 6.
10. Hajjar R, Hodgkin T (2007) The use of wild relatives in crop improvement: a survey of developments in the last 20 years. Euphytica 156:1–13.
11. Schierenbeck KA, Norman CE (2009) Hybridization and the evolution of invasiveness in plants and other organisms. Biological invasions 11: 1093–1105.
12. Nevo E, Chen G (2010) Drought and salt tolerances in wild relatives for wheat and barley improvement. Plant, cell & environment 33: 670–685.
13. Schoen DJ, Brown AH (1993) Conservation of allelic richness in wild crop relatives is aided by assessment of genetic markers. Proceedings of the national academy of sciences 90:10623–10627.
14. Pyhäjärvi T, García-Gil RM, Knürr T, Mikkonen M, Wachowiak W, et al. (2007) Demographic history has influenced nucleotide diversity in European *Pinus sylvestris* populations. Genetics 177: 1713–1724.
15. Morrell PL, Clegg MT (2007) Genetic evidence for a second domestication of barley (*Hordeum vulgare*) east of the fertile crescent. Proceedings of the national academy of sciences 104: 3289–3294.
16. Allaby RG (2008) The rise of plant domestication: life in the slow lane. Biologist 55:94–99.
17. Olsen KM, Schaal BA (2001) Microsatellite variation in cassava (*Manihot esculenta, Euphorbiaceae*) and its wild relatives: further evidence for a southern Amazonian origin of domestication. American journal of botany 88: 131–142.
18. Friesen ML, von Wettberg EJ (2010) Adapting genomics to study the evolution and ecology of agricultural systems. Current opinion in plant biology 13: 119–125.
19. van der Maesen LJG (1980) India is the native home of the pigeonpea. Liber gratulatorius in honorem HCD de Wit Misc Paper 19 Wageningen, the Netherlands pp 257–262.
20. van der Maesen LJG (1990) Pigeonpea origin, history, evolution, and taxonomy In: Nene YL, . Halls D, . Sheila VK, eds The Pigeonpea UK: CAB International, pp 15–46.
21. Bohra A, Dubey A, Saxena RK, VarmaPenmetsa R, Poornima KN, et al. (2011) Analysis of BAC-end sequences (BESs) and development of BES-SSR markers for genetic mapping and hybrid purity assessment in pigeonpea (*Cajanus* spp). BMC plant biology11: 56.
22. Saxena KB (2008) Genetic improvement of pigeon pea–a review. Tropical plant biology 1: 159–178.
23. Ratnaparkhe MB, Gupta VS, Ven Murthy MR, Ranjekar PK (1995) Genetic fingerprinting of pigeonpea [*Cajanus cajan* (L) Millsp] and its wild relatives using RAPD markers. Theoretical and applied genetics 91: 893–898.
24. Yang S, Pang W, Ash G, Harper J, Carling J, et al. (2006) Low level of genetic diversity in cultivated pigeonpea compared to its wild relatives is revealed by diversity arrays technology. Theoretical and applied genetics 113: 585–595.
25. Varshney RK, Nayak SN, May GD, Jackson SA (2009) Next-generation sequencing technologies and their implications for crop genetics and breeding. Trends in biotechnology 27: 522–530.
26. Cuesta-Marcos A, Szűcs P, Close T, Filichkin T, Muehlbauer G, et al. (2010) Genome-wide SNPs and re-sequencing of growth habit and inflorescence genes in barley: implications for association mapping in germplasm arrays varying in size and structure. BMC genomics 11: 707.
27. Cavanagh CR, Shiaoman C, Shichen W, Bevan EH, Stuart S, et al. (2013) Genome-wide comparative diversity uncovers multiple targets of selection for improvement in hexaploid wheat landraces and cultivars. Proceedings of the national academy of sciences 110: 8057–8062.
28. Ganal MW, Gregor D, Andreas P, Aurelie B, Buckler ES, et al. (2011) A large maize (*Zea mays* L) SNP genotyping array: development and germplasm genotyping, and genetic mapping to compare with the B73 reference genome. PloS one 6: e28334.
29. Durstewitz G, Polley A, Plieske J, Luerssen H, Graner EM, et al. (2010) SNP discovery by amplicon sequencing and multiplex SNP genotyping in the allopolyploid species *Brassica napus*. Genome 53: 948–956.
30. Hyten DL, Choi IKY, Song Q, Specht JE, Carter TE, et al. (2010) A high density integrated genetic linkage map of soybean and the development of a 1536 universal soy linkage panel for quantitative trait locus mapping. Crop science 50: 960–968.
31. Muchero W, Diop NN, Bhat PR, Fenton RD, Wanamaker S, et al. (2009) A consensus genetic map of cowpea [*Vigna unguiculata* (L) Walp] and synteny based on EST-derived SNPs. Proceedings of the national academy of sciences 106: 18159–18164.
32. Deulvot C, Charrel H, Marty A, Jacquin F, Donnadieu C, et al. (2010) Highly-multiplexed SNP genotyping for genetic mapping and germplasm diversity studies in pea. BMC genomics 11:468.

33. Kumari N, Brown-Guedira G, Huang L (2013) Development and validation of a breeder-friendly KASPar marker for wheat leaf rust resistance locus Lr21. Molecular breeding 31: 233–237.

34. Cortés AJ, Chavarro MC, Blair MW (2011) SNP marker diversity in common bean (*Phaseolus vulgaris* L.). Theoretical and applied genetics 123: 827–845.

35. Hiremath PJ, Kumar A, VarmaPenmetsa R, Farmer A, Schlueter JA, et al. (2012) Large-scale development of cost-effective SNP marker assays for diversity assessment and genetic mapping in chickpea and comparative mapping in legumes. Plant biotechnology journal 10: 716–732.

36. Khera P, Upadhyaya HD, Pandey M, Roorkiwal M, Sriswathi M, et al. (2013) Single nucleotide polymorphism–based genetic diversity in the reference set of peanut (*Arachis* spp) by developing and applying cost-effective kompetitive allele specific polymerase chain reaction genotyping assays. The plant genome 6:1–11.

37. Cuc LM, Mace ES, Crouch JH, Quang VD, Long TD, et al. (2008) Isolation and characterization of novel microsatellite markers and their application for diversity assessment in cultivated groundnut (*Arachis hypogaea*). BMC plant biology 8: 55.

38. Kudapa H, Bharti AK, Cannon SV, Farmer AD, Mulaosmanovic B, et al. (2012) A comprehensive transcriptome assembly of pigeonpea (*Cajanus cajan* L) using sanger and second-generation sequencing platforms. Molecular plant 5:1020–1028.

39. Peakall R, Smouse PE (2006) GENALEX 6: Genetic analysis in excel population genetic software for teaching and research. Molecular ecology notes 6: 288–295.

40. Pritchard JK, Stephens P, Donnelly P (2000) Inference of population structure using multilocus genotype data. Genetics 155: 945–959.

41. Earl DA (2012) STRUCTURE HARVESTER: a website and program for visualizing STRUCTURE output and implementing the Evanno method. Conservation genetics resources 4: 359–361.

42. Evanno G, Regnaut S, Goudet J (2005) Detecting the number of clusters of individuals using the software STRUCTURE: a simulation study. Molecular ecology 14: 2611–2620.

43. Tamura K, Peterson D, Peterson N, Stecher G, Nei M, et al. (2011) MEGA5: molecular evolutionary genetics analysis using maximum likelihood, evolutionary distance, and maximum parsimony methods. Molecular biology and evolution 28: 2731–2739.

44. Gutenkunst RN, Hernandez RD, Williamson SH, Bustamante CD (2009) Inferring the joint demographic history of multiple populations from multidimensional SNP frequency data. PLoS genetics 5: e1000695.

45. Bohra A, Mallikarjuna N, Saxena KB, Upadhyaya HD, Vales I, et al. (2010) Harnessing the potential of crop wild relatives through genomics tools for pigeonpea improvement. Journal of plant biology 37: 83–98.

46. Panguluri SK, Janaiah K, Govil JN, Kumar PA, Sharma PC (2006) AFLP fingerprinting in pigeonpea (*Cajanus cajan* (L) Millsp) and its wild relatives. Genetic resources and crop evolution 53: 523–531.

47. Yang S, Saxena RK, Kulwal PL, Ash GJ, Dubey A, et al. (2011) First genetic map of pigeonpea based on Diversity Array Technology (DArT) markers. Journal of genetics 90:103–109.

48. Saxena RK, Cui X, Thakur V, Walter B, Close TJ, et al. (2011) Single feature polymorphisms (SFPs) for drought tolerance in pigeonpea (*Cajanus* spp). Functional & integrative genomics 11: 651–657.

49. Roorkiwal M, Sawargaonkar SL, Chitikineni A, Thudi M, Saxena RK, et al. (2013) Single nucleotide polymorphism genotyping for breeding and genetics applications in chickpea and pigeonpea using the BeadXpress platform. The plant genome 6: 2.

50. Mudaraddi B, Saxena KB, Saxena RK, Varshney RK (2013) Molecular diversity among wild relatives of *Cajanus cajan* (L) Millsp. African journal of biotechnology 12: 3797–3801.

51. Varshney RK, Chen W, Li Y, Bharti AK, Saxena RK, et al. (2012) Draft genome sequence of pigeonpea (*Cajanus cajan*), an orphan legume crop of resource-poor farmers. Nature biotechnology 30: 83–89.

Genetic Diversity and Ecological Niche Modelling of Wild Barley: Refugia, Large-Scale Post-LGM Range Expansion and Limited Mid-Future Climate Threats?

Joanne Russell[1], Maarten van Zonneveld[2], Ian K. Dawson[1]*, Allan Booth[1], Robbie Waugh[1], Brian Steffenson[3]

1 Cell and Molecular Sciences, The James Hutton Institute, Invergowrie, Scotland, United Kingdom, 2 Regional Office for the Americas, Bioversity International, Cali, Colombia, 3 Department of Plant Pathology, University of Minnesota, Saint Paul, Minnesota, United States of America

Abstract

Describing genetic diversity in wild barley (*Hordeum vulgare* ssp. *spontaneum*) in geographic and environmental space in the context of current, past and potential future climates is important for conservation and for breeding the domesticated crop (*Hordeum vulgare* ssp. *vulgare*). Spatial genetic diversity in wild barley was revealed by both nuclear- (2,505 SNP, 24 nSSR) and chloroplast-derived (5 cpSSR) markers in 256 widely-sampled geo-referenced accessions. Results were compared with MaxEnt-modelled geographic distributions under current, past (Last Glacial Maximum, LGM) and mid-term future (anthropogenic scenario A2, the 2080s) climates. Comparisons suggest large-scale post-LGM range expansion in Central Asia and relatively small, but statistically significant, reductions in range-wide genetic diversity under future climate. Our analyses support the utility of ecological niche modelling for locating genetic diversity hotspots and determine priority geographic areas for wild barley conservation under anthropogenic climate change. Similar research on other cereal crop progenitors could play an important role in tailoring conservation and crop improvement strategies to support future human food security.

Editor: Mingliang Xu, China Agricultural University, China

Funding: These authors have no support or funding to report.

Competing Interests: The authors have declared that no competing interests exist.

* E-mail: iankdawson@aol.com

Introduction

Ecological niche modelling of the distributions of crop wild relatives in present, past and future climates can provide important insights into the past evolution and future trajectories of crop progenitors and domesticates [1]. Non-overlaps between current and predicted future distributions may reveal populations at particular threat from anthropogenic climate change [2]. At the same time, it has been suggested that overlaps between modelled past and present distributions may indicate refugial areas rich in genetic diversity, although this is a theory that requires wider validation [3–5]. In both instances, distributional differences may indicate wild genetic resources of particular importance for conservation and for breeding of the domesticated crop, in order to respond to new environmental pressures [6].

Wild barley (*Hordeum vulgare* ssp. *spontaneum*), the progenitor of the agriculturally important domesticated *H. vulgare* ssp. *vulgare* [7], provides an excellent opportunity to explore the utility of ecological niche modelling for supporting conservation and use. One reason is that its extensive natural distribution, which covers a range of environments across the Fertile Crescent and Central Asia [8], has been widely sampled for seed. This seed has been made available for genotyping and is an important resource for characterising genetic variation, to assist the cultivated crop to respond to anthropogenic climate change and other production challenges [9,10]. Another reason for wild barley's utility is that a wide range of molecular tools are available to describe genetic diversity. These tools include single nucleotide polymorphisms (SNPs [11]), nuclear simple sequence repeats (nSSRs [12]) and chloroplast simple sequence repeats (cpSSRs [13]) that were developed initially for studying cultivated barley, but can also be used to characterise the wild resource. If centres of diversity in wild barley (as described by these tools) are spatially coincident with habitat common to past and present modelled distributions, then this would support the utility of niche modelling for locating genetic refugia. On the other hand, if centres of maximum variation are outside areas of common past-present habitat, then the utility of niche modelling for locating genetic diversity would be weakened.

In this paper, we explore this issue by combining ecological niche modelling, based on the MaxEnt procedure [14], with a spatial analysis of SNP, nSSR and cpSSR data sets, using various geographic information systems [15,16,17]. Our intention is to build a greater understanding of the impacts of climate change on wild barley, and to provide information to help manage natural stands better in the context of environmental change. In turn, this will support breeding to adapt the domesticated barley crop to future climate. Our analysis is based on a range-wide, fully geo-referenced collection of 256 wild barley accessions sampled from 19 countries, and involves distribution modelling under three conditions: current climate, climate at the Last Glacial Maximum (the LGM) and future climate for the 2080s under anthropogenic scenario A2. We begin to explore the possible utility of linkage disequilibrium analysis for discriminating between alternative hypotheses for describing and explaining spatial genetic structure, and discuss the merits and limitations of the methods we employ. An approach similar to that described in this paper could be used to improve the management of other progenitors of domesticated cereals originating in the Fertile Crescent and Central Asia. This is important for supporting future global human food security in the context of anthropogenic climate change [18].

Materials and Methods

The Wild Barley Collection

The Wild Barley Diversity Collection (WBDC) is the most comprehensive geo-referenced collection of *H. vulgare* ssp. *spontaneum* currently subject to wide characterisation [19]. Our sampling of 256 individuals from the WBDC included 19 countries and was designed to cover as much of the accepted natural distribution of the taxon as possible [8]. Sampling extended from North Africa through the Fertile Crescent into Central Asia (Fig. 1 and Table S1). Sampling did not include Tibet, where wild barley very different from that to the west is found [20,21], as very few geo-referenced samples (as required for ecological niche modelling purposes) are available from there. Most accessions originated from the International Center for Agricultural Research in the Dry Areas (ICARDA), Aleppo, Syria, and were assembled by the former gene bank curator there, Dr Jan Valkoun. The majority of accessions were sampled from the wild in the fifty-year period 1953 to 2002, especially in the ten-year periods of 1983 to 1992 and 1993 to 2002 (79 and 101 accessions, respectively). These collection periods correspond well with when the weather station data that are used to support the interpolation of bioclimatic variables for ecological niche modelling were obtained (see below, [22]). For some accessions with early collection dates, latitudes and longitudes used in the current study are based on the interpretation of passport site-description data rather than actual given GPS coordinates. These accessions are therefore likely to be less precisely located.

Assembling Molecular Marker Data Sets

Three molecular marker data sets were analysed in the current study. First, SNP data derived from two Illumina barley oligonucleotide pool assay platforms were used (see [11,23] for a description of these platforms, referred to as BOPAs 1 and 2 or collectively as BOPA SNPs). Here, a subset of 2,505 mostly chromosome-position-mapped BOPA SNPs from an existing study on the WBDC ([24], to investigate disease resistance traits) was used. 'Ascertainment bias' can confound the interpretation of BOPA SNP results when comparing domesticated and wild barley genetic resources [25]. In the current study, however, which only involved comparing different portions of wild barley's range, no significant confounding effect is expected (see discussion in the study by Russell et al. [26], which compared landrace and wild barleys in the Fertile Crescent using BOPA SNPs). Second, we characterised variation *de novo* at 24 of the barley nSSR loci described by Ramsay et al. [12], using the methods given there. Third, we determined variation *de novo* at five of the cpSSR loci designed for *Hordeum* by Provan et al. [13], using the methods of Comadran et al. [27]. A list of all 2,534 loci used in the current study is given in Table S1.

Analysing Molecular Marker Data

Spatial autocorrelation analysis. Spatial autocorrelation analysis using SPAGeDi [28] was undertaken to assess the relationship between inter-individual genetic identities of the 256 tested wild barley accessions and geographic distances. Separate analyses were carried out for BOPA SNPs, nSSRs and cpSSRs. Ritland's [29] kinship coefficient was employed to quantify average pairwise genetic identity based on 20 geographic distance classes of equal sample size. Whether or not individual kinship values were different from expectations (under a random spatial distribution of genetic variation) was assessed by a randomisation test with 1,000 permutations. Kinship values were regressed against the natural logarithm of distance classes to estimate the overall extent of spatial genetic structure. The significance of the regression slope was determined by 1,000 random permutations of locations.

STRUCTURE analysis. STRUCTURE analysis was not designed for predominantly inbreeding species such as barley, but it has been widely applied to cultivated and wild barley populations to reveal interesting genetic features (see discussion in [26]). Here, BOPA SNP and nSSR data sets were each analysed with STRUCTURE 2.2 [30] to assign accessions to one of K groups for different values of K. Each analysis was based on 25,000 'burn-in' replications and a further 25,000 Markov chain Monte Carlo steps (initial trial runs indicated that these numbers of replications were sufficient to ensure the convergence of key parameters). After trial runs, K was set at five because log $\Pr(X/K)$ values in STRUCTURE had started to plateau at this point [30]. (Note that for our purposes it is more important to capture the major genetic divisions within data sets than to determine an 'absolute' value for K.) The 'no admixture' model in STRUCTURE was used to assign single states to individuals. Other analysis options were kept at default settings. STRUCTURE was run five times for each data set and the most common group assignments used as the basis for the interpretation of results (most accessions placed in the same groups in separate runs).

Circular neighbourhood analysis. To overlay genetic diversity onto geographic maps we employed DIVA-GIS 7.3 [15] (www.diva-gis.org) and ArcGIS 10 [16] (www.esri.com/software/arcgis/). Two approaches were used, the first based on allelic (or haplotype) richness and the second based on K groupings. In the first, allelic (BOPA SNP and nSSR) and haplotype (cpSSR) richness estimates were calculated for groups of accessions. Groups were circumscribed using a circular neighbourhood diameter of four degrees and a grid size of 30 minutes (method described in [6]). This allowed us to capture sufficient collection sites within neighbourhoods to estimate genetic parameters with some confidence. To account for varying sampling intensity in geographic space, which otherwise affects diversity estimates [31], rarefaction to a sample size of 10 individuals in neighbourhoods was undertaken using ADZE [17]. In the second approach, K groupings ($K=5$) revealed by STRUCTURE for BOPA SNPs and nSSRs were used instead of allelic/haplotype richness estimates, in order to reveal genetic differentiation at a local geographic scale. Apart from this, the method of analysis was the same as applied in the first approach.

Circular neighbourhood analysis with rarefaction, as conducted in both the above approaches, has the advantage of allowing unbiased comparisons of genetic diversity across geographic space. However, it necessarily excludes accessions from analysis where sampling intensity is low. In the current study, a total of 38 accessions were thereby excluded (including from North Africa, southwestern Iran, Afghanistan and Azerbaijan).

Ecological Niche Modelling

Although MaxEnt [14] has a number of well-documented limitations [32], it is reported to predict the natural distributions of

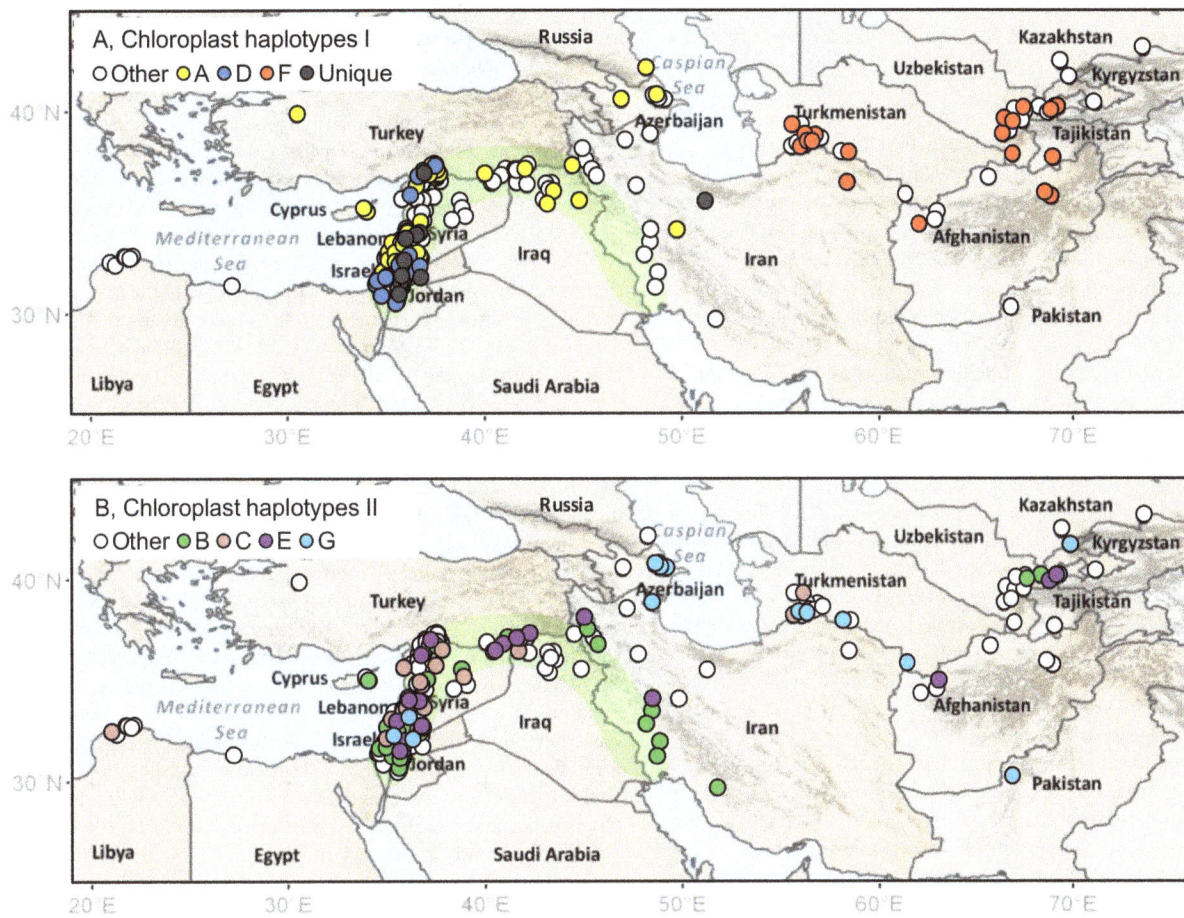

Figure 1. 256 wild barley accessions sampled for genetic analysis and ecological niche modelling. Sampling covered 19 countries and much of the geographic range of wild barley. Superimposed on the positions of accessions are cpSSR haplotype designations for seven common haplotypes (frequency ≥0.05 across all accessions). A, distribution of three common, clearly geographically-differentiated, haplotypes. The distribution of 10 unique haplotypes is also shown. B, distribution of the other four common haplotypes. In both A and B, other sampled accessions are indicated by white circles ('Other'). In total, 31 chloroplast haplotypes were revealed, as described in the Results and Discussion. The approximate dimensions of the Fertile Crescent, a region considered crucial in the development of agriculture and where dense stands of wild barley can occur [8], are indicated by green shading for reference purposes (see [26] for a discussion of barley domestication). The coordinates of sampled accessions and full chloroplast haplotype data are given in Table S1.

plants well when based on 'presence only' location data compared to other ecological niche modelling approaches [33–35]. This is especially so when modelling is based on location data from a limited number of sites (less than 700 [34]). We therefore employed MaxEnt 3.3.1 to model geographic distributions for wild barley under current, past and future climates, in a manner similar to van Zonneveld et al. [36]. This involved extracting data for 19 bioclimatic variables for each of the 256 wild barley accession collection sites from WorldClim [22] (www.worldclim.org/) (extracted values for variables listed in Table S1). These 19 variables are derived from monthly temperature and rainfall values and include seasonality and limiting environmental factors. Our modelling was bounded by longitudes of 18.63 and 80.42 degrees east, latitudes of 21.58 and 47.96 degrees north. The output of MaxEnt is a grid map with each cell assigned a probability of taxon presence [37]. Modelled distribution areas were restricted to the threshold suitability value of maximum training sensitivity plus specificity recommended by Liu et al. [38]. In total, 13 accessions were excluded from modelling because they were identified to occur at 'outlier' sites (see Table S1).

For the LGM (~21,000 years before present), modelling was based on CCSM and MIROC models (available at WorldClim, [2]; the results of models were averaged to provide overall estimates. The LGM is believed to have been an influential period in determining contemporary patterns of genetic variation in many plant species and much modelling of past distributions has therefore been based on it [2,3,5]. For future climate, modelling was based on the 2080s period (2070 to 2099) and the medium- to high-emission trajectory A2 for anthropogenic global warming. The 2080s A2 scenario has been widely used in modelling to provide insights on a timescale and threat level that is useful for planning purposes [39,40]. Nineteen general circulation models (GCM) were used for future climate (again, results were averaged across models). Data on future climate projections were provided by the CGIAR Climate Change, Agriculture and Food Security Research Programme (CCAFS) and downscaled with the Delta method [39].

For current, past and future distribution modelling, 2.5-minute downscaled climate layers were employed, which is the same resolution as used by Waltari et al. [2] for past-climate distribution projections. Based on the overall geographic scale of our sample range, we consider this degree of resolution sufficient for our study.

Results and Discussion

Isolation-by-Distance Alone Does Not Explain the Observed Genetic Structure in Wild Barley

Individual scores for all 256 wild barley accessions for 2,505 BOPA SNPs, 24 nSSRs and five cpSSRs are provided in Table S1. The overall quality of the data was high, with the mean level of missing data (including ambiguous calls) across all markers less than 1%. 2,363 BOPA SNPs were polymorphic and all 24 nSSRs and 5 cpSSRs. A mean of 16.7 alleles per locus (ranging from 3 to 54) was revealed at nSSRs. Length variation at cpSSR products indicated a series of single base differences, with combined data (summing differences across separate products) revealing 31 haplotypes. Seven haplotypes could be defined as common (A to G, occurring at a frequency ≥0.05), while 10 were unique. Compared to haplotype A, haplotypes B, C, D and E differed by a single nucleotide length at one cpSSR product, while haplotypes F and G differed by single nucleotide length differences at two products. Of the common haplotypes, A, D and F showed very clear geographic structuring but the others did not (compare Figs. 1A and B). Haplotype A occurred throughout the Fertile Crescent but not further east, D occurred only in the Eastern Mediterranean region of the Fertile Crescent, while F occurred only in Turkmenistan and further east.

Spatial autocorrelation analysis has been widely applied to assess genetic structure in plant species and to describe deviations from isolation-by-distance expectations (e.g. [78,79]). It can also be a useful method for comparing different molecular marker data sets compiled on the same taxon, as we do here for wild barley (Fig. 2). In our analysis, a degree of geographic-based genetic structure is evident for BOPA SNPs, nSSRs and cpSSR haplotypes ($P<0.01$ in a test for overall structure in each case). The decrease in similarity observed with geographic distance is, however, not a simple trend for any of our three data sets. Differences in profiles are also observed between marker types. For both BOPA SNPs and nSSRs, an increase in similarity at a distance class of around 1,000 km is observed, after which similarity declines again. For cpSSR haplotypes, an obvious increase in similarity is also observed at a distance class of around 500 km. Spatial autocorrelation analysis therefore indicates that a simple isolation-by-distance model does not fully explain genetic structure across the geographic range of wild barley tested (as indicated also, e.g., by the distribution of accessions among BOPA SNP STRUCTURE groups, as shown in Fig. 3). In such situations, climate change-related expansions and contractions in range could have a role in determining patterns of variation [3,5].

The differences we observed for spatial autocorrelation analysis profiles for nuclear BOPA SNPs and nSSRs compared to maternally-inherited cpSSRs may indicate the more restricted role of seed when compared to pollen in gene flow (even though wild barley is predominantly self-pollinated and so pollen-mediated gene flow is expected to be relatively low [41]). The smaller effective population size of the organellar genome compared to the nuclear genome may also be a factor in determining the differences observed [42].

The Spatial Distribution of Genetic Variation Corresponds With Niche Modelling in Locating Diversity Hotspots and is Consistent With Post-LGM Expansion in Central Asia

Geographic information systems are underutilised in genetic diversity studies, but they can be very effective in expressing variation in geographic and environmental space [43–45]. Our analysis is the first on wild barley to use circular neighbourhoods with rarefaction to account for differences in sampling intensity across geographic space. These differences otherwise skew the visualisation and interpretation of genetic diversity, as illustrated by nSSR analyses of cacao (*Theobroma cacao* [5]) and the cherimoya fruit tree (*Annona cherimola* [6]) in South America. The results of our analyses of wild barley are given in Figures 4 and 5. Figure 4, based on allelic/haplotype richness estimates, demonstrates that BOPA SNPs, nSSRs and cpSSRs all provide similar profiles of diversity across geographic space. In each case, higher sample-size-corrected values of richness were observed in the Eastern Mediterranean region than in Central Asia. Circular neighbourhood analysis based on chloroplast haplotypes therefore clearly corresponds with the distribution of unique haplotypes shown in Figure 1 (nine of 10 unique haplotypes occurred in the Eastern Mediterranean region). Figure 5 (A, B), which shows levels of BOPA SNP and nSSR diversity based on K group richness, also indicates higher diversity (greater genetic differentiation at a local level) in the Eastern Mediterranean than in Central Asia (as also evident from individual K group assignments in Fig. 3).

Our findings are consistent with the limited previous molecular marker research (uncorrected for sampling intensity) comparing wild barley from the Eastern Mediterranean region and environs with Central Asia. For example, Volis et al. [46] measured lower variation in wild barley in Central Asia (samples from Turkmenistan only) than in the Eastern Mediterranean using isozymes, while Fu and Horbach [47] found the same based on nSSRs. Our analysis provides comprehensive evidence to reinforce these observations and confirms the status of the Eastern Mediterranean wild barley stands as important resources for conservation and evaluation [9,10].

Our intention in this study is to compare patterns of genetic variation in wild barley with the modelled distributions of the taxon under all three conditions of current, past and future climates, something which to our knowledge has not been undertaken before for any member of the genus *Hordeum*. The results of our ecological niche modelling are presented in Figure 6, from which several interesting observations can be drawn in relation to the genetic data, as set out in this section for the current-past climate comparison and in the next section for the current-future climate comparison.

Considering first the modelled present-day geographic distribution of wild barley, visual assessment confirms that the accessions included in our study for genetic analysis provide good coverage of most of the taxon's range in the Fertile Crescent and Central Asia (compare Fig. 6A with Fig. 1). The most obvious exception is the 'peak' of the Fertile Crescent in southern Turkey, where it is known that important stands of wild barley occur [8]. Although not sampled in this study, these Turkish stands should be incorporated in future work. A comparison of the modelled present-day geographic distribution of wild barley with that projected for the LGM (Figs. 6B, D) suggests that since the LGM suitable habitat has been lost in areas that include southeastern Iran and northern Saudi Arabia. At the same time, the comparison indicates that at the LGM wild barley was, just as it is now, widely present in the region bordering the Eastern Mediterranean coast. A comparison of current and past modelled distributions also indicates habitat gains since the LGM. Particular areas identified in this regard include the northern Iraq portion of the Fertile Crescent and, especially, a large part of Central Asia.

The apparent relatively recent range expansion of wild barley in Central Asia as revealed by ecological niche modelling is consistent with our findings of lower levels of genetic diversity in the region

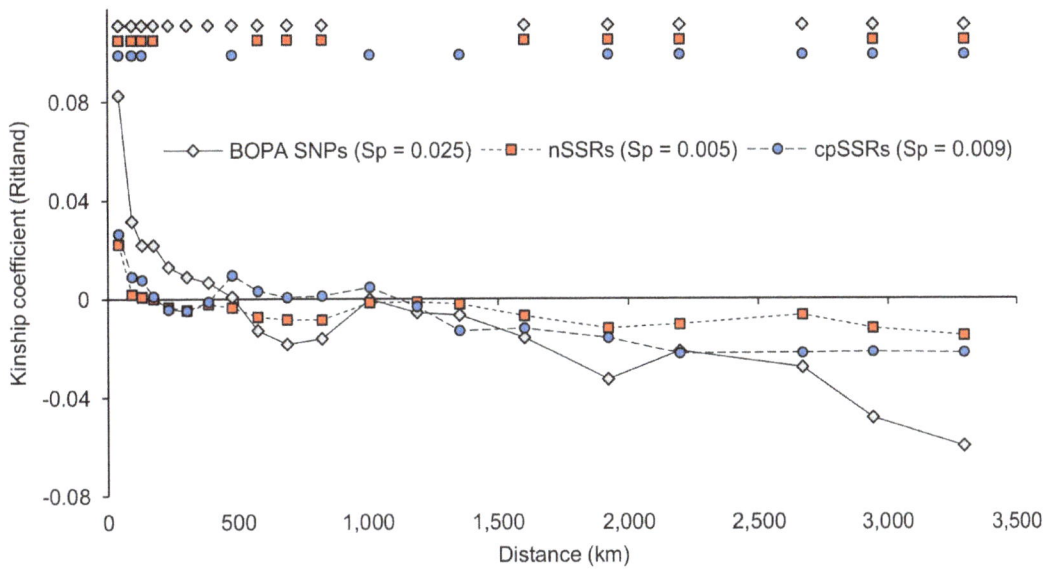

Figure 2. Spatial autocorrelation analysis profiles for wild barley accessions based on BOPA SNPs, nSSRs and cpSSRs. Geographic distances on the x-axis are the mean values of distance classes. The symbols at the top of the figure mark observations significantly larger or smaller ($P \leq 0.01$) than the average for distance classes. Values for the Sp statistic, calculated from the regression slope of the graph and the kinship coefficient of the first distance class [76], are also shown. Placing all three data sets on the same graph allows profiles to be compared. Increases in similarity at a distance class of around 1,000 km, and an earlier additional increase for cpSSRs at around 500 km, illustrate that a simple isolation-by-distance model is not sufficient to describe genetic variation in wild barley.

compared to the Eastern Mediterranean, in which latter region it can be postulated that a continuous presence of the taxon has led to the accumulation of genetic diversity (as shown in Figs. 4, 5A and B) there. The correspondence between past-present ecological niche modelling and our analysis of spatial genetic diversity in wild barley has important implications, as it supports the utility of niche modelling as a tool for identifying genetically diverse and potentially refugial areas. There are, however, other possible reasons why different levels of genetic diversity are observed across wild barley's range. The Eastern Mediterranean region (as represented, e.g., by altitudinal variation for the accessions included in the current study, see Fig. 5C) is, for example, particularly environmentally heterogeneous. This may have

allowed more genetic variation to develop and accumulate there compared to Central Asia without recourse to an explanation based upon post-LGM macro-geographic range adjustment. We are not able to distinguish between these alternatives (or, indeed, to understand whether a combination of both range expansion and environmental heterogeneity are important) for determining the current pattern of spatial genetic diversity observed in wild barley. Nevertheless, one interesting feature of our data that deserves further exploration in this regard is the level of linkage disequilibrium (LD) between chromosome-position-mapped BOPA SNP markers in different parts of wild barley's range, as we relate below.

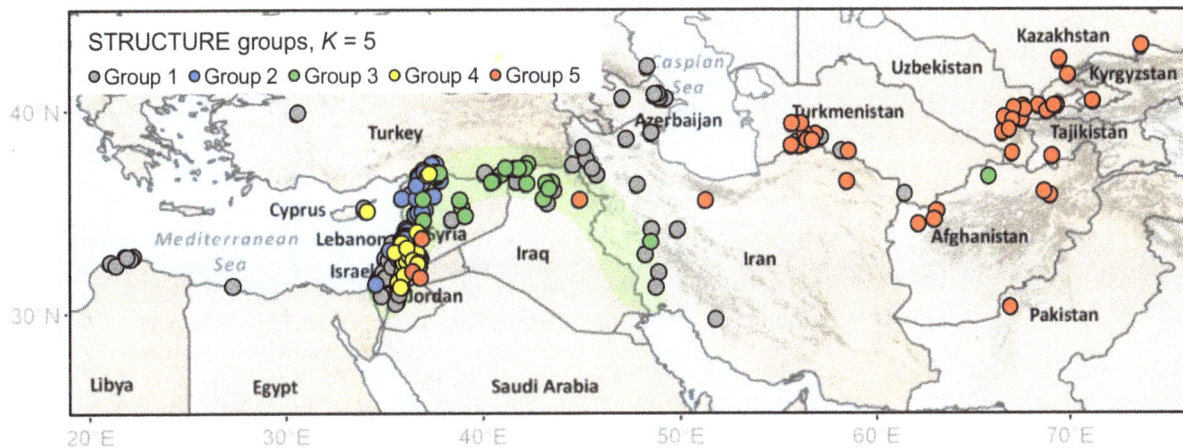

Figure 3. STRUCTURE group assignments for individual wild barley accessions. The results shown are based on $K = 5$ and for BOPA SNPs. The results for nSSRs (not shown) were similar. Results correspond with spatial autocorrelation analysis (Fig. 2) in describing a more complex genetic structure in wild barley than might be expected with a simple isolation-by-distance model.

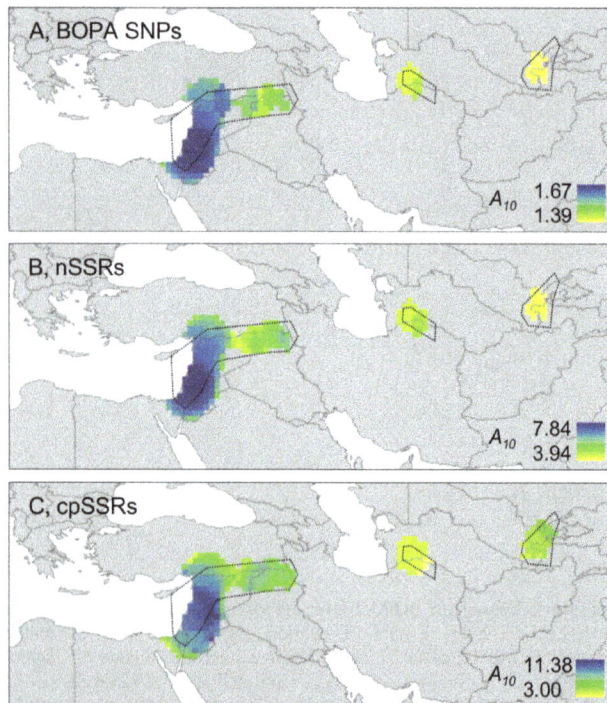

Figure 4. Allelic (A and B) and haplotype (C) richness (A_{10}) maps for wild barley. BOPA SNPs, nSSRs and cpSSRs all indicate the Eastern Mediterranean region as more diverse (highly diverse areas = dark blue) than Central Asia. As expected, nSSRs with high allelic diversity and cpSSRs with multiple haplotypes reveal relatively higher richness values within neighbourhoods (A_{10} as high as 7.84 and 11.38, respectively) than biallelic BOPA SNPs (maximum A_{10} = 1.67). Not all of the original sample range could be included in analysis because of the required minimum sampling intensity to calculate a standardised diversity value (see Materials and Methods; compare the current figure with Fig. 1). Accessions included in analysis in a particular geographic area are circumscribed by a dotted line. A, 2,426 from 2,505 BOPA SNPs used in calculations (SNPs excluded with ≥25% missing data in one or more grid cells); B, all 24 nSSRs used in calculations; C, all cpSSR haplotypes used in calculations.

Figure 5. STRUCTURE group richness (A and B, K_{10}) and 'altitude richness' (C, Alt_{10}) maps for wild barley. A and B, richness estimates for BOPA SNPs and nSSRs, respectively, K = 5 in STRUCTURE analysis. Both marker sets indicate the Eastern Mediterranean region as more diverse (highly diverse areas = dark brown) than Central Asia. C, 'altitude richness' of wild barley sample sites, based on five altitude categories (<200 m, 200 to 600 m, 600 to 1,000 m, 1,000 to 1,400 m, >1,400 m). Altitude data provide an indication of environmental heterogeneity and were downloaded from WorldClim (www.worldclim.org/; values given in Table S1). Unlike the 19 bioclimatic variables used elsewhere in the current study, altitude data are actual values rather than interpolations from weather station records, so they are particularly appropriate for assessing real environmental heterogeneity [22,77]. Altitude richness estimates indicate sample points in the Eastern Mediterranean region as more diverse than those in Central Asia. Not all of the original sample range could be included in analyses because of the required minimum sampling intensity to calculate standardised diversity values (see Materials and Methods; compare the current figure with Figs. 1 and 3 [individual STRUCTURE K group assignments], see also Fig. 4). Accessions included in analyses in a particular geographic area are circumscribed by a dotted line. The analysis to generate 'altitude richness' was carried out in the same way as for STRUCTURE group richness, except 'altitude category' substituted for 'STRUCTURE group'.

To explore LD, we undertook a further analysis based on two sub-samples of our wild barley accessions taken to represent the Eastern Mediterranean and Central Asia regions, as shown in Figure 7. These sub-samples, each of 40 individuals, represent approximately balanced sets of material (as explained in the legend to Fig. 7) for LD comparison. Compared to the Eastern Mediterranean sub-sample, the Central Asian sub-sample is of lower genetic diversity, comes from a more uniform environment and has a much higher proportion of accessions collected from habitat established (apparently) since the LGM. For these sub-samples of accessions, we then calculated LD for pairs of chromosome-mapped BOPA SNPs falling into different centimorgan (cM) distance categories along each of barley's seven chromosomes. Estimates for LD were based on 487 SNPs (the number of mapped SNPs on each chromosome ranged from 57 to 80) with a minimum minor allele frequency of 0.1 in both sub-samples, while the chromosome distance interval for making pairwise comparisons was set at five cM (so comparisons for paired SNPs 0 to 5 cM apart, 5 to 10 cM apart, etc.). The level of LD was estimated with r^2 values (the squared correlation of allele frequencies [48,49]) using DNASP 5.00.07 [50] with all SNPs assigned homozygous status (i.e., no intra-locus component in

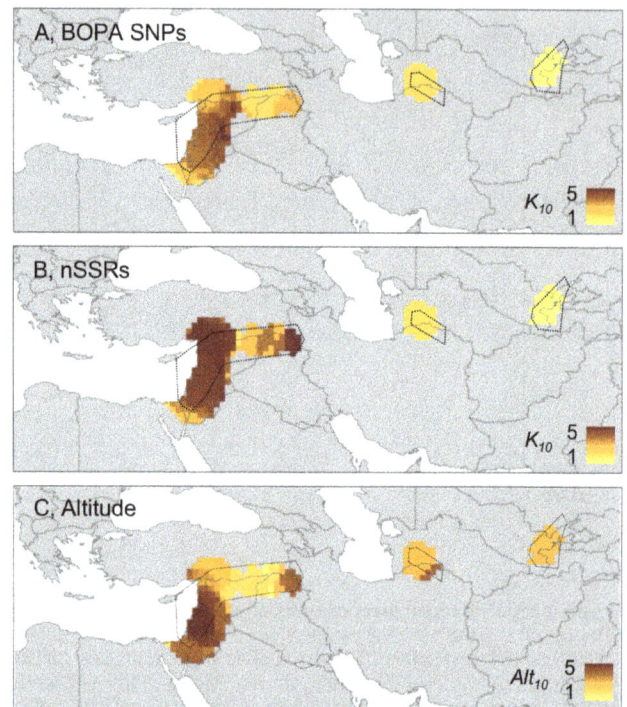

analysis). Finally, once r^2 values were generated, they were compiled into mean values for chromosome distance categories (a minimum of 10 observations for a distance interval were required before assigning a mean value) for each chromosome, and then averaged across chromosomes, in EXCEL. Results were then expressed in graphical form comparing LD estimates across sub-samples (mean r^2 Central Asia sub-sample/mean r^2 Eastern Mediterranean sub-sample) (Fig. 8).

Our comparison of LD estimates indicates that at shorter chromosome distance intervals values of LD are relatively higher in Central Asia than in the Eastern Mediterranean, but that with increased distance along chromosomes values become more equal. Linkage disequilibrium values are difficult to interpret because of the many influencing factors, including population structure, selection pressures, mating patterns and changes in population size

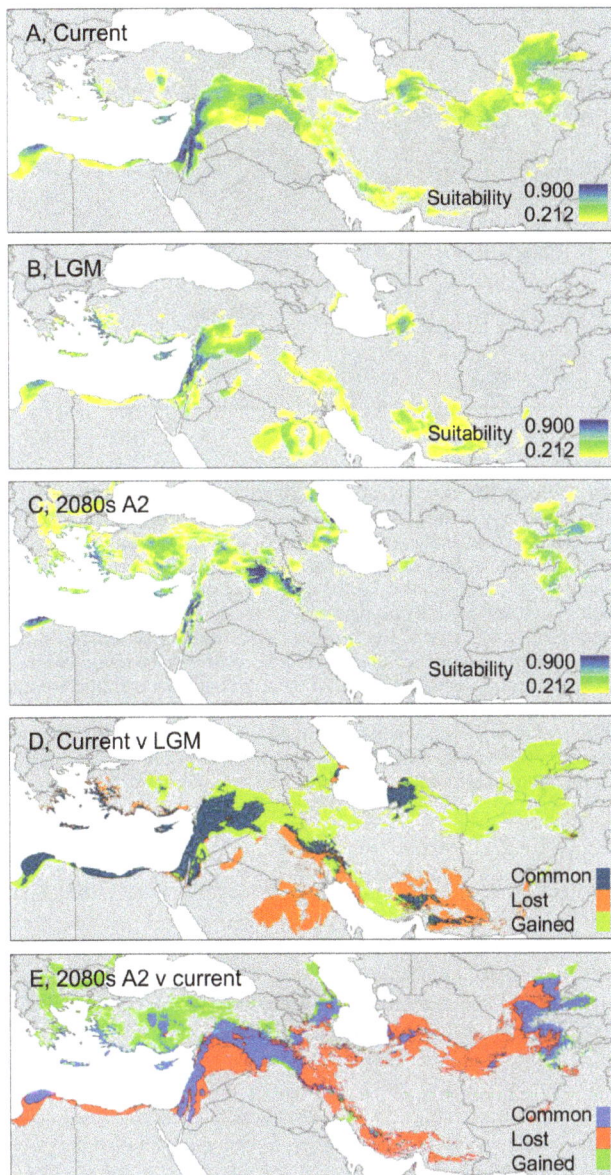

Figure 6. Potential wild barley distributions in current (A), past (B) and future (C) climates. Distributions are based on ecological niche modelling using MaxEnt (see Materials and Methods). D, differences between current and past modelled distributions, including areas lost and gained since the LGM. E, differences between future and current modelled distributions, including areas expected to be lost and gained by the 2080s. Note that past and future distribution maps take no account of rises or falls in sea levels or of other water bodies, and that these distributions are shown superimposed on current country boundaries.

and range [26,51–53]. In the absence of confounding factors, however, our observations are consistent with a relatively recent geographic expansion in Central Asia that has not allowed for enough time for recombination between proximate paired markers to result in equilibrium (convergence) between them [54–56]. Although interesting, we stress that this interpretation of the results of our LD analysis is speculative and must be treated with caution. For example, the selfing rate for wild barley can vary across populations (see [41] and [57] for Israeli and Jordanian populations, respectively), and if wild barley stands in Central

Asia were more highly selfed [58], this could also explain why LD for proximate markers was higher there than in the Eastern Mediterranean. Clearly, more research on this topic is required.

Niche Modelling Suggests Small, But Statistically Significant, Losses of Genetic Diversity in Wild Barley Under Mid-Term Anthropogenic Climate Change

A comparison of present-day plant distributions with predictions for the 2080s is useful for devising responses to anthropogenic climate change [1,32]. For example, areas of predicted habitat loss may be targets for the collection of seed that can then be stored in gene banks. In addition, locations where habitat is likely to be retained may be priorities for *in situ* conservation measures [59,60]. Of most interest may be locations where habitat is predicted to be retained in geographic regions of general habitat loss. Our ecological niche modelling comparing the current distribution of wild barley with that predicted for the 2080s under the A2 emission scenario is shown in Figure 6 (C, E). The comparison suggests that suitable habitat will be lost in particular in large areas of Iran, northern Syria and in the border region of Afghanistan and Turkmenistan. At the same time, potential habitat will be gained most notably in parts of Turkey.

In order to calculate the possible losses in range-wide genetic diversity in wild barley associated with future habitat loss, we compared allelic richness at nSSRs for accessions predicted to be in shared future and current habitat ($N = 155$, i.e., excluding accessions in 'lost' habitat, as indicated in Table S1) with accessions in the current distribution ($N = 243$). We chose nSSRs as the estimator for this analysis because of their high allelic variability and hence sensitivity in describing diversity differences. Our analysis indicated a relatively small reduction in allelic richness under climate change (shared habitat, $A = 14.88$; current habitat, $A = 15.88$; estimates calculated in FSTAT 2.9.4 [61] and corrected by rarefaction to a sample size of 135 complete genotypes across all nSSRs). Although small, the difference was statistically significant ($P = 0.012$ based on a two-tailed t-test of individual locus allelic richness values undertaken in EXCEL). Further assessment, based on BOPA SNPs and cpSSRs, revealed a relatively modest 108 SNP alleles (4.6% of SNPs in the comparison) and four chloroplast haplotypes unique to 2080s 'lost' habitat. Our analysis therefore suggests that, overall, mid-term future losses in genetic diversity due to climate change are expected to be relatively low. A comparison of modelled current, past and future distributions (Fig. 6) suggests that in part this is because much predicted future habitat loss is in areas of putative post-LGM range expansion, where contemporary genetic variation is relatively low (e.g., in the border region of Afghanistan and Turkmenistan). On the other hand, shared future-current habitat includes much of the more genetically diverse putative LGM refugial regions. Of concern, though, could be future habitat loss in parts of northern Syria, where habitat is in common in the past-current comparison, but not in the future-current comparison (compare Figs. 6D, E).

It is important to consider a number of provisos when interpreting our findings. First, we have not considered in our analysis that existing wild barley populations in habitat that will be lost under climate change could migrate to (newly) environmentally-matched sites. Such migration will presumably be easier in areas with greater micro-geographic environmental heterogeneity, as the distances to be moved are then smaller. This would suggest migration is more feasible in the Eastern Mediterranean region (see Fig. 5C). Whether migrations are possible also depends on the level of human activity in wild barley's habitat. Human disturbance that provides opportunities for establishment could

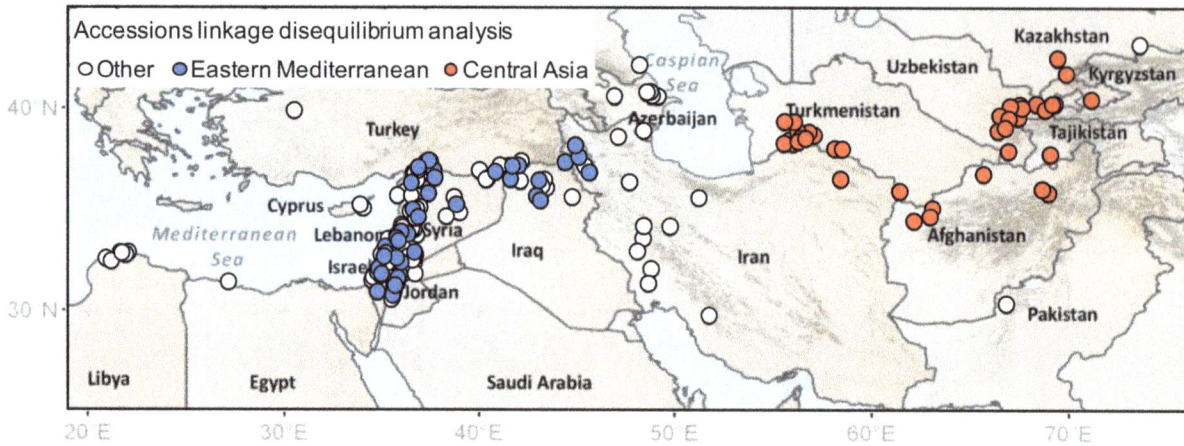

Figure 7. Locations of wild barley individuals sub-sampled from two regions for testing of linkage disequilibrium (LD). For LD assessment, forty accessions were chosen at random from the Eastern Mediterranean and Central Asia regions, across an approximately equal-dimensioned geographic area to minimise confounding sample size and dimensional effects in analysis (a significant issue in LD calculations [26]). In the case of the Eastern Mediterranean region, to ensure similar geographic coverage to Central Asia, sampling was extended eastward away from the coast below the peak of the Fertile Crescent into northwestern Iran. The coordinates of the accessions sampled for LD analysis are given in Table S1. Compared to the Eastern Mediterranean sub-sample, that from Central Asia had a less variable environment across accession collection sites (see Fig. 5C and the bioclimatic variables given in Table S1). Furthermore, climate modelling suggested a much greater proportion of accessions in the Central Asian sub-sample to be associated with range expansion since the LGM (see Fig. 6D and Table S1). Consistent with genetic diversity levels expressed on maps (Figs. 4, 5A, 5B), the latter sub-sample also had lower nuclear diversity according to FSTAT 2.9.4 [61] calculations (nSSR allelic richness for the Eastern Mediterranean sub-sample = 10.44, for Central Asia = 7.44, corrected by rarefaction to a sample size of 36 complete genotypes across all nSSRs; $P<0.001$ based on a two-tailed t-test of individual locus allelic richness values undertaken in EXCEL). Another factor that can confound LD comparisons is differences in allele frequency distributions between samples. We therefore tested allele frequency profiles for our two sub-samples, and found them to be similar (proportion of markers with a minimum minor allele frequency between 0.1 and 0.3 was 0.493 and 0.499 for the Eastern Mediterranean and Central Asia areas, respectively).

be beneficial, while modern agricultural practices that intensify crop production (excluding other plants from fields) and fragment wild habitat could be detrimental [62]. Second, in common with most other studies that compare present-day and potential future

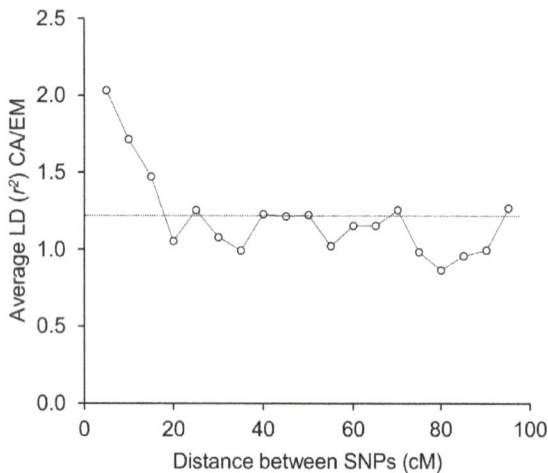

Figure 8. Comparison of mean linkage disequilibrium (LD) (r^2) values across all wild barley chromosomes. CA = Central Asia, EM = Eastern Mediterranean sub-samples. Comparisons are for BOPA SNPs at five cM intervals (SNPs 0 to 5 cM apart, 5 to 10 cM apart, etc.). The dotted line indicates the average value of all plotted comparisons. The graph indicates that when compared to longer pairwise SNP distances, LD estimates for shorter pairwise SNP distances are relatively higher in the Central Asia sub-sample than in the Eastern Mediterranean sub-sample. The difference between sub-samples appears to be lost after about 15 cM.

distributions to make conservation predictions (e.g., [5,63]), our analysis does not consider the possibilities for the local adaptation of plant populations to new climatic conditions. There is little relevant research on this topic for wild barley, which ideally requires multiple time-interval-based monitoring of wild stands [64]. Nevo et al. [65], however, did assess genetic variation and phenotypic traits in the same 10 natural stands of wild barley in Israel sampled first in 1980 and then again in 2008. The authors observed some change in the distribution of nSSR alleles and larger changes in flowering times (accessions sampled in 2008 flowered significantly earlier under greenhouse conditions). These changes could indicate responses to climate change, although other explanations are also possible. Interestingly, the changes in wild barley nSSR composition observed by Nevo et al. [65] were much smaller than those found in wild emmer wheat (*Triticum dicoccoides*) populations included in the same study and sampled at the same dates. This suggests different responses to climate change by different cereals in the Eastern Mediterranean region. Third and finally, our current assessment was based on genetic markers that are presumably (mostly) neutral with regard to phenotype, so we are not able to determine whether or not there will be important losses in functional genetic diversity under anthropogenic climate change.

Final Considerations

Our analyses are consistent with the view that climate change has played a role in determining the levels of present-day genetic variation observed in wild barley in different portions of its natural range. Our data support the utility of ecological niche modelling of current and past plant distributions for predicting geographic areas of high genetic diversity, and suggest limited future losses of genetic diversity in wild barley under mid-term future climate change. We have explored the use of chromosome-position-mapped SNPs for discriminating between different hypotheses to explain diversity patterns in wild barley, but more research is required on this topic, ideally using SNPs that have been physically positioned in the genome. With the promise soon of a complete genome sequence for barley (building on the current sequence assembly [66,67]), the physical distances between the SNPs used in the current study will soon be available. This will allow more formal LD analysis of possible range expansions and contractions in relation to climate change. Further research on wild barley should also explore the ensemble forecasting of distributions based on both the differences between GCM and the multiple statistical methods available for species modelling [68,69]. Modelling should also investigate the possible further downscaling of environmental data in predictions [70], which could provide greater accuracy [34,71].

The wider utility of past-present ecological niche modelling for locating centres of genetic diversity in the Fertile Crescent and Central Asia regions could be tested by examining the wild progenitors of other important cereals located there [8]. Molecular marker data sets are available for comparison purposes (e.g., for einkorn wheat [*Triticum monococcum*] and emmer wheat [72,73]), although more systematic assessments of genetic diversity are required based on fully geo-referenced samples. It would be interesting to model the distributions of different wild cereals at the time it is proposed that humans began to manage them significantly. How modelled distributions correspond with putative sites of first cultivation and first domestication [74] could then be explored. For example, modelling to understand distributions over

the transition to the Younger Dryas (~12,000 years ago) would be useful. This was a relatively cold and unfavourable period for humans in the Fertile Crescent region that is believed to be associated with early cultivation events, leading eventually to domestications [75]. Distribution modelling of wild barley combined with geographically coincident sampling and genetic analysis of wild and landrace accessions of the taxon (and of other cereals) throughout the region will provide important insights into domestication processes [26].

Acknowledgments

We thank Shiaoman Chao, Nicola Cook, Linzi Jorgensen, Andrzej Kilian and Joy Roy who helped assemble the molecular marker data sets used in the current study, and Carlos Navarro and Denis Marechal who helped prepare future climate data. Dave Miller and Ian Williamson helped in the preparation of figures. Anonymous review greatly improved the manuscript. MvZ thanks the CGIAR Research Programme on Climate Change, Agriculture and Food Security (CCAFS) for support.

Author Contributions

Conceived and designed the experiments: JR MvZ IKD RW BS. Performed the experiments: JR IKD AB BS. Analyzed the data: JR MvZ IKD RW BS. Wrote the paper: JR MvZ IKD AB RW BS.

References

1. Warren R, VanDerWal J, Price J, Welbergen JA, Atkinson I, et al. (2013) Quantifying the benefit of early climate change mitigation in avoiding biodiversity loss. Nature Climate Change 3: 678–682.
2. Waltari E, Hijmans RJ, Peterson AT, Nyári ÁS, Perkins SL, et al. (2007) Locating Pleistocene refugia: comparing phylogeographic and ecological niche model predictions. PLoS ONE 2: e563. Available: www.plosone.org/
3. Provan J, Bennett KD (2008) Phylogeographic insights into cryptic glacial refugia. Trends in Ecology and Evolution 23: 564–571.
4. Nogués-Bravo D (2009) Predicting the past distribution of species climatic niches. Global Ecology and Biogeography 18: 521–531.
5. Thomas E, van Zonneveld M, Loo J, Hodgkin T, Galluzzi G, et al. (2012) Present spatial diversity patterns of *Theobroma cacao* L. in the neotropics reflect genetic differentiation in Pleistocene refugia followed by human-influenced dispersal. PLoS ONE 7: e47676. Available: www.plosone.org/
6. van Zonneveld M, Scheldeman X, Escribano P, Viruel MA, Van Damme P, et al. (2012) Mapping genetic diversity of cherimoya (*Annona cherimola* Mill.): application of spatial analysis for conservation and use of plant genetic resources. PLoS ONE 7: e29845. Available: www.plosone.org/
7. Newton AC, Flavell AJ, George TS, Leat P, Mullholland B, et al. (2011) Crops that feed the world 4. Barley: a resilient crop? Strengths and weaknesses in the context of food security. Food Security 3: 141–178.
8. Harlan JR (1975) Crops and man. Madison, Wisconsin, , USA: The American Society of Agronomy and the Crop Science Society of America.
9. Feuillet C, Langridge P, Waugh R (2008) Cereal breeding takes a walk on the wild side. Trends in Genetics 24: 24–32.
10. Nevo E, Chen G (2010) Drought and salt tolerances in wild relatives for wheat and barley improvement. Plant, Cell and Environment 33: 670–685.
11. Rostoks N, Mudie S, Cardle L, Russell J, Ramsay L, et al. (2005) Genome-wide SNP discovery and linkage analysis in barley based on genes responsive to abiotic stress. Molecular Genetics and Genomics 274: 515–527.
12. Ramsay L, Macaulay M, degli Ivanissevich S, MacLean K, Cardle L, et al. (2000) A simple sequence repeat-based linkage map of barley. Genetics 156: 1997–2005.
13. Provan J, Russell JR, Booth A, Powell W (1999) Polymorphic chloroplast simple sequence repeat primers for systematic and population studies in the genus *Hordeum*. Molecular Ecology 8: 505–511.
14. Elith J, Graham CH, Anderson RP, Dudík M, Ferrier S, et al. (2006) Novel methods improve prediction of species' distributions from occurrence data. Ecography 29: 129–151.
15. Hijmans RJ, Guarino L, Mathur P (2012) DIVA-GIS version 7.5 manual. California, , USA: University of California Davis.
16. Mitchell A (2005) The ESRI guide to GIS analysis, volume 2: spatial measurements and statistics. California, , USA: ESRI press.
17. Szpiech ZA, Jakobsson M, Rosenberg NA (2008) ADZE: a rarefaction approach for counting alleles private to combinations of populations. Bioinformatics 24: 2498–2504.
18. Groves CR, Game ET, Anderson MG, Cross M, Enquist C, et al. (2012) Incorporating climate change into systematic conservation planning. Biodiversity and Conservation 21: 1651–1671.
19. Steffenson BJ, Olivera P, Roy JA, Jin Y, Smith KP, et al. (2007) A walk on the wild side: mining wild wheat and barley collections for rust resistance genes. Australian Journal of Agricultural Research 58: 532–544.
20. Dai F, Nevo E, Wu D, Comadran J, Zhou M, et al. (2012) Tibet is one of the centers of domestication of cultivated barley. Proceedings of the National Academy of Sciences of the USA 109: 16969–16973.
21. Ren X, Nevo E, Sun D, Sun G (2013) Tibet as a potential domestication center of cultivated barley of China. PLoS ONE 8: e62700. Available: www.plosone.org/
22. Hijmans RJ, Cameron SE, Parra JL, Jones PG, Jarvis A (2005) Very high resolution interpolated climate surfaces for global land areas. International Journal of Climatology 25: 1965–1978.
23. Close TJ, Bhat PR, Lonardi S, Wu Y, Rostoks N, et al. (2009) Development and implementation of high-throughput SNP genotyping in barley. BioMed Central Genomics 10: 582. Available: www.biomedcentral.com
24. Roy JK, Smith KP, Muehlbauer GJ, Chao S, Close TJ, et al. (2010) Association mapping of spot blotch resistance in wild barley. Molecular Breeding 26: 243–256.
25. Moragues M, Comadran J, Waugh R, Milne I, Flavell AJ, et al. (2010) Effects of ascertainment bias and marker number on estimations of barley diversity from high throughput SNP genotype data. Trends in Applied Genetics 120: 1525–1534.
26. Russell JR, Dawson IK, Flavell A, Steffenson B, Weltzien E, et al. (2011) Analysis of >1000 single nucleotide polymorphisms in geographically matched samples of landrace and wild barley indicates secondary contact and chromosome-level differences in diversity around domestication genes. New Phytologist 191: 564–578.
27. Comadran J, Thomas W, van Eeuwijk F, Ceccarelli S, Grando S, et al. (2009) Patterns of genetic diversity and linkage disequilibrium in a highly structured *Hordeum vulgare* association-mapping population for the Mediterranean basin. Theoretical and Applied Genetics 119: 175–187.
28. Hardy OJ, Vekemans X (2002) SPAGeDi: a versatile computer program to analyse spatial genetic structure at the individual or population levels. Molecular Ecology Notes 2: 618–620.

29. Ritland K (1996) Estimators for pairwise relatedness and individual inbreeding coefficients. Genetical Research 67: 175–185.

30. Falush D, Stephens M, Pritchard JK (2007) Inference of population structure using multilocus genotype data: dominant markers and null alleles. Molecular Ecology Notes 7: 574–578.

31. Leberg PL (2002) Estimating allelic richness: effects of sample size and bottlenecks. Molecular Ecology 11: 2445–2449.

32. Davis AP, Gole TW, Baena S, Moat J (2012) The impact of climate change on indigenous arabica coffee (*Coffea arabica*): predicting future trends and identifying priorities. PLoS ONE 7: e47981. Available: www.plosone.org/

33. Elith J, Phillips SJ, Hastie T, Dudik M, Chee YE, et al. (2011) A statistical explanation of MaxEnt for ecologists. Diversity and Distributions 17: 43–57.

34. Aguirre-Gutiérrez J, Carvalheiro LG, Polce C, van Loon EE, Raes N, et al. (2013) Fit-for-purpose: species distribution model performance depends on evaluation criteria – Dutch hoverflies as a case study. PLoS ONE 8: e63708. Available: www.plosone.org/

35. Merow C, Smith MJ, Silander Jr JA (2013) A practical guide to MaxEnt for modeling species' distributions: what it does, and why inputs and settings matter. Ecography 36: 1058–1069.

36. van Zonneveld M, Jarvis A, Dvorak W, Lema G, Leibing C (2009) Climate change impact predictions on *Pinus patula* and *Pinus tecunumanii* populations in Mexico and Central America. Forest Ecology and Management 257: 1566–1576.

37. Phillips SJ, Dudik M (2008) Modeling of species distributions with Maxent: new extensions and a comprehensive evaluation. Ecography 31: 161–175.

38. Liu C, Berry PM, Dawson TP, Pearson RG (2005) Selecting thresholds of occurrence in the prediction of species distributions. Ecography 28: 385–393.

39. Ramirez J, Jarvis A (2010) Downscaling global circulation model outputs: the delta method. Decision and Policy Analysis Working Paper No. 1. Cali, Columbia: Centro Internacional de Agricultura Tropical (CIAT).

40. Vermeulen SJ, Challinor AJ, Thornton PK, Campbell BM, Eriyagama N, et al. (2013) Addressing uncertainty in adaptation planning for agriculture. Proceedings of the National Academy of Sciences of the USA 110: 8357–8362.

41. Brown AHD, Zohary D, Nevo E (1978) Outcrossing rates and heterozygosity in natural populations of *Hordeum spontaneum* Koch in Israel. Heredity 41: 49–62.

42. Petit RJ, Duminil J, Fineschi S, Hampe A, Salvini D, et al. (2005) Comparative organization of chloroplast, mitochondrial and nuclear diversity in plant populations. Molecular Ecology 14: 689–701.

43. Kozak KH, Graham CH, Wiens JJ (2008) Integrating GIS-based environmental data into evolutionary biology. Trends in Ecology and Evolution 23: 141–148.

44. Scheldeman X, van Zonneveld M (2010) Training manual on spatial analysis of plant diversity and distribution. Rome, Italy: Bioversity International.

45. van Etten J, Hijmans RJ (2010) A geospatial modelling approach integrating archaeobotany and genetics to trace the origin and dispersal of domesticated plants. PLoS ONE 5: e12060. Available: www.plosone.org/

46. Volis S, Mendlinger S, Turuspekov Y, Esnazarov U, Abugalieva S, et al. (2001) Allozyme variation in Turkmenian populations of wild barley, *Hordeum spontaneum* Koch. Annals of Botany 87: 435–446.

47. Fu Y-B, Horbach C (2012) Genetic diversity in a core subset of wild barley germplasm. Diversity 4: 239–257.

48. Hill WG, Robertson A (1968) Linkage disequilibrium in finite populations. Theoretical and Applied Genetics 38: 226–231.

49. Kelly JK (1997) A test of neutrality based on interlocus associations. Genetics 146: 1197–1206.

50. Librado P, Rozas J (2009) DNASP version 5, a software for comprehensive analysis of DNA polymorphism data. Bioinformatics 25: 1451–1452.

51. Slatkin M (2008) Linkage disequilibrium – understanding the evolutionary past and mapping the medical future. Nature Reviews Genetics 9: 477–485.

52. Excoffier L, Foll M, Petit RE (2009) Genetic consequences of range expansions. Annual Review of Ecology, Evolution, and Systematics 40:481–501.

53. Siol M, Wright SI, Barrett SCH (2010) The population genomics of plant adaptation. New Phytologist 188: 313–332.

54. McVean GAT (2002) A genealogical interpretation of linkage disequilibrium. Genetics 162: 987–991.

55. Caldwell KS, Russell J, Langridge P, Powell W (2006) Extreme population-dependent linkage disequilibrium detected in an inbreeding plant species, *Hordeum vulgare*. Genetics 172: 557–567.

56. Tenesa A, Navarro P, Hayes BJ, Duffy DL, Clarke GM, et al. (2007) Recent human effective population size estimated from linkage disequilibrium. Genome Research 17: 520–526.

57. Abdel-Ghani AH, Parzies HK, Omary A, Geiger HH (2004) Estimating the outcrossing rate of barley landraces and wild barley populations collected from ecologically different regions of Jordan. Theoretical and Applied Genetics 109: 588–595.

58. Morrell PL, Toleno DM, Lundy KE, Clegg MT (2006) Estimating the contribution of mutation, recombination and gene conversion in the generation of haplotypic diversity. Genetics 173: 1705–1723.

59. Keppel G, Van Niel KP, Wardell-Johnson GW, Yates CJ, Byrne M, et al. (2012) Refugia: identifying and understanding safe havens for biodiversity under climate change. Global Ecology and Biogeography 21: 393–404.

60. Shoo LP, Hoffmann AA, Garnett S, Pressey RL, Williams YM, et al. (2013) Making decisions to conserve species under climate change. Climatic Change 119: 239–246.

61. Goudet J (1995) FSTAT version 1.2: a computer program to calculate F statistics. Journal of Heredity 86: 485–486.

62. Bishaw Z, Struik PC, van Gastel AJG (2011) Wheat and barley seed system in Syria: farmers, varietal perceptions, seed sources and seed management. International Journal of Plant Production 5: 323–347.

63. Bálint M, Domisch S, Engelhardt CHM, Haase P, Lehrian S, et al. (2011) Cryptic biodiversity loss linked to global climate change. Nature Climate Change 1: 313–318.

64. Hansen MM, Olivieri I, Waller DM, Nielsen EE, the GeM Working Group (2012) Monitoring adaptive genetic responses to environmental change. Molecular Ecology 21: 1311–1329.

65. Nevo E, Fu Y-B, Pavlicek T, Khalifa S, Tavasi M, et al. (2012) Evolution of wild cereals during 28 years of global warming in Israel. Proceedings of the National Academy of Sciences of the USA 109: 3412–3415.

66. IBSC (The International Barley Genome Sequencing Consortium) (2012) A physical, genetic and functional sequence assembly of the barley genome. Nature 491: 711–717.

67. Mascher M, Richmond TA, Gerhardt DJ, Himmelbach A, Leah C, et al. (2013) Barley whole exome capture: a tool for genomic research in the genus *Hordeum* and beyond. The Plant Journal 76: 494–505.

68. Araujo MB, New M (2007) Ensemble forecasting of species distributions. Trends in Ecology and Evolution 22: 42–47.

69. Buisson L, Thuiller W, Casajus N, Lek S, Grenouillet G (2010) Uncertainty in ensemble forecasting of species distribution. Global Change Biology 16: 1145–1157.

70. Franklin J, Davis FW, Ikegami M, Syphard AD, Flint LE, et al. (2013) Modeling plant species distributions under future climates: how fine scale do climate projections need to be? Global Change Biology 19: 473–483.

71. Conlisk E, Syphard AD, Franklin J, Flint L, Flint A, et al. (2013) Uncertainty in assessing the impacts of global change with coupled dynamic species distribution and population models. Global Change Biology 19: 858–869.

72. Brown TA, Jones MK, Powell W, Allaby RG (2008) The complex origins of domesticated crops in the Fertile Crescent. Trends in Ecology and Evolution 24: 103–109.

73. Burger JC, Chapman MA, Burke JM (2008) Molecular insights into the evolution of crop plants. American Journal of Botany 95: 113–122.

74. Willcox G (2005) The distribution, natural habitats and availability of wild cereals in relation to their domestication in the Near East: multiple events, multiple centres. Vegetation History and Archaeobotany 14: 534–541.

75. Fuller DQ (2007) Contrasting patterns in crop domestication and domestication rates: recent archaeobotanical insights from the Old World. Annals of Botany 100: 903–924.

76. Vekemans X, Hardy OJ (2004) New insights from fine-scale spatial genetic structure analyses in plant populations. Molecular Ecology 13: 921–935.

77. Farr TG, Rosen PA, Caro E, Crippen R, Duren R, et al. (2007) The Shuttle Radar Topography Mission. Reviews of Geophysics 45: RG2004. doi: 10.1029/2005RG000183

78. Llaurens V, Castric V, Austerlitz F, Vekemans X (2008) High paternal diversity in the self-incompatible herb *Arabidopsis halleri* despite clonal reproduction and spatially restricted pollen dispersal. Molecular Ecology 17: 1577–1588.

79. Ismail SA, Ghazoul J, Ravikanth G, Shaanker RU, Kushalappa CG, et al. (2012) Does long-distance pollen dispersal preclude inbreeding in tropical trees? Fragmentation genetics of *Dysoxylum malabaricum* in an agro-forest landscape. Molecular Ecology 21: 5484–5496.

Exploring Germplasm Diversity to Understand the Domestication Process in *Cicer* spp. Using SNP and DArT Markers

Manish Roorkiwal[1], Eric J. von Wettberg[2,3], Hari D. Upadhyaya[1], Emily Warschefsky[2,3], Abhishek Rathore[1], Rajeev K. Varshney[1]*

1 International Crops Research Institute for the Semi-Arid Tropics (ICRISAT), Hyderabad, Andhra Pradesh, India, **2** Department of Biological Sciences, Florida International University, Miami, Florida, United States of America, **3** Center for Tropical Plant Conservation, Fairchild Tropical Botanic Garden, Miami, Florida, United States of America

Abstract

To estimate genetic diversity within and between 10 interfertile *Cicer* species (94 genotypes) from the primary, secondary and tertiary gene pool, we analysed 5,257 DArT markers and 651 KASPar SNP markers. Based on successful allele calling in the tertiary gene pool, 2,763 DArT and 624 SNP markers that are polymorphic between genotypes from the gene pools were analyzed further. STRUCTURE analyses were consistent with 3 cultivated populations, representing kabuli, desi and pea-shaped seed types, with substantial admixture among these groups, while two wild populations were observed using DArT markers. AMOVA was used to partition variance among hierarchical sets of landraces and wild species at both the geographical and species level, with 61% of the variation found between species, and 39% within species. Molecular variance among the wild species was high (39%) compared to the variation present in cultivated material (10%). Observed heterozygosity was higher in wild species than the cultivated species for each linkage group. Our results support the Fertile Crescent both as the center of domestication and diversification of chickpea. The collection used in the present study covers all the three regions of historical chickpea cultivation, with the highest diversity in the Fertile Crescent region. Shared alleles between different gene pools suggest the possibility of gene flow among these species or incomplete lineage sorting and could indicate complicated patterns of divergence and fusion of wild chickpea taxa in the past.

Editor: Tianzhen Zhang, Nanjing Agricultural University, China

Funding: This study was funded by grants from Australian Indo Strategic Research Fund (AISRF) of Ministry of Science & Technology and CGIAR Generation Challenge Programme (GCP), Mexico. This work has been undertaken as part of the CGIAR Research Program on Grain Legumes. ICRISAT is a member of CGIAR Consortium. Thanks are also due to several colleagues at ICRISAT and partners in collaborating centres. EvW received support from HHMI award #52006924 to FIU's discipline-based education research group, and USDA-NIFA- Hispanic Serving Institutions Grant 2011-38422-30804 to FIU's agroecology program. EW is supported by the Fairchild Challenge of Fairchild Tropical Botanic Garden. The funders had no role in study design, data collection and analysis, decision to publish, or preparation of the manuscript.

Competing Interests: The authors have declared that no competing interests exist.

* Email: r.k.varshney@cgiar.org

Introduction

Many crops that are grown across multiple regions have limited genetic diversity due to bottlenecks from domestication, selective breeding and in some taxa, natural processes [1–4]. Recurrent selection of improved cultivars over multiple generations results in an increasingly narrow genetic base for a crop, making it more vulnerable to disease and limiting its adaptability. Such genetically depauperate crops could have disastrous consequences in the face of emerging diseases and climate change [5,6]. Recent applications of genome mapping suggest that the genetic diversity stored in germplasm banks can be utilized with a much higher level of efficiency than previously imagined [6,7]. This is particularly true for self-pollinated crops like chickpea (*Cicer arietinum*). During the past few decades, our understanding of the importance of plant genetic resources and the need to conserve them has grown [8], and wild relatives are now commonly seen as a key source of genetic diversity that can be used to increase diversity in breeding

material [7,9]. Diversity estimates of germplasm collections have not been universally performed to assess the scope of diversity available in existing collections. Such estimates are critical for providing insight into efforts to introgress wild germplasm into elite lines, and for guiding future collections of wild germplasm [10].

In order to make more efficient use of wild relatives, we need improved classifications of their relationship to crop material and to other wild species [11]. Characterizing patterns of diversity within the secondary and tertiary gene pools [12] can provide insight into which subdivisions of germplasm collections contain wild material that is most likely to increase diversity and can guide the use of wild material in breeding efforts. Although wild material is rarely used in breeding programs due to agronomically poor traits, it remains a chief reservoir for many disease and abiotic stress resistance traits. Effective characterization of wild material can facilitate its more effective use [13].

Chickpea is an important crop in semi-arid tropical regions such as South Asia and Eastern & Southern Africa, Mediterranean regions, and cool temperate areas [14]. Globally, chickpea is the second most widely consumed legume after beans (*Phaseolus*) [15]. Lack of genetic diversity has long been a critical problem for chickpea breeding [16], limiting efforts to improve resistance to diseases like *Ascochyta* blight and *Fusarium* wilt, pod borer insects, and tolerance to abiotic stresses like terminal drought, high and low temperatures [17,18]. Chickpea reference set has also been used to understand the available diversity for stress responsive genes [19]. Widening the genetic diversity of cultivated chickpea is dependent on the introduction of alleles controlling the traits of interest from wild germplasm [1]. Currently chickpea's immediate ancestor, *C. reticulatum*, and its interfertile sister species *C. echinospermum*, is the main source of new variation, although introgression is possible from the more distantly related gene pools with greater effort [20].

Cultivated chickpea first appears in the archaeological record some 6.6–7.2 thousand years ago in Syria [21,22]. The immediate wild relatives (*C. reticulatum* and *C. echinospermum*) of chickpea are restricted to southeastern Turkey [1]. Domestication is thought to have happened earlier, as much as 10.5 thousand years ago, concurrent with or soon after the domestication of other Fertile Crescent crops such as wheat, barley, pea, and lentil. Domesticated chickpea was likely brought to Syria about 7,000 years ago, while records for the dates of introduction into East Africa and the Indian subcontinent are limited [22]. Abbo and co-workers [1,23] have speculated that chickpea is particularly genetically depauperate because it may have gone through four distinct bottlenecks: modern breeding, domestication, a shift early in its cultivation from a winter annual phenology to a spring phenology, and wild relatives (particularly *C. reticulatum* and *C. echinospermum*) that have a narrow geographic distribution compared to other crops domesticated in the Fertile Crescent. The shift in phenology may have accompanied the introduction of other crops such as sesame and sorghum that are summer annuals [24]. Breeding for preferred phenotypes, such as seed colour and shape, may exacerbate chickpea's narrow genetic base and may be one of the key reasons for slow progress in yield improvement and increased tolerance to various biotic and abiotic stresses. Based on seed shape, size and colour, chickpea is classified into two seed types, kabuli and desi. The kabuli chickpea is characterized by a larger, cream-coloured seed with a thin seed coat, while the desi seed type has a smaller, darker coloured seed with a thick seed coat. In addition, a third seed type, designated as intermediate or pea-shaped, is characterized by medium to small size and round, pea-shaped seeds [25].

Single nucleotide polymorphism (SNP) markers have become the markers of choice for various genome wide analyses because they are widespread across genomes, accurate and reproducible, and well suited to automated detection [26]. A range of low- to high-throughput SNP genotyping platforms have become available to make SNP genotyping cost-effective such as BeadXpress, KBioscience Competitive Allele-Specific Polymerase chain reaction (KASPar) assays, and GoldenGate assays from Illumina Inc. [27,28]. In addition, another high-throughput marker system, Diversity arrays technology (DArT), has proven useful for screening large numbers of loci in crops with low genetic diversity, and DArT markers for chickpea have recently been developed [29].

The present study is focused on the assessment of relationships in a diversity panel of chickpea which includes breeding material from the three seed types (kabuli, desi, and pea-shaped) and wild species from the primary, secondary, and tertiary gene pools using KASPar technology and hybridization based DArT arrays for high-throughput SNP genotyping. We examined the level of genetic differentiation among these groups of genotypes and assessed how segregating variation is spread across the genome of chickpea.

Materials and Methods

Germplasm and DNA isolation

A diverse set of 94 chickpea genotypes (Table S1) including 66 cultivars and landraces (23 desi, 41 kabuli, and 2 pea-shaped seed type genotypes) and 28 genotypes from 9 wild species including genotypes from primary, secondary and tertiary gene pool was selected as a diversity panel for assessment from the ICRISAT germplasm collection [30].

Total genomic DNA was isolated from 10–12 leaves of two week old plants following a modified CTAB protocol as described in Cuc et al. [31]. Only one plant per accession was used for DNA isolation. DNA quality and quantity for each sample was assessed on 0.8% agarose gel.

Genotyping

SNPs were identified using four different approaches: Solexa/Illumina sequencing, mining of Sanger Expressed Sequence Tags (ESTs), allele-specific sequencing of candidate genes, and allele-specific sequencing of tentative orthologous genes (TOGs) as described by Hiremath et al. [28]. In total, 2,486 SNPs were used for validation and development of KASPar assays by KBioscience, of which 2,005 (80.6%) assays could be validated and designated as Chickpea KASPar Assay Markers (CKAMs) [28]. A subset of highly polymorphic 651 CKAMs was used for genotyping using KASPar assays. In addition, this diverse set was also genotyped with high-density DArT array with 15,360 DArT clones as described in Thudi et al. [29].

Data Analysis

The germplasm was divided into three different clusters based on geographical origin, namely the Fertile Crescent, Central and South Asia, and Ethiopian Highlands (Figure 1). Additionally, germplasm was classified based on gene pools (primary, secondary, and tertiary) [32], seed type (desi, kabuli, and pea-shaped) and wild vs. cultivated species. The purpose of these different divisions of the data was to determine the scale over which genetic variation is present in the germplasm collection. In order to assess hierarchical levels of variation within and between different sub-groups, DArT and SNP genotyping data were analyzed separately. AMOVA was conducted on the DArT markers based on the hierarchical model and permutational procedures of Excoffier et al. [33] to assess the level of variation among these wild and domesticated groups. We implemented AMOVA in GenAlEx 6.5 [34,35] and Arlequin [36]. AMOVA analysis with populations nested within regions was also performed to examine the distribution of variation and differential connectivity among populations (PhiPT; an analogue of Fst, i.e., genetic diversity among populations). In addition, Shannon information index (measure of species diversity in a population) was calculated for all the population using GenAlEx 6.5. This index provides important information about rarity and commonness of species in a community by taking relative abundances of different species into account [34,37].

A separate AMOVA was performed on the SNP data to assess variation within and among desi, kabuli, and pea-shaped seed types. In both AMOVAs, we assessed genetic variation within groups (Fct), within populations (Fst), between populations within a group (Fsc), population polymorphism, and Nei's genetic distance and gene flow (Nm) using GenAlEx v.6.41 [34,35] and

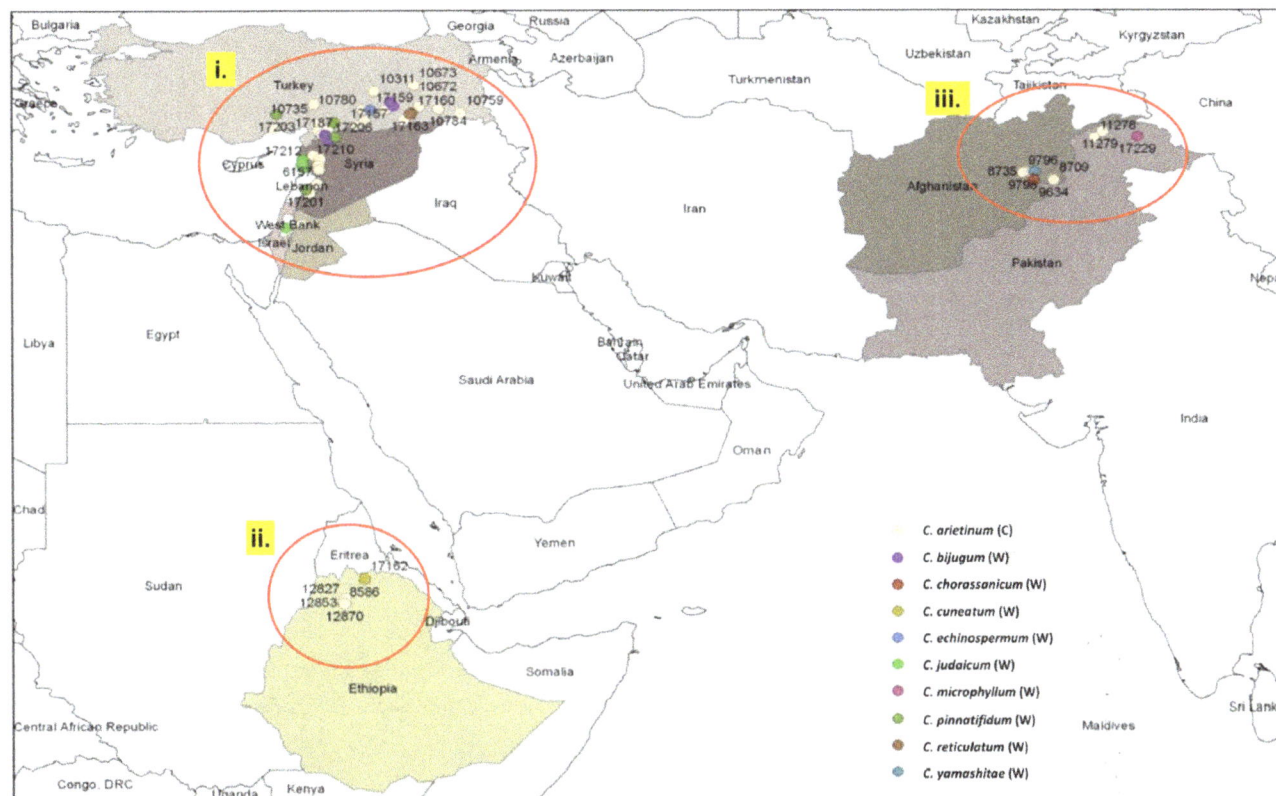

Figure 1. Geographic locations of cultivated and wild *Cicer* species collection sites (C: Cultivated; W: Wild) i. Fertile Crescent; ii. Ethiopia; iii. Central Asia.

Arlequin [36]. For each group presence of private alleles (np), percentage of polymorphic loci (%p), the average number of alleles per locus (k), the expected heterozygosity (He), and unbiased expected heterozygosity (UHe) across different subgroups (i.e., wild species *vs* cultivated with the DArT markers and seed type with the SNP markers) was calculated. The polymorphism information content (PIC) values for SNP and DArT markers across 94 diverse genotypes were calculated by using Power-Marker software [38].

STRUCTURE 2.3 [39] was used to estimate the number of natural genetic groups (K), the distribution of individuals among these groups, and to assign individual genotypes to a specified number of groups "K" based on membership coefficients calculated from the genotype data. This approach is an important complement to the hierarchical division of the germplasm (see above), as it can determine the number of groups best supported by the DArT and SNP data. DArT data was converted in to psuedo-diploid format by assigning a row of missing data to each individual so that it could be analysed with STRUCTURE. We assessed a range of population numbers from K = 1 to K = 15 using a burn-in period of 50,000 steps followed by 500,000 MCMC (Monte Carlo Markov Chain) replicates with 3X iterations, assuming admixture and correlated allele frequencies. Due to missing SNP calls in the wild material, data from wild material was separated from that of cultivated material and a separate STRUCTURE analysis of cultivated material alone was performed using SNP markers. In order to compliment the STRUCTURE analyses, pair-wise genetic differentiation between individuals was calculated from the DArT markers, which was used in principal coordinate analysis (PCoA), implemented in GenAlEx

6.5. These analyses labelled the material based on its source region: the Fertile Crescent, Central Asia, and the Ethiopian highlands.

A complementary approach to assessing relationships among taxa is a phylogenetic analysis. Distance-based phylogenetic analysis of SNP data was performed using the software package Geneious v. 7.0.6 (Biomatters) (http://www.geneious.com). A cladogram was produced using unweighted pair-group method with arithmetic mean (UPGMA) cluster analysis under the Jukes-Cantor genetic distance model with 100 bootstrap replications. The consensus tree was then rooted with the clade of individuals from the tertiary gene pool.

Results

Marker attributes

In total, 651 SNP markers using KASPar assays and DArT arrays were used for genotyping the set of 94 diverse chickpea genotypes. This set includes 66 cultivated chickpea genotypes and 27 wild relatives representing eight wild *Cicer* species from primary, secondary, and tertiary gene pools along with one perennial wild chickpea genotype. The genotypes were carefully selected to represent geographical areas with the most phenotypic diversity: the Fertile Crescent, Central Asia, and the Ethiopian highlands (Figure 1). SNP markers were highly polymorphic across this diverse set and a total of 611 SNPs were found polymorphic. The polymorphic information content (PIC) value ranged from 0.02 to 0.50 across these 94 genotypes with mean PIC value of 0.23 (Figure 2a). Although these SNPs were highly polymorphic, in many cases SNPs could not be called for wild chickpea genotypes (Table S2). SNPs were developed using cultivated chickpea and

a.

b.

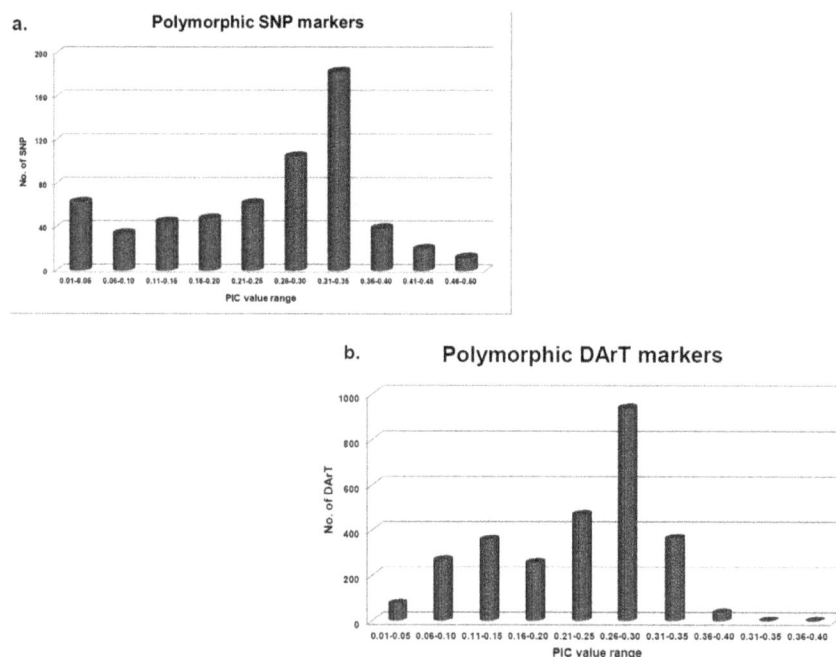

Figure 2. Polymorphism information content (PIC) value of markers used in study. a. PIC value of SNP markers used for diversity analysis. b. PIC value of DArT markers used for diversity analysis.

later used for genotyping the wild species, which may account for the greater number of missing loci in the tertiary gene pool and the bimodal distribution of PIC values.

To overcome the issue of missing data in the wild material and to compliment the SNP data, the set was genotyped using high density DArT arrays with 15,360 clones [29]. A total, 5,257 DArT markers were polymorphic across 94 lines. Of these, a subset of 2,763 markers was selected for use in the present study based on the presence of the allele in wild chickpea (tertiary gene pool). PIC for these 2,763 DArT markers ranged from 0.02 to 0.37, with an average of 0.22 across the 94 genotypes (Figure 2b) (Table S3).

Differences among the wild species and cultivated germplasm

The chickpea diversity panel used in the present study is comprised of 94 genotypes from 9 wild species (8 annual and 1 perennial) and cultivated species (*C. arietinum*). DArT data was used to understand the diversity and genetic architecture of the germplasm. As expected, wild species genotypes had higher levels of polymorphic markers (99.60%) compared to cultivated genotypes (35.79%) (Table 1). A UPGMA tree was constructed based on pairwise genetic distances using the SNP markers to understand the relationships between the genotypes from wild and cultivated species (Figure 3). Two major groups were identified by this analysis, separating wild from cultivated genotypes. Cultivated and wild species genotypes from the primary gene pool were grouped in one cluster (Figure 3). However, genotypes from the chickpea ancestor, *C. reticulatum*, were interspersed with those from cultivated individuals, consistent with a close relationship between ancestral and cultivated chickpea. Genotypes from the secondary gene pool species were found to cluster together, as were genotypes from the tertiary gene pool.

In parallel, STRUCTURE was also used to understand the clustering between cultivated and wild species genotypes. With the DArT data, STRUCTURE resolved four clusters using the

Evanno method (Figure 4a). This grouping indicates a substantial difference between wild and cultivated material, as well as major differences within the wild material. These results suggest that there are three major groups among the wild material (Figure 4a), corresponding to different gene pools. Individuals in the tertiary gene pool are represented largely as one cluster with admixture; although these individuals represent several species (with the capacity to hybridize) and are certainly not a homogenous group, they do cluster together. The perennial species in the tertiary gene pool, *C. microphyllum*, appears admixed with the primary gene pool. However, this could be due to its closer phylogenetic relationship to *C. reticulatum* or accidental gene flow in the germplasm collection. The secondary gene pool, with the closely related and interfertile species of *C. pinnatifidum*, *C. bijugum* and *C. judaicum* formed one tight cluster. The immediate ancestors of the crop, *C. reticulatum* and *C. echinospermum*, show up as a group with substantial admixture with the cultivated individuals. This could represent the derivation of the crop, and could also represent introgression from the crop to the wild species (or artefacts of maintenance in germplasm facilities). The cultivated accessions of *C. arietinum* showed little admixture with the wild material in this analysis.

In addition, principal coordinate analysis, which was performed as a complementary approach to display clustering of genotypes, separated cultivated genotypes from wild species genotypes. Few genotypes of the wild chickpea clustered with cultivated material. Those wild genotypes that did cluster were *C. reticulatum*, the likely progenitor of cultivated chickpea (Figure 4b). The PCoA showed substantial differences among the wild material; *C. reticulatum* and *C. echinospermum* genotypes clustered with closely related cultivated material (Figure 4b). However, the closely related species from the secondary gene pool clustered individually rather than all clustering together. Furthermore, genotypes from a species in the tertiary gene pool, *C. yamashatae*, clustered more closely with the primary gene pool than did the species of the secondary gene pool. AMOVA partitioned 39% of variation between wild and

Table 1. Assessment of genetic diversity across groups of wild and cultivated chickpea using DArT markers.

	Polymorphic marker (%)	N	Na	Ne	I	He	UHe
Cultivated	35.79	63.401±0.04	1.219±0.013	1.096±0.004	0.113±0.004	0.068±0.002	0.068±0.002
Wild	99.60	27.143±0.023	1.996±0.001	1.766±0.005	0.607±0.002	0.421±0.002	0.429±0.002
Mean	67.70	45.272±0.245	1.607±0.008	1.431±0.005	0.360±0.004	0.244±0.003	0.249±0.003

No. of polymorphic alleles (N), No. of Different Alleles (Na), No. of Effective Alleles (Ne, = 1/(Sum pi^2)), Shannon's Information Index (I = −1 * Sum (pi * Ln (pii)), Expected Heterozygosity (He = 1−Sum pi^2) and Unbiased Expected Heterozygosity (UHe = (2N/(2N−1)) * He).

cultivated groups and 61% of variation segregating within groups (Figure 4c).

Genetic diversity among the genotypes from wild chickpea

The present study included analysis of 28 chickpea genotypes from nine wild species including genotypes from primary, secondary, tertiary gene pools and one individual of a perennial species, *C. microphyllum*. Genotyping using SNP markers resulted in high rates of failed SNP allele calls and null alleles. We therefore used DArT data to estimate the genetic diversity and relationships among the cultivated and wild species genotypes for primary, secondary and tertiary gene pools. AMOVA of wild species genotypes indicated that 31% of variation was found among the species while 69% of variation was observed within the species. Genetic distance between populations (primary, secondary and tertiary) was calculated based on Nei's genetic distance. As expected, higher similarity was observed between the primary and secondary gene pools (Nei's genetic distance 0.15), while greater distance was observed between primary and tertiary gene pools (Nei's genetic distance 0.69). Furthermore, a greater distance was observed between the secondary and tertiary gene pools than between the primary and secondary gene pools, which suggests that genotypes from the primary and secondary gene pools are more closely related to each other than to the tertiary gene pool. Across all wild material, numbers of effective alleles and values of heterozygosity were much higher than in the crop material. Within the wild material, the secondary gene pool had the greatest diversity, with highest effective allele estimates and highest heterozygosity (Table 2).

In the PCoA of the wild material alone (Figure S1a), a few genotypes from the primary gene pool clustered with the tertiary gene pool genotypes. Other genotypes from the primary gene pool clustered with the secondary gene pool. In parallel, we performed a STRUCTURE analysis on the 28 wild species genotypes using DArT markers. The STRUCTURE results complemented the observation from PCoA and diversity analysis (Figure S1b). We selected K = 2 based on Evanno method. The first cluster corresponds to the primary gene pool, while the second cluster corresponds to the secondary gene pool. The tertiary gene pool was admixed, likely representing the great diversity in those disparate species.

Genetic diversity among phenotypic classes of cultivated chickpea

Diversity among the 66 cultivated genotypes was assessed using both the DArT and SNP markers. These 66 genotypes were classified in three sub-groups based on seed type, *i.e.* desi, kabuli and pea-shaped. SNP markers were used in the program STRUCTURE to resolve differences among phenotypic classes of cultivated chickpea. Three groups of the cultivated material (K = 3) were observed, with most individuals demonstrating substantial admixture (Figure S2a). Genetic diversity among the phenotypic classes was also assessed using DArT and SNP markers (Table 3). The number of effective alleles (Ne) and heterozygosity (He) were very similar among the phenotypic classes (with overlapping standard deviations around their means), and all values were low (i.e., <1.1 for Ne, and <0.1 for He). Hierarchical AMOVA using both SNP and DArT data provided similar results. More than 90% of variation was observed within these phenotypic classes, while only about 10% variation was reported among these different populations (Figure S2b).

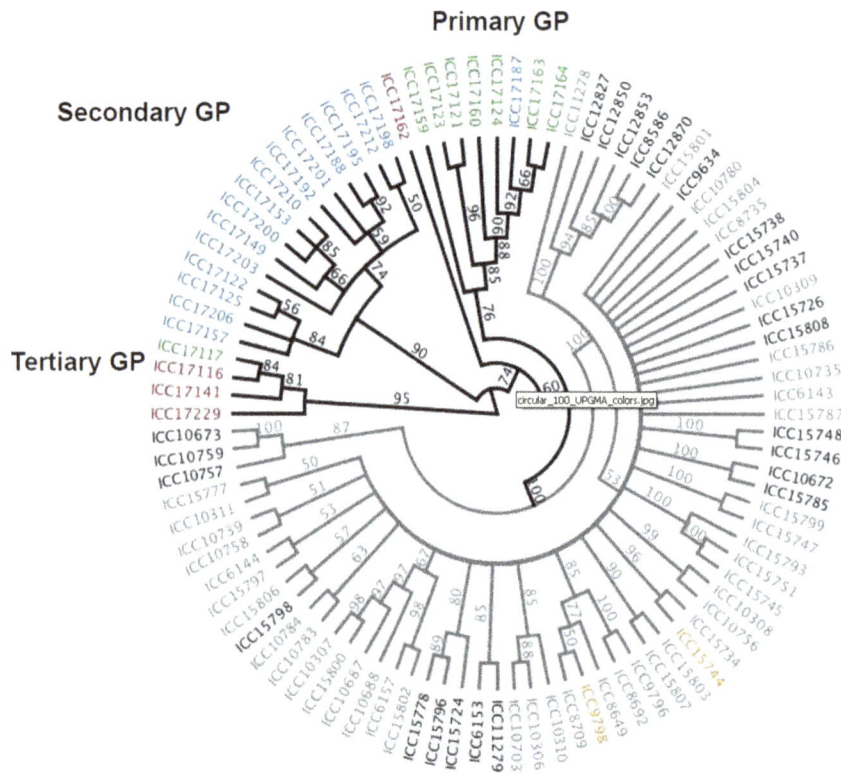

Figure 3. UPGMA tree of pairwise relatedness of cultivated (grey branches) and wild (black branches) chickpea. Genepools and seed types are represented by the following colors: primary, green; secondary, blue; tertiary, red; pea-shaped, orange; kabuli, grey; and desi, black.

Genetic diversity among the cultivars from different geographic regions

To understand the diversity in chickpea cultivars from different regions, an analysis was also performed based on the geographical distribution of cultivated and wild species genotypes. Based on geographical origin, germplasm was divided in three clusters: the Fertile Crescent, Central Asia, and the Ethiopian highlands. Substantial geographic variation was observed, with the greatest

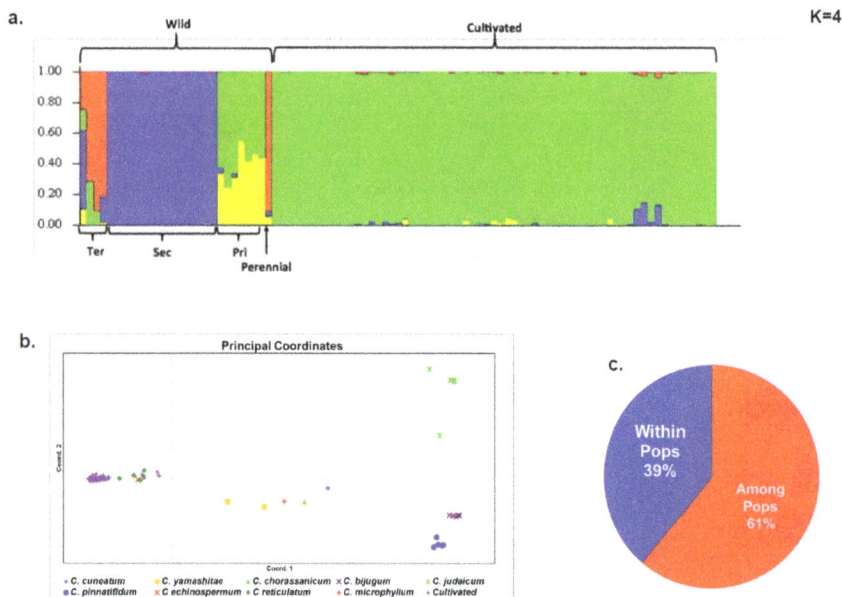

Figure 4. Population structure analysis using STRUCTURE of *Cicer* accessions. a. Structure showing distinct group of wild and cultivated species; wild further classified in primary (Pri), secondary (Sec) and tertiary (Ter) gene pool species. b. Principal coordinates analysis among wild and cultivated species. c. Analysis of molecular variance between and among wild and cultivated species genotypes.

Table 2. Assessment of genetic diversity across wild germplasm using DArT markers.

	Polymorphic marker (%)	N	Na	Ne	I	He	UHe
Primary GP	74.34	6.915±0.006	1.743±0.008	1.521±0.007	0.429±0.005	0.294±0.004	0.317±0.004
Secondary GP	95.04	15.519±0.014	1.935±0.006	1.63±0.007	0.521±0.004	0.353±0.003	0.365±0.003
Tertiary GP	10.82	3.754±0.011	0.86±0.011	1.072±0.004	0.061±0.003	0.041±0.002	0.048±0.003
Mean	60.07	8.729±0.055	1.513±0.007	1.407±0.004	0.337±0.003	0.229±0.002	0.243±0.002

No. of polymorphic alleles (N), No. of Different Alleles (Na), No. of Effective Alleles (Ne, = 1/(Sum pi^2)), Shannon's Information Index (I = −1* Sum (pi * Ln (pi)), Expected Heterozygosity (He = 1−Sum pi^2) and Unbiased Expected Heterozygosity (UHe = (2N/(2N−1)) * He).

Table 3. Assessment of genetic diversity across chickpea germplasm based on seed type.

Marker Type	Seed type	Polymorphic marker (%)	N	Na	Ne	I	He	UHe
SNP	kabuli	15.22	41.747±0.026	1.152±0.014	1.046±0.006	0.052±0.006	0.031±0.004	0.032±0.004
	desi	4.81	23.897±0.014	1.048±0.009	1.031±0.006	0.025±0.005	0.017±0.003	0.018±0.003
	pea	2.56	1.901±0.012	1.026±0.006	1.026±0.006	0.018±0.004	0.013±0.003	0.017±0.004
	Total	7.53	22.515±0.377	1.075±0.006	1.034±0.004	0.032±0.003	0.02±0.002	0.022±0.002
DArT	kabuli	18.64	39.101±0.028	0.987±0.012	1.06±0.003	0.065±0.003	0.04±0.002	0.041±0.002
	desi	26.89	22.348±0.019	1.094±0.013	0.11±0.004	0.109±0.004	0.069±0.003	0.071±0.003
	pea	0.54	1.952±0.004	0.725±0.009	1.004±0.001	0.003±0.001	0.002±0.001	0.003±0.001
	Total	15.36	21.134±0.167	0.936±0.007	1.058±0.002	0.059±0.002	0.037±0.001	0.038±0.001

No. of polymorphic alleles (N), No. of Different Alleles (Na), No. of Effective Alleles (Ne, = 1/(Sum pi^2)), Shannon's Information Index (I = −1* Sum (pi * Ln (pi)), Expected Heterozygosity (He = 1−Sum pi^2) and Unbiased Expected Heterozygosity (UHe = (2N/(2N−1)) * He).

diversity found in the Fertile Crescent and much lower diversity in the Ethiopian highlands and central Asia (Table 4). In parallel, PCoA was also performed (Figure 5). Outside of the Fertile Crescent, wild and cultivated material did not cluster together, which is consistent with a single domestication in the Fertile Crescent followed by dispersal to Central and South Asia and the East African highlands and subsequent divergence (Figure 5).

Discussion

Chickpea is believed to have been domesticated 10,000 years ago in southeastern Turkey and adjoining Syria [40–42]. The crop suffers from a narrow genetic base among the cultivated germplasm, which may be due to four population bottlenecks the crop has experienced [1]. This low genetic diversity makes the crop more susceptible to a range of diseases and pests [1,17]. Recently, Varshney et al. [43] also confirmed the problem of narrow diversity in elite chickpea using whole genome re-sequencing of 90 chickpea lines. Wild relatives of chickpea could serve an important role in enhancing the genetic base of cultivated material. In an effort to understand the genetic diversity available in cultivated and wild gene pools, the present study was undertaken using SNP and DArT markers. Genetic diversity was analyzed for these loci across a panel of domesticated and wild germplasm in the ICRISAT collection [30].

Understanding the available genetic diversity in the germplasm collection is a pre-requisite to adopt effective conservation and management strategies to use these genetic resources in crop improvement. Understanding patterns of genetic diversity can complement efforts to match collections from differing climatic regions to planting zones differing in climate [24]. The present study focuses on exploration of the genetic diversity and population structure of this diverse set of chickpea that includes cultivated and wild species genotypes ranging from primary to tertiary gene pools [12]. Global research efforts have resulted in the development of a large number of markers (SSR, SNPs, DArT) and genotyping platforms that can be used to study genetic diversity and explore the diverse germplasm for the traits to use in chickpea improvement programs [44]. KASPar assay from KBiosciences (Hertfordshire, UK) (http://www.kbioscience.co.uk) provides flexibility in use and have been proven successful for molecular breeding applications involving only few markers for genotyping a large number of segregating lines [45–47]. In the case of chickpea, more than 2,000 KASPar assay [28] and high density DArT array with 15,360 DArT clones have been developed [29]. The present study used a subset of 651 SNPs along with DArT arrays for genotyping. SNP genotyping data was used for cultivated germplasm as alleles could not be called for most of the wild species genotypes. SNPs used in the present study were designed from cultivated chickpea, which may be the reason they could not be amplified in wild species and could contribute to the biomodel PIC values. SNPs, although powerful as a marker due to their declining costs and high number [28], can be biased by being developed from a small number of individuals. This bias can skew the pool towards older and more intermediate frequency SNPs [48,49]. The benefit remains the large number of low cost markers. We minimized any effect of SNP bias by restricting its usage in the wild *Cicer* material where it lacks the information needed to separate patterns of relationships and complemented our analysis with the inclusion of independent DArT data that lacks such bias. In particular, focusing our analysis of the wild material on the DArT data should avoid the skew that SNP data can introduce.

Table 4. Genetic variation across the three primary regions of diversity: Fertile Crescent, Central Asia, and the Ethiopian highlands using DArT markers.

	Polymorphic marker (%)	N	Na	Ne	I	He	UHe
Ethiopia	49.84%	5.895±0.007	1.467±0.011	1.206±0.004	0.232±0.004	0.145±0.003	0.175±0.003
Central Asia	54.11%	12.840±0.008	1.527±0.010	1.277±0.006	0.267±0.005	0.174±0.003	0.189±0.004
Fertile Crescent	98.62%	71.810±0.047	1.986±0.002	1.415±0.004	0.439±0.003	0.277±0.002	0.281±0.002

No. of polymorphic alleles (N), No. of Different Alleles (Na), No. of Effective Alleles (Ne, = 1/(Sum pi^2)), Shannon's Information Index (I = −1* Sum (pi * Ln (pi)), Expected Heterozygosity (He = 1 − Sum pi^2)) and Unbiased Expected Heterozygosity (UHe = (2N/(2N−1)) * He).

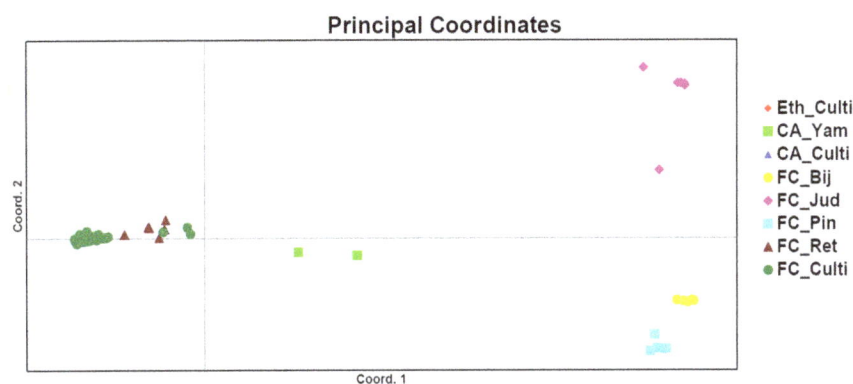

Figure 5. Principal coordinates analysis of wild and cultivated species of chickpea based on their geographical distribution (Eth_Culti: Cultivated chickpea from Ethiopia; CA_Yam: *Cicer yamashatae* from Central Asia; CA_culti: Cultivated chickpea from Central Asia; FC_Bij:, *C. bijugum;* from Fertile Crescent; FC_Jud: *C. judaicum* from Fertile Crescent; FC_Pin: *C. pinnatifidum* from Fertile Crescent; FC_Ret: *C. reticulatum* from Fertile Crescent and FC_Culti: Cultivated chickpea from Fertile Crescent).

In many crops that are deficient in genetic variation, wild relatives remain a critical resource. As is the case in other crops [4,47,50], higher levels of genetic variation were observed across all of the wild species. Significant genetic variation was observed in *C. reticulatum*, the immediate progenitor of cultivated chickpea, but genotypes of this species were less diverse than other *Cicer* species. Our results will allow the most genetically distinct of the existing accessions of these species to be used in breeding to maximize the diversity introgression into cultivated forms. However, as international germplasm collections contain only 18 unique *C. reticulatum* accessions [51], our results suggest that further collecting of *C. reticulatum*, particularly beyond the Mardin region of southeastern Anatolia where most existing collections were made, would be greatly beneficial. Relatively higher levels of genetic variation were present in the wild species of the secondary and tertiary gene pools, which span a far greater ecological range than *C. reticulatum*, which is restricted to oak savannas and disturbed pastures in southeastern Anatolia. However, the levels of genetic variation were still not all that high, consistent with the high probabilities on the assignment tests and the primarily selfing reproductive system of most *Cicer* species. Traits of wild species that are beneficial in a Mediterranean climate, such as vernalization, can hinder efforts to breed chickpea for cultivation in subtropical climates. Therefore, wild species from different regions, such as the African highlands or Central Asia could provide climatically adaptive traits for chickpea production in non-Mediterranean climates. For instance, species from outside the Fertile Crescent, such as *C. cuneatum* from Ethiopia and *C. microphyllum* from Central Asia (Pakistan and Afghanistan) could be exploited as sources of adaptive variation for those regions. Furthermore, wild species from more arid environments, such as *C. judaicum* and *C. pinnatifidum*, could be useful in expanding the resistance of cultivated chickpea to important biotic stresses like *Ascochyta*, *Helicoverpa*, *Fusarium* and *Botrytis* Gray Mold [20].

Based on seed type, chickpea has been subdivided in to three groups: desi, kabuli and pea-shaped. Significant differentiation among desi and kabuli seed type cultivars was observed, although far less than exists between wild species. The distinction could be due to a relatively recent evolution of kabuli seed type from a desi seed type ancestor that closely resembled the wild species, as previously speculated [16], but could just as easily represent artificial population structure generated by breeders [52]. Regardless, the division between the phenotypic classes of seed type appears to be weak and likely of recent origin. The dearth of

desi seed type genotypes from the Fertile Crescent could suggest that kabuli seed types were favoured in this region, potentially as a means to prevent introgression from *C. reticulatum* and *C. echinospermum*, which have seed and flower colours similar to desi seed types.

Germplasm collections contain relatively low numbers of wild relatives of crops [6]. Although often several individual lines of a wild species are available, rarely has collecting been aimed at understanding patterns of variation in populations of wild relatives [53,54]. Our results indicate that collecting diverse population samples of several *Cicer* species spanning ecologically meaningful gradients in abiotic or biotic factors such as moisture, soil fertility or pathogen distribution would be extremely useful. Analysis of variation across these gradients in wild relatives could show how natural selection has adapted populations of wild relatives to these localized conditions, giving us natural targets for breeding.

Supporting Information

Figure S1 a. Principal coordinates analysis of wild species of chickpea based on primary, secondary and tertiary gene pool. b. Population structure analysis across wild chickpea accessions to understand the distribution of primary, secondary and tertiary gene pool species.

Figure S2 a. Population structure analysis across cultivated chickpea accessions based on seed type. b. Analysis of molecular variance within and among cultivated population based on seed type.

Author Contributions

Conceived and designed the experiments: RKV. Performed the experiments: RKV MR. Analyzed the data: MR EvW EW AR. Contributed reagents/materials/analysis tools: RKV HDU. Wrote the paper: RKV MR EvW EW.

References

1. Abbo S, Berger J, Turner NC (2003) Evolution of cultivated chickpea: four bottlenecks limit diversity and constraint adaptation. Funct Plant Biol 30: 1081–1087.
2. Gepts P (2004) Who owns biodiversity, and how should the owners be compensated? Plant physiology 134: 1295–1307.
3. Fu YB (2006) Impact of plant breeding on genetic diversity of agricultural crops: searching for molecular evidence. Plant Genetic Resources: Characterization and Utilization 4: 71–78.
4. Gross BL, Olsen KM (2010) Genetic perspectives on crop domestication, Trends in plant science 15: 529–537.
5. Tanksley SD, McCouch SR (1997), Seed banks and molecular maps: unlocking genetic potential from the wild. Science 277: 1063–1066.
6. McCouch S, Baute GJ, Bradeen J, Bramel P, Bretting PK, et al. (2013) Agriculture: Feeding the future. Nature 499: 23–24.
7. Hajjar R, Hodgkin T (2007) The use of wild relatives in crop improvement: a survey of developments in the last 20 years. Euphytica 156: 1–13.
8. Roa V, Hodgkin T (2002) Genetic diversity and conservation and utilization of plant genetic resources. Plant Cell Tiss Org 68: 1–19.
9. Dempewolf H, Eastwood RJ, Guarino L, Khoury CK, Müller JV, et al. (2014) Adapting agriculture to climate change: A global initiative to collect, conserve, and use crop wild relatives. Agroecology and Sustainable Food Systems 38: 369–377.
10. Tester M, Langridge P (2010) Breeding technologies to increase crop production in a changing world. Science 327: 818–822.
11. Miflin B (2000) Crop improvement in the 21st century. J Exp Bot 51: 1–8.
12. Harlan JR, Wet JMJd (1971) Toward a rational classification of cultivated plants. Taxon 20: 509–517.
13. Haussmann BIG, Parzies HK, Presterl T, Sušic Z, Miedaner T (2004) Plant genetic resources in crop improvement. Plant Genetic Resources 2: 3–21.
14. Berger JD, Milroy SP, Turner NC, Siddique KHM, Imtiaz M, et al. (2011) Chickpea evolution has selected for contrasting phenological mechanisms among different habitats. Euphytica 180: 1–15.
15. Akibode S, Maredia M (2011) Global and regional trends in production, trade and consumption of food legume crops. CGIAR Draft Report. Available: http://impact.cgiar.org/sites/default/files/images/Legumetrendsv2.pdf.
16. Ladizinsky G (1985) Founder effect in crop plant evolution. Economic Botany 39: 191–199.
17. Millan T, Clarke HJ, Siddique KHM, Buhariwalla HK, Gaur PM, et al. (2006) Chickpea molecular breeding: new tools and concepts. Euphytica 147: 81–103.
18. Berger JD, Turner NC (2007) The Ecology of Chickpea. *In:* Chickpea breeding and management, Yadav SS, Redden RJ, and Sharma (eds). CAB International, Oxfordshire, UK.
19. Roorkiwal M, Nayak SN, Thudi M, Upadhyaya HD, Brunel D, et al. (2014) Allelic diversity and association analysis for candidate abiotic stress responsive genes with drought tolerance in chickpea. Front Plant Sci 5: 248 doi:10.3389/fpls.2014.00248.
20. Mallikarjuna N, Coyne C, Cho S, Rynearson S, Rajesh PN, et al. (2011) *Cicer.* Kole C (ed.), Wild Crop Relatives: Genomic and Breeding Resources, Legume Crops and Forages, DOI 10.1007/978-3-642-14387-8_4, Springer-Verlag Berlin Heidelberg.
21. Hillman GC (1975) The plant remains from Tell Abu Hureya in Syria: a preliminary report. In: Moore AMT (ed.) The Excavation of Tell Abu Hureya in Syria: A Preliminary Report. Proceedings of the Prehistory Society 41: 70–73.
22. Redden RJ, Berger JD (2007) History and Origin of Chickpea. In *Chickpea breeding and management*, Yadav SS, Redden RJ, and Sharma (eds). CAB International, Oxfordshire, U.K.
23. Abbo S, Saranga Y, Peleg Z, Kerem Z, Lev-Yadun S, et al. (2009) Reconsidering domestication of legumes versus cereals in the ancient near east. Q Rev Biol 84: 29–50.
24. Berger JD, Hughes S, Snowball R, Redden B, Bennett SJ, et al. (2013) Strengthening the impact of plant genetic resources through collaborative collection, conservation, characterisation, and evaluation: a tribute to the legacy of Dr Clive Francis. Crop and Pasture Science 64: 300–311.
25. Sharma S, Upadhyaya HD, Roorkiwal M, Varshney RK, Gowda CLL (2013) Chickpea. *In:* Genetic and Genomic Resources of Grain Legume Improvement to production. doi:http://dx.doi.org/10.1016/B978-0-12-397935-3.00001-3.
26. Varshney RK, Hiremath PJ, Lekha P, Kashiwagi J, Balaji J, et al. (2009) A comprehensive resource of drought- and salinity- responsive ESTs for gene discovery and marker development in chickpea (*Cicer arietinum* L.). BMC Genomics 10: 523.
27. Roorkiwal M, Sawargaonkar SL, Chitikineni A, Thudi M, Saxena RK, et al. (2013) Single nucleotide polymorphism genotyping for breeding and genetics applications in chickpea and pigeonpea using the BeadXpress platform. The Plant Genome 6 doi:10.3835/plantgenome2013.05.0017.
28. Hiremath PJ, Kumar A, Penmetsa RV, Farmer A, Schlueter JA, et al. (2012) Large-scale development of cost-effective SNP marker assays for diversity assessment and genetic mapping in chickpea and comparative mapping in legumes. Plant Biotechnol J 10: 716–732.
29. Thudi M, Bohra A, Nayak SN, Varghese N, Shah TM, et al. (2011) Novel SSR markers from BAC-end sequences, DArT arrays and a comprehensive genetic map with 1,291 marker loci for chickpea (*Cicer arietinum* L.). PLoS One 6: e27275.
30. Gowda CLL, Upadhyaya HD, Sharma S, Varshney RK, Dwivedi SL (2013) Exploiting genomic resources for efficient conservation and use of chickpea, groundnut, and pigeonpea collections for crop improvement. The Plant Genome 6: doi:10.3835/plantgenome2013.05.0016.
31. Cuc LM, Mace ES, Crouch JH, Quang VD, Long TD, et al. (2008) Isolation and characterization of novel microsatellite markers and their application for diversity assessment in cultivated groundnut (*Arachis hypogaea*). BMC Plant Biology 8: 55.
32. Ladizinsky G, Adler A (1976) Genetic relationships among the annual species of *Cicer* L. Theor Appl Genet 48: 197–203.
33. Excoffier L, Smouse PE, Quattro JM (1992) Analysis of molecular variance inferred from metric distances among DNA haplotypes: application to human mitochondrial DNA restriction data. Genetics 131: 479–491.
34. Peakall R, Smouse PE (2006) GENALEX 6: genetic analysis in Excel. Population genetic software for teaching and research. Molecular Ecology Notes 6: 288–295.
35. Peakall R, Smouse PE (2012) GenAlEx 6.5: genetic analysis in Excel. Population genetic software for teaching and research-an update. Bioinformatics 28: 2537–2539.
36. Excoffier L, Estoup A, Cornuet JM (2005) Bayesian analysis of an admixture model with mutations and arbitrarily linked markers. Genetics 169: 1727–1738.
37. Shannon CE (1948) A mathematical theory of communication. The Bell System Technical Journal 27: 379–423 and 623–656.
38. Liu K, Muse SV (2005) PowerMarker: an integrated analysis environment for genetic marker analysis. Bioinformatics 21: 2128–2129.
39. Pritchard JK, Stephens M, Donnelly P (2000) Inference of population structure using multilocus genotype data. Genetics 155: 945–959.
40. van der Maesen LJG (1987) Origin, history and taxonomy of chickpea. In: Saxena MC, Singh RB (eds) The chickpea. CABI, Wallingford, UK, 139–156.
41. Lev-Yadun S, Gopher A, Abbo S (2000) Archaeology. The cradle of agriculture. Science 288: 1602–1603.
42. Zohary D, Hopf M (2000) Domestication of plants in the old world, 3rd edn. Oxford University Press, New York, USA.
43. Varshney RK, Song C, Saxena RK, Azam S, Yu S, et al. (2013) Draft genome sequence of chickpea (*Cicer arietinum*) provides a resource for trait improvement. Nat Biotechnol 31: 240–246.
44. Varshney RK, Mohan SM, Gaur PM, Gangarao NV, Pandey MK, et al. (2013) Achievements and prospects of genomics-assisted breeding in three legume crops of the semi-arid tropics. Biotechnol Adv 31: 1120–1134.
45. Allen AM, Barker GL, Berry ST, Coghill JA, Gwilliam R, et al. (2011) Transcript-specific, single-nucleotide polymorphism discovery and linkage analysis in hexaploid bread wheat (*Triticum aestivum* L.). Plant Biotechnol J 9: 1086–1099.
46. Cortés AJ, Chavarro MC, Blair MW (2011) SNP marker diversity in common bean (*Phaseolus vulgaris* L.). Theor Appl Genet 123: 827–845.
47. Saxena RK, von Wettberg E, Upadhyaya HD, Sanchez V, Songok S, et al. (2014) Genetic Diversity and Demographic History of *Cajanus* spp. Illustrated from Genome-Wide SNPs. PLoS One 9(2): e88568.
48. Albrechtsen A, Nielsen FC, Nielsen R (2010) Ascertainment biases in SNP chips affect measures of population divergence. Mol Biol Evol 27: 2534–2547.
49. Lachance J, Tishkoff SA (2013) SNP ascertainment bias in population genetic analyses: why it is important, and how to correct it. BioEssays 35: 780–786.
50. Kassa MT, VarmaPenmetsa R, Carrasquilla-Garcia N, Sarma BK, Datta S, et al. (2012) Genetic patterns of domestication in pigeonpea (*Cajanus cajan* (L) Millsp) and wild *Cajanus* relatives. PloS one 7: 6.
51. Berger J, Abbo S, Turner NC (2003) Ecogeography of annual wild *Cicer* species: The poor state of the world collection. Crop Sci 41: 1976–1090.
52. Moreno MT, Cubero JI (1978) Variation in *Cicer arietinum*. Euphytica 27: 465–485.
53. Gepts P, Famula TR, Bettinger RL, Brush SB, Damania AB, et al. (eds) (2012) Biodiversity in agriculture: domestication, evolution, and sustainability. Cambridge University Press, Cambridge, U.K.
54. Hufford MB, Xu X, van Heerwaarden J, Pyhäjärvi T, Chia JM, et al. (2012) Comparative population genomics of maize domestication and improvement. Nat Genet 44: 808–811.

5

Mapping Genetic Diversity of Cherimoya (*Annona cherimola* Mill.): Application of Spatial Analysis for Conservation and Use of Plant Genetic Resources

Maarten van Zonneveld[1,2]*, **Xavier Scheldeman**[1], **Pilar Escribano**[3], **María A. Viruel**[3], **Patrick Van Damme**[2,4], **Willman Garcia**[5], **César Tapia**[6], **José Romero**[7], **Manuel Sigueñas**[8], **José I. Hormaza**[3]

1 Bioversity International, Regional Office for the Americas, Cali, Colombia, 2 Ghent University, Faculty of Bioscience Engineering, Gent, Belgium, 3 Instituto de Hortofruticultura Subtropical y Mediterránea, (IHSM-UMA-CSIC), Estación Experimental La Mayora, Algarrobo-Costa, Málaga, Spain, 4 World Agroforestry Centre (ICRAF), GRP1 - Domestication, Nairobi, Kenya, 5 PROINPA, Oficina Regional Valle Norte, Cochabamba, Bolivia, 6 Instituto Nacional Autónomo de Investigaciones Agropecuarias (INIAP) Panamericana sur km1, Quito, Ecuador, 7 Naturaleza y Cultura Internacional (NCI), Loja, Ecuador, 8 Instituto Nacional de Innovación Agrícola (INIA), La Molina, Lima, Peru

Abstract

There is a growing call for inventories that evaluate geographic patterns in diversity of plant genetic resources maintained on farm and in species' natural populations in order to enhance their use and conservation. Such evaluations are relevant for useful tropical and subtropical tree species, as many of these species are still undomesticated, or in incipient stages of domestication and local populations can offer yet-unknown traits of high value to further domestication. For many outcrossing species, such as most trees, inbreeding depression can be an issue, and genetic diversity is important to sustain local production. Diversity is also crucial for species to adapt to environmental changes. This paper explores the possibilities of incorporating molecular marker data into Geographic Information Systems (GIS) to allow visualization and better understanding of spatial patterns of genetic diversity as a key input to optimize conservation and use of plant genetic resources, based on a case study of cherimoya (*Annona cherimola* Mill.), a Neotropical fruit tree species. We present spatial analyses to (1) improve the understanding of spatial distribution of genetic diversity of cherimoya natural stands and cultivated trees in Ecuador, Bolivia and Peru based on microsatellite molecular markers (SSRs); and (2) formulate optimal conservation strategies by revealing priority areas for *in situ* conservation, and identifying existing diversity gaps in *ex situ* collections. We found high levels of allelic richness, locally common alleles and expected heterozygosity in cherimoya's putative centre of origin, southern Ecuador and northern Peru, whereas levels of diversity in southern Peru and especially in Bolivia were significantly lower. The application of GIS on a large microsatellite dataset allows a more detailed prioritization of areas for *in situ* conservation and targeted collection across the Andean distribution range of cherimoya than previous studies could do, i.e. at province and department level in Ecuador and Peru, respectively.

Editor: Pär K. Ingvarsson, University of Umeå, Sweden

Funding: This study has been carried out within the context of the CHERLA project funded by the International Cooperation with Developing Countries (INCO-DEV) Sixth Framework Programme (Contract 015100) of the European Commission. Additional financial support was provided by the Spanish Ministry of Education (Project Grants AGL2010-15140), the Instituto Nacional de Investigación y Tecnología Agraria y Alimentaria (INIA) from Spain (RF2009-00010), Junta de Andalucía (FEDER AGR2742) and by the INIA-Spain financed project 'Strengthening Regional Collaboration in Conservation and Sustainable Use of Forest Genetic Resources in Latin America and Sub-Saharan Africa.' The funders had no role in study design, data collection and analysis, decision to publish, or preparation of the manuscript.

Competing Interests: The authors have declared that no competing interests exist.

* E-mail: mvanzonneveld@cgiar.org

Introduction

Many useful tropical and subtropical tree species, even those commonly cultivated, are still in incipient stages of domestication, with their genetic resources often principally or exclusively, present *in situ*, i.e. on farm in home gardens or orchards and/or in natural populations. The local diversity of these tree species could offer yet-unknown traits of high value to further domestication [1]. For many outcrossing species, such as most tropical tree species, this genetic diversity is important to sustain local production as many of these species are vulnerable to inbreeding depression [2]. Diversity is also a key factor for adaption to environmental changes [2]. However, tree species are increasingly vulnerable to losses of genetic diversity, referred to as genetic erosion, due to decreased population sizes resulting from land use changes and land degradation, and due to changes in local climate that may select against some genotypes [3]. Therefore, there is a growing call to assess the conservation status of the genetic resources of tree species [4].

The formulation of effective and efficient conservation strategies requires a thorough understanding of spatial patterns of genetic diversity [5]. A better knowledge of areas of high genetic diversity is also important in optimizing the use of genetic resources, as the likelihood to find interesting materials for breeding is higher where levels of genetic diversity are maximal [6], [7]. Initiatives to prioritize research on global plant genetic resources, such as those lead by the Food and Agriculture Organization of the United Nations (FAO), include calls for more inventories and surveys to increase understanding of variation in plant genetic resources, explicitly referring to the application of molecular tools in such assessments [8], [9].

This study focuses on cherimoya (*Annona cherimola* Mill.), an underutilized fruit tree species that belongs to the Annonaceae, a family included within the Magnoliales in the Eumagnoliid clade among the early-divergent angiosperms [10]. This Neotropical tree species still is in its initial stages of domestication [11] and it is considered at high risk of losing valuable genetic material from its genepool [12]. Cherimoya fruits are widely praised for their excellent organoleptic characteristics, and the species is therefore considered to have a high potential for commercial production and income generation for both small and large-scale producers in subtropical climates [13]. Cherimoya presents protogynous dichogamy, i.e. it has hermaphroditic flowers wherein female and male parts do not mature simultaneously, which favors outcrossing in its native range [14]. For commercial production, hand pollination with pollen and stamens is common practice due to lack in overlap of the female and male stages and absence of pollinating agents outside its native range [14]. At present, advanced commercial production is found in Spain, the world's largest cherimoya producer, with around 3000 ha of plantations, while small-scale cultivation occurs throughout the Andes, Central America and Mexico.

Most early chroniclers and scientists proposed the Andean region, and more specifically, the valleys of southern Ecuador and northern Peru, as cherimoya's centre of origin [12], [15], [16]. The existence of natural cherimoya forest patches, which are scattered across the inter-Andean valleys in Ecuador and northern Peru, supports this hypothesis. Nonetheless the possibility that these are feral populations cannot be excluded. This phenomenon has been observed in the case of several fruit tree species, such as olives [17]. An alternative hypothesis for the centre of origin of cherimoya is Central America [18], which would imply that the area of northern Peru and southern Ecuador is a secondary centre of diversity. Most relatives of cherimoya are native to Central America and southern Mexico, which is an argument in favor of this alternate hypothesis (H. Rainer, Institute of Botany, University of Vienna, 2011, pers. comm.). In any case, cherimoya fruits were consumed in the Andean region in antiquity [12] and the movement of germplasm across Mesoamerica, southern Mexico and the Andes probably took place in pre-Columbian times.

The conservation status of cherimoya genetic resources has improved considerably in recent years. Due to increasing commercial prices for cherimoya at local markets, Andean farmers are stimulated to conserve *in situ* the cherimoya trees growing in their backyards. Indeed, trees established in home gardens and orchards are common throughout the Andean region in Bolivia, Ecuador and Peru, which usually originate from planted local seeds or chance seedlings [11], and among them some individuals show promising traits for future breeding programs [19]. In Peru, the local cultivar 'Cumbe' is already fetching retail prices significantly above the prices of unselected cherimoya fruit types [20]. In contrast to most tropical and subtropical underutilized fruit tree species, cherimoya genetic resources are also well conserved *ex situ*. Several field collections have been established in Spain, Peru and Ecuador, preserving over 500 different accessions [11], [21]. The Spanish collection based at la Estación Experimental La Mayora in Malaga, which holds over 300 accessions (190 of them collected in the Andean region), is currently used as a source of materials for the Spanish cherimoya breeding program and has been thoroughly analyzed using isozymes [22]–[24] and microsatellite markers [11], [25]–[27].

The recent development of new molecular tools in combination with new spatial methods and increased computer capacity has created opportunities for new applications of genetic diversity analyses [28]–[30]. Whereas neutral molecular markers are considered a sound tool to measure patterns and trends in the use and conservation of plant genetic resources [31], Geographic Information Systems (GIS) provide opportunities to carry out spatial analyses of genetic diversity patterns identified with these markers [32]. GIS can be used to interpolate genetic parameters between sampled populations (e.g. [33]–[35]), to apply re-sampling of georeferenced samples within a defined buffer zone [36], [37], or to develop grid-based genetic distance models [38], [39]. GIS are also an acknowledged tool to prioritize areas for conservation of plant genetic resources [40]. Several studies have used spatial analysis to develop conservation strategies for plant genetic resources based on molecular marker data (e.g. [36], [41]). Moreover, results obtained using GIS can be presented in a clear way on maps, which facilitates the incorporation of these findings into the formulation of conservation strategies and the implementation of conservation measures [42].

In this article we further explore the possibilities of incorporating molecular marker data into GIS to better visualize and understand spatial patterns of genetic diversity, as a key input to optimize conservation and enhance use of local plant diversity, based on a case study of cherimoya. The specific objectives of this article are to (1) apply innovative spatial analysis to improve understanding of the geographic distribution of cherimoya 's genetic diversity in its putative native range, based on microsatellite molecular markers (SSRs); and (2) formulate optimal conservation strategies by prioritizing areas for *in situ* conservation and identifying existing diversity gaps in *ex situ* collections. Based on the outcomes, we discuss how these spatial analyses can be used to define possible strategies that guarantee the long term conservation of cherimoya genetic resources and how these analyses can be applied to improve conservation and use of tree and crop genetic resources in general.

Results

A total of 1504 trees were analyzed in this study, i.e. 395 from Bolivia, 351 from Ecuador and 758 from Peru. Of those, 502 are currently conserved in *ex situ* collections (either in Ecuador, Peru or Spain) whereas the remainder trees were sampled *in situ*. The molecular analysis included a core set of nine microsatellite loci [27] resulting in 71 different alleles. In all analyses of α-diversity and β-diversity (also referred to as divergence) we applied circular neighborhood re-sampling technique resulting in a total dataset of 48,128 trees (Figure 1). This technique facilitates analysis of patterns in genetic variation across extensive distribution ranges while maintaining high-resolution grids.

Allelic richness

Allelic richness is a straightforward measure of genetic diversity that is commonly used in studies based on molecular markers that aim at selecting populations for conservation [5], [43]. Figure 2 presents the distribution of the average number of alleles per locus found in the study area. It clearly shows that a higher number of alleles is present in the northern part of the study area, specifically in northern Peru, around Cajamarca Department, while other areas of high diversity are located on the border zone between Ecuador (Loja Province) and Peru (Piura Department), in the northern part of Ecuador around its capital Quito and in the northern part of the Lima Department in Peru.

Despite the effort to implement a similar sampling density throughout the study area, some areas (often locations with a higher abundance of traditionally managed cherimoya trees and stands) have been sampled more intensively than others (Figure 1),

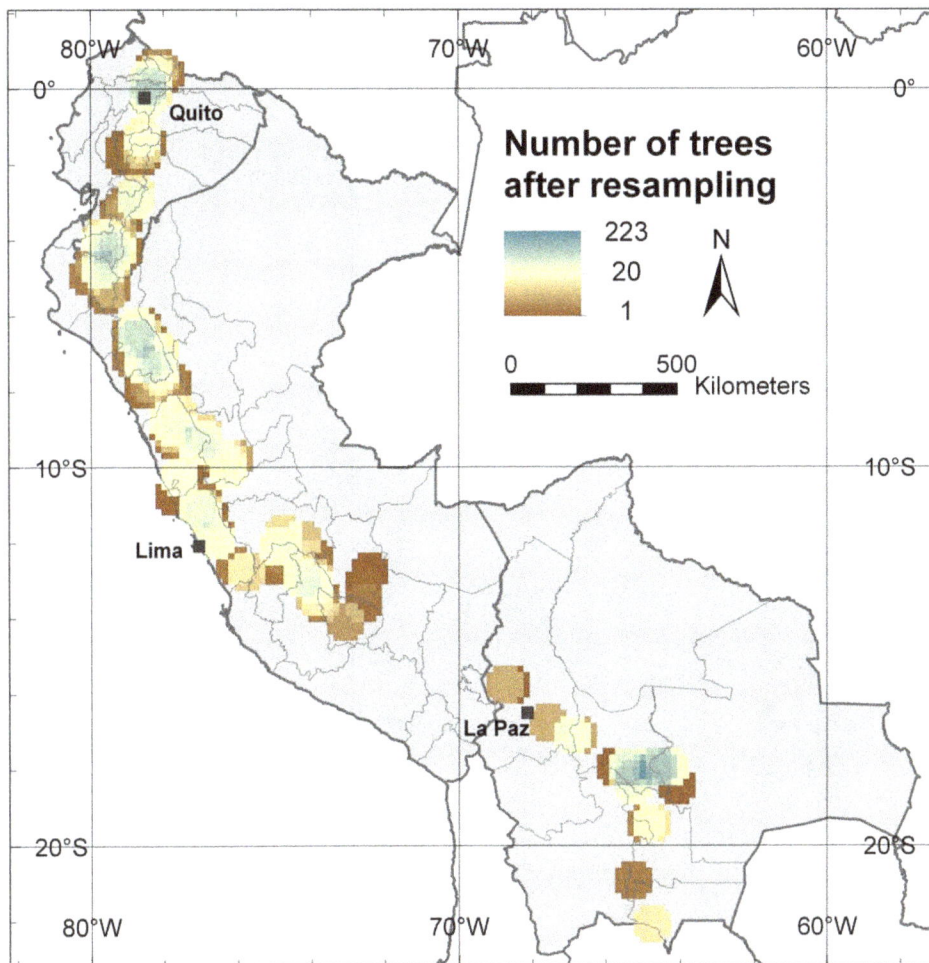

Figure 1. Number of trees per grid after re-sampling. This map is made with a 10-minutes grid applying a one-degree circular neighborhood.

generating a sampling bias [44]. The rarefaction methodology corrects this sampling bias by recalculating allelic richness in each grid cell to a minimum sample size [5]. Figure 3 shows only the grid cells where 20 or more trees were present after applying a one-degree circular neighborhood, and for which allelic richness was corrected following the rarefaction methodology to a minimum sample size of 20 trees. The Cajamarca Department in northern Peru remains the area with the highest diversity, up to an average of 5.18 different alleles per locus. After correction by rarefaction, diversity in Ecuador, especially around Quito, is reduced, whereas the same seems to happen in the northern part of the Lima Department, in Peru, indicating the presence of a sampling bias around the capitals of both countries. The area around the Peruvian capital Lima, an important commercial cherimoya cultivation area, shows the lowest allelic richness within Peru, probably due to the widespread cultivation of a vegetatively propagated cultivar, 'Cumbe'. Another striking result is that allelic richness in Bolivia, already low in the uncorrected analysis, is even lower with correction of sampling bias, resulting in an even higher contrast between cherimoya genetic diversity in Bolivia and that found in Peru and Ecuador.

Locally common alleles

Priority for conservation should be given to populations that retain locally common alleles; these are alleles that occur in high frequency in a limited area, and can indicate the presence of genotypes adapted to specific environments [43]. Figure 4 shows the richness of locally common alleles per locus in the study area. The high diversity levels found in the Cajamarca Department in northern Peru are reconfirmed. Besides harboring the highest number of different alleles, it also contains the highest number of locally common alleles. This makes this area a priority for *in situ* conservation, both of cultivated trees on farm and of natural stands. The border region between Peru and Ecuador (Piura Department and Loja Province) is another area where a high concentration of locally common alleles has been observed and may, therefore, be a second area to prioritize *in situ* conservation efforts. To a lesser extent, the area around Quito in Ecuador and the northern part of the Lima Department in Peru also present locally common alleles.

Expected Heterozygosity (He), Fixation Index (F) and Genetic Distance (GD)

In situ conservation should focus on viable populations, where inbreeding and subsequent loss of alleles are minimal. Parameters that allow assessment of inbreeding are expected heterozygosity (He) and the fixation index (F). The fixation index (F) was used to detect areas subjected to high inbreeding depression and, as the inverse to that, excess in heterozygosity [45]. Figure 5 shows the values for He in the study area, again confirming Cajamarca

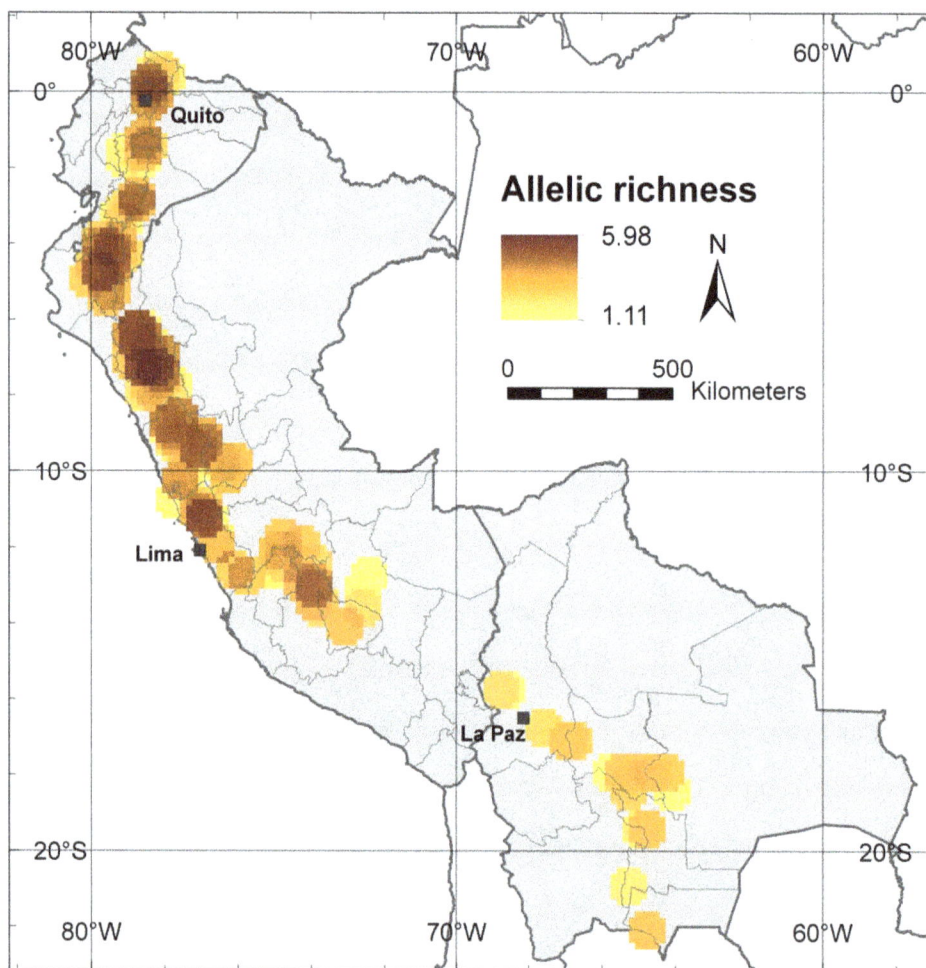

Figure 2. Allelic richness. This map shows the average number of alleles per locus in all 10-minutes grid cells applying a one-degree circular neighborhood.

Department in northern Peru as the area with the highest genetic diversity. High He values, however, radiate towards the south (as opposed to the higher diversity towards the north found in the allelic richness analyses) indicating higher levels of diversity in terms of heterozygosity in central Peru compared to Ecuador. Figure 6 shows the values for the fixation index, with F values close to 0 in the Cajamarca Department indicating that natural and cultivated cherimoya tree stands in this area have not experienced inbreeding. The highest values for F are observed in central Ecuador, suggesting that the level of inbreeding is highest in that part of cherimoya's Andean distribution range.

The most important Peruvian commercial cherimoya cultivation area, located near the Capital Lima, particularly shows negative F values, i.e. excess of heterozygosity. Most of the cherimoyas cultivated in this area are vegetatively propagated clones of the cultivar 'Cumbe' which resulted in highly heterozygous values from the molecular analysis, i.e. the 'Cumbe' accession conserved in the Spanish genebank is heterozygote for eight of the nine microsatellite loci analyzed in this study (Ho value of 0.89). An analysis of the average genetic distance, between the 'Cumbe' accession and the genotypes in each grid cell with 20 or more re-sampled trees in the study area, clearly shows lowest genetic distance values near the Peruvian capital, Lima, indicating that the cherimoya trees in this area are very similar to the cultivar 'Cumbe' (Figure 7). This area clearly differs from the rest of the

cherimoya distribution area in our study, which is more likely to be a product of natural gene flow patterns.

β-diversity (divergence)

Besides α-diversity parameters, aimed at identifying those areas with highest allelic richness and balanced allele frequencies, *in situ* conservation also needs to take into account allelic composition (β-diversity or divergence) as it is possible that populations with low allelic richness possess unique allele compositions, different from those of populations in other areas of the range, which would warrant their *in situ* conservation [5]. Applying the Structure software (see [46]) and using the statistic parameter ΔK [47] to define the number of clusters with genetically similar trees present in the study area, we differentiated two main populations. Figure 8 shows the differentiation of the populations among distribution areas in cluster A and B, respectively. Cluster A has the highest presence in the areas previously identified as those with the highest allelic richness (Cajamarca Department in northern Peru, border zone between Ecuador and Peru and the area around Quito in Ecuador), whereas cluster B is mainly confined to southern Peru and Bolivia. Bolivian cherimoya trees are almost exclusively assigned to cluster B. Particular areas that did not show a strong linkage to either of the two clusters included the surroundings of the city of Lima and Loja Province in southern Ecuador.

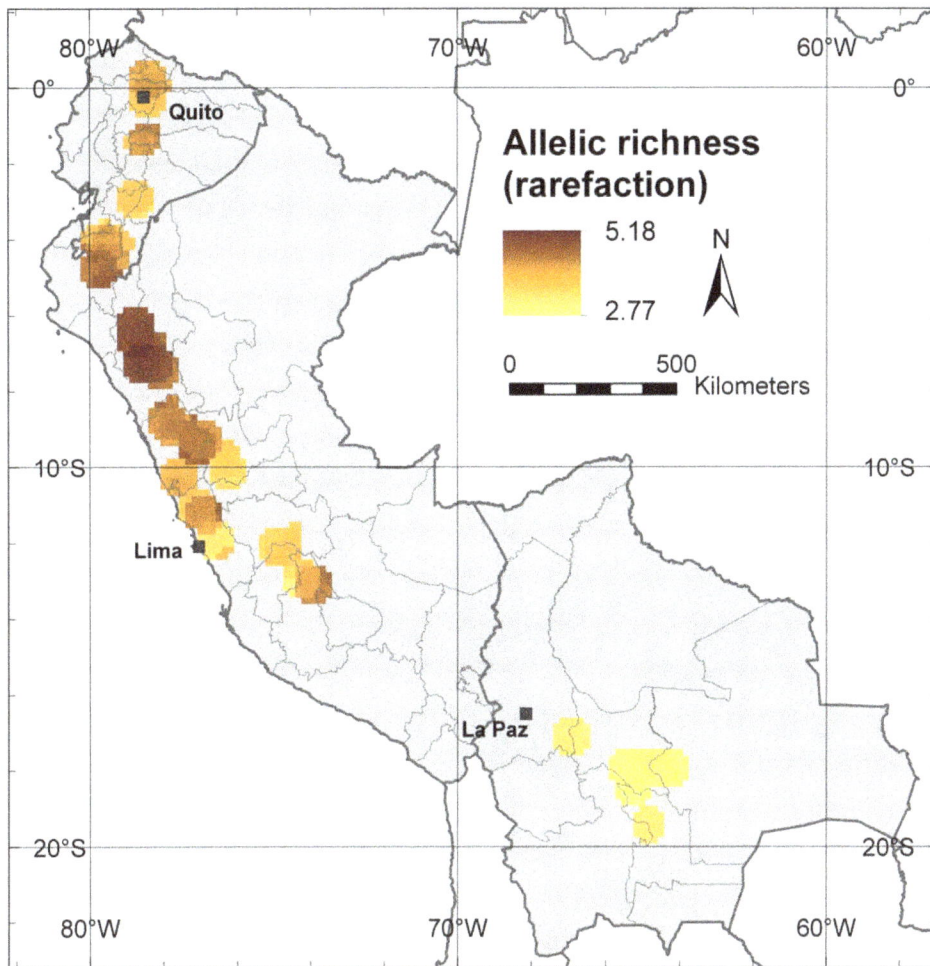

Figure 3. Allelic richness corrected for sample size by using rarefaction. This map shows the average number of alleles per locus in the 10-minutes grid cells applying a one-degree circular neighborhood and a correction by rarefaction to a minimum sample size of 20 trees.

Ex situ conservation status

Of the 1504 trees included in this study, 502 genotypes are currently conserved in *ex situ* collections (either in Ecuador, Peru or Spain). Only eight alleles, corresponding to 11% of the total of 71 alleles that have been found in the study area, are not represented in any accession of these collections. Figure 9 shows the distribution of the missing alleles. There is only a small area with a significant portion of missing alleles (3 in total), i.e. in southern Ecuador (Azuay Province). Natural cherimoya forest patches and areas of traditional cherimoya cultivation in this province should be prioritized for future cherimoya collection missions. With almost 90% of alleles found to be present in *ex situ* collections, it can be concluded that, in general, cherimoya diversity from the countries analyzed is fairly well conserved *ex situ*.

Distribution range of cherimoya in the Andes

The above results and subsequent conclusions are obviously only of practical use if the sampling performed was indeed representative for the distribution of cherimoya in the study area. Maxent species distribution modeling software was applied to model cherimoya's distribution range in Ecuador, Peru and Bolivia based on the climatic niche in which the 1504 sampled trees of our study were located. The modeled distribution was then compared with the sampled areas in these countries.

Cross-validation, to evaluate the quality of the distribution model, returned an Area Under Curve (AUC) value of 0.9, which indicates good model performance [48]. AUC is a commonly used parameter in the validation of distribution models. Another measure of validation, the Kappa value, returned a value of 0.799 indicating the model performed even excellent [49].

In general, sampling covered most of the cherimoya-modeled distribution (Figure 10); 46% of the modeled distribution area is covered by grid cells with 20 or more re-sampled trees (Figure 10, dark blue areas). In 24.5% of the potential area of cherimoya occurrence less than 20 trees were re-sampled (light blue areas) whereas 29.5% of the modeled range was not sampled (red areas) and are considered sample gaps. The largest sample gaps are located in northern Peru in the transition zone between the Peruvian Andes and the Amazon (in the Departments of San Martin and Amazonas) and in southern Peru (in the Departments of Junín, Pasco, Huancavelica, Ayacucho and Puno). The Andean-Amazon transition zone should be priority for future complementary cherimoya collection trips because it is adjacent to an area where already high levels of diversity have been found, i.e. Cajamarca Department in northern Peru.

Cherimoya was predicted absent by the distribution model in a significant area of southern Peru, indicating that the environmental conditions in substantial parts of that region are not suitable for

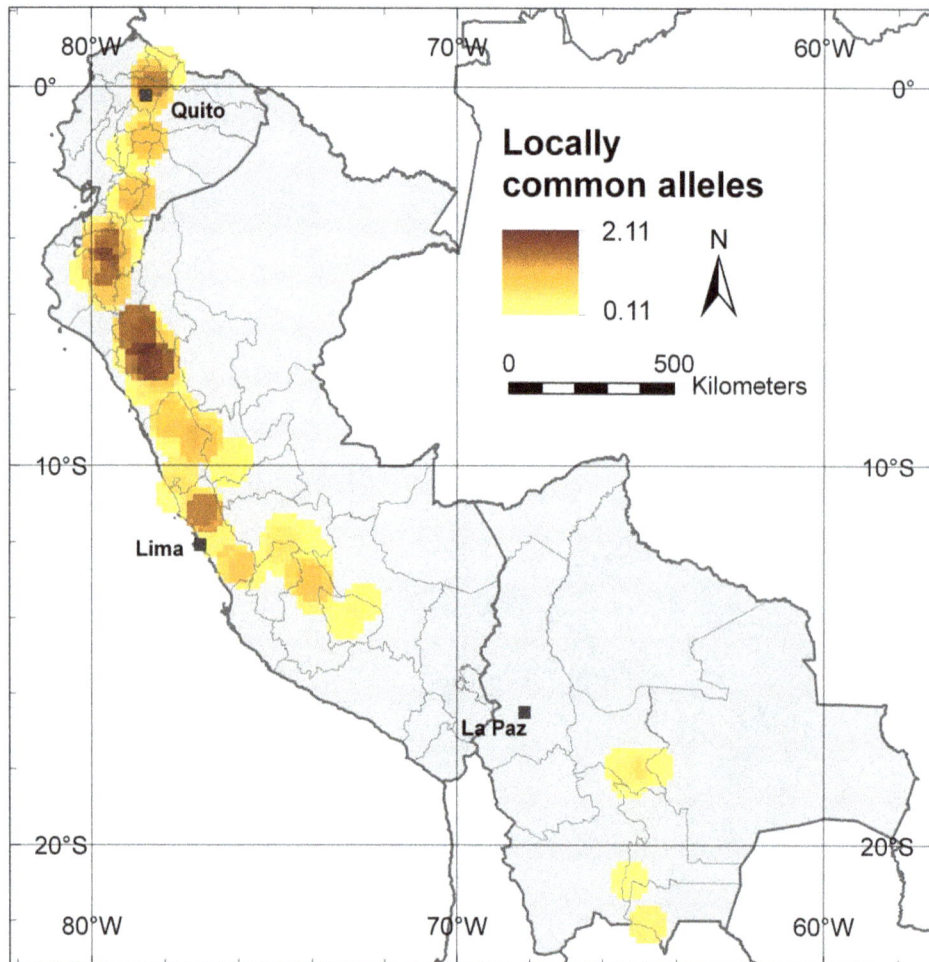

Figure 4. Locally common alleles. This map shows the average number of alleles per locus in a 10-minutes grid cell that are relatively common (occurring with a frequency higher that 5%) in a limited area (in 25% or less of the grid cells) applying a one-degree circular neighborhood.

cherimoya cultivation (Figure 10). This explains why no trees have been sampled in that area.

Discussion

Areas of high diversity in the cherimoya centre of origin

Our results are in line with a previous genetic study of the Spanish cherimoya collection that also distinguished populations in Ecuador and northern Peru from those in southern Peru [11], and corroborate with results from isozyme markers that showed high genetic variation present in Peru and Ecuador [24]. However, our study is based on a much higher number of samples and, therefore, provides much more detail for prioritizing areas for *in situ* conservation and germplasm collection.

At the allele level, our analysis confirms that, within our study area, the highest allelic richness as well as the highest number of locally common alleles are found in the area of southern Ecuador and northern Peru, i.e. the putative centre of origin of cherimoya. Northern Peru, and more specifically the Cajamarca Department, shows the highest levels of genetic diversity.

The highest values of the fixation index, which is an indication of inbreeding, were found in Ecuador. Inbreeding may occur because of reduction and fragmentation of natural stands and cultivated areas, increasing the risk of allele loss, which eventually

leads to genetic erosion [50]. Our results do not allow us to determine how much genetic erosion has taken place in Ecuador in comparison to Peru and Bolivia, but the high inbreeding values in Ecuador could explain why currently allelic richness is lower in this country than in northern Peru.

At the population level, significant differences can be observed between the cherimoya germplasm present in the area with highest diversity (where genotypes belonging to cluster A are predominant) and genotypes found in areas with lower diversity, i.e. in southern Peru and Bolivia (represented by cluster B). Cluster A seems likely to represent material that is genetically closer to the "wild" cherimoya type. No natural cherimoya stands have been observed in Bolivia, and this probably explains why no genotypes pertaining to cluster A have been recorded there. Cluster B probably corresponds to a genepool that is genetically different from most of the wild or semi-domesticated cherimoya found in northern Peru and Ecuador and that could have formed the basis for Bolivian cherimoya cultivation. Looking at the areas of high cluster B dominance, Bolivian germplasm probably originates from southern Peru.

Although most early chroniclers and scientists proposed southern Ecuador and northern Peru to be cherimoya's centre of origin [12], [15], [16], [51], the possibility of that area being a secondary centre of origin cannot be discarded. A diversity study

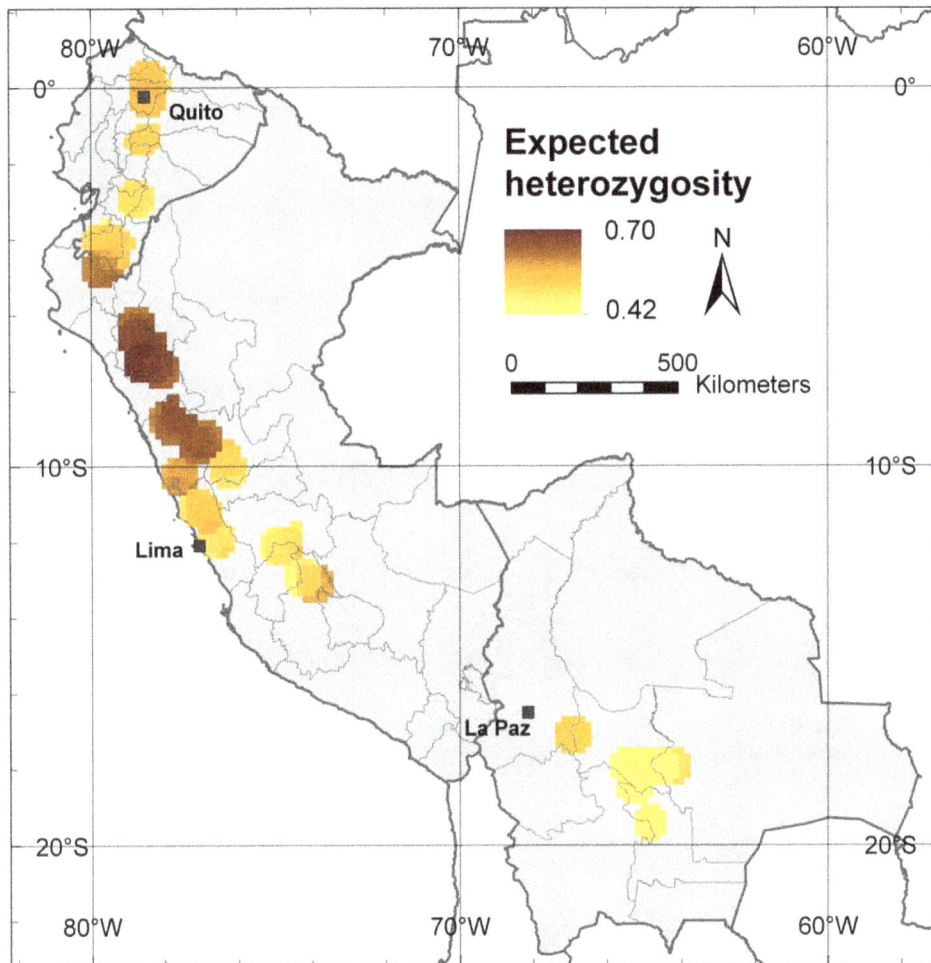

Figure 5. Expected heterozygosity (*He*). This map shows the average *He* value in each 10-minutes grid cell with 20 or more trees applying a one-degree circular neighborhood.

similar to the one described in this study, but including cherimoya genotypes from Central America and Mexico, would shed light on the genetic variation across the complete pre-Columbian distribution range of cherimoya and provide additional clues on the primary centre of origin and diversification of this species.

Ex situ and *in situ* conservation of cherimoya genetic resources in the Andean region

Most alleles identified in our study are represented in one or more of the existing *ex situ* collections in Ecuador, Peru and Spain. The results obtained suggest that the highest priority for further collection should be the Azuay Province in Ecuador, since cherimoya stands in this area harbor most alleles not yet included in genebanks. It is also one of the areas with the highest risk of allele loss because of the high observed levels of inbreeding, compared to other parts of the study area. An additional priority area for germplasm collection is the transition zone from the Andes to the Amazon in Peru (in the higher elevation areas of the Departments of San Martin and Amazonas), which was not sampled in this study. According to the distribution model there is a high probability of finding cherimoya stands in this region, which probably is also high in genetic diversity, because it is adjacent to the area with the highest diversity found in this study, i.e. the Cajamarca Department in northern Peru.

A priority for *in situ* conservation should be the Cajamarca Department, the area with the highest levels of genetic diversity. A second area of priority should be the Loja Province in southern Ecuador, an area with a high number of locally common alleles. Both areas are assigned mostly to cluster A. Since trees predominantly assigned to cluster B have a particular allelic composition in comparison to trees predominantly grouped in cluster A, genotypes of cluster B should also be considered in conservation activities. The part of Lima Department north of the Peruvian capital, which is assigned mostly to cluster B, could be prioritized for *in situ* conservation of genotypes from this cluster. In contrast to the low levels of allelic richness around Lima city in the southern part of the Lima Department, the northern part of this Department contains a fair number of locally common alleles.

The long-term conservation of cherimoya genetic resources is far from guaranteed. As commercial prices for fruits can fluctuate, short-term incentives for farmers to maintain cherimoya as a profitable crop are reduced and a decline in commercial interest may lead to the replacement of cherimoya trees by other crops, increasing the risks of genetic erosion. Around Quito, for example, most of the traditional cherimoya cultivation is being replaced by avocado plantations, which are commercially more attractive (X. Scheldeman, pers. obs.). An increase in commercial prices for cherimoya products will not necessarily promote the conservation

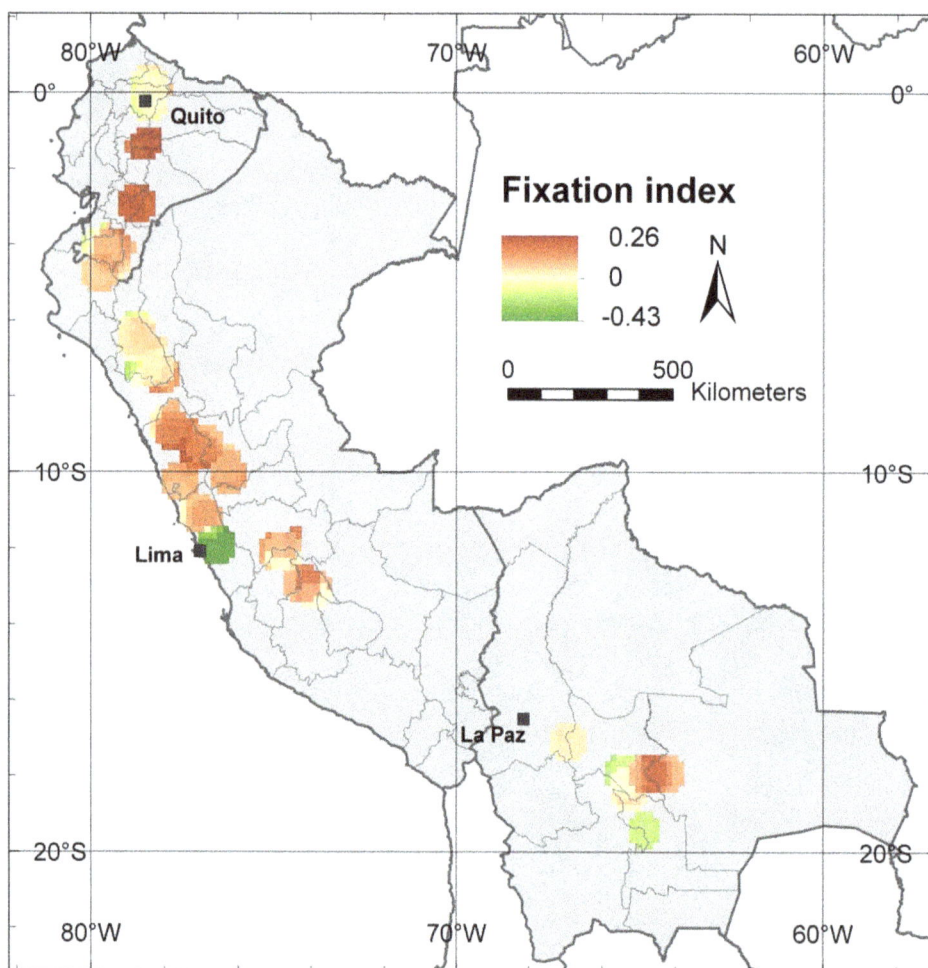

Figure 6. Fixation *index (F)*. This map shows the average *F* value in each 10-minutes cell with 20 or more trees applying a one-degree circular neighborhood. Yellow areas indicate cherimoya stands where observed heterozygosity is as expected, red areas indicate stands where observed heterozygosity is lower than expected (indicating inbreeding) whereas observed heterozygosity is higher than expected in green areas.

of the existing genetic diversity. Indeed, in our study we found low levels of genetic diversity around the Peruvian capital, Lima where the clonally propagated cultivar 'Cumbe' is widely cultivated, because it fetches higher prices in the market.

A promising strategy to enhance *in situ* conservation on farm is through the promotion of seed or bud-for-grafting exchange between farmers [52]. During the CHERLA project, cherimoya fairs, which facilitate exchange of plant material, were organized in different areas of this study, including the Cajamarca and Piura Departments in Peru, Loja Province in Ecuador and various departments in Bolivia. Seed and bud exchange can also be a way to conserve local races from unfavorable alterations in the local environment due to climate change, by re-distributing them in new areas with suitable climate conditions [53]. Another way to ensure conservation of genetic resources of tree species while their use is stimulated could be the establishment of local clonal seed orchards if and when adequate propagation techniques, to enable the multiplication of clones, are made available as well [1], [54]. This is the case for cherimoya, as demonstrated by the successful clonal propagation of the cultivar 'Cumbe' around the city of Lima.

Ideally, each area targeted for *in situ* conservation - where the existing cherimoya stands and forest patches can evolve within the local environment - should be backed up by *ex situ* conservation of

germplasm (which currently is the case for cherimoya), and be monitored periodically to assess the dynamics in diversity use and risks of genetic erosion. *Ex situ* collections of fruit tree species often consist of living trees, such as the cherimoya collections. This allows conservation of superior combination of alleles that can be propagated vegetatively through grafting. Additional reasons include the following: (1) many tropical and subtropical trees (including cherimoya) have seeds with recalcitrant or intermediate behavior, which cannot be stored for long-term conservation; and (2) pollen, fruits and seeds can be collected continuously for characterization, evaluation and genetic improvement once trees have reached the reproductive stage. Nevertheless, the high costs for research institutions to maintain field genebanks of woody perennial species, can be a reason to minimize *ex situ* collections and focus especially on *in situ* conservation [55]. In that case, it is important to screen the existing accessions through morphological, biochemical and/or molecular characterization to maximize the conservation of genetic diversity and potentially interesting functional attributes in a reduced collection [6]. This approach has already successfully been used in the cherimoya collection la Mayora, Malaga, Spain [27]. *Ex situ* conservation may particularly be important for areas that suffer from inbreeding -an indicator for high rates of allelic loss and genetic erosion- such as central Ecuador in the case of cherimoya, whereas *in situ* conservation

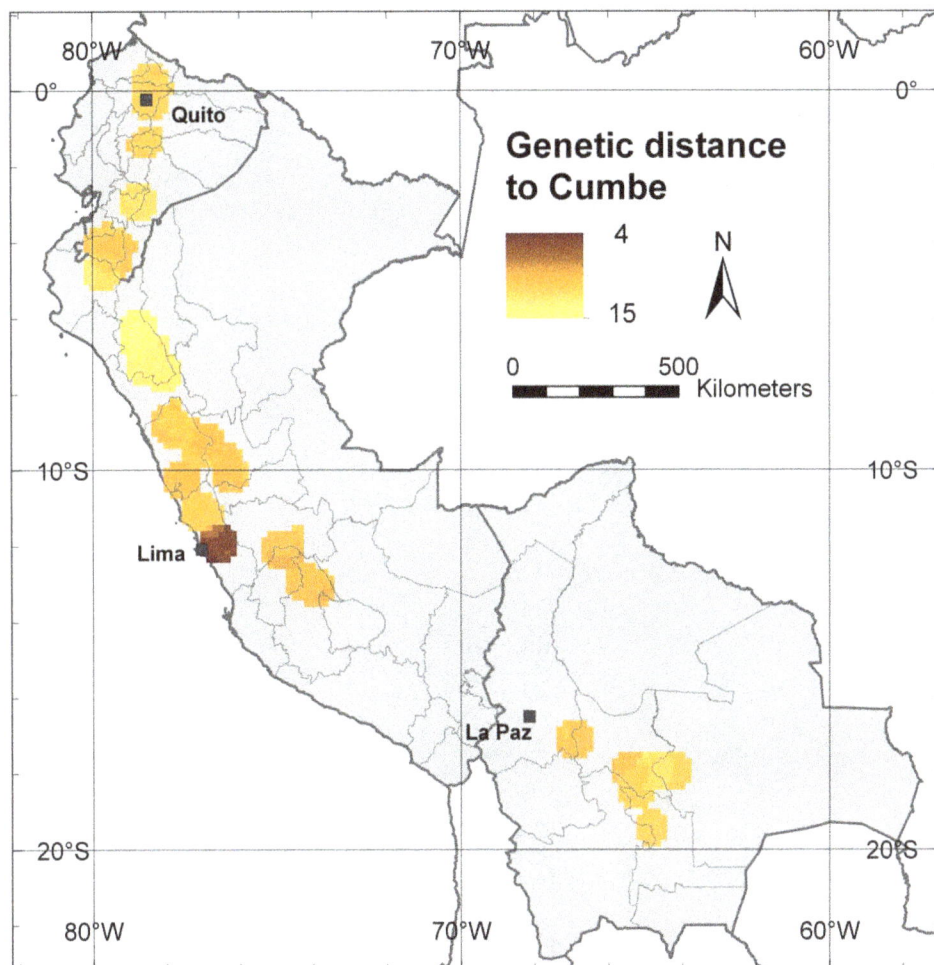

Figure 7. Genetic distance to the Peruvian cultivar 'Cumbe'. This maps shows the average genetic distance (GD) to the cultivar 'Cumbe', in each 10-minutes cell with 20 or more trees applying a one-degree circular neighborhood. As reference of the cultivar, the 'Cumbe' accession from the collection la Mayora, Malaga, Spain, was used.

may be most successful in areas of high diversity where still low rates of inbreeding are observed such as in the cherimoya stands from northern Peru.

Use of GIS and molecular markers to enhance conservation and use of plant genetic resources

Despite the advances in new computational applications and the use of molecular tools, spatial analyses are still underutilized in efforts to conserve plant diversity [56]. With respect to targeting collection sites and prioritizing the conservation of plant genetic resources, spatial analyses of diversity have been carried out mainly at the species level for crop genepools (e.g. [57]–[59]). Only a few studies have mapped intraspecific diversity to enhance the conservation of genetic resources of specific crops and trees (e.g. [36], [41]). Kiambi et al. [41] grouped samples using a grid to compare diversity between geographic areas of similar size, whereas Lowe et al. [36] applied re-sampling to enable the calculation of diversity estimates with high degrees of confidence. However, these studies were carried out with fewer than 100 individuals per species, which limits the type of spatial analysis that can be carried out over the geographic distribution range of species. Our analysis combines both techniques on a large dataset (1504 trees), which can be conceptualized as a continuous distribution of plant individuals, in which each individual is connected to its neighboring trees because

they share the same seed system, and/or breed with each other. Based on this concept, trees have been sampled in this study following a scattered distribution to calculate, across the Andean distribution range of cherimoya, several diversity estimates important to prioritize areas for conservation, including two recommended parameters: allelic richness [5] and the number of locally common alleles [43]. Since the application of molecular tools is becoming cheaper, intraspecific diversity studies with large datasets will probably be more common in the near future, allowing for studies of this sort on other tree species and annual crops.

The size of the grid cells and width of the circular neighborhood for this type of spatial analysis depends on how many plant individuals have been collected across the landscape, and the minimum number of plant individuals that is considered sufficient to make confident estimates of genetic parameters per grid cell. Application of circular neighborhood provides an effective way to decrease grid cell size, which facilitates detection of spatial patterns in genetic variation across an extensive distribution range. By re-sampling the trees in the landscape, it generates a high number of grid cells with a sufficient number of trees to make confident calculations of genetic parameters per grid cell. It also makes analyses less sensitive to grid origin definition and enables the inclusion of isolated trees in the calculation of the genetic parameters, i.e. together with their closest neighboring trees.

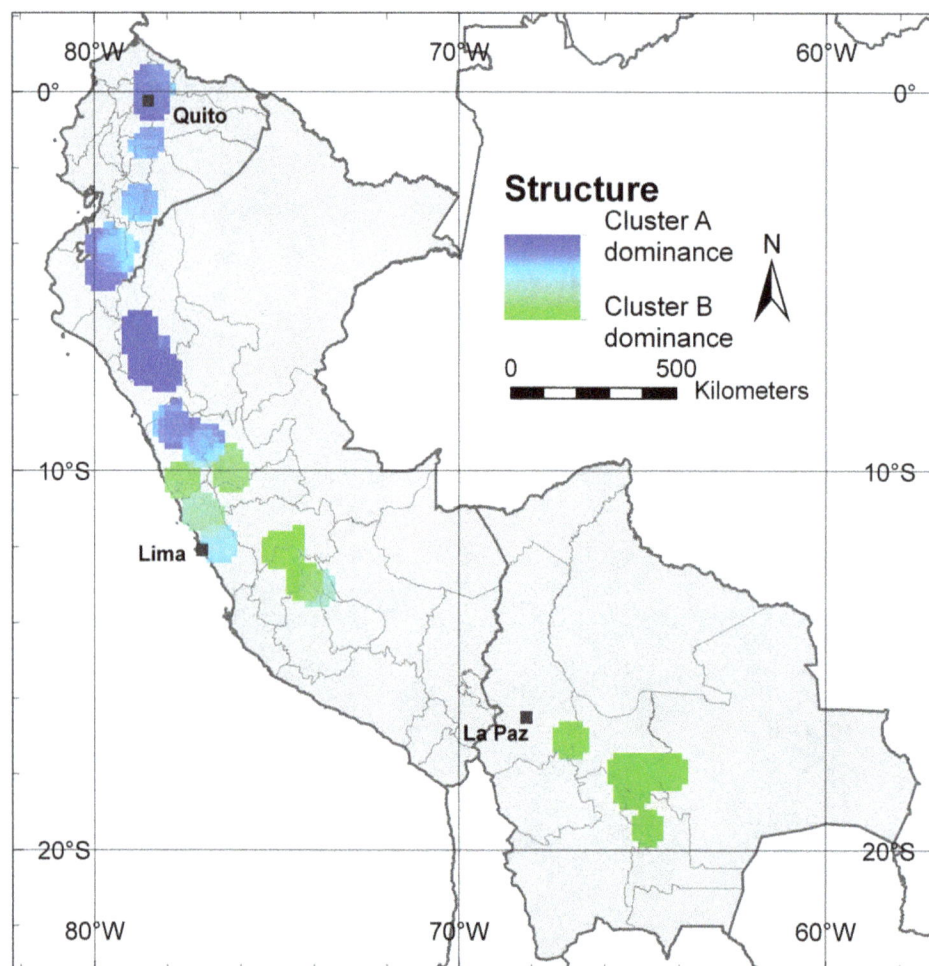

Figure 8. Genetic structure of Andean cherimoya distribution in Population clusters A and B. This map shows in each 10-minutes cell with 20 or more trees applying a one-degree circular neighborhood, the average probability of finding a cherimoya tree belonging to cluster A or B. Dark blue areas show a higher probability of finding trees belonging to cluster A whereas dark green areas show a higher probability of finding trees belonging to cluster B. Light blue colored areas are not clearly assigned to any of the two clusters.

Ideally, the sampling strategy for this type of analysis should be identified based on a pre-defined grid, aiming at measuring the same number trees per grid cell. However, due to logistical constraints and because a species simply may be more abundant in some areas than in others, in practice, sampling will always be sub-optimal to a certain degree. Of all the genetic parameters, allelic richness is most sensitive to uneven sampling and, accordingly, we have corrected sample size by rarefaction [5]. Repeated subsampling of a minimum number of tree individuals per grid cell is another possibility to correct for sampling bias [60]. This technique could also be used to correct other genetic parameters than allelic richness for sampling bias, such as expected heterozygosity, although these are less sensitive to uneven sampling [61].

Given the sampling distribution in our study area and the fact that for the calculation of most genetic parameters, we maintained a minimum of 20 re-sampled trees per grid cell, we defined a cell size of 10 minutes and a circular neighborhood with a diameter of one degree, which enabled us to detect spatial patterns of genetic variation at administrative level one in Ecuador, Peru and Bolivia (provinces and departments). For studies of plant species, in which individuals are sampled in a more clumped distribution compared to our scattered sampling distribution and/or in lower densities

across the landscape, larger grid cells and/or a larger width of circular neighborhood could be applied, always assuring a sufficient number of trees per grid cell. The overall resolution of the study will obviously be lower.

Following Frankel et al. [6], we hypothesized that areas with high diversity measured by neutral molecular markers (like our microsatellite loci) have a high probability to contain genetic material that will also show diversity in functional traits, including traits of agronomic interest. Molecular markers are considered an appropriate indicator to quantify patterns and trends in the use and conservation of plant genetic resources [31]. However, while neutral molecular marker surveys are suitable for diversity studies, direct measurement of traits in field trials may be more desirable to evaluate the genetic health and adaptive capacity of tree populations [50]. Nevertheless, molecular marker studies representative of the whole genome provide a less expensive and scientifically sound alternative to assess the genetic resource status of tree species, for which, in comparison to annual crops, field trials are particularly expensive because of the long generation times [62]. Markers of DNA sequences related to phenotypic traits, including expressed sequence tagged (EST) markers and markers in specific genes, could be of interest to include in spatial analysis of patterns and trends in plant genetic resources. More

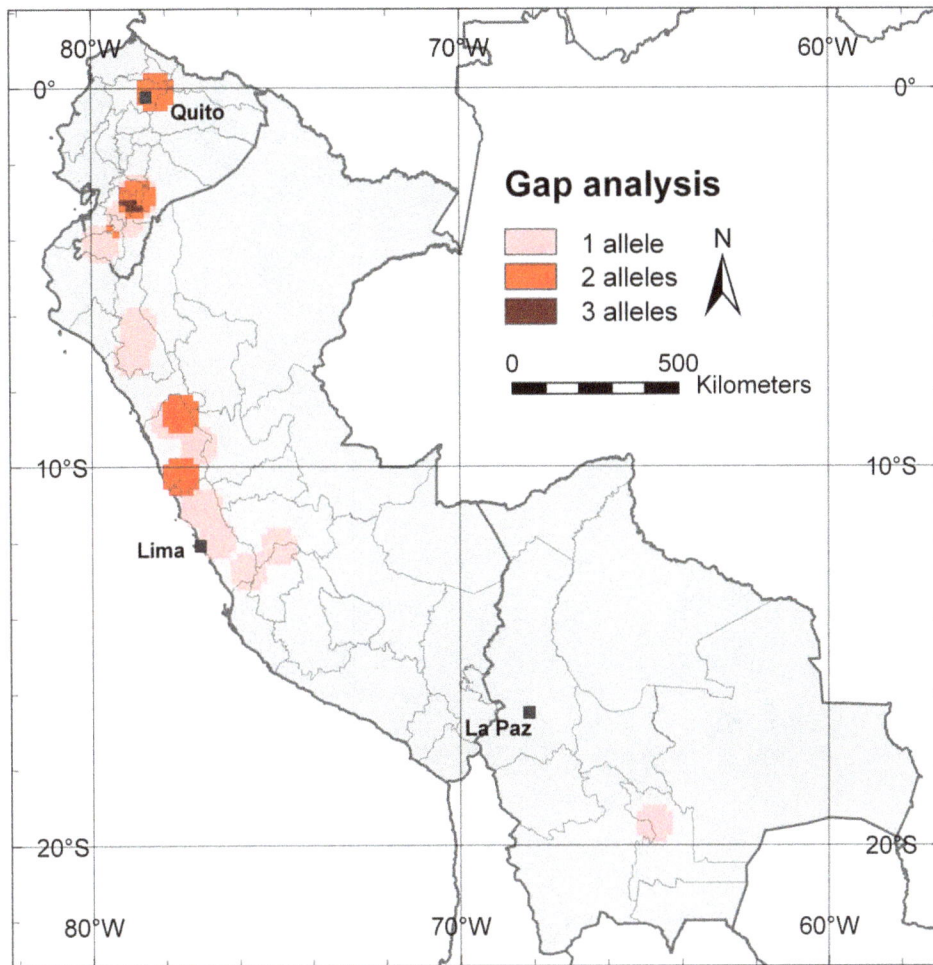

Figure 9. Gap analysis of alleles not found in *ex situ* collections. Richness analysis of alleles (eight alleles out of the total of 71 observed alleles) that are not found in any *ex situ* collection based on 10-minutes grid with a one-degree circular neighborhood.

and more are becoming available, especially for important crops where sequencing programs have been performed or will be carried out in the near future. An example in cherimoya is a recently described gene involved in seedlessness in a sister species, *Annona squamosa* [63]. However these markers are less polymorphic than neutral ones, such as those that have been used in our study, so the use of neutral markers to study spatial patterns of genetic diversity is still necessary.

It is difficult to compare our results with those of Lowe et al. [36] and Kiambi et al. [41] because of the differences in methodology used. To examine molecular marker studies on the same species, minimum standard sets of markers have already been suggested [64]. Standardization of methodologies in studies on different species would improve comparability of results and also would facilitate Meta-analyses, for example to better understand how well genetic diversity of tropical and subtropical tree species is protected on farm and in protected areas.

In our study we only examined spatial patterns of genetic variation without relating them to other spatial attributes. GIS can also be used to link genetic data to available spatial information relevant to conservation of plant genetic resources, for instance to reveal short-term threats such as accessibility and long-term threats such as climate change. With this type of analysis, hotspots of diversity under threat could be identified following Myers et al. [65] but instead of looking at species level, this could be done at the

intraspecific level, to ensure the conservation of priority populations of specific crops and useful tree species. Spatial information on the patterns and characteristics of human societies can be used to understand the drivers behind threats. In a study on changes in cassava diversity in the Peruvian Amazon, GIS was used to correlate cassava diversity data with biotic and socio-economic spatial data to identify possible drivers behind diversity and genetic erosion [66]. This can be useful information in the development of adequate policies and measures to promote *in situ* conservation of plant genetic resources on farms and in natural populations.

Materials and Methods

Sampling and SSR analysis: A total of 1504 cherimoya accessions have been analyzed in this study, 395 from Bolivia, 351 from Ecuador and 758 from Peru. DNA was extracted from young leaves after [67]. Based on polymorphism, a set of nine SSRs has been selected from those previously developed in cherimoya [26]. A 15 μL of reaction solution containing 16 mM (NH4)2SO4, 67 mM Tris-ClH pH 8.8, 0.01% Tween20, 2 mM MgCl2, 0.1 mM each dNTP, 0.4 μM each primer, 25 ng genomic DNA and 0.5 units of BioTaq[TM] DNA polymerase (Bioline, London, UK) was used for amplification on an I-cycler (Bio-Rad Laboratories, Hercules, CA, USA) thermocycler using the following temperature profile: an initial step of 1 min at 94°C,

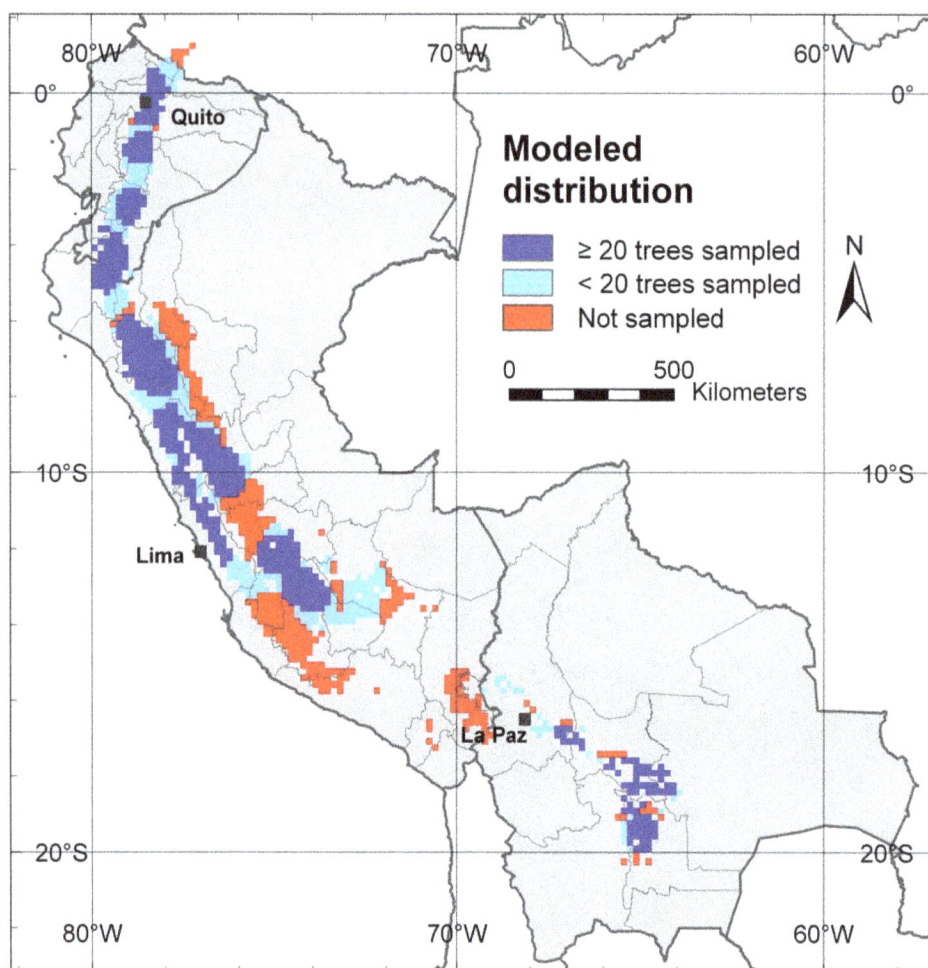

Figure 10. Modeled distribution of cherimoya. Areas of the modeled distribution in dark blue are covered by the 10-minutes grid cells with 20 or more trees applying circular neighborhood. Light blue areas of modeled distribution coincide with grid cells that contain less than 20 trees after re-sampling. Red areas indicate potential areas for cherimoya occurrence and cultivation that have not been in sampled.

35 cycles of 30 s at 94°C, 30 s at 45°C–55°C and 1 min at 72°C, and a final step of 5 min at 72°C. Forward primers were labeled with a fluorescent dye on the 5′ end. The PCR products were analyzed by capillary electrophoresis in a CEQ^{TM} 8000 capillary DNA analysis system (Beckman Coulter, Fullerton, CA, USA). Samples were denaturalized at 90°C during 120 s, injected at 2.0 kV, 30 s and separated at 6.0 kV during 35 min. Each reaction was repeated twice and the Spanish cultivar Fino de Jete was used as control in each run to ensure size accuracy and to minimize run-to-run variation.

Data cleaning: The coordinates of the respective tree locations were checked in DIVA-GIS (www.diva-gis.org) on erroneous points based on passport data at administrative level one (e.g. departments, provinces) with a buffer of 20 minutes (approx 30 km), and outliers based on climate data derived from the Worldclim data set [68] (two or more of the 19 bioclim variables according the Reverse jackknife method [69]). Based on these analyses, two points were excluded. The cleaned dataset thus included microsatellite data of 1504 georeferenced trees. Taking into account that nine SSR markers were analyzed, this results in a total of 27,072 georeferenced alleles.

Spatial analysis – Circular neighborhood: Grids for all genetic parameters were generated in DIVA-GIS and are based on a grid with a cell size of 10 minutes (which corresponds to

approximate 18 km in the study area) applying a circular neighborhood with a diameter of one degree (corresponding to approximate 111 km) constructed in Excel. The circular neighborhood is used to re-sample the allelic composition of a single tree to all surrounding grid cells, in this case, 32 cells with a size of 10 minutes, within a diameter of one degree around its location. In this way, the allelic composition of each sampled tree is representative for the area within the defined buffer zone. Applying the circular neighborhood re-sampling technique resulted in a total dataset of 48,128 trees and 866,304 alleles.

Spatial analysis – α-diversity: After applying circular neighborhood to all trees, genetic parameters were calculated in GenAlEx per 10-minutes grid cell, for all trees present in each cell after re-sampling. Genetic parameters included the average number of alleles per locus (Na), the number of locally common alleles per locus (alleles occurring with a frequency higher than 5% in 25% or less of the grid cells), average expected heterozygosity per locus (He), fixation index (F) and genetic distance (GD) (see [45]). Na and the number of locally common alleles per locus were presented for all grid cells with trees included. Na was corrected by rarefaction to a minimum sample size of 20 trees per cell with the HP-RARE software (see [70]); consequently, this parameter was only calculated for grid cells with 20 or more re-sampled trees. This minimum sample size was also used as a threshold of the number of trees per grid cell to get interpretable results for the parameters *He*, *F* and *GD*. *GD*, which was used to calculate distance in allelic composition of each cherimoya genotype to the commercial variety 'Cumbe', was calculated in GenAlEx using the *GD* option for codominant markers (see [71]). Final *GD* value per grid cell was the average *GD* for all re-sampled trees present in each cell. The reference tree was the accession 'Cumbe' from the Spanish cherimoya genebank in Malaga.

Spatial analysis - β-diversity: Population structure was defined by running the software Structure (see [46]) on all 1504 samples applying a 10,000 burn-in period, 10,000 Markov chain Monte Carlo (MCMC) repetitions after burn-in, and 20 iterations. Optimal K was selected after [47] by running Structure for K values between one and 10 and defining the final number of clusters where value of ΔK was highest. This was at K = 2, hence a map was developed for these two clusters, which we named respectively A and B. We used the probabilities of each tree belonging to cluster A and B to visualize the clusters on a map. Mapping of probabilities was done based on the average value of all trees per 10-minutes cell for those grid cells with 20 or more re-sampled trees after applying the one-degree circular neighborhood.

Spatial analysis - *Ex situ* conservation status: The private alleles function in GenAlEx (PAS) was used to identify the alleles exclusively found in trees that were sampled *in situ*. To visualize patterns in these alleles that are not included in any genebank, a point-to-grid richness analysis, using a 10-minutes grid, was carried out in DIVA-GIS based on the one-degree circular neighborhood re-sampled tree grid.

Spatial analysis - distribution modeling: To identify how well the sampling covered the Andean distribution range of cherimoya, and thus to identify potential collection gaps, we modeled the distribution (presence only) of cherimoya in the study area using the distribution modeling program Maxent (see [72], [73]). With this technique, potential distribution areas are identified as those areas where similar environmental conditions prevail as those at the sites where the species has already been observed. The data required to identify these areas include species presence points as well as layers of environmental variables covering the study area. Maxent is a species distribution modeling tool for which the applied algorithm has been evaluated as performing very well, in comparison to other ecological niche modeling software [74], [75]. Therefore, it was selected for this study's distribution modeling analysis. The coordinates in the passport data of the sampled trees were used for the presence point input. For environmental layer input, we used the 10-minutes grids of 19 bioclimatic variables (see [76]), derived from the Worldclim dataset [68]. The modeled distribution area was restricted using the 10 percentile training presence threshold, which indicates the probability value at which 10% of the presence points falls outside the potential area. The modeled distribution was generated in Maxent with 80% of the points (training data) and was cross-validated in DIVA-GIS with 20% of the remaining tree observations (test data). Besides 20% of the presence points, test data included randomly generated points in 0.1× the bounding box of the presence points as a proxy for absence points (5 times the number of presence points). Based on the cross-validation, the Area Under Curve (AUC) and Kappa value were calculated in DIVA-GIS as measures of model performance.

All maps were edited in ArcMap.

Acknowledgments

We thank Jorge Rojas and his team from PROINPA for DNA extraction and Bernardo Guzmán from PROINPA for field prospection and sampling in Bolivia. We also thank the personnel from INIA for the DNA extraction and field prospection in Peru and from INIAP Ecuador. Doris Chalampuente, Fernando Paredes, Marcelo Tacán, Eddie Zambrano and Edwin Naranjo. Laura Snook and Evert Thomas from Bioversity, and an anonymous reviewer provided useful comments on an early version of the manuscript.

Author Contributions

Conceived and designed the experiments: XS JIH WG CT JR MS MAV. Performed the experiments: PE MV JIH. Analyzed the data: MvZ XS. Contributed reagents/materials/analysis tools: PE MAV JIH WG CT JR MS MvZ XS. Wrote the paper: MvZ XS PVD JIH.

References

1. Ræbild A, Larsen AS, Jensen JS, Ouedraogo M, De Groote S, et al. (2011) Advances in domestication of indigenous fruit trees in the West African Sahel. New Forests 41: 297–315.

2. Dawson IK, Lengkeek A, Weber JC, Jamnadass R (2009) Managing genetic variation in tropical trees: linking knowledge with action in agroforestry ecosystems for improved conservation and enhanced livelihoods. Biodiversity and Conservation 18: 969–986.

3. Dawson IK, Vinceti B, Weber JC, Neufeldt H, Russell J, et al. (2011) Climate change and tree genetic resource management: maintaining and enhancing the productivity and value of smallholder tropical agroforestry landscapes. A review. Agroforestry Systems 81: 67–78.

4. Palmberg-Lerche C (2008) Thoughts on the conservation of forest biological diversity and forest tree and shrub genetic resources. Journal of Tropical Forest Science 20: 300–312.

5. Petit RJ, El Mousadik A, Pons O (1998) Identifying populations for conservation on the basis of genetic markers. Conservation Biology 12: 844–855.

6. Frankel OH, Brown AHD, Burdon J (1995) The conservation of cultivated plants. In: The conservation of plant biodiversity, Cambridge University Press, UK. First edition. pp 79–117.

7. Tanksley SD, McCouch SR (1997) Seed banks and molecular maps: unlocking genetic potential from the wild. Science 227: 1063–1066.

8. FAO (2010) The second report on the state of the world's plant genetic resources for food and agriculture. Rome.

9. FAO (2011) Draft updated global plan of action for the conservation and sustainable utilization of plant genetic resources for food and agriculture. Fifth session of the Intergovernmental Technical Working Group on Plant Genetic Resources for Food and Agriculture, Rome, 27–29 April 2011.

10. Bremer B, Bremer K, Chase MW, Fay MF, Reveal JL, et al. (2009) An update of the Angiosperm Phylogeny Group classification for the orders and families of flowering plants: APG III. Botanical Journal of the Linnean Society 161: 105–121.

11. Escribano P, Viruel MA, Hormaza JI (2007) Molecular analysis of genetic diversity and geographic origin within an ex situ germplasm collection of cherimoya by using SSRs. Journal of the American Society for Horticultural Science 132: 357–367.

12. Popenoe H, King SR, León J, Kalinowski LS, Vietmeyer ND, et al. (1989) Cherimoya. In: Lost crops of the Incas: Little-known plants of the Andes with promise for worldwide cultivation, National Academy Press, Washington, D.C. pp 228–239.

13. Van Damme P, Scheldeman X (1999) Promoting cultivation of cherimoya in Latin America. Unasylva 198: 43–47.

14. Lora J, Hormaza JI, Herrero M (2010) The progamic phase of an early-divergent angiosperm, *Annona cherimola* (Annonaceae). Annals of Botany 105: 221–231.

15. Popenoe W (1921) The native home of the cherimoya. Journal of Heredity 12: 331–336.

16. Guzman VL (1951) Informe del viaje de exploración sobre la cherimoya y otros frutales tropicales. Ministerio de Agricultura, Centro Nacional de Investigación y Experimentación Agrícola La Molina, Lima, Peru. 25 p.

17. Gepts P (2003) Crop domestication as a long-term selection experiment. In: Janick J, ed. Plant breeding reviews 24 Part 2: Long-term Selection: Crops, Animals, and Bacteria. pp 1–44.

18. Pozorski T, Pozorski S (1997) Cherimoya and guanabana in the archaeological record of Peru. Journal of Ethnobiology 17: 235–248.

19. Scheldeman X, Van Damme P, Ureña Alvarez JV, Romero Motoche JP (2003) Horticultural potential of Andean fruit crops exploring their centre of origin. Acta Horticulturae 598: 97–102.

20. Vanhove W, Van Damme P (2009) Marketing of cherimoya in the Andes for the benefit of the rural poor and as a tool for agrobiodiversity conservation. Acta Horticulturae 806: 497–504.

21. CHERLA (2008) *Inventory of current ex situ germplasm collections.* Deliverable 7, Project no. 015100, INCO sixth framework programme.

22. Pascual L, Perfectti F, Gutierrez M, Vargas AM (1993) Characterizing isozymes of Spanish cherimoya cultivars. HortScience 28: 845–847.

23. Perfectti F, Pascual L (1998) Characterization of cherimoya germplasm by isozyme markers. Fruit Varieties Journal 52: 53–62.

24. Perfectti F, Pascual L (2005) Genetic diversity in a worldwide collection of cherimoya cultivars. Genetic Resources and Crop Evolution 52: 959–966.

25. Escribano P, Viruel MA, Hormaza JI (2004) Characterization and cross-species amplification of microsatellite markers in cherimoya (Annona cherimola Mill. Annonaceae). Molecular Ecology Notes 4: 746–748.

26. Escribano P, Viruel MA, Hormaza JI (2008) PERMANENT GENETIC RESOURCES: Development of 52 new polymorphic SSR markers from cherimoya (Annona cherimola Mill.). Transferability to related taxa and selection of a reduced set for DNA fingerprinting and diversity studies. Molecular Ecology Resources 8: 317–321.

27. Escribano P, Viruel MA, Hormaza JI (2008) Comparison of different methods to construct a core germplasm collection in woody perennial species with simple sequence repeat markers. A case study in cherimoya (Annona cherimola, Annonaceae), an underutilised subtropical fruit tree species. Annals of Applied Biology 153: 25–32.

28. Manel S, Schwartz MK, Luikart G, Taberlet P (2003) Landscape genetics: combining landscape ecology and population genetics. Trends in Ecology and Evolution 18: 189–197.

29. Holderegger R, Buehler D, Gugerli F, Manel S (2010) Landscape genetics of plants. Trends in Plant Science 15: 675–683.

30. Scheldeman X, van Zonneveld M (2010) Training manual on spatial analysis of plant diversity and distribution. Bioversity International, Rome, Italy.

31. Eaton D, Windig J, Hiemstra SJ, van Veller M, Trach NX, et al. (2006) Indicators for livestock and crop biodiversity. Report.2006/05. CGN/DLO Foundation, Wageningen UR, Wageningen.

32. Kozak KH, Graham CH, Wiens JJ (2008) Integrating GIS-based environmental data into evolutionary biology. Trends in Ecology and Evolution 23: 141–148.

33. Degen B, Scholz F (1998) Spatial genetic differentiation among populations of European beech (Fagus sylvatica L.) in western Germany as identified by geostatistical analysis. Forest Genetics 5: 191–199.

34. Hanotte O, Bradley DG, Ochieng JW, Verjee Y, Hill EW, et al. (2002) African pastoralism: genetic imprints of origins and migrations. Science 296: 336–339.

35. Hoffmann MH, Glaß AS, Tomiuk J, Schmuths H, Fritsch RM, et al. (2003) Analysis of molecular data of Arabidopsis thaliana (L.) Heynh. (Brassicaceae) with Geographical Information Systems (GIS). Molecular Ecology 12: 1007–1019.

36. Lowe AJ, Gillies ACM, Wilson J, Dawson IK (2000) Conservation genetics of bush mango from central/west Africa: implications from random amplified polymorphic DNA analysis. Molecular Ecology 9: 831–841.

37. Vigouroux Y, Glaubitz JC, Matsuoka Y, Goodman MM, Sánchez GJ, et al. (2008) Population structure and genetic diversity of New World maize races assessed by DNA microsatellites. American Journal of Botany 95: 1240–1253.

38. McRae BH (2006) Isolation by resistance. Evolution 60: 1551–1561.

39. van Etten J, Hijmans RJ (2010) A geospatial modelling approach integrating archaeobotany and genetics to trace the origin and dispersal of domesticated plants. PLoS ONE 5: e12060. doi:10.1371/journal.pone.0012060.

40. Guarino L, Jarvis A, Hijmans RJ, Maxted N (2002) Geographic Information Systems (GIS) and the conservation and use of plant genetic resources. In: Engels JMM, Ramanatha Rao V, Brown AHD, Jackson MT, eds. Managing plant genetic diversity, International Plant Genetic Resources Institute (IPGRI) Rome, Italy. 2002. pp 387–404.

41. Kiambi DK, Newbury HJ, Maxted N, Ford-Lloyd BV (2008) Molecular genetic variation in the African wild rice Oryza longistaminata A. Chev. et Roehr. and its association with environmental variables. African Journal of Biotechnology 7: 1446–1460.

42. Jarvis A, Touval JL, Castro Schmitz M, Sotomayor L, Hyman GG (2010) Assessment of threats to ecosystems in South America. Journal for Nature Conservation 18: 180–188.

43. Frankel OH, Brown AHD, Burdon J (1995) The genetic diversity of wild plants. In: Frankel OH, Brown AHD, Burdon J, eds. The conservation of plant biodiversity, Cambridge University Press, UK. First edition. pp 10–38.

44. Hijmans RJ, Garrett KA, Huamán Z, Zhang DP, Schreuder M, et al. (2000) Assessing the geographic representativeness of genebank collections: the case of Bolivian wild potatoes. Conservation Biology 14: 1755–1765.

45. Peakall R, Smouse PE (2006) GENALEX 6: genetic analysis in Excel. Population genetic software for teaching and research. Molecular Ecology Notes 6: 288–295.

46. Pritchard JK, Stephens M, Donnelly P (2000) Inference of Population Structure Using Multilocus Genotype Data. Genetics 155: 945–959.

47. Evanno G, Regnaut S, Goudet J (2005) Detecting the number of clusters of individuals using the software STRUCTURE: a simulation study. Molecular Ecology 14: 2611–2620.

48. Araújo MB, Pearson RG, Thuiller W, Erhard M (2005) Validation of species-climate impact models under climate change. Global Change Biology 11: 1504–1513.

49. Fielding AH, Bell JF (1997) A review of methods for the assessment of prediction errors in conservation presence/absence models. Environmental Conservation 24: 38–49.

50. Lowe AJ, Boshier D, Ward M, Bacles CFE, Navarro C (2005) Genetic resource impacts of habitat loss and degradation; reconciling empirical evidence and predicted theory for neotropical trees. Heredity 95: 255–273.

51. Bonavia D, Ochoa CM, Tovar SO, Palomino RC (2004) Archaeological evidence of cherimoya (Annona cherimolia Mill.) and guanabana (Annona muricata L.) in ancient Peru. Economic Botany 58: 509–522.

52. Tapia ME (2000) Mountain agrobiodiversity in Peru. Seed fairs, seed banks, and mountain-to-mountain exchange. Mountain Research and Development 20: 220–225.

53. Mercer KL, Perales HR (2010) Evolutionary response of landraces to climate change in centers of crop diversity. Evolutionary Applications 3: 480–493.

54. Cornelius JP, Clement CR, Weber JC, Sotelo-Montes C, van Leeuwen J, et al. (2006) The trade-off between genetic gain and conservation in a participatory improvement programme: the case of peach palm (Bactris gasipaes Kunth). Forest, Trees and Livelihoods 16: 17–34.

55. van Leeuwen J, Lleras Pérez E, Clement CR (2005) Field genebanks may impede instead of promote crop development: Lessons of failed genebanks of "promising" Brazilian palms. Agrociencia 9: 61–66.

56. Escudero A, Iriondo JM, Torres ME (2003) Spatial analysis of genetic diversity as a tool for plant conservation. Biological Conservation 113: 351–365.

57. Hijmans RJ, Spooner DM (2001) Geographic distribution of wild potato species. American Journal of Botany 88: 2101–2112.

58. Jarvis A, Ferguson ME, Williams DE, Guarino L, Jones PG, et al. (2003) Biogeography of wild Arachis: assessing conservation status and setting future priorities. Crop Science 43: 1100–1108.

59. Scheldeman X, Willemen L, Coppens D'eeckenbrugge G, Romeijn-Peeters E, Restrepo MT, et al. (2007) Distribution, diversity and environmental adaptation of highland papayas (Vasconcellea spp.) in tropical and subtropical America. Biodiversity and Conservation 16: 1867–1884.

60. Leberg PL (2002) Estimating allelic richness: Effects of sample size and bottlenecks. Molecular Ecology 11: 2445–2449.

61. Lowe A, Harris S, Ashton P (2004) Genetic diversity and differentiation. In: Lowe A, Harris S, Ashton P, eds. Ecological genetics: design, analysis, and application, Blackwell Publishing, UK. First edition. pp 50–105.

62. Rajora OP, Mosseler A (2001) Challenges and opportunities for conservation of forest genetic resources. Euphytica 118: 197–212.

63. Lora J, Hormaza JI, Herrero M, Gasser CS (2011) Seedless fruits and the disruption of a conserved genetic pathway in angiosperm ovule development. Proceedings of the National Academy of Sciences USA 108: 5461–5465.

64. Van Damme V, Gómez-Paniagua H, de Vicente MC (2011) The GCP molecular marker toolkit, an instrument for use in breeding food security crops. Molecular breeding 28: 597–610.

65. Myers N, Mittermeier RA, Mittermeier CG, da Fonseca GAB, Kent J (2000) Biodiversity hotspots for conservation priorities. Nature 403: 853–858.

66. Willemen L, Scheldeman X, Soto Cabellos V, Salazar SR, Guarino L (2007) Spatial patterns of diversity and genetic erosion of traditional cassava (Manihot esculenta Crantz) cultivation in the Peruvian Amazon: An evaluation of socio-economic and environmental indicators. Genetic Resources and Crop Evolution 54: 1599–1612.

67. Viruel MA, Hormaza JI (2004) Development, characterization and variability analysis of microsatellites in lychee (Litchi chinensis Sonn., Sapindaceae). Theoretical and Applied Genetics 108: 896–902.

68. Hijmans RJ, Cameron SE, Parra JL, Jones PG, Jarvis A (2005) Very high resolution interpolated climate surfaces for global land areas. International Journal of Climatology 25: 1965–1978.

69. Chapman AD (2005) Principles and methods of data cleaning – primary species and species-occurrence data, version 1.0. Report for the Global Biodiversity Information Facility, Copenhagen.

70. Kalinowski ST (2005) HP-RARE 1.0: a computer program for performing rarefaction on measures of allelic richness. Molecular Ecology Notes 5: 187–189.

71. Smouse PE, Peakall R (1999) Spatial autocorrelation analysis of individual multiallele and multilocus genetic structure. Heredity 82: 561–573.

72. Phillips SJ, Anderson RP, Schapire RE (2006) Maximum entropy modeling of species geographic distributions. Ecological Modeling 190: 231–259.

73. Elith J, Phillips SJ, Hastie T, Dudík M, Chee YE, et al. (2011) A statistical explanation of MaxEnt for ecologists. Diversity and Distributions 17: 43–57.

74. Elith J, Graham CH, Anderson RP, Dudík M, Ferrier S, et al. (2006) Novel methods improve prediction of species' distributions from occurrence data. Ecography 29: 129–151.

75. Hernandez PA, Graham CH, Master LL, Albert DL (2006) The effect of sample size and species characteristics on performance of different species distribution modeling methods. Ecography 29: 773–785.

76. Busby JR (1991) BIOCLIM a bioclimatic analysis and prediction system. In: Margules CR, Austin MP, eds. Nature Conservation: Cost Effective Biological Surveys and Data Analysis, CSIRO, Canberra. pp 64–68.

Impact of Transgenic Wheat with *wheat yellow mosaic virus* Resistance on Microbial Community Diversity and Enzyme Activity in Rhizosphere Soil

Jirong Wu[1,2,3◐], Mingzheng Yu[1,2,3◐], Jianhong Xu[1,2,3], Juan Du[1,2,3], Fang Ji[1,2,3], Fei Dong[1,2,3], Xinhai Li[4*], Jianrong Shi[1,2,3*]

1 Institute of Food Safety and Detection, Jiangsu Academy of Agricultural Sciences, Nanjing, China, 2 Key Lab of Food Quality and Safety of Jiangsu Province—State Key Laboratory Breeding Base, Nanjing, China, 3 Jiangsu Center for GMO evaluation and detection, Nanjing, China, 4 Institute of Crop Sciences, Chinese Academy of Agricultural Sciences, Beijing, China

Abstract

The transgenic wheat line N12-1 containing the *WYMV-Nib8* gene was obtained previously through particle bombardment, and it can effectively control the wheat yellow mosaic virus (WYMV) disease transmitted by *Polymyxa graminis* at turngreen stage. Due to insertion of an exogenous gene, the transcriptome of wheat may be altered and affect root exudates. Thus, it is important to investigate the potential environmental risk of transgenic wheat before commercial release because of potential undesirable ecological side effects. Our 2-year study at two different experimental locations was performed to analyze the impact of transgenic wheat N12-1 on bacterial and fungal community diversity in rhizosphere soil using polymerase chain reaction-denaturing gel gradient electrophoresis (PCR-DGGE) at four growth stages (seeding stage, turngreen stage, grain-filling stage, and maturing stage). We also explored the activities of urease, sucrase and dehydrogenase in rhizosphere soil. The results showed that there was little difference in bacterial and fungal community diversity in rhizosphere soil between N12-1 and its recipient Y158 by comparing Shannon's, Simpson's diversity index and evenness (except at one or two growth stages). Regarding enzyme activity, only one significant difference was found during the maturing stage at Xinxiang in 2011 for dehydrogenase. Significant growth stage variation was observed during 2 years at two experimental locations for both soil microbial community diversity and enzyme activity. Analysis of bands from the gel for fungal community diversity showed that the majority of fungi were uncultured. The results of this study suggested that virus-resistant transgenic wheat had no adverse impact on microbial community diversity and enzyme activity in rhizosphere soil during 2 continuous years at two different experimental locations. This study provides a theoretical basis for environmental impact monitoring of transgenic wheat when the introduced gene is derived from a virus.

Editor: Newton C M Gomes, University of Aveiro, Portugal

Funding: This work was supported by the National Special Transgenic Project (2014ZX08011-003), Natural Science Foundation of Jiangsu Province, China (BK20130721) and Jiangsu Agriculture Science and Technology Innovation Fund [cx(11)4064]. The funders had no role in study design, data collection and analysis, decision to publish, or preparation of the manuscript.

Competing Interests: The authors have declared that no competing interests exist.

* E-mail: lixinhai@caas.cn (XL); shiji@jaas.ac.cn (JS)

◐ These authors contributed equally to this work.

Introduction

Since the first successful genetically engineered (GE) plant was reported in 1983 [1], the planting area of transgenic crops has increased rapidly [2]. The global area cultivated commercially with transgenic crops has increased from 1.7 million ha in 1996 to 170.3 million ha in 2012 [3]. With the continued release and use of transgenic crops, there is a growing concern about their impact on the biota and soil microbial processes, such as nutrient cycling, and the potential risk of gene transfer from transgenic crops to indigenous soil microbes [4–5]. The microbes in rhizosphere soil play an important role in plant growth and development [6–7]. Transgenic crops planted in soil will inevitably interact with microorganisms such as bacteria, fungi, and actinomycetes [8–10]. Thus, transgenic crops may affect soil microbial population structure and quantity [11–13]. Additionally, root exudates have marked effects on soil microbial diversity and spatial distribution

[14–15]. At this time, most studies of environmental risk assessment focused on transgenic Bt crops such as transgenic cotton, rice and maize containing the *Bt* gene [16–18]; these studies provided basic methods for environmental risk assessment for other crops.

Enzymes in the rhizosphere soil derived from animal, plant roots and soil microbial cell secretion and decomposition of residues are an important component of the soil ecosystem [19]. They play an important role in soil biochemical processes and directly affect soil fertility [19]. Urease is associated with nitrogen transformation in the soil, while sucrase is associated with soil organic matter, nitrogen and phosphorus contents, and dehydrogenase is associated with the redox ability of the soil [19]. Previous studies showed that transgenic plants might affect enzyme activities in rhizosphere soil [11,20–21]. Therefore, it is important to investigate the impact of transgenic crops on rhizosphere soil enzyme activity when performing environmental safety risk assessments.

The first report of transgenic plants with virus resistance, expressing the coat protein of the *tobacco mosaic virus* (TMV) and delaying the development of disease, appeared in 1986 [22]. The same strategy was subsequently used to create resistance to a range of other viruses [23–24]. The exogenous genes of the transgenic virus-resistant crops are generally derived from the virus itself, including genes encoding coat protein and replicase [22–24]. Sequences derived from the genomes of plant viruses have been used to generate viral resistance in transgenic crop plants, but potential safety issues have been raised due to the environmental risks of transgenic plants with virus resistance, including hetero-encapsidation, virus recombination, gene flow, synergism and effects on non-target organisms [25,26].

Wheat yellow mosaic disease, caused by the wheat yellow mosaic virus (WYMV) at turngreen stage, is a serious illness affecting wheat in the middle and lower reaches of the Yangtze River region in China [27–28]. Disease-resistant variety breeding is one of the most cost-effective ways to control this disease through conventional wheat breeding. In recent years, conventional wheat breeding in combination with genetic engineering techniques has been applied to address wheat yellow mosaic disease, and some disease-resistant wheat lines have been cultivated. Using the particle bombardment method, genes from WYMV encoding replicase WYMV-Nib8 were transferred to the disease-sensitive variety Yangmai158 (Y158), and the disease-resistant transgenic wheat line named N12-1 was obtained by successive backcross with Y158 [29]. N12-1 showed stable and effective resistance to wheat yellow mosaic disease in a previous study [30].

Considering the above risks, transgenic virus-resistant wheat may affect the microbial community diversity in rhizosphere soil and change the population structure. Exogenous insertion of genes may also cause changes in the metabolic pathways of genetically modified crops and alter the composition of root exudates, resulting in changes in soil enzyme activity [31]. Thus, further studies on the impact on soil microbial community diversity and enzyme activities should be performed. In this study, environmental risk assessment of N12-1 was performed during 2 consecutive years of wheat cultivation under field conditions at two different experimental stations. The research involved primarily: (i) differences in soil microbial (bacterial and fungal) diversity in rhizosphere soil between N12-1 and Y158 using polymerase chain reaction–denaturing gradient gel electrophoresis (PCR-DGGE) and (ii) the activity of enzymes (urease, sucrase and dehydrogenase) in rhizosphere soil. In this report, we provide a theoretical basis for environmental transgenic wheat monitoring.

Materials and Methods

Ethics statement

In our study, the research samples were rhizosphere soils in the presence of transgenic and non-transgenic wheat. This presented no ethical issue.

Plant materials and field trial

Transgenic wheat line N12-1 and its recipient Yangmai158 (Y158) provided by the Chinese Academy of Agricultural Sciences (CAAS) were applied in this study. N12-1, which contains the *WYMV-Nib8* gene from wheat yellow mosaic virus, can effectively control the WYMV disease transmitted by *Polymyxa graminis* at turngreen stage. Y158 was one of the most popular varieties in the middle and lower reaches of the Yangtze River region in China.

However, it is sensitive to WYMV disease and the yield decreased significantly due to effects of this severe disease.

This study was performed at Luhe experimental station for transgenic crop, Jiangsu Academy of Agricultural Sciences (Luhe) and Xinxiang experimental station for transgenic crop, Henan Academy of Agricultural Sciences (Xinxiang). The physical and chemical properties of the soil are provided in Table 1. pH value, water content, available nitrogen, phosphorus potassium and organic matter content were determined by potentiometry method, alkali solution diffusion method, sodium bicarbonate method, ammonium acetate extraction method, potassium dichromate method, respectively [32]. The experiment was conducted in two successive growth seasons of wheat (October 2010-June 2011 and October 2011-June 2012) in the same field in which transgenic crops had never been planted. Each variety (line) had four blocks, each of which was 10×6 m. The materials were planted in a row with a row length of 6 m and row spacing of 0.3 m. Distance between plants was 3 cm within a row. Completely random design was applied to arrange the experiment performed in the field, and the wheat was subjected to conventional field management, that was 375 kg/ha of compound fertilizer (N:P_2O_5:K_2O = 1:0.4:1) as base fertilizer and 225 kg/ha of urea as topdressing at seedling stage.

Soil sampling

Rhizosphere soil samples were collected in both years at Luhe and Xinxiang at four growth stages [seeding stage (SS), turngreen stage (TS), grainfilling stage (GS), maturing stage (MS)]. Rhizosphere soil was defined as the soil still attached to the roots after the roots were shaken by hand. For each sampling site, five wheat plants were selected to collect rhizosphere soil and each block contains five sampling site. Rhizosphere soil from the five sampling sites per block was mixed as a composite rhizosphere soil sample. The soil samples were sieved using a 20-mesh sieve and then stored at 4°C until further use, usually within one month before DNA extraction.

Soil DNA extraction

Total community DNA was extracted from 0.5 g of rhizosphere soil using an UltraClean Soil DNA Isolation Kit (MoBio Lab, USA). DNA extraction was performed according to the manufacturer's protocol.

PCR amplification of 16S and18S rDNA fragments for DGGE analysis

The 16S rDNA fragments of bacteria were amplified by using the primer pair GC338f (5′-CGCCGCGCGCGGCGGGCG-GGGCGGGGGCACGGGGGGGACTCCTACGGGAGGCAGC-AG-3′, the sequence underlined was the GC clamp) and 518r (5′-ATTACCGCGGCTGCTGG -3′) as described by Bakke et al. [33]. High fidelity polymerase of KOD-Plus-Neo (Toyobo, Japan) was applied to perform PCR amplification and avoid mutations in the PCR product. Briefly, the reaction mixture consisted of 1 μl of template DNA (1–5 ng), 5 μl 10×PCR Buffer, 5 μl of 2 mM dNTPs, 3 μl of 25 mM $MgSO_4$, 0.5 μl of 10 μM forward primer, 0.5 μl of 10 μM reverse primer, and 1 U of DNA polymerase, after which ddH_2O was added to a final volume of 50 μl. The thermal cycling program was performed with an initial denaturation at 94°C for 5 min, followed by 35 cycles at 95°C for 15 sec, 58°C for 15 sec, and 68°C for 30 sec before the final extension at 68°C for 10 min. Products were checked by electrophoresis in 1% (wt/vol) agarose gels followed by ethidium bromide staining.

Table 1. Main physical and chemical properties of the soil from two experiment locations before planting.

Experiment station	Physical and chemical properties					
	pH	water (%)	available nitrogen (mg/kg)	available phosphorus (mg/kg)	available potassium (mg/kg)	organic matter (%)
Luhe	5.8	20.55	110.16	90.81	857.99	1.44
Xinxiang	8.5	4.92	70.39	28.26	863.69	0.68

The 18S rDNA fragments of fungi were amplified by using the primer pair (GC-Fungi: 5′-<u>CGCCCGCCGCGCCCCGCGCCC</u><u>GGCCCGCCGCCCCCGCCCC</u>ATTCCCCGTTACCCGTT-G-3′; NS1: 5′- GTAGTCATATGCTTGTCTC -3′, the sequence underlined was the GC clamp) as described by Das et al. [34]. The protocol for PCR amplification was similar as above. All products were purified before electrophoresis using a Cycle Pure Kit (Omega, USA).

PCR-DGGE

DGGE analysis for 16S rDNA and 18S rDNA products was performed with the DCode System (Bio-Rad, USA). Polyacrylamide gels were composed of a denaturing gradient of 50–65% (bacteria) and 30–38% (fungi) urea, 0.17% (vol/vol) TEMED, 0.047% (wt/vol) ammonium persulfate, 6% acrylamide-N,N_-methylenebisacrylamide (37.5:1) and 1×TAE. PCR products (up to 50 μl) were applied to the gel. DGGE was performed at 50 V in 1×TAE at 60°C for 12 h (bacteria) and at 50 V in 1×TAE at 60°C for 20 h (fungi), respectively. A silver staining method was used for the detection of DNA in DGGE gels.

Migration and intensity of DGGE bands were analyzed using Quantity One according to the manual. The bands that shared identical migration positions were considered to be the same species. Shannon's diversity index (H) of bacterial and fungal DGGE profiles was calculated with the following formula [35]:

$$H = -\sum_{i=1}^{S} \frac{n_i}{N} \ln \frac{n_i}{N}$$

Simpson's diversity index (D) was calculated with the following formula:

$$D = \sum_{i=1}^{S} \left[\frac{n_i}{N}\right]^2$$

Evenness (E) was calculated with the following formula:

$$E = \frac{H}{\ln S}$$

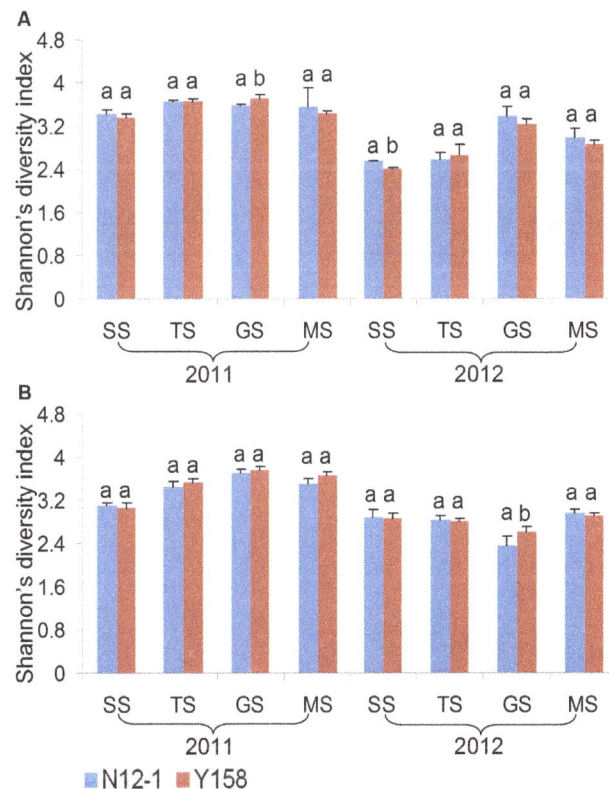

Figure 2. Shannon's index of bacterial communities at different growth stages. Error bars indicate standard errors (n = 4). Different letters above bars denote a statistically significant difference between the means of the fields. A: Luhe; B: Xinxiang. SS: seeding stage; TS: turngreen stage; GS: grainfilling stage; MS: maturing stage.

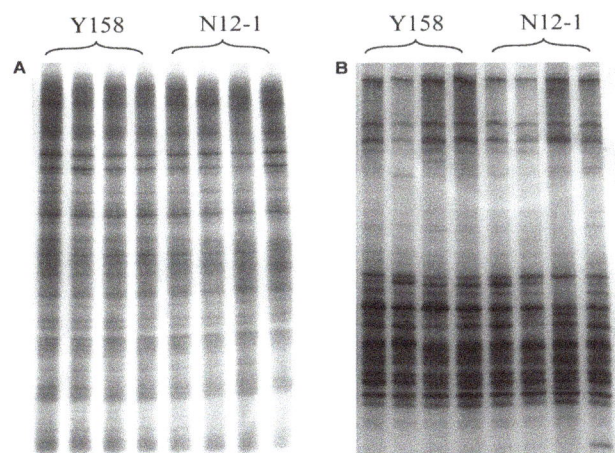

Figure 1. DGGE profiles of 16S rDNA and 18S rDNA fragments amplified from DNA extracted from rhizosphere soil of N12-1 and Y 158 at turngreen stage from Luhe experiment station in 2011. A: bacteria; B: fungus.

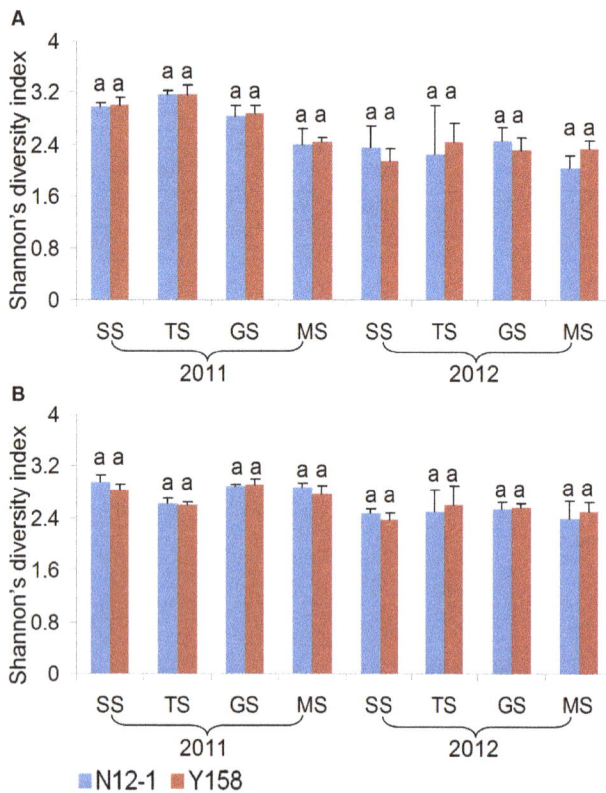

Figure 3. Shannon's index of fungi communities at different growth stages. Error bars indicate standard errors (n = 4). Different letters above bars denote a statistically significant difference between the means of the fields. A: Luhe; B: Xinxiang. SS: seeding stage; TS: turngreen stage; GS: grainfilling stage; MS: maturing stage.

n_i represented the square of individual peaks detected by Quantity One; N represented the square of all peaks in the same lane; S represented the number of bands in the same lane.

Band sequencing

Visible bands in the fungi DGGE gel were picked with sterile tips and transferred into a 200 μl tube. Sterile ddH$_2$O (50 μl) was added to the tube and the gel was pounded to pieces. The tubes with broken gels were incubated at room temperature overnight and then centrifuged at 12000 rpm for 5 min. The supernatant solution was used as the template for PCR, which was performed as described above or for fungi using primers without GC-clamps. The PCR products were purified (Omega, USA), ligated into pM19-T vector (Takara, Japan) and transformed into competent cells (*E coli* DH5α, Takara, Japan) according to the instructions of the manufactures and plated on LB solid medium with ampicillin. Positive clones were selected by PCR with primer pair of NS1 and Fungi (GC-Fungi without GC-clamps) and plasmids were extracted for sequencing (Invitrogen, Shanghai). All the sequences that have been sequenced successfully were submitted to GenBank (Accession numbers: KJ755390-KJ755404).

Enzyme activity analysis

Activities of urease, sucrose and dehydrogenase were analyzed in this study. Urease and sucrose activities in soil were assayed using the method of Guan [36]: urease activity was determined by measuring the release of NH$_3$ as mg.(g.d)$^{-1}$, and sucrose activity was determined based on 3,5-dinitrosalicylic acid colorimetry as mg.(g.d)$^{-1}$. Dehydrogenase activity was determined based on the reduction of triphenyltetrazolium chloride (TTC) to triphenylformazan (TPF), as described by Serra-Wittling et al. [37] with minor modifications, which was expressed as μg.(g.d)$^{-1}$. The data were subjected to analysis of variance, and the means and standard deviations of four replicates were calculated.

Table 2. Simpson's index and Evenness of bacterial community.

Experiment station	Growth stage	Variety(line)	Simpson's index		Evenness	
			2011	2012	2011	2012
Luhe	SS	N12-1	0.04±0.00a	0.10±0.00a	0.95±0.01a	0.88±0.01a
		Y158	0.04±0.00a	0.12±0.00b	0.95±0.01a	0.86±0.02a
	TS	N12-1	0.03±0.00a	0.09±0.01a	0.84±0.01a	0.92±0.03a
		Y158	0.03±0.00a	0.08±0.02a	0.95±0.01a	0.93±0.01a
	GS	N12-1	0.03±0.00a	0.04±0.01a	0.95±0.00a	0.93±0.01a
		Y158	0.03±0.00b	0.05±0.01a	0.96±0.01a	0.93±0.02a
	MS	N12-1	0.04±0.00a	0.06±0.02a	0.99±0.10a	0.93±0.03a
		Y158	0.04±0.00a	0.07±0.00a	0.95±0.01a	0.90±0.02a
Xinxiang	SS	N12-1	0.05±0.01a	0.06±0.01a	0.93±0.03a	0.98±0.01a
		Y158	0.05±0.01a	0.06±0.01a	0.95±0.01a	1.00±0.01a
	TS	N12-1	0.04±0.01a	0.07±0.01a	0.93±0.02a	0.93±0.02a
		Y158	0.04±0.00a	0.07±0.01a	0.93±0.00a	0.94±0.02a
	GS	N12-1	0.03±0.00a	0.11±0.02a	0.95±0.01a	0.92±0.03a
		Y158	0.03±0.00a	0.08±0.01b	0.95±0.01a	0.96±0.01a
	MS	N12-1	0.04±0.01a	0.07±0.00a	0.93±0.02a	0.90±0.01a
		Y158	0.03±0.00a	0.07±0.01a	0.96±0.01b	0.89±0.02a

SS: seeding stage; TS: turngreen stage; GS: grainfilling stage; MS: maturing stage. The alphabets after the value represented the significance level of the index.

Table 3. Simpson's index and Evenness of fungi community.

Experiment station	Growth stage	variety(line)	Simpson's index		Evenness	
			2011	2012	2011	2012
Luhe	SS	N12-1	0.06±0.01a	0.11±0.03a	0.92±0.02a	0.94±0.02a
		Y158	0.06±0.01a	0.14±0.02a	0.94±0.01a	0.91±0.03a
	TS	N12-1	0.05±0.00a	0.04±0.06a	0.92±0.00a	0.78±0.23a
		Y158	0.05±0.01a	0.11±0.05a	0.91±0.01a	0.97±0.08a
	GS	N12-1	0.07±0.01a	0.10±0.02a	0.93±0.01a	0.95±0.02a
		Y158	0.07±0.01a	0.11±0.03a	0.93±0.01a	0.93±0.02a
	MS	N12-1	0.12±0.02a	0.16±0.03a	0.88±0.03a	0.86±0.04a
		Y158	0.11±0.01a	0.12±0.02b	0.89±0.04a	0.92±0.04a
Xinxiang	SS	N12-1	0.06±0.01a	0.10±0.01a	0.93±0.03a	0.87±0.02a
		Y158	0.07±0.00b	0.10±0.01a	0.93±0.01a	0.86±0.03a
	TS	N12-1	0.08±0.01a	0.09±0.03a	0.95±0.01a	0.97±0.02a
		Y158	0.08±0.00a	0.08±0.02a	0.95±0.01a	0.97±0.00a
	GS	N12-1	0.07±0.00a	0.08±0.01a	0.94±0.02a	0.96±0.01a
		Y158	0.07±0.01a	0.08±0.01a	0.93±0.01a	0.96±0.01a
	MS	N12-1	0.07±0.01a	0.12±0.04a	0.89±0.03a	0.88±0.04a
		Y158	0.07±0.01a	0.10±0.02a	0.91±0.02a	0.89±0.03a

SS: seeding stage; TS: turngreen stage; GS: grainfilling stage; MS: maturing stage. The alphabets after the value represented the significance level of the index.

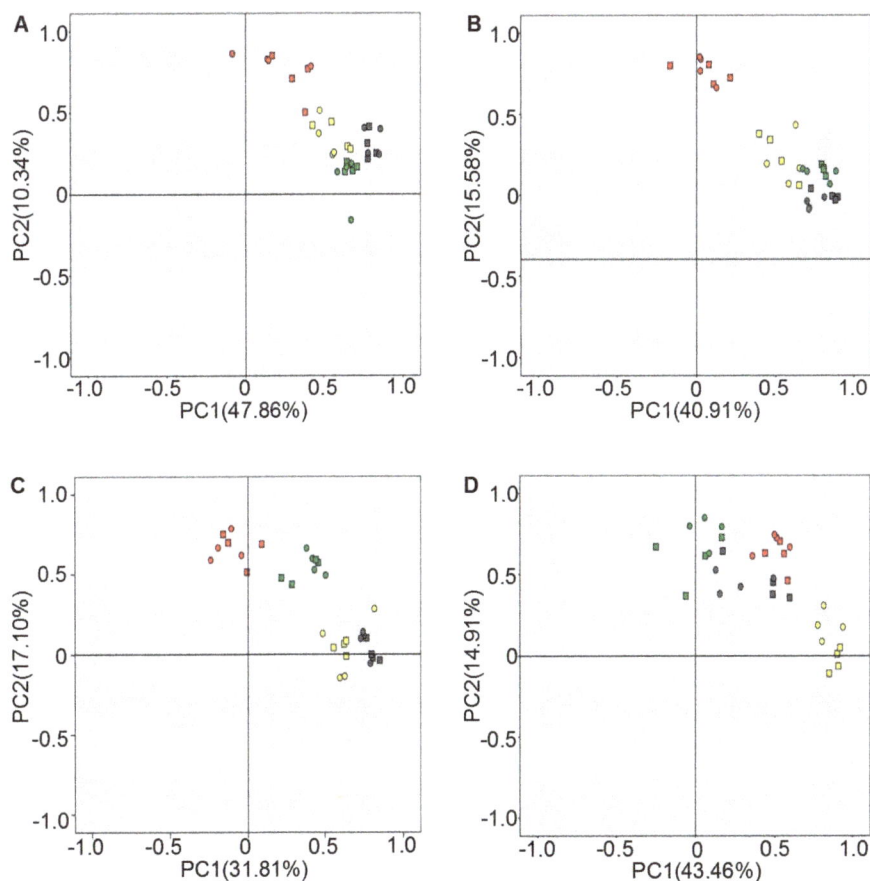

Figure 4. Principal component analysis of bacterial community diversities in rhizosphere soil. A: Luhe in 2011; B: Luhe in 2012; C: Xinxiang in 2011; D: Xinxiang in 2012. Square: N12-1; Round: Y158. Gray: seeding stage; Green: turngreen stage; Red: grainfilling stage; Yellow: maturing stage. Band position and presence (presence/absence) were used to carry out PCA analyses.

Statistical analysis

SPSS 16.0 was applied to determine whether the indices and enzyme activities differed between years, varieties and growth stages by ANOVA. PCA analyses were carried out based on band position and presence (presence/absence), and then the correlation matrix principal component analysis was performed by SPSS 16.0 [35]. Microsoft Excel 2003 was used to construct column diagrams.

Results

Impact of transgenic wheat on bacterial and fungal community diversity

One of the DGGE profiles of 16S rDNA and 18S rDNA fragments amplified from DNA extracted from rhizosphere soil was presented as figure 1. Three diversity indices (Shannon's, Simpson's, evenness) were used to analyze the bacterial and fungal DGGE profiles of the soil samples from Luhe and Xinxiang at four different growth stages in 2011 and 2012. For bacteria, the effect of wheat line on DGGE diversity indices was insignificant, except GS stage in 2011, SS in 2012 at Luhe and GS stage in 2012 at Xinxiang for Shannon's diversity index (Fig. 2). The Simpson's diversity index showed the same results as Shannon's diversity index (Table 2). For evenness, only one difference was found at the MS stage at Xinxiang in 2011 (Table 2). For fungi, the effect of wheat line on DGGE diversity indices was insignificant, except for

SS in 2011 at Xinxiang and MS in 2012 at Luhe for Simpson's index (Fig. 3; Table 3).

Principal component analysis of bacterial community diversity

Principal components analysis (PCA) using both band position and presence/absence as parameters were performed to further analyze DGGE fingerprint profiles. For experiments conducted at Luhe, the contribution rates of the two principal components were 47.86% and 10.34% in 2011 (Fig. 4A) and 40.91% and 15.58% in 2012 (Fig. 4B), respectively. Different growth stages showed a distinct separation along the principal components axes, whereas different replications of experimental materials formed a cluster at the same growth stage. This was consistent with the result of Shannon's diversity analysis. In 2011, the first principal component axis clearly separated the GS and SS stage (Fig. 4A), but separated the GS and MS stage in 2012 (Fig. 4B). The second principal component axis clearly distinguished the GS stage in 2011 (Fig. 4A) and the GS stage in 2012 (Fig. 4B).

For experiments conducted at Xinxiang, the contribution rates of the two principal components were 31.81% and 17.10% in 2011 (Fig. 4C) and 43.46% and 14.91% in 2012 (Fig. 4D), respectively. Different growth stages also showed a distinct separation along the principal components axes, whereas different replications of experimental materials clustered together at the same growth stage. In 2011, the first principal component axis

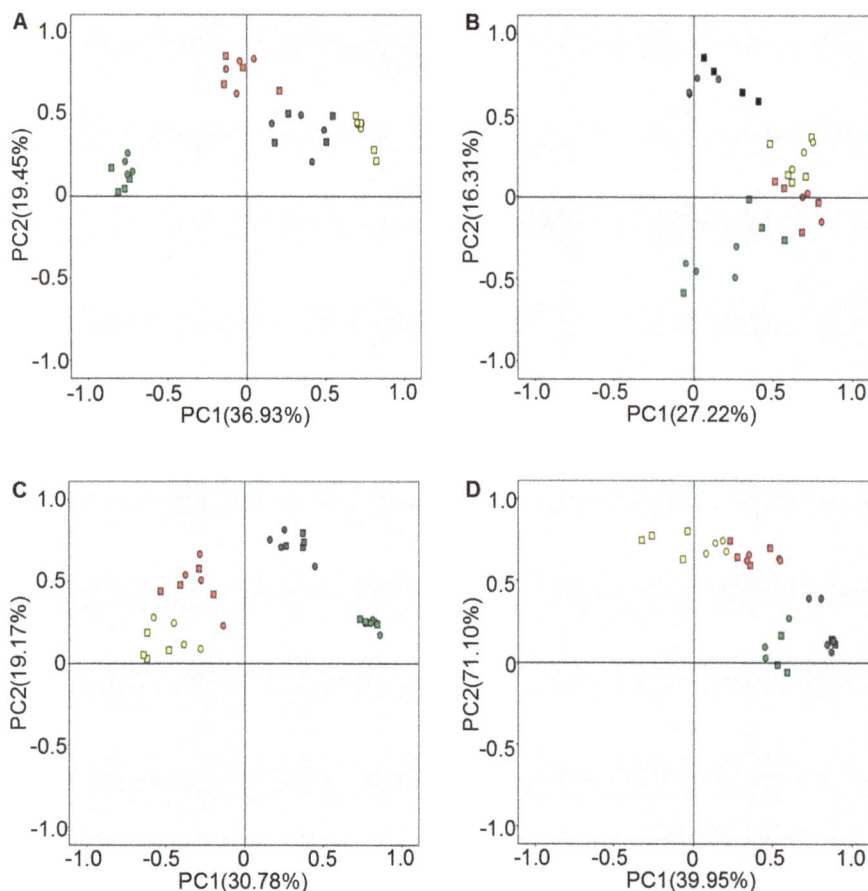

Figure 5. Principal component analysis of fungi communities diversity in rhizosphere soil. A: Luhe in 2011; B: Luhe in 2012; C: Xinxiang in 2011; D: Xinxiang in 2012. Square: N12-1; Round: Y158. Gray: seeding stage; Green: turngreen stage; Red: grainfilling stage; Yellow: maturing stage. Band position and presence (presence/absence) were used to carry out PCA analyses.

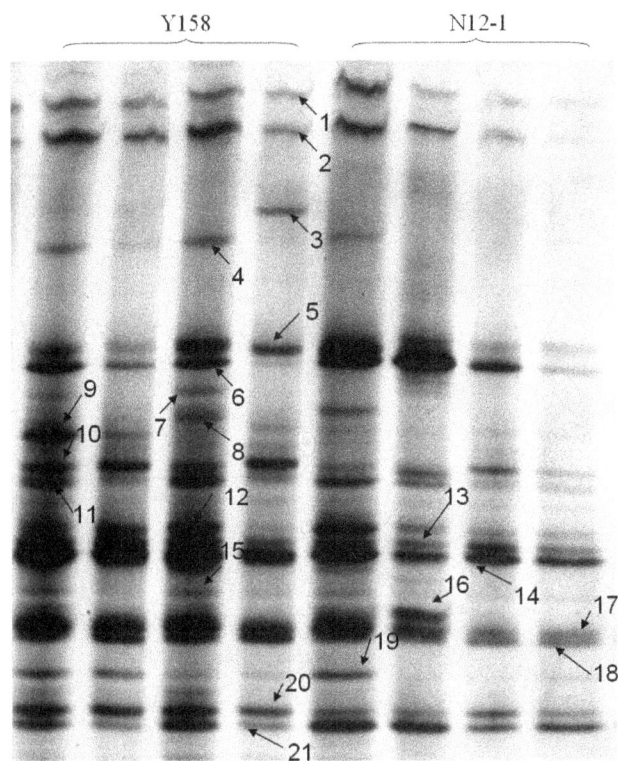

Y158   N12-1

Figure 6. PCR-DGGE gel profile of fungi communities used for band sequencing. The numbers means different bands picked for sequencing.

clearly separated the four growth stages (Fig. 4C), but separated the MS stage from the other three stages in 2012 (Fig. 4D). The second principal component axis clearly distinguished the TS and GS stages from the SS and MS stages in 2011 (Fig. 4C), and distinguished the MS stage in 2012 (Fig. 4D).

These PCA analysis results showed that growth stage played an important role in bacterial community diversity, rather than the presence of transgenic and non-transgenic wheat.

Principal component analysis of fungal community diversity

For experiments at Luhe, the contribution rates of the two principal components were 36.93% and 19.45% in 2011 (Fig. 5A) and 27.22% and 16.31% in 2012 (Fig. 5B), respectively. Different sampling times showed a distinct separation along the principal components axes, whereas different replications of experiment materials formed a cluster at the same sampling time. In 2011, the first principal component axis clearly separated the four growth stages (Fig. 5A), but separated SS and TS from GS and MS in 2012 (Fig. 5B). The second principal component axis clearly distinguished the TS stage in 2011 (Fig. 5A), and the SS and TS stage in 2012 (Fig. 5B).

For experiments at Xinxiang, the contribution rates of the two principal components were 30.78% and 19.17% in 2011 (Fig. 5C) and 39.95% and 17.10% in 2012 (Fig. 5D). Different sampling times also showed a distinct separation along the principal components axes, whereas different replications of experimental materials formed a cluster at the same sampling time. In 2011, the first principal component axis clearly separated the SS and TS stages (Fig. 5C), but separated the SS and MS stages in 2012 (Fig. 5D). The second principal component axis could not clearly

Table 4. Blast results of the bands from the DGGE gels of fungI community analysis.

No. of bands	Accession No.	Blast result	identity
1	GU214699.1	*Septoria dysentericae* strain CPC 12328 18S ribosomal RNA gene	100%
2	GQ330624.1	Uncultured *Mucorales* clone PR3 4E 28 18S ribosomal RNA gene	95%
3		Cannot be amplified	
4	AJ515922.1	Uncultured soil ascomycete partial 18S rDNA gene	100%
5	EU120944.1	Uncultured *Cystofilobasidiales* (aff. Guehomyces) clone Y9 18S ribosomal RNA gene	100%
6	AJ515941.1	Uncultured soil ascomycete partial 18S rDNA gene	99%
7		Cannot be amplified	
8	AY789390.1	*Peziza varia* strain ZW-Geo94-Clark 18S small subunit ribosomal RNA gene	99%
9	FJ176814.1	*Saccobolus dilutellus* isolate AFTOL-ID 1299 18S small subunit ribosomal RNA gene	97%
10	FO181499.1	Balen uncultured eukaryote partial 18S ribosomal RNA	80%
11	AY771600.1	Polyozellus multiplex isolate AFTOL-ID 677 18S small subunit ribosomal RNA gene	99%
12	GU190186.1	*Cochliobolus* sp. Enrichment culture clone NJ-F5 18S small subunit ribosomal RNA gene	100%
13		Cannot be amplified	
14	AJ515948.1	Uncultured soil ascomycete partial 18S rDNA gene	99%
15		Cannot be amplified	
16	AJ301992.1	*Myrothecium leucotrichym* 18S RNA gene	99%
17	JX159444.1	Uncultured *Filobasidium* clone Cegs 957 18S ribosomal RNA gene	99%
18	KC171701.1	Uncultured fungus isolate DGGE gel band f10 18S ribosomal RNA gene	100%
19		Cannot be amplified	
20		Cannot be amplified	
21	EU120947.1	Uncultured *Ascobolus* clone Y12 18S ribosomal RNA gene	99%

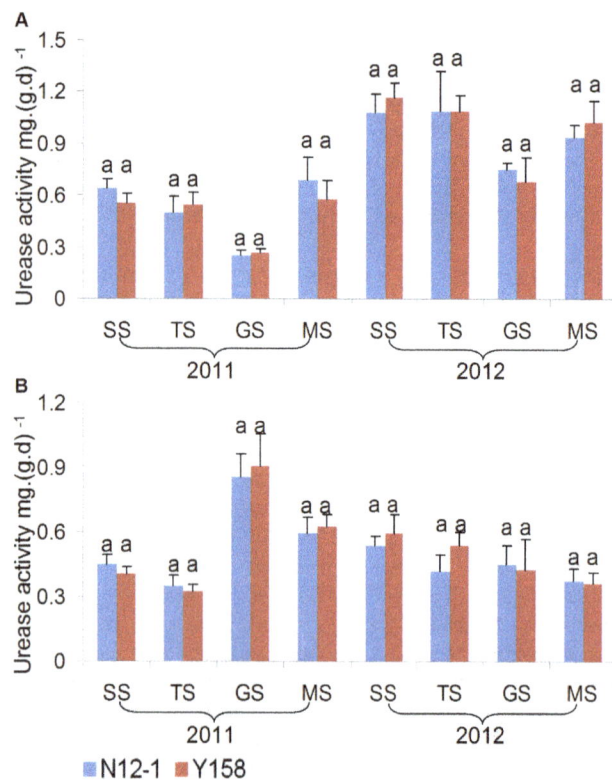

Figure 7. Urease activity in rhizosphere soil at different growth stages. Error bars indicate standard errors (n = 4). Different letters above bars denote a statistically significant difference between the means of the fields. A: Luhe; B: Xinxiang. SS: seeding stage; TS: turngreen stage; GS: grainfilling stage; MS: maturing stage.

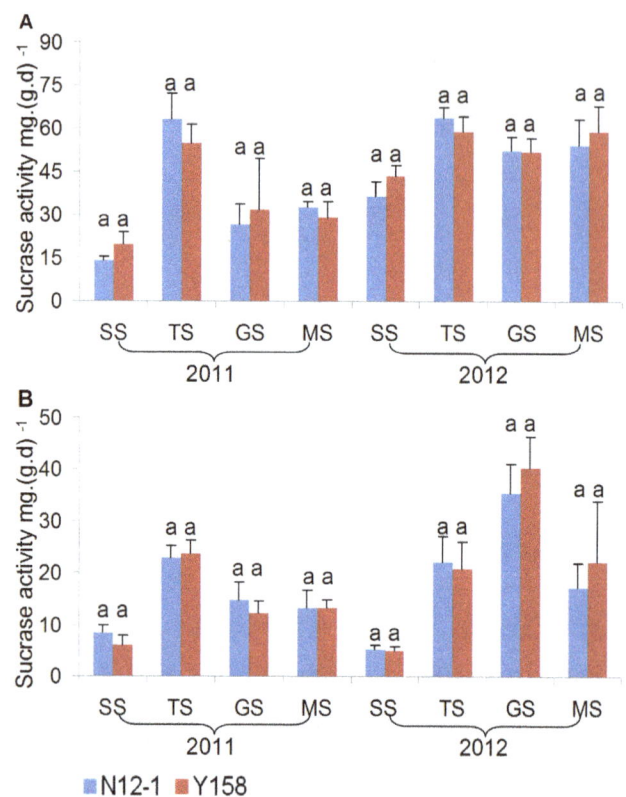

Figure 8. Sucrase activity in rhizosphere soil at different growth stages. Error bars indicate standard errors (n = 4). Different letters above bars denote a statistically significant difference between the means of the fields. A: Luhe; B: Xinxiang. SS: seeding stage; TS: turngreen stage; GS: grainfilling stage; MS: maturing stage.

distinguish any growth stage in 2011 (Fig. 5C), but could distinguish the MS and GS stages from the SS and TS stages in 2012 (Fig. 5D).

These PCA analysis results showed that fungal communities exhibited marked diversity at different growth stages, rather than between the transgenic line and non-transgenic wheat recipient.

Band sequencing

A total of 21 visible bands from the DGGE gel of fungi from Luhe in 2011 were subjected to sequencing (Fig. 6), and 15 were sequenced successfully. Using NCBI BLAST, we found that most of the sequenced bands represented uncultured fungi. Others were partial 18S rRNA sequences of *Septoria dysentericae*, *Peziza varia*, *Saccobolus dilutellus*, *Polyozellus*, *Cochliobolus*, and *Myrothecium leucotrichym* (Table 4).

Enzyme activity analysis

Urease, sucrase, and dehydrogenase activities in rhizosphere soil were applied as indicators for environmental risk assessment of transgenic wheat N12-1 in this study.

In general, there was no consistent significant difference in the enzyme activity between soils of transgenic wheat N12-1 and its recipient Y158 within the same growth stage during the 2 years. Only one significant difference in activity was observed; for dehydrogenase at the MS stage at Xinxiang in 2011. In 2011, the dehydrogenase activity in soil of N12-1 was significantly ($p<0.05$) higher than in soil of its recipient Y158 (Figs. 7–9). Significant differences were observed between years ($p<0.01$) and among

growth stages ($p<0.001$) at both Luhe and Xinxiang, with the exception of dehydrogenase among growth stages at Xinxiang ($p<0.25$) (Table 5). These results showed that N12-1 had a minor impact on soil enzyme activities.

Discussion

With the cultivation of more varieties of virus-resistant transgenic plants and large-scale planting, environmental impact monitoring after commercial release has attracted increasing attention from the scientific community and public [38,39]. In soil, there are high microbial population densities and large numbers of microbial species that interact with the plants and surrounding environment and have an effect on the function of the soil ecosystem, such as the enzyme activity and physicochemical properties.

Soil microbial analysis has been used widely to evaluate the impact of various exogenous chemical or environmental pollutants (such as herbicides, fertilizers, heavy metals, et al.) on soil fertility and crop yields [7]. Therefore, monitoring changes in soil microbial populations will increase our understanding of the potential risks of introduction of exogenous genes to soil [4,7]. In our study, two years and two locations of field research was performed to compare the impact of transgenic wheat with genes encoding replicase from WYMV on microbial population diversity in agricultural systems. One of the major outcomes was that transgenic insertion did not significantly alter bacterial or fungal population diversity at each growth stage; however, growth stage and planting year had important effects on microbial diversity.

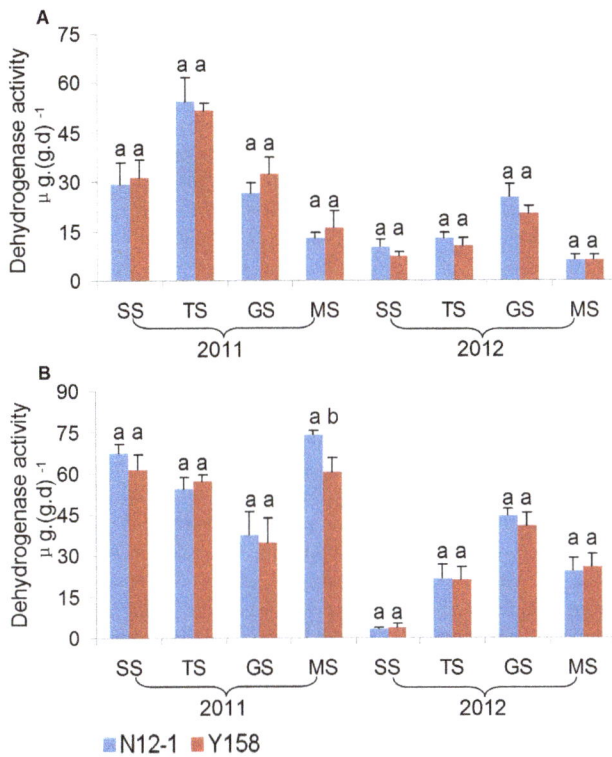

Figure 9. Dehydrogenase activity in rhizosphere soil at different growth stages. Error bars indicate standard errors (n = 4). Different letters above bars denote a statistically significant difference between the means of the fields. A: Luhe; B: Xinxiang. SS: seeding stage; TS: turngreen stage; GS: grainfilling stage; MS: maturing stage.

This result was similar to the findings of Meyer et al. [40]. In that study, the authors found that the effects of GM wheat on plant-beneficial root-colonizing microorganisms are minor and not of ecological importance. Lupwayi et al. reported that glyphosate-resistant wheat–canola rotations under low-disturbance direct seeding and conventional tillage did not affect the functional diversity of rhizosphere soil bacteria in 18 of 20 site-years [20]. The observation that certain growth stages (mainly SS and GS) showed differences between transgenic and non-transgenic wheat may be due to inconsistencies in the soil at seeding time, and at later growth stages the temperature and humidity increased rapidly. The differences between years and growth stages indicated that the diversity of bacteria and fungi might be affected by various environmental factors, such as temperature, humidity, and light. In studies of GM crops against virus, there was no significant difference between microbial communities with transgenic or non-transgenic watermelon resistant or cucumber green mottle mosaic virus (CGMMV), but significant changes in the microbial community were observed during the growing season [41]. Transgenic tomato resistant to cucumber mosaic virus (CMV) had no effect on the variation of soil microbial communities, in which soil position and environmental factors played more dominant roles [42]. Also, non-environmental factor, such as root exudates, may also play an important role for diversity changes of bacteria and fungi between years and growth stages [14–15]. Fang et al. thought that bacterial communities differed due to changes in root exudates quantity and composition by developing corn plant, which select different bacterial groups during root colonization [43]; Donegan et al. have speculated that the reason for the different in the communities of genetically modified plants is due to differences in the root exudates patterns of these plants [44]. However, Wei et al. reported the opposite result [45]. In a transgenic alfalfa study performed using the cultivation-dependent plating method, statistically significant differences in densities of rhizospheric bacteria between transgenic and non-transgenic

Table 5. Generalized Linear Mixed Model results for overall effects on enzyme activity.

Location	Enzyme	Effect	F Value	p Value
Luhe	Urease	Years	282.21	0.00
		Growth stage	36.89	0.00
		Variety (Line)	0.01	0.91
	Sucrase	Years	74.66	0.00
		Growth stage	33.54	0.00
		Variety (Line)	0.05	0.83
	Dehydrogenase	Years	82.20	0.00
		Growth stage	16.23	0.00
		Variety (Line)	0.03	0.87
Xinxiang	Urease	Years	6.38	0.01
		Growth stage	7.97	0.00
		Variety (Line)	0.103	0.75
	Sucrase	Years	11.45	0.00
		Growth stage	23.67	0.00
		Variety (Line)	0.10	0.75
	Dehydrogenase	Years	73.88	0.00
		Growth stage	1.40	0.25
		Variety (Line)	0.11	0.74

alfalfa clones were observed for ammonifying bacteria, cellulolytic bacteria, rhizobial bacteria, denitrifying bacteria and *Azotobacter* spp. [46]. These results indicated that transgenic crops containing a viral gene conferring resistance to viral disease had little effect on soil microbial diversity (excluding a small number of studies) compared with non-transgenic crops. Transgenic wheat also had no adverse effects on soil biological indicators, such as *Folsomia candida* [47] and earthworm [48]. Duc et al. found that GM wheat with race-specific antifungal resistance against powdery mildew (Pm3b), and two with nonspecific antifungal resistance, had no impact on the soil fauna community (mites, springtails, annelids, and diptera). However, sampling date and location significantly influenced the soil fauna community and decomposition processes [49].

Soil enzymes in the soil nutrient cycle and energy transfer play an important role in soil ecology, and are derived mainly from soil microbial populations. Many studies have used soil enzymes as indicators of soil microbial activity and fertility [50–52]. In our study, urease, sucrase and dehydrogenase were used as indicators of the impact of transgenic wheat on soil quality. The results showed no significant difference in enzyme activity in rhizosphere soil between transgenic and non-transgenic wheat at each growth stage at two locations in 2 years, excluding dehydrogenase during the maturing stage at Xinxiang in 2011. In other studies of transgenic crops, there was no consistent significant difference in soil enzymes between transgenic and non-transgenic plants, but there were differences among seasons and crop varieties [53–54]. These results are consistent with our study. In other studies, some enzymes showed significant differences between transgenic and non-transgenic plants [11,20–21]. There have been no previous studies of soil enzyme activities of transgenic wheat. Thus, our results should be confirmed in future studies and at more experimental locations. Additionally, other types of transgenic wheat, such as insect-resistant and stress-tolerant varieties, should be used to perform risk assessments.

Due to the complexity of DGGE profiles, several bands can be difficult to identify visually, and different bands represent different microbes. These issues make it difficult to compare varieties. Thus, the combination of DGGE and cloning sequencing methods is often used to investigate the impact of transgenic plants on microorganisms in rhizosphere soil [55,56]. In our study, most of the bands from fungi DGGE gels represented uncultured fungal taxa. This is in agreement with the fact that only ~1% of microbes in soil can be artificially cultured and identified [57].

With the development of sequencing technology, the way we study microbial communities has been changed. Traditionally, the study of genes from natural environments included cloning DNA into a vector, inserting that vector into a host, screening, and Sanger sequencing. Sequence-by-synthesis methods provide faster, cheaper, and simpler methods for (meta)genome sequence that bypass the PCR amplification bias, cloning bias and labor-intensive Sanger method [58]. Currently, massively parallel high-throughput pyrosequencing methods can process hundreds of thousands of sequences simultaneously [59]. Fierer et al. used metagenomic and small subunit rRNA analyses to study the genetic diversity of bacteria, archaea, fungi, and viruses in soil [60]; Uroz et al. used functional assays and metagenomic analyses to reveal difference between the microbial communities [61]. Li et al. analyzed the impact on bacterial community in midguts of the asian corn borer larvae by transgenic *Trichoderma* strain overexpressing a heterologous *chit42* gene with chitin-binding domain by using 16s rRNA library. All above studies have used the next generation sequencing technology [62]. Now, this technology is being adopted to study the microbial community in rhizosphere soil of transgenic plants gradually [62].

In conclusion, our study has produced weak evidence for the effect of virus-resistant transgenic wheat on soil microbial community diversity and enzyme activities. The community structure was markedly affected by natural variations in the environment related to wheat growth stage and planting year. Little difference was observed in bacterial and fungal communities in the presence of the wild-type Y158 or the transgenic line N12-1. This requires further investigation using extended field observations involving more varieties for more years. Based on this information, we can determine whether the altered composition is attributable to the presence of transgenic crops, or is simply part of the variation driven by the presence of different genotypes [63]. These studies should also involve more soil types and longer-term monitoring to account for the variability of the natural environment.

Acknowledgments

We thank everyone that helped with the fieldwork and Dr. YZ from CAAS for providing seeds of transgenic and non-transgenic wheat.

Author Contributions

Conceived and designed the experiments: JW MY JX XL JS. Performed the experiments: JW MY JX JD FJ FD. Analyzed the data: JW MY. Wrote the paper: JW MY JS.

References

1. Horsch RB, Fraley RT, Rogers SG, Sanders PR, Lloyd A, et al. (1984) Inheritance of functional foreign genes in plants. Science 223: 496–498

2. Vauramo S, Pasonen HL, Pappinen A, Setälä H (2006) Decomposition of leaf litter form chitinase transgenic birch (*Betula pendula*) and effects on decomposer populations in a field trial. Appl Soil Ecol 32: 338–349.

3. James C (2013) Global Status of Commercialized Biotech/GM Crops: 2012. ISAAA Brief No. 44. ISAAA, Ithaca, New York.

4. McGregor AN, Turner MA (2000) Soil effects of transgenic agriculture: biological processes and ecological consequences. NZ Soil News 48(6):166–169

5. Sengeløv G, Kristensen KJ, Sørensen AH, Kroer N, Sørensen SJ (2001) Effect of genomic location on horizontal transfer of a recombinant gene cassette between Pseudomonas strains in the rhizosphere and spermosphere of barley seedlings. Cur Microbiol 42:160–167

6. Gyaneshwar P, Naresh Kumar G, Parekh LJ, Poole PS (2002) Role of soil microorganisms in improving P nutrition of plants. Plant Soil 245: 83–93

7. Kent AD, Triplett EW (2002) Microbial communities and their interactions in soil and rhizosphere ecosystems. Annu Rev Microbiol 56:211–236

8. O'Callaghan M, Glare TR. (2001) Impacts of transgenic plants and microorganisms on soil biota. New Zeal Plant Prot 54: 105–110.

9. Andow DA, Zwahlen C (2006) Assessing environmental risks of transgenic plants. Ecology Letters 9:196–214

10. Liu B, Zeng Q, Yan F, Xu H, Xu C (2005) Effects of transgenic plants on soil microorganisms. Plant Soil 271:1–13

11. Chen ZH, Chen LJ, Zhang YL, Wu ZJ (2011) Microbial properties, enzyme activities and the persistence of exogenous proteins in soil under consecutive cultivation of transgenic cottons (*Gossypium hirsutum* L.). Plant Soil Environ, 57: 67–74

12. Blackwood CB, Buyer JS (2004) Soil microbial communities associated with Bt and non-Bt corn in three soils. J Environ Qual 33:832–836

13. Donegan KK, Palm CJ, Fieland VJ, Porteous LA, Ganio LM, et al. (1995) Changes in levels, species and DNA fingerprints of soil microorganisms associated with cotton expressing the *Bacillus thuringiensis* var. *kurstaki* endotoxin. Appl Soil Ecol 2:111–124

14. Bais HP, Weir TL, Perry LG, Gilroy S, Vivanco JM (2006) The role of root exudates in rhizosphere interactions with plants and other organisms. Annu Rev Plant Biol 57:233–266

15. Saxena D, Flores S, Stotzky G (1999) Insecticidal toxin in root exudates from Bt corn. Nature 402: 480–481

16. Raybould A, Higgins LS, Horak MJ, Layton RJ, Storer NP, et al. (2012) Assessing the ecological risks from the persistence and spread of feral populations of insect-resistant transgenic maize. Transgenic Res 21:655–664

17. Devarc MH, Jones CM, Thies JE (2004) Effect of Cry3Bb transgenic corn and tefluthrin on the soil microbial community: biomass, activity, and diversity. J Environ Qual 33: 837–843

18. Lu H, Wu W, Chen Y, Zhang X, Devare M, et al. (2010) Decomposition of *Bt* transgenic rice residues and response of soil microbial community in rapeseed–rice cropping system. Plant Soil 336: 279–290

19. Burns RG (1982) Enzyme activity in soil: Location and a possible role in microbial ecology. Soil Biol Biochem 14: 423–427

20. Lupwayi NZ, Hanson KG, Harker KN, Clayton GW, Blackshaw RE, et al. (2007) Soil microbial biomass, functional diversity and enzyme activity in glyphosate-resistant wheat–canola rotations under low-disturbance direct seeding and conventional tillage. Soil Biol Biochem 39: 1418–1427

21. Sun CX, Chen LJ, Wu ZJ, Zhou LK, Shimizu H (2007) Soil persistence of *Bacillus thuringiensis* (Bt) toxin from transgenic Bt cotton tissues and its effect on soil enzyme activities. Biol Fertil Soils 43: 617–620

22. Abel PP, Nelson RS, De B, Hoffman N, Rogers SG, et al. (1986) Delay of disease development in transgenic plants that express the tobacco mosaic virus coat protein. Science 232, 738–43.

23. Beachy RN, Loesch-Fries S, Tumer NE (1990) Coat-protein mediated resistance against virus infection. Annu Rev Phytopathol 28, 451–474.

24. Fuchs M, Gonsalves D (2007) Safety of virus-resistant transgenic plants two decades after their introduction: lessons from realistic field risk assessment studies. Annu Rev Phytopathol 45: 173–202

25. Tepfer M (2002) Risk assessment of virus-resistant transgenic plants. Annu Rev Phytopathol 40: 467–491

26. Robinson DJ (1996) Environmental risk assessment of release of transgenic plants containing virus-derived inserts. Transgenic Res 5: 359–362

27. Chen J (1993) Occurrence of fungally transmitted wheat mosaic viruses in China. Ann Appl Biol 123: 55–61

28. Han C, Li D, Xing Y, Zhu K, Tian Z, et al. (2000) *Wheat yellow mosaic virus* widely occurring in wheat (*Triticum aestivum*) in China. Plant Dis 84: 627–630

29. Xu H, Pang J, Ye X, Du L, Li L, et al. (2001) Study on the gene transferring of Nib8 into wheat for It's resistance to the yellow mosaic virus by bombardment. Acta Agronomica Sinica 27: 688–693 (in Chinese with English abstract)

30. Wu H, Zhang B, Gao D, Xu H, Cheng S (2006) Disease resistance test of transgenic wheat lines with *WYMV-Nib8* gene and their application in breeding. J Triticeae Crops 26: 11–14 (in Chinese with English abstract)

31. Conner AJ, Glare TR, Nap J (2003) The release of genetically modified crops into the environment. Part II. Overview of ecological risk assessment. The Plant J 33: 19–46

32. Lu RK (1999) Methods of Soil Agrochemical analysis. Beijing: China Agricultural Science and Technology Press

33. Bakke I, Schryver PD, Boon N, Vadstein O (2011) PCR-based community structure studies of bacteria associated with marine organisms: a simple PCR strategy to avoid co-amplification of eukaryotic DNA. J Microbiol Meth 84: 349–351

34. Das M, Royer TV, Leff LG (2007) Diversity of fungi, bacteria, and actinomycetes on Leaves decomposing in a stream. Appl Environ Microbiol 73: 756–767

35. Liu W, Lu HH, Wu W, Wei QK, Chen YX, et al. (2008) Transgenic Bt rice does not affect enzyme activities and microbial composition in the rhizosphere during crop development. Soil Biol Biochem 40: 475–486

36. Guan SY (1986) Soil Enzyme and Its Research Methods. Beijing: China Agricultural Press, 62–142

37. Serra-Wittling C, Houot S, Barriuso E (1995) Soil enzymatic response to addition of municipal solid-waste compost. Biol Fert Soils 20: 226–236

38. Graef F, Züghart W, Hommel B, Heinrich U, Stachow U, et al. (2005) Methodological scheme for designing the monitoring of genetically modified crops at the regional scale. Environ Monit Assess 111: 1–26

39. Züghart W, Benzler A, Berhorn F, Sukopp U, Graef F (2008) Determining indicators, methods and sites for monitoring potential adverse effects of genetically modified plants to the environment: the legal and conceptional framework for implementation. Euphytica 164: 845–852.

40. Meyer JB, Song-Wilson Y, Foetzki A, Luginbühl C, Winzeler M, et al. (2013) Does wheat genetically modified for disease resistance affect root-colonizing pseudomonads and arbuscular mycorrhizal fungi? PLoS ONE, 8(1): e53825. doi:10.1371/journal.pone.0053825

41. Yi H, Kin H, Kim C, Harn CH, Kim HM, et al. (2009) Using T-RFLP to assess the impact on soil microbial communities by transgenic lines of watermelon rootstock resistant to cucumber green mottle mosaic virus (CGMMV). J Plant Biol 52: 577–584

42. Lin C, Pan T (2010) PCR-denaturing gradient gel electrophoresis analysis to assess the effects of a genetically modified cucumber mosaic virus-resistant tomato plant on soil microbial communities. Applied Environ Microbiol 76: 3370–3373

43. Fang M, Kremer RJ, Motavalli PP, Davis G (2005) Bacterial diversity in rhizospheres of nontransgenic and transgenic corn. Appl Environ Microb 71:4132–4136

44. Donegan KK, Seidler RJ, Doyle JD, Porteous LA, Digiovanni G, et al. (1999) A field study with genetically engineered alfalfa inoculated with recombinant *Sinorhizobium meliloti*: effects on the soil ecosystem. J Appl Ecol 36:920–936

45. Wei XD, Zou HL, Chu LM, Liao B, Ye CM, et al. (2006) Field released transgenic papaya affects microbial communities and enzyme activities in soil. Plant Soil 285:347–358

46. Faragova N, Gottwaldova K, Farago J (2011) Effect of transgenic alfalfa plants with introduced gene for alfalfa mosaic virus coat protein on rhizosphere microbial community composition and physiological profile. Biologia 66: 768–777

47. Romeis J, Battini M, Bigler F (2003) Transgenic wheat with enhanced fungal resistance causes no effects on *Folsomia candida* (Collembola: Isotomidae). Pedobiologia 47: 141–147

48. Lindfield A, Nentwig W (2012) Genetically engineered antifungal wheat has no detrimental effects on the key soil species *Lumbricus terrestris*. The Open Ecology J 5:45–52

49. Duc C, Nentwig W, Lindfeld A (2011) No adverse effect of genetically modified antifungal wheat on decomposition dynamics and the soil fauna community – a field study. PLoS ONE: e25014. doi:10.1371/journal.pone.0025014

50. Weaver RW, Angle JS, Bottomiey PS (1994) Methods of soil analysis. Part 2. Microbiological and Biochemical properties, No. 5. Soil Sci Soc Am, Madison

51. Alef K, Nannipieri P (1995) Methods in Applied Soil Microbiology and Biochemistry. San Diego, CA: Academic Press

52. Dick RP, Breakwel DP, Tureo RF (1996) Soil enzyme activities and biodiversity measurements as integrating microbiological indicators. In: Doran JW, Jones AJ. ed. Methods for Assessing Soil Quality. Soil Sci Soc Am, Madison, WI, 247–272

53. Icoz I, Saxena D, Andow DA, Zwahlen C, Stotzky G (2008) Microbial populations and enzyme activities in soil In Situ under transgenic corn expressing Cry proteins from *Bacillus thuringiensis*. J. Environ. Qual 37: 647–662

54. Shen RF, Cai H, Gong WH (2006) Transgenic Bt cotton has no apparent effect on enzymatic activities of functional diversity of microbial communities in rhizosphere soil. Plant Soil 285:149–159

55. Tan F, Wang J, Feng Y, Chi G, Kong H, et al. (2010) Bt corn plants and their straw have no apparent impact on soil microbial communities. Plant Soil 329: 349–364

56. Weiner N, Meincke R, Gottwald C, Radl V, Dong X, Schloter M, et al. (2010) Effects of genetically modified potatoes with increased zeaxanthin content on the abundance and diversity of rhizobacteria with in vitro antagonistic activity do not exceed natural variability among cultivars. Plant Soil 326:437–452

57. Kowalchuk GA, Bruinsma M, Van Veen JA (2003) Assessing responses of soil microorganisms to GM plants. Trends Ecol and Evol 18: 403–410

58. Cardenas E, Tiedje JM (2008) New tools for discovering and characterizing microbial diversity. Curr Opin Biotech 19: 544–549

59. Hirsch PR, Mauchline TH, Clark IM (2010) Culture-independent molecular techniques for soil microbial ecology. Soil Bio Biochem 42: 878–887

60. Fierer N, Breitbart M, Nulton J, Peter S, Lozupone C, et al. (2007) Metagenomic and small-subunit rRNA analyses reveal the genetic diversity of bacteria, archaea, fungi, and viruses in soil. Appl Environ Microbiol 73: 7059–7066

61. Uroz S, Ioannidis P, Lengelle J, Cébron A, Morin E, et al. (2013) Functional assays and metagenomic analyses reveals differences between the microbial communities inhabiting the soil horizons of a Norway spruce plantation. PLoS ONE 8(2): e55929. doi:10.1371/journal.pone.0055929

62. Li Y, Fu K, Gao S, Wu Q, Fan L, et al. (2013) Impact on bacterial community in midguts of the asian corn borer larvae by transgenic *Trichoderma* Strain overexpressing a heterologous chit42 gene with chitin-binding domain. PLoS ONE 8(2): e55555. doi:10.1371/journal.pone.0055555

63. Chun YJ, Kim DY, Kim H, Park KW, Jeong S, et al. (2011) Do transgenic chili pepper plants producing viral coat protein affect the structure of a soil microbial community? Applied Soil Ecol 51: 130–1

Single-Nucleotide Polymorphism Markers from De-Novo Assembly of the Pomegranate Transcriptome Reveal Germplasm Genetic Diversity

Ron Ophir[1]*, Amir Sherman[1]*, Mor Rubinstein[1], Ravit Eshed[1], Michal Sharabi Schwager[1], Rotem Harel-Beja[2], Irit Bar-Ya'akov[2], Doron Holland[2]

1 Department of Fruit Tree Sciences, Agricultural Research Organization, Volcani Center, Bet Dagan, Israel, 2 Department of Fruit Tree Sciences, Agricultural Research Organization, Newe Ya'ar Center, Ramat Yishai, Israel

Abstract

Pomegranate is a valuable crop that is grown commercially in many parts of the world. Wild species have been reported from India, Turkmenistan and Socotra. Pomegranate fruit has a variety of health-beneficial qualities. However, despite this crop's importance, only moderate effort has been invested in studying its biochemical or physiological properties or in establishing genomic and genetic infrastructures. In this study, we reconstructed a transcriptome from two phenotypically different accessions using 454-GS-FLX Titanium technology. These data were used to explore the functional annotation of 45,187 fully annotated contigs. We further compiled a genetic-variation resource of 7,155 simple-sequence repeats (SSRs) and 6,500 single-nucleotide polymorphisms (SNPs). A subset of 480 SNPs was sampled to investigate the genetic structure of the broad pomegranate germplasm collection at the Agricultural Research Organization (ARO), which includes accessions from different geographical areas worldwide. This subset of SNPs was found to be polymorphic, with 10.7% loci with minor allele frequencies of (MAF<0.05). These SNPs were successfully used to classify the ARO pomegranate collection into two major groups of accessions: one from India, China and Iran, composed of mainly unknown country origin and which was more of an admixture than the other major group, composed of accessions mainly from the Mediterranean basin, Central Asia and California. This study establishes a high-throughput transcriptome and genetic-marker infrastructure. Moreover, it sheds new light on the genetic interrelations between pomegranate species worldwide and more accurately defines their genetic nature.

Editor: Randall P. Niedz, United States Department of Agriculture, United States of America

Funding: The authors acknowledge funding from the Chief Scientist Office, Ministry of Agriculture & Rural Development, http://www.science.moag.gov.il/, Agricultural Biotechnology 203-0753. The funders had no role in study design, data collection and analysis, decision to publish, or preparation of the manuscript.

Competing Interests: The authors have declared that no competing interests exist.

* E-mail: ron@agri.gov.il (RO); asherman@agri.gov.il (AS)

Introduction

Pomegranate (*Punica granatum* L.) belongs to the family Lythraceae and is grown in many regions worldwide. Wild pomegranates (*P. granatum* L. or *Punica protopunica*) have been reported as native to India [1], Turkmenistan [2], Iran [3] and Socotra [4]. The pomegranate fruit is considered to be highly valuable due to its health-beneficial effects, appealing taste and pleasing esthetics. Recent scientific publications have focused mainly on its health-beneficial characteristics and mechanisms of action in human and animal model disease systems. The health-promoting activity attributed to various parts of the pomegranate fruit, flower and bark included anticancer, antidiabetic, antiviral and antimicrobial functions [5–7]. In addition, consumption of pomegranate juice has been shown to reduce blood pressure and serum fatty acid concentration [8,9]. Studies of the metabolites produced in pomegranate include mainly ellagitannins and gallotannins, ellagic acid derivatives, flavonoids, organic acids, fatty acids, triglycerides, sterols, terpenoids and alkaloids [10,11].

In contrast to the vast knowledge about its health-beneficial activities, genetic studies of the pomegranate tree are based on insufficient information. The genetic diversity of pomegranate has been studied by several groups from Turkey [12], Iran [3,13,14], India [15], Tunisia [16,17], Spain [18] and China [19,20]. Those studies suggested classifying pomegranate accessions based on simple-sequence repeats (SSRs) [21], random amplification of polymorphic DNA [12,22,23], amplified fragment length polymorphisms [19,24], 28S rDNA [18] or internal transcribed spacers [3]. Most of the analyses were performed on local accessions except for one report which included a limited number of accessions introduced from different countries [21]. Two studies [15,16] that used broader collections showed that plant material from different regions is genetically distinguishable. Other studies have compared wild, semi-wild and domesticated pomegranate accessions [3,15]. Overall, these studies identified substantial polymorphism among pomegranate accessions and were successful in splitting the local collections into several closely related subgroups. However, this was not based on a broad genomic infrastructure. Progress in the genomic area has been shown by a recent study that produced partial transcriptome data from pomegranate fruit peel mRNA using deep sequencing with the Illumina platform [25].

The currently available genomic resources for pomegranate cannot support modern breeding efforts such as marker-assisted selection or elaborate molecular studies. Next-generation sequencing (NGS) methods provide powerful tools for non-model organisms. In particular, 454-GS-FLX Titanium technology provides long reads along with considerable yield [26]. Transcriptome sequencing is one solution for large (>0.5 Gb) genomes that enables reducing complexity by focusing on genic regions. Sequencing the transcriptome using NGS technologies rapidly generates large sets of genetic markers linked to genes. Therefore, many studies have used this technology to investigate the behavior of wide gene-expression profiles in non-model organisms such as *Ammopiptanthus mongolicus*, bitter melon fruit and root, peony, watermelon and carrot [27–32]. Moreover, the overwhelming information provides an opportunity to characterize a crop's transcriptome in terms of its genetic variation and repertoire of gene functionalities [33–36]. The most useful markers are SSRs and single-nucleotide polymorphisms (SNPs). These genetic markers are abundant and distributed throughout the genome. On average, there is 1 SNP per 200–500 bp [37,38] and 1 SSR per 1.5–5 Kbp [39]. Therefore, the integration of SSR and SNP discovery in cDNA sequencing using NGS results in an overwhelming number of useful markers. In the previously published transcriptome from pomegranate, only 115 SSR markers were identified but no SNP markers[25].

In this work, we used high-throughput sequencing to discover major parts of the pomegranate transcriptome. We used this genomic information to compile a genetic-variation resource. A sample of the discovered genetic variation was utilized to learn about the genetic basis of the broad pomegranate germplasm at the Agricultural Research Organization (ARO) [40].

Materials and Methods

Plant material

The ARO's pomegranate germplasm collection is located at the Newe Ya'ar Research Center in northern Israel (http://igb.agri.gov.il/main/index.pl?page = 22). In the present study we analyzed 105 pomegranate accessions (Table S1) originating from different geographical locations all over the world and in particular, Israel [40]. Two *P. granatum* accessions with differing phenotypes were chosen to reconstruct the pomegranate transcriptome: 'Nana' (P.G.233–244) is a *P. granatum* var. Nana seedling characterized as a dwarf, conditional dormant pomegranate. 'Nana' has a very small and sour fruit with hard seeds and a green to red peel. Because 'Nana' is so distinct, it was recognized in one study as a third species of the genus *Punica* (*P. nana* L.) [41]. 'Black' (P.G.127–28), is a domesticated cultivar characterized as a deciduous normal-sized tree with a very distinct deep-purple peel. The 'Black' accession has a fruit of medium size, with sweet taste and soft seeds.

Accession phenotyping

Phenotypic classification was based on a phenotypic description of the ARO collection at the Newe Ya'ar Research Center. The collection was established in 1978 and new accessions have been added to the orchard ever since. Once each of the accessions reached the fruiting stage, the tree and fruit were characterized for at least 5 years. Characterization included, among other traits, date of maturity (when fruit reach edible quality), fruit size (weight and diameter), peel and aril colors (visual description) and taste (organoleptic description). Five mature fruits were harvested from two different trees of each accession, from different parts of each tree. Mature fruits were chosen by their combination of aril and peel color and by tasting for astringency and sourness. Fruit size, weight and diameter of the fruits were averaged. Colors and taste were described by observing and tasting the fruit.

RNA extraction for 454-GS-FLX sequencing

Total RNA was extracted from 3 g of ground tissue according to Meisel et al. [42], followed by two additional sodium acetate/ethanol precipitations. RNA was extracted from leaves, roots, flowers (petals, anthers, ovaries) and fruit at developmental (Stage 3; [43]). For each accession, equal amounts of total RNA from each tissue were mixed.

DNA extraction

The DNA-extraction protocol was based on Porebski et al. [44] with a few modifications. Young leaves from 105 pomegranate accessions were used for DNA preparation, 0.5 g resuspended in 6 ml extraction buffer. The chloroform–octanol solution was replaced with chloroform–isoamyl alcohol. DNA was precipitated with sodium acetate instead of sodium chloride.

De novo transcriptome assembly and functional annotation

Raw files of sequence reads were preprocessed by "SFF_extract" (http://bioinf.comav.upv.es/sff_extract/) and arguments for removing the adaptors and clipping the poly-A were applied. Reads from 454-FLX GS Titanium were assembled by a stable version of MIRA, v3.2 [45]. For the MIRA run, we used the "denovo,est,normal,454" set of "'Do-What-I-Mean" (DWIM) parameters. The complete set of sequence reads of 'Black' and 'Nana' were uploaded to SRA (SRS516503 and SRS516504, respectively).

All contigs were searched for open reading frames (ORFs) by the "getorf" program from the EMBOSS package[46]. The longest ORF with start and stop codons was chosen for each contig with a minimum cutoff of 50 amino acids. ORFs were run against *Eucalyptus grandis* proteins downloaded from Phytozome v9 [47]. Estimates of the homologous-segment fraction between pomegranate ORFs and eucalyptus proteins were calculated by cover index (CI) as CI = 2*HSP/(QL+HL) where HSP is the high-scoring segment pair (i.e., the blast alignment), QL is the query length (i.e., pomegranate ORF) and HL is the hit length (i.e., the eucalyptus protein).

A sequence-similarity search of contigs was run against the nonredundant protein (nr) database using blastx with a filter of e-value $<10^{-5}$. Best hits were further mapped to GO-slim by Blast2GO [48] and only hits with Blast2GO annotation score >55 were scored. Mapping the pomegranate fruit peel transcripts to the transcripts of the pooled tissues in this study was performed by blast search for all transcripts of peel against pool and vice versa, and selecting the reciprocal best hits.

SNP and SSR discovery

The SNP position and the coverage of each nucleotide allele were derived from the MIRA contigs output file. The coverage was only counted for nucleotides with PHRED-scale quality >30 [49]. Only positions whose nucleotide allele had a coverage of ≥3 were considered valid SNPs.

SSR scanning was performed on the 67,532contigs. MIcroSAtellite (MISA) identification tools and SciRoKo were run with default parameters.

SNP subset for genetic analysis

A subset of 480 SNPs included SNPs with the highest coverage restricted to a single SNP per contig. SNP assays for all 480 SNPs were developed by Fluidigm Corporation (http://http://www.fluidigm.com/snp-genotyping.html) based on variation data between cultivars 'Nana' and 'Black'. The SNP assays were used to screen the 105 accessions' DNA samples by running on FR48.48 arrays of the EP1 Fluidigm platform according to the manufacturer's instructions (http://www.fluidigm.com). To exclude bad samples and failed marker assays, samples that had more than 10% "No Call" and assays with more than 30% "No Call" were removed. The remaining subset was submitted for the downstream analysis.

The polymorphism information content was calculated as [50]

$$PIC = 1 - \sum p_i^2 \qquad (1)$$

where i is the ith allele.

Germplasm accession classification and diversity

To assess the relationship between pomegranate accessions, we estimated the genetic distance as D = 1-proportion of shared alleles (PSA). PSA was calculated as

$$PSA = \frac{\sum_{i=1}^{L} PS_i}{2 * L} \qquad (2)$$

where PS is the proportion of shared alleles for each locus and L is the total number of loci [51].

Hierarchical clustering was performed on a pairwise D distance matrix and the "ward" agglomerative method [52] was applied. The confidence limits of the tree topology were calculated by applying bootstrap method (1,000 resampling of loci). To count the number of bipartitions fit to the tree we used the "ape" R-package [53,54] and presented the bootstrap values as percentages.

The subpopulation structure underlying the germplasm collection was estimated by running a simulation of STRUCTURE software v2.3.3 [55] with 5,000 burn-in periods and 50,000 repetitions. The number of populations, K, was inferred by running the simulation of K = 1 to K = 10 (20 runs for each K) and using the likelihood method of ΔK [56].

The fixation indices F_S and F_{ST} [57] were calculated as

$$F_S = \frac{H_{\exp} - H_{obs}}{H_{\exp}} \qquad (3)$$

where F_S is the fixation index of each subpopulation, H_{obs} is the observed heterozygous types and $H_{\exp}$ is the estimated heteozygosity under Hardy-Weinberg equilibrium (HWE),

$$F_{ST} = \frac{H_S - H_T}{H_T} \qquad (4)$$

where F_{ST} is the genetic differentiation of a subpopulation due to genetic drift, H_S is the weighted average of all subpopulations' expected heterozygozity, and H_T is the expected heterozygosity in the entire population (germplasm collection).

Results

Pomegranate transcriptome reconstruction

We extracted mRNA from two *P. granatum* L. cultivars: 'Nana' (P.G.233–244) and 'Black' (P.G.127–28), from the ARO pomegranate collection. These accessions are very different phenotypically and are therefore assumed to be genetically different as well. 'Nana' is characterized as a dwarf pomegranate that has a temperature-conditional dormancy period. 'Black' is an edible deciduous cultivar with purple (almost black) peel color. To avoid over-representation of tissue-specific gene expression, mRNA samples of leaves, roots, flower parts (petals and reproductive organs) and fruits at developmental stage 3 [43] were pooled together. For each cultivar accession, cDNA pooled from these tissues was sequenced by the 454-GS-FLX Titanium platform, a pool per accession per half plate. The sequence results yielded a total of 755,519 and 728,665 reads for 'Nana' and 'Black', respectively, in which most of the reads (80.08–82.62%) from both samples were successfully assembled (Table 1). Half of the contigs were longer than 707 and 719 bp, respectively. The joint assembly of reads of both accessions yielded a median 714 bp, suggesting no preference for separate assemblies. The skewness to the right (positive skewness values) of contig-length distributions indicated that these distributions are asymmetric, i.e., there is a tail of long contigs (Figure 1). The skewness values of the joint assembly, the 'Nana' accession assembly and the 'Black' accession assembly were 2.16, 1.16 and 1.18, respectively, indicating that the joint assembly generated longer contigs. We therefore focused on the joint assembly as a reference for further analysis.

Functional annotation of pomegranate transcriptome

To explore the gene repertoire in the pomegranate transcriptome, a DNA–protein similarity search (blastx) against the nr database was performed. Mapping blast hits to gene ontology (GO) and downstream annotation analysis were performed using Blast2GO [48,58]. Out of 67,532 contigs, 58,473 (86%) included an ORF, 54,838 (81%) had a significant hit (e-value $<10^{-5}$) and 45,187 (67%) passed the minimum blast2 GO-annotation score of 55, which means that they were mapped to one of the GO categories (Table S2). Most of the homologous protein hits in GenBank were plants (99%) (Figure 2), with 81.27% of the hits being proteins of *Vitis vinifera*, *Ricinus communis*, *Populus trichocarpa*, and *Glycine max*. This suggests that most of the functional annotation derived from the plant-homologous hits, and that the DNA sample was not contaminated.

Annotation relies on sequence similarity of the mRNA products to homologous proteins with functional descriptions. The joint assembly produced long, but not essentially more informative contigs. Therefore, we estimated whether the proteins derived from the ORFs of the pomegranate assembly include most of the coding sequence. A blast search was run against 46,315 proteins of *Eucalyptus grandis*, a sister taxon in the Myrtales clade, and the ratio of blast alignment to length of the pomegranate query and eucalyptus hit was calculated (see Materials and Methods). The ratio was notated as CI. Half of the contigs included ORFs with CI ≥0.86, and 90% of the contigs included ORFs with CI ≥0.41. In comparison, in the pomegranate transcriptome from the peel [25], half of the contigs included ORFs with CI ≥0.38 and only 10% of the contigs included ORFs with CI ≥0.85.

The cDNA sequencing was performed with multi-tissue samples. Therefore, the knowledge that could be derived on gene functionality was essentially non-tissue-specific. However, it would be interesting to investigate whether pomegranate has a bias toward specific functions. The pomegranate contigs mapped to 323,654 GO categories. Where the contigs were mapped to more than one category, the most specific category was chosen to convert the "one to many" to "one to one" mapping. Many studies use the GO-Slim set of GO categories [28,29,59–61]. Plant

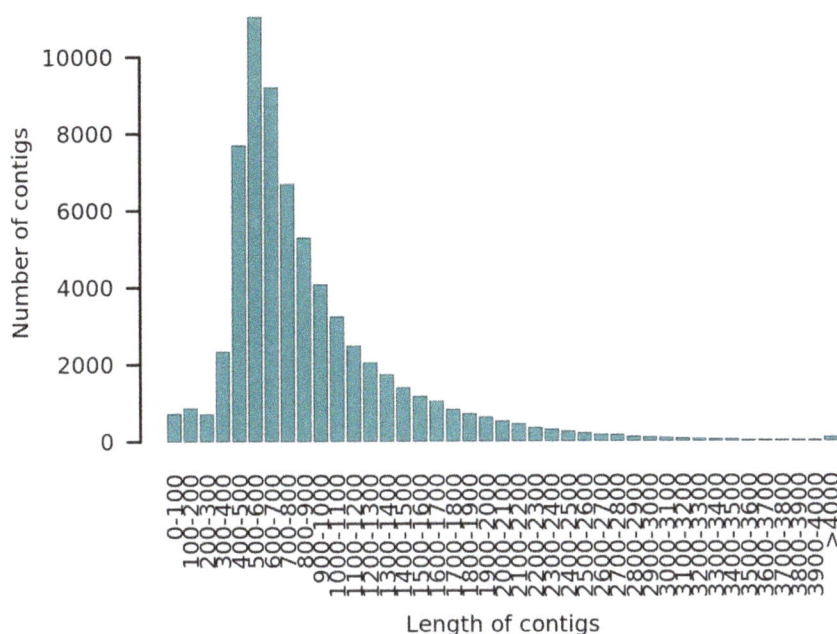

Figure 1. Distribution of contig lengths from both accession assemblies. Reads of cDNA sequencing by 454-GS-FLX Titanium technology were assembled by MIRA assembly program. The distribution of consensus contig lengths is drawn as 100-bp long bins.

GO-Slim represents a set of GO categories which is most relevant to all plants. Contigs mapped to the root of the ontologies can be considered unknown. Notably, the root category was not the most frequent on any of the ontology graphs (biological processes, molecular functions, and cellular components) (Figure 3), indicating successful annotation. The most frequent (30%) informative biological processes were various types of metabolic processes, cellular organization, and responses to abiotic and biotic stimuli. The most frequent molecular functions were various binding functions (41.05%), and hydrolase (16.43%), catalytic (15.93%) and transferase (12.05%) activities.

A comparison between the transcripts from the pooled pomegranate tissues and the transcripts from the pomegranate peel [25] enabled us to outline the differential scheme of biological processes and molecular functionalities between peel and other tissues (root, leaf, flower and fruit at a developmental stages). The intersection of the transcripts from the pool with those from the peel was mapped by reciprocal blast hit of the contig sequences, resulting in 5,943 corresponding transcripts. Similar to the annotation of transcripts from the pool of several tissues, full ORFs derived from the peel transcripts were submitted to the Blast2GO pipeline against "nr" (Figure 3). Peel seemed to lack transcripts corresponding to flower development, embryo development, pollination, photosynthesis, death, response to extracellular stimulus, and epigenetics. This illustrates the advantage of a pool in which flower tissues and fruit development stages contribute genes that are not abundant or are absent in the peel. Furthermore, plastid genes dominated the peel tissue, whereas cytoplasmic and membranal genes were contributed by the pooled tissues.

SSR discovery

Although SNPs are more abundant than SSRs and more economic per marker, the latter are still valuable for genetic studies when focusing on a small number of markers [62]. We therefore screened for SSRs in all contigs by running two different SSR-detection programs: MISA [63] and SciRoKo [64]. Whereas the former is faster and more popular, the latter is expected to be more robust for SNPs in the SSRs. The number of SSRs found in the entire set of contigs (67,532) was 10,330 and 12,309 for MISA and SciRoKo, respectively. The robustness of SciRoKo is reflected in a higher proportion of SSRs with four and five motifs. However, we assumed that the intersection of the two approaches for SSR identification would decrease the number of false SSRs. We therefore crossed the lists of SSRs based on the SSR motif and its start position in a contig. The intersection list included 7,155 putative SSRs (Table S3). The SSR motifs consisted of mono- to hexanucleotides. However, most were di- and trinucleotide motifs, totaling 3,910 (54%) and 2,876 (40%), respectively, whereas tetranucleotide motifs were an order of magnitude less abundant

Table 1. Assembly parameters for 'Nana' and 'Black' accession reads.

Assembly	Number of reads	Reads filtered out (%)	Number of contigs	Contig length (N50)	Contig length (average)	%GC (N50)
Black	728,665	19.15	43,027	719	862.5	47.32
Nana	755,519	17.38	46,734	707	849.5	47.33
Both	1,484,183	18.25	67,532	714	880.2	47.36

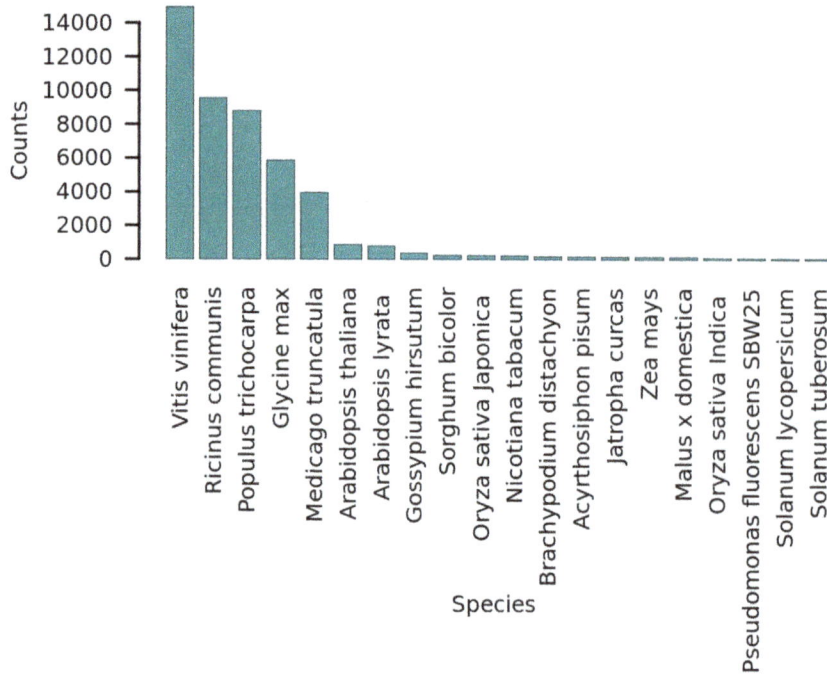

Figure 2. Distribution of top homologs according to their taxonomy. The assembled contigs were submitted to sequence-similarity search (blastx) against the nonredundant protein (nr) GenBank database. The frequency of best-hit species was counted and is illustrated in a bar plot. Only species with more than 100 hits are presented.

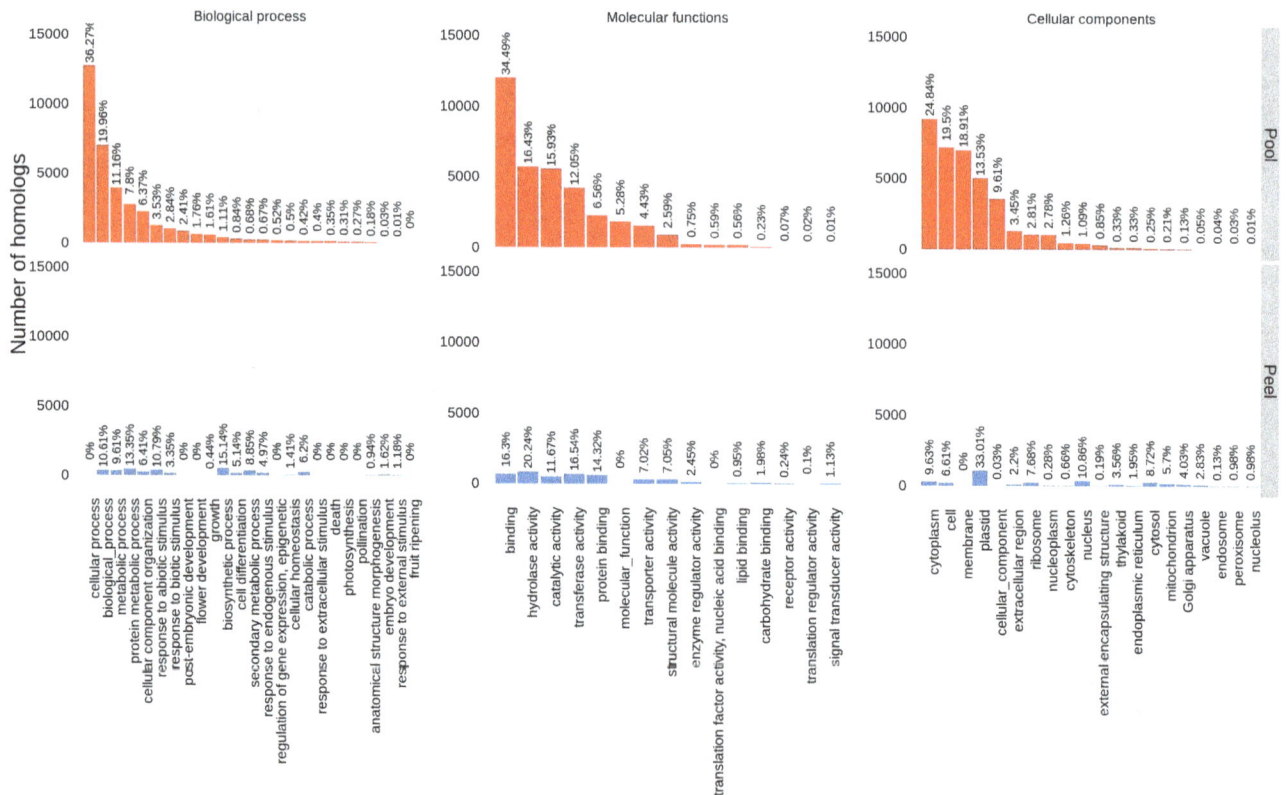

Figure 3. Pomegranate gene ontology categories. Contigs were annotated by blast hits against the nr database and mapped to GO categories by Blast2GO. The distribution of contigs into the GO categories biological processes, molecular functions, and cellular components is plotted for a pool of tissues (root, leaf, flower and fruit developmental stage 3; red bars) and from peel (turquoise bars).

(201; 3%) (Figure 4A). The preponderance of (AG/TC)n and (GA/CT)n for dinucleotide motifs, and of (GAA/TTC)n, (AGA/TCT)n and (AGG/CCT)n for trinucleotide motifs is altogether unique to pomegranate (Figure 4B and 4C) and is consistent with a previous study in which 117 pomegranate SSR markers were reported in AG/TC-rich regions [21]. Moreover, the (AG/TC)n motif was found to be the most frequent among 115 SSRs in the pomegranate fruit-peel transcriptome [25]. Nevertheless, the (AAG/CTT)n motif is most abundant in many closely related species such as *Prunus* spp. [65], *Eucalyptus grandis* [66] and Cucurbitaceae species [59], as well as Brassicaceae speceis [67]. In summary, the uniqueness of pomegranate lies in the order of trinucleotide-motif frequency: (GAA/TTC)n was most abundant, (AGA/TCT)n was the second and (AGG/CCT)n was the third most abundant trinucleotide motif. In melon, for example, (AAG/CTT)n is the most abundant motif whereas in *Prunus* species, (AAC/GTT)n is the most frequent. In *E. grandis*, (AAG/CTT)n is the second most abundant while the (GGC/CCG)n motif is the most frequent.

SNP discovery

SNP discovery has become routine in high-throughput parallel sequencing [33,34,68,69]. We investigated the polymorphism of SNPs between the two accessions where each accession was homozygous for a different allele, and within each accession. To avoid sequencing errors, we relied on minimal resequencing coverage at the SNP locus. Therefore, the coverage of each allele was only counted for good-quality base calls and we filtered out base-call variants if one of the alleles was covered (resequenced), using 3X coverage as the cutoff. The number of SNPs for 'Nana' and 'Black' was 2,336 and 2,436, respectively, whereas the number of SNPs between the two accessions was 1,728. For all three sets of SNPs, transition substitutions were more frequent than transversion substitutions. The highest transition-to-transversion (t/v) ratio was estimated between the accessions (3.5), whereas the t/v ratios within 'Nana' and within 'Black' were 1.8 and 2.2, respectively. Transitions are expected to be more frequent than transversions [70,71]. However, this t/v ratio was higher for SNPs between accessions rather than within accessions. The increase in the ratio was due to decreasing transversion substitutions of these SNPs, rather than increased transitions (Figure 5). This underlies the fact that the SNPs were sampled mostly from coding sequences and the selection on transversion substitutions in a coding region is higher than on transition substitutions. Moreover, this effect increased when comparing the t/v ratio between the two cultivars.

Germplasm diversity

Our major interest was to understand the genetic structure of the ARO's pomegranate germplasm collection at Newe Ya'ar Research Center (Table S3;[40]). Thus, 480 SNPs were selected from the transcriptomes of the 'Nana' and 'Black' accessions and used to survey the genotypes in the ARO germplasm collection by Fluidigm-EP platform. To ensure even genome distribution of high-confidence markers, SNPs with the highest read coverage in each contig were selected. Because our criteria were not biased toward a specific type of SNP, we expected no change in the t/v ratio. Indeed, the proportion of transition SNPs was 0.6. First, the marker diversity was estimated by calculating the polymorphism information content (PIC) for each SNP. Half of the SNPs were highly diverse (PIC ≥0.43), indicating that the selected SNPs are highly diverse in the ARO pomegranate germplasm. Only 10.7% of the SNPs showed minor allele frequencies (MAF <0.05), consistent with previous studies [19,21], suggesting that the pomegranate genome is highly diverse. MAF SNPs are an

uninformative subset and were therefore filtered out, as were SNPs whose genotype called as heterozygous for all germplasm accessions. Consequently, a set of 346 SNPs was retained for further analysis.

Relationship among germplasm accessions

The ARO germplasm collection includes local accessions and several accessions introduced from foreign countries, including India, China, Central Asia, USA, Spain, Turkey and the Mediterranean region (Table S1; [40]). However, the genetic relationships among the 105 examined accessions and their genuine geographical origins are still uncertain. Thus, classifying the accessions only by their origin might be misleading. To classify the accessions into groups of kinship based on their genetics, we calculated the pairwise genetic distances as 1- PSA [51,72] among all accessions, and a hierarchical classification tree was reconstructed (Figure 6). Confidence in the tree topology was obtained by bootstrap method (1,000 resampling). As a complement, the STRUCTURE program was used to reveal the number of subpopulations composing the genetic structure of the ARO germplasm collection. Therefore, we applied the δK likelihood method [56]. The maximum ΔK corresponded to K = 2, suggesting that there are two major subpopulations [56]. The proportion of subpopulations (Q) is presented in Figure 6 in comparison to the dendrogram, which classifies the collection into groups of closely related genetic accessions (hereafter called groups and subgroups).

The classification tree (i.e., dendrogram) revealed two major groups of accessions (G1 and G2); G2 was further divided into two subgroups (G2.1 and G2.2). Strictly speaking, the dendrogram implied altogether three statistically significant groups (bootstrap value >90%). Group G1 (P.G.160-61 to P.G.102-3) includes ornamental, inedible pomegranate accessions such as *P. granatum* var. Nana seedlings (e.g., P.G.149-50 and P.G.167-68), as well as accessions from India, China and Iran (ICI; e.g., P.G.145-46, P.G.202-213, P.G.102-3; brown bar, Figure 6). The G2 group (P.G.129-30 to P.G.117-18) includes accessions from the Mediterranean region, Central Asia and California. This group includes domesticated pomegranate cultivars and is further divided into two main subgroups: subgroup G2.1 (P.G.129-30 to P.G.105-6; yellow bar, Figure 6) consists mostly of the "Wonderful-like" (WL) accessions and seems to be genetically and phenotypically uniform. Most of the WL accessions were collected in Israel and contain divergent landraces and mutants [40]; subgroup G2.2 (P.G.211-222 to P.G.214-225) includes accessions originated from the Mediterranean basin, Central Asia and California and was split into two further subgroups: G2.2.1 (P.G.221-222 to P.G.165-66) and G2.2.2 (P.G.218-229 to P.G.117-18), although this separation was statistically insignificant. Nevertheless, G2.2.1 includes accessions introduced from Spain ["Mollar" (M); green bar, Figure 6] and accessions introduced from Central Asia and California (CAC; orange bar, Figure 6), both of which are more homogeneous than the rest of the G.2 members. In contrast, G2.2.2 includes slightly more of an admixture of accessions. In this subgroup, the exception is a few accessions that are genetically homogeneous and have similar phenotypic characteristics. For example, the "Early Red Sweet" (ERS) accessions (P.G.234-245 to P.G.154-55), which are characterized by dark red color of the arils and peel, sweet taste and soft seeds, clustered together with high statistical confidence. A world view of the geographical location representing the suggested origin of each of the subgroups described in the dendrogram of germplasm accessions is presented in Figure 7. The groups' location on the world map illustrates that

A

B

C

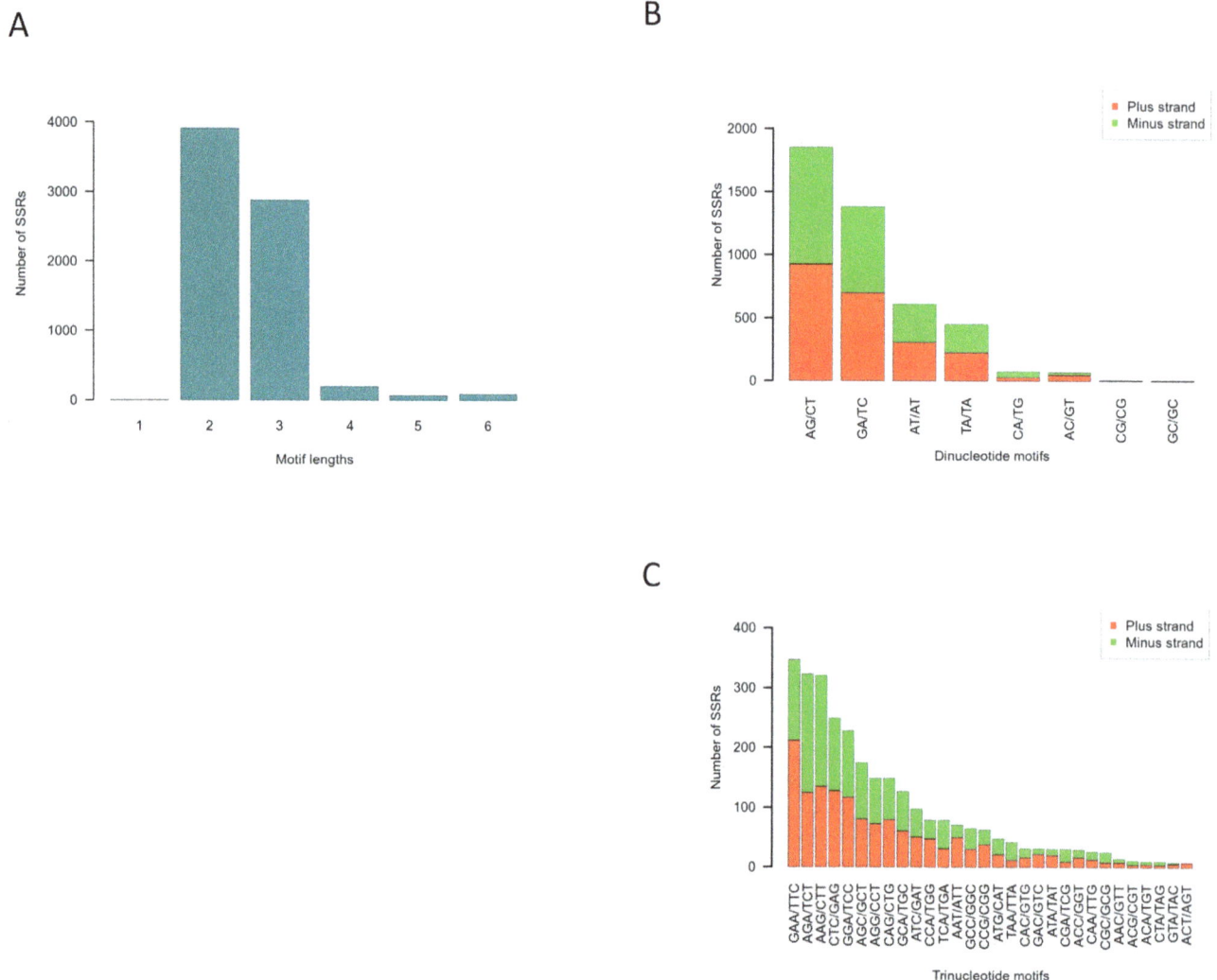

Figure 4. SSR length and motif distribution. The number of mono- to hexanucleotide SSR motifs was counted (A). The nucleotide compositions of the most frequent motifs (di- and trinucleotide motifs) was determined for each type and is illustrated in a bar plot for dinucleotide (B) and trinucleotide (C) motifs. Motifs that are reverse-complementary were plotted one above the other as "plus strand" (red) and "minus strand" (green).

the two major groups in the dendrogram are separated by their origin.

Genetic group diversity

The genetic flow within the germplasm collection can be retrieved by estimating the medians of the fixation index of a subpopulation (F_S) for all SNP loci. We estimated the median F_S of each of the three subgroups defined based on the dendrogram (Figure 6). The F_S values deviated from the HWE with positive values of 0.41, 0.33, and 0.32 for G1, G2.1, and G2.2, respectively. These results indicate that the subgroups are much less heterozygous than expected under the HWE, probably due to a long period of cultivation. The median F_S value for G1 was surprising as it is expected to be composed of more ancient accessions. One might assume that they would be more diverse because they had more time to evolve. Our results seem to suggest that they were evolved from different founders that were not intercrossed. Moreover, subgroup G2.2.1 includes three different geographical origins. The calculation of F_S for the Spanish M accession and for CAC-originated accessions revealed that G2.2.1 is a mix of two separate subgroups. The median F_S of the M

accession was 0.52 and that of the CAC accessions was 0.19. Next, the estimation of genetic differentiation among the three subgroups was calculated as F_{ST}. The loci's F_{ST} values deviated only slightly from the HWE. Most of the loci (90%) deviated less than 5% and the median was only 1.2%. This suggests that the subgroup differentiation occurred recently.

Discussion

The primary objective of this work was to generate a comprehensive genomic infrastructure of pomegranate (*P. granatum* L.). This was done by reconstructing the transcriptome from several tissues that originated from two genetically divergent pomegranate accessions. Once a solid genomic infrastructure was established, a sample of the genetic variation was selected to explore the genetic structure of our pomegranate germplasm collection.

Transcriptome assembly quality

We used 454-GS-Titanium as a strategy for the de novo assembly of a transcriptome from a non-model organism. This

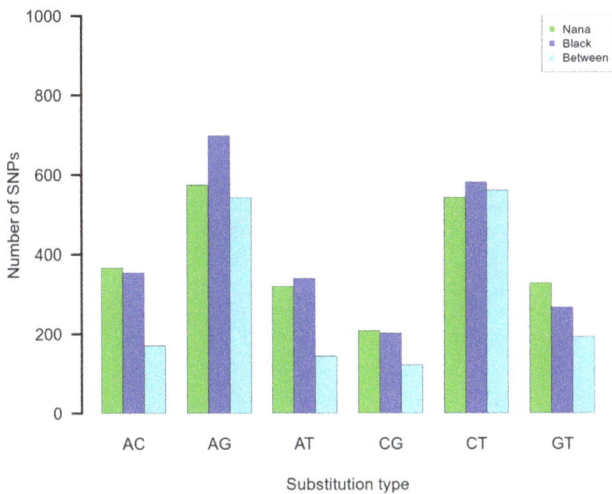

Figure 5. Distribution of SNP substitutions within and between 'Nana' and 'Black' accessions. The frequency of nucleotide substitutions was counted in 'Nana' (green) and 'Black' (blue). Where 'Nana' was homozygous for one allele and 'Black' was homozygous for another, it was counted as a between-accession SNP (turquoise).

approach has been commonly used in the last few years [31–33,35]. In the present study, this strategy was used to create a resource for SNP mining [33], in order to reveal the genetic diversity of the ARO pomegranate collection. The use of two putatively genetically different accessions was presumed to enable discovering the genetic variations reflected as SNPs and SSRs. For this purpose, mRNA from a variety of tissues from each of the accessions was extracted and pooled, as has been done previously in melon and eucalyptus [34,69]. Aside from the fact that the transcriptome assembly in this study resulted in a high frequency of long contigs, and most (70%) of the contigs included ORFs resembling eucalyptus homologs (CI ≥0.7), 45,187 (67%) contigs were fully annotated by Blast2GO [48]. This suggests the establishment of a high-quality genomic infrastructure. The annotation process was performed against the nr GenBank protein database which contains a broad set of proteins from a variety of organisms. As expected, most homologous hits were plants, mainly rosids. However, the most frequent hits were from *Vitis vinifera* rather than the more closely phylogenetically related species *Eucalyptus grandis* and *Populus trichocarpa* [73]. A possible explanation for this is that the proportion of *V. vinifera* sequences in the nr database is higher than those of *Eucalyptus* and *Populus*.

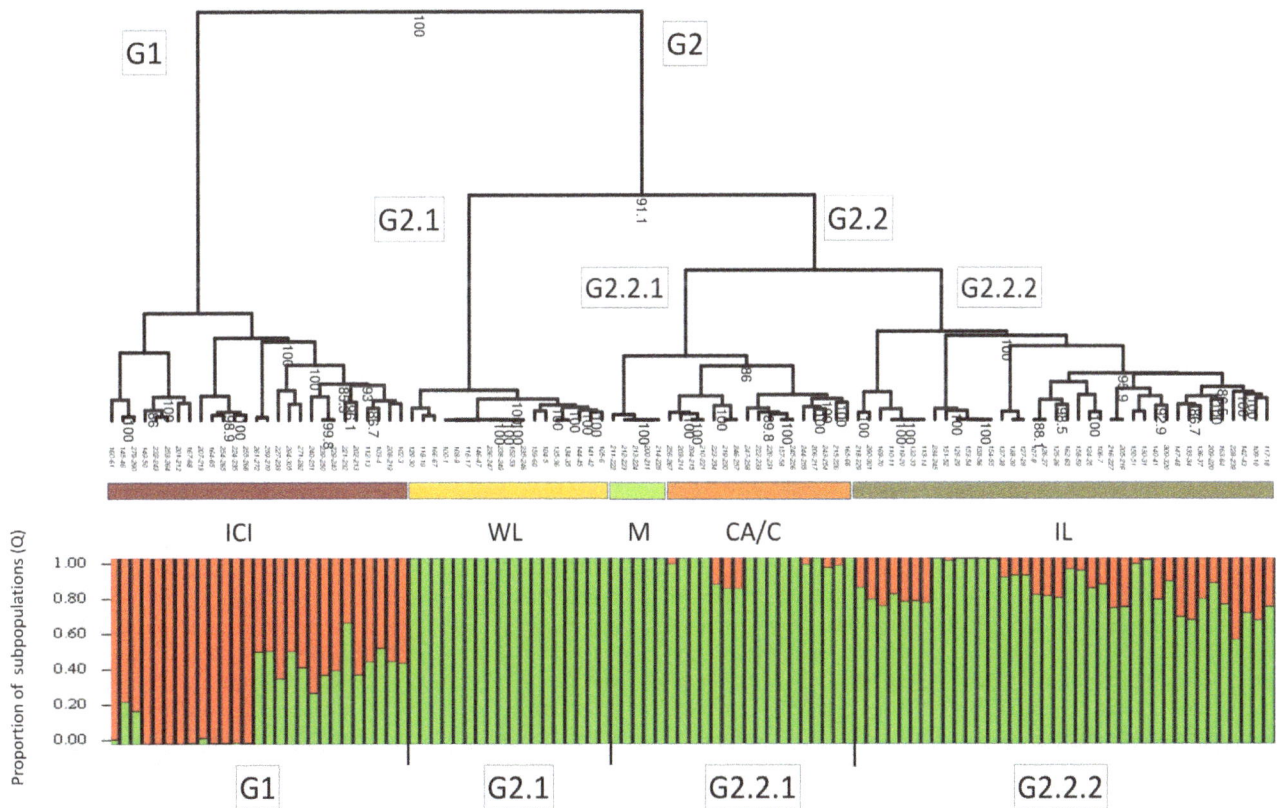

Figure 6. Dendrogram and genetic structure of 105 accessions in the ARO pomegranate germplasm collection. Genotyping of 105 pomegranate accessions from the ARO collection was performed with 346 SNPs. The genotyping results were used to classify the accessions into subpopulations and reveal their genetic structure. Classification was performed by drawing a dendrogram based on 1-PSA as genetic distance. Only confident branches with bootstrap values above 85 were assigned. The two major groups are notated as G1 and G2, and G2 was further divided into G2.1 and G2.2, giving altogether three groups of related accessions. The colored horizontal bars correspond to phenotypically homogeneous pomegranate groups of accessions: India-China-Iran (ICI), "Wonderful-like" (WL), Mollar (M) Central Asia and California (CAC), and Israel (IL). Genetic structure was revealed by STRUCTURE program with K = 2 as found by simulation and δK likelihood method. The two subpopulations (K = 2) are represented as green and red bars. The division of STRUCTURE's Q-value bar plot into four (vertical lines) corresponds to the four major significant clusters in the dendrogram.

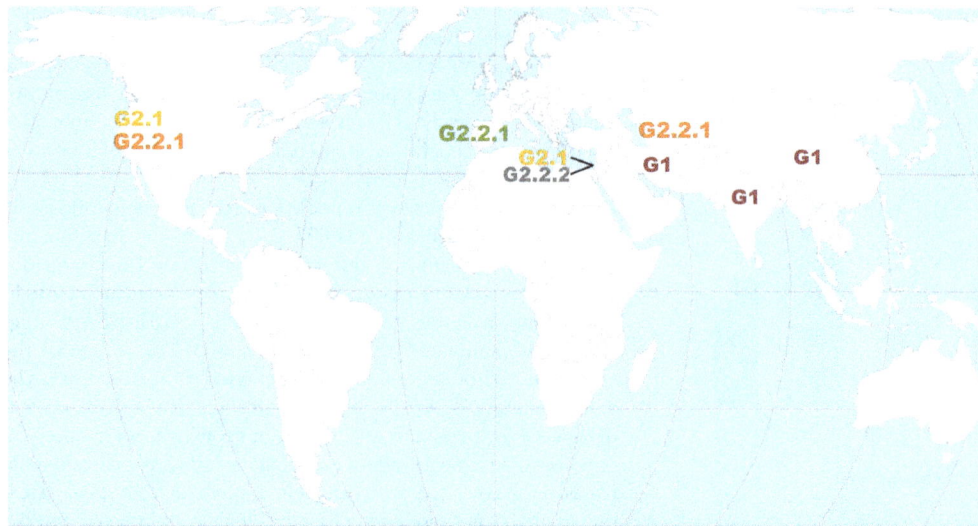

Figure 7. Worldwide distribution of the pomegranate genetic groups. The genetic groups as clustered together by the genetic dendrogram were located on a world map by geographical regions. Genetic groups and subgroups are colored as in Figure 6.

Pomegranate transcriptome

Pomegranate is a fruit tree that belongs to the family Lythraceae and is part of the order Myrtales. A comprehensive genomic study was recently performed on its relative tree, *Eucalyptus*, which is part of the Myrtales, and peach (*Prunus persica*) and pear (*Pyrus pyrifolia*), which are part of the family Rosaceae. It is difficult to compare functionality among transcriptome studies for three reasons: 1) the source of the mRNA is not uniform and tissue-specific expression can bias the comparison; 2) the annotation method can affect the functionality results (e.g. use of different GO levels); 3) corresponding results may be due to lack of specific information. For example, "metabolic processes" is one of the most frequent GO categories, regardless of species or type of GO annotation method [59,74–76]. Therefore, we could not conclude on the uniqueness of a functionality related to pomegranate. However, in pomegranate, like in pear (~20%) [75], reproductive processes and pollination are enriched as compared to other trees such as peach (~2%) [76], eucalyptus (0%) [69] and olive (6%) [74]. We did, however, compare the transcriptome from a pool of tissues in this study and from peel in a previous study [25], and showed that the strategy of pooling different tissues can enrich the repertoire of genes.

SSR markers from transcribed regions are commonly used in crops for genetic analysis [77–79] and for cultivar classification of crops [80–82]. In this work, the resource of such markers for pomegranate was immensely broadened. The combination of frequent SSR motifs is known to be species-specific [83]. However, as expected, some of the frequent SSR motifs are common to closely phylogenetically related species. In pomegranate, for example, the most frequent dinucleotide motif is (AG/TC)n, which is common to *Prunus* and *Eucalyptus*, whereas the high frequency of (GAA/TTC)n is unique to pomegranate and is barely found in these other species. The resemblance of most SSR motifs to closely related species supports the validity of the long list of SSR markers obtained in this study.

SNP markers are more abundant within a genome and are cost-effective for genotyping. However, the discovery of SNPs is primarily dependent on the sequencing coverage [84]. This coverage is not evenly distributed when sequencing cDNA [85] and therefore, when using the allele coverage as a cutoff for

selecting true SNPs, the number of SNPs drops to less than the number of SSRs. Thus a SNP-discovery study of a non-model organism that lacks either a genome or transcriptome reference should integrate de novo assembly followed by resequencing of deep sequencing [86]. One of the pitfalls of SNP discovery from de novo assembly is false identification of SNPs that are paralogous, i.e., that result from misassembly of reads from repetitive sequences in the genome. This might cause an overestimation of SNP frequency. Another pitfall might stem from the detection of SNPs in cDNA. Allelic-specific expression might cause extensive coverage by one allele but not by the other. Thus, setting high allele coverage may cause an underestimation of SNP frequency. In pomegranate, the t/v ratio is approximately 2 as in peach coding sequences [76]. The decrease in the transversion SNPs between the accessions is typical to genic regions [70,71]. Transversions cause nonsynonymous substitutions which are subjected to deleterious selection in the coding sequence. This supports the assumption that 'Nana' and 'Black' accessions are genetically different and were separately subjected to recurrent selection.

Germplasm accession kinship and diversity

Understanding the genetic structure of the pomegranate germplasm is crucial for studying the inheritance and breeding relevance of important agricultural traits. It is essential to understand the pomegranate collection's structure to avoid stratification and admixture when performing genome-wide association studies [87], as well as to avoid redundancy when crossing accessions. After filtering out the bad-call and noninformative SNPs, 71% informative markers were left. The dendrogram which was established in this study is based on the analysis of those 346 SNPs and surveys 105 accessions from the ARO pomegranate germplasm collection. This collection compiles many accessions originating from Israel and from several other countries, such as the USA, India, China, Iran and Turkey.

The genetic classification divides the germplasm collection into two statistically significantly distinct genetic groups: G1 and G2. The G1 group is composed of *P. granatum* var. Nana seedlings and its descendant accessions, in addition to accessions of Indian, Chinese and Iranian origin. It includes the Indian cultivar Bhagwa and all of the conditional dormant (evergreen) accessions. This group is defined as the ICI group (red bar, Figure 6). The division into G1 and G2 groups separates accessions originating in India from those originating in Israel, California, Spain and Turkey, as has been reported previously [21]. However, in this study, G1 includes additional accessions from Iran and China. The dendogram separates the pomegranate accessions into two general geographical regions (Figure 7). One branch spreads from the suggested origin of the pomegranate species toward the Far East, while the other spreads toward the West. One important aspect is the classification of wild pomegranate. A clear definition of the term "wild" in *P. granatum* species is missing. Some studies refer to 'Daru' pomegranates that grow in the forests on Himalayan slopes as wild pomegranates (e.g., [88]). These pomegranates are characterized by thorny bushes with very low quality sour fruits which resemble the fruits of 'Nana' [89].

In general, the G1 group is more admixed than the G2 group, which includes the rest of the accessions. The division that is close to the root of the dendrogram (i.e., the two major groups G1 and G2) corresponds to the subpopulations revealed by the STRUC-TURE program analysis and global geographical origins, whereas the clusters that are closer to the leaves correspond the phenotypic and local geographical origins. The G2 group can be subdivided into two subgroups (G2.1 and G2.2), and further divided into four subgroups (WT, M, CAC and IL). Some of these groups are still highly distinct, both phenotypically and genotypically. The WL (G.2.1) group is characterized by large fruit with a sour–sweet taste and red peel and arils. Some accessions do not correspond to this description, but those are most probably mutations of WL types as they have been reported to be sports of 'Wonderful' (e.g. sweet P.G.105–6). 'Wonderful' is present in a large number of landraces in the Mediterranean region. Moreover, there are reports of phenotypes similar to that of 'Wonderful' that have been described in the Mediterranean region for many years, such as 'Red Lufani' [90]. This suggests that the Mediterranean region is the geographical origin of this group. The genetic structure of the WL group is homogeneous (green in STRUCTURE analysis), in contrast to most other accessions in the second subtree of the G2 group. This suggests that the WL group underwent a process of stratification. Some of the subclusters of G2.2 branch have a prominent geographical profile or phenotypic characteristics. Two distinct subgroups make up G2.2.1: a subgroup from Spain ('Mollar' types) characterized by soft seeds with light pink arils and peel, and the CAC subgroup from Central Asia and California. The IL subgroup (G.2.2.2) has a common origin but is a phenotypically mixed group which splits into subgroups with strong phenotypic characteristics. Among these are the ERS subgroup characterized by red arils and peel, sweet taste and soft seeds, the "Black" subgroup (P.G.137–38 to P.G.127–28) with characteristic deep purple skin, and the "Hassas" subgroup characterized by pink arils and a weakly colored peel. The accessions in this latter subgroup were collected mostly from northern Israel and two accessions originated from Turkey (P.G.163–64 and P.G.209–220). This subgroup shares geographical origin as well as the aforementioned phenotypic characteristics.

Conclusions

The significance of this study lies in creating an infrastructure for the elaboration of breeding strategies and for genetic mapping. It was aimed at establishing a global view of genetic interrelations in the pomegranate germplasm. As a result, it potentially orients the origins of, and suggests interrelations among pomegranate accessions all over the world as the accessions analyzed included some introduced from distant geographical locations. Although several studies on the genetic structure of pomegranate germplasm have already been performed [16,73], they were based on a small set of genetic markers and accessions. Moreover, the broad information revealed on the pomegranate transcriptome in this study provides a source for further research into gene-function identification and metabolic pathways.

Author Contributions

Conceived and designed the experiments: RO AS. Performed the experiments: RO AS RE IBY RHB. Analyzed the data: RO MR MSS. Contributed reagents/materials/analysis tools: DH. Wrote the paper: RO.

References

1. Rana JC, Pradheep K, Verma V (2007) Naturally occurring wild relatives of temperate fruits in Western Himalayan region of India: an analysis. Biodiversity and conservation 16: 3963–3991.

2. Levin GM (1981) Wild pomegranate (*Punica granatum* L.) in Turkmenistan. [title translated from Russian]. Izvestiia Akademii Nauk Turkmenskoi SSR Seriia Seriia Biologicheskikh Nauk (2): 60–64.

3. Hajiahmadi Z, Talebi M, Sayed-Tabatabaei BE (2013) Studying Genetic Variability of Pomegranate (*Punica granatum* L.) Based on Chloroplast DNA and Barcode Genes. Mol Biotechnol 10.1007/s12033-013-9676-2.

4. Balfour IB (1888) Punica. In: Balfour IB, editor.Botany of Socotra. Edinburgh: Robert Grant & Sons.pp. 93–96.

5. Neurath AR, Strick N, Li YY, Debnath AK (2004) *Punica granatum* (Pomegranate) juice provides an HIV-1 entry inhibitor and candidate topical microbicide. BMC Infect Dis 4: 41.

6. Rocha A, Wang L, Penichet M, Martins-Green M (2012) Pomegranate juice and specific components inhibit cell and molecular processes critical for metastasis of breast cancer. Breast Cancer Res Treat 136: 647–658.

7. Wang L, Ho J, Glackin C, Martins-Green M (2012) Specific pomegranate juice components as potential inhibitors of prostate cancer metastasis. Transl Oncol 5: 344–355.

8. Asgary S, Sahebkar A, Afshani MR, Keshvari M, Haghjooyjavanmard S, et al. (2013) Clinical Evaluation of Blood Pressure Lowering, Endothelial Function Improving, Hypolipidemic and Anti-Inflammatory Effects of Pomegranate Juice in Hypertensive Subjects. Phytother Res 10.1002/ptr.4977.

9. Aviram M, Dornfeld L (2001) Pomegranate juice consumption inhibits serum angiotensin converting enzyme activity and reduces systolic blood pressure. Atherosclerosis 158: 195–198.

10. Seeram NP, Henning SM, Zhang Y, Suchard M, Li Z, et al. (2006) Pomegranate juice ellagitannin metabolites are present in human plasma and some persist in urine for up to 48 hours. The Journal of Nutrition 136: 2481–2485.

11. Dafny-Yalin M, Glazer I, Bar-Ilan I, Kerem Z, Holland D, et al. (2010) Color, Sugars and Organic Acids Composition in Aril Juices and Peel Homogenates Prepared from Different Pomegranate Accessions. J Agric Food Chem 58: 43424352.

12. Ercisli S, Gadze J, Agar G, Yildirim N, Hizarci Y (2011) Genetic relationships among wild pomegranate (*Punica granatum*) genotypes from Coruh Valley in Turkey. Genet Mol Res 10: 459–464.

13. Sarkhosh A, Zamani Z, Fatahi R, Ebadi A (2006) RAPD markers reveal polymorphism among some Iranian pomegranate (*Punica granatum* L.) genotypes. Scientia Horticulturae 111: 24–29.

14. Pirseyedi SM, Valizadehghan S, Mardi M, Ghaffari MR, Mahmoodi P, et al. (2010) Isolation and Characterization of Novel Microsatellite Markers in Pomegranate (*Punica granatum* L.). Int J Mol Sci 11: 2010–2016.

15. Ranade SA, Rana TS, Narzary D (2009) SPAR profiles and genetic diversity amongst pomegranate (*Punica granatum* L.) genotypes. Physiol Mol Biol Plants 15: 61–70.

16. Hasnaoui N, Buonamici A, Sebastiani F, Mars M, Zhang D, et al. (2012) Molecular genetic diversity of *Punica granatum* L. (pomegranate) as revealed by microsatellite DNA markers (SSR). Gene 493: 105–112.

17. Hasnaoui N, Mars M, Chibani J, Trifi M (2010) Molecular Polymorphisms in Tunisian Pomegranate (*Punica granatum* L.) as Revealed by RAPD Fingerprints. Diversity 2: 107–114.

18. Melgarejo P, Martcnez JJ, HerncLndez F, Martcnez R, Legua P, et al. (2009) Cultivar identification using 18S–28S rDNA intergenic spacer-RFLP in pomegranate (*Punica granatum* L.). Scientia Horticulturae 120: 500–503.

19. Yuan Z, Yin Y, Qu J, Zhu L, Li Y (2007) Population genetic diversity in Chinese pomegranate (*Punica granatum* L.) cultivars revealed by fluorescent-AFLP markers. J Genet Genomics 34: 1061–1071.

20. Zhang YP, Tan HH, Cao SY, Wang XC, Yang G, et al. (2012) A novel strategy for identification of 47 pomegranate (*Punica granatum*) cultivars using RAPD markers. Genet Mol Res 11: 3032–3041.

21. Soriano JM, Zuriaga E, Rubio P, LlcLcer G, Infante R, et al. (2011) Development and characterization of microsatellite markers in pomegranate (*Punica granatum* L.). Molecular Breeding 27: 119–128.

22. Zamani Z, Sarkhosh A, Fatahi R, Ebadi A (2007) Genetic relationships among pomegranate genotypes studied by fruit characteristics and RAPD markers. Journal of Horticultural Science & Biotechnology 82: 11–18.

23. Ercisli S, Agar G, Orhan E, Yildirim N, Hizarci Y (2007) Interspecific variability of RAPD and fatty acid composition of some pomegranate cultivars (*Punica granatum* L.) growing in Southern Anatolia Region in Turkey. Biochemical Systematics and Ecology 35: 764–769.

24. Ercisli S, Kafkas E, Orhan E, Kafkas S, Dogan Y, et al. (2011) Genetic characterization of pomegranate (*Punica granatum* L.) genotypes by AFLP markers. Biol Res 44: 345–350.

25. Ono NN, Britton MT, Fass JN, Nicolet CM, Lin D, et al. (2011) Exploring the Transcriptome Landscape of Pomegranate Fruit Peel for Natural Product Biosynthetic Gene and SSR Marker Discovery. J Integr Plant Biol 10.1111/j.1744-7909.2011.01073.x.

26. Imelfort M, Edwards D (2009) De novo sequencing of plant genomes using second-generation technologies. Brief Bioinform 10: 609–618.

27. Natarajan P, Parani M (2011) De novo assembly and transcriptome analysis of five major tissues of *Jatropha curcas* L. using GS FLX titanium platform of 454 pyrosequencing. BMC Genomics 12: 191.

28. Guo S, Liu J, Zheng Y, Huang M, Zhang H, et al. (2011) Characterization of transcriptome dynamics during watermelon fruit development: sequencing, assembly, annotation and gene expression profiles. BMC Genomics 12: 454.

29. Gai S, Zhang Y, Mu P, Liu C, Liu S, et al. (2012) Transcriptome analysis of tree peony during chilling requirement fulfillment: assembling, annotation and markers discovering. Gene 497: 256–262.

30. Su CL, Chao YT, Alex Chang YC, Chen WC, Chen CY, et al. (2011) De novo assembly of expressed transcripts and global analysis of the *Phalaenopsis aphrodite* transcriptome. Plant Cell Physiol 52: 1501–1514.

31. Yang P, Li X, Shipp MJ, Shockey JM, Cahoon EB (2010) Mining the bitter melon (*momordica charantia* l.) seed transcriptome by 454 analysis of non-normalized and normalized cDNA populations for conjugated fatty acid metabolism-related genes. BMC Plant Biol 10: 250.

32. Zhou Y, Gao F, Liu R, Feng J, Li H (2012) De novo sequencing and analysis of root transcriptome using 454 pyrosequencing to discover putative genes associated with drought tolerance in *Ammopiptanthus mongolicus*. BMC Genomics 13: 266.

33. Barbazuk WB, Emrich SJ, Chen HD, Li L, Schnable PS (2007) SNP discovery via 454 transcriptome sequencing. Plant J 51: 910–918.

34. Blanca J, Esteras C, Ziarsolo P, Perez D, Ferna Ndez-Pedrosa V, et al. (2012) Transcriptome sequencing for SNP discovery across *Cucumis melo*. BMC Genomics 13: 280.

35. Iorizzo M, Senalik DA, Grzebelus D, Bowman M, Cavagnaro PF, et al. (2011) De novo assembly and characterization of the carrot transcriptome reveals novel genes, new markers, and genetic diversity. BMC Genomics 12: 389.

36. Jhanwar S, Priya P, Garg R, Parida SK, Tyagi AK, et al. (2012) Transcriptome sequencing of wild chickpea as a rich resource for marker development. Plant Biotechnol J 10: 690–702.

37. Clark RM, Schweikert G, Toomajian C, Ossowski S, Zeller G, et al. (2007) Common sequence polymorphisms shaping genetic diversity in *Arabidopsis thaliana*. Science 317: 338–342.

38. Rostoks N, Mudie S, Cardle L, Russell J, Ramsay L, et al. (2005) Genome-wide SNP discovery and linkage analysis in barley based on genes responsive to abiotic stress. Mol Genet Genomics 274: 515–527.

39. Sonah H, Deshmukh RK, Sharma A, Singh VP, Gupta DK, et al. (2011) Genome-wide distribution and organization of microsatellites in plants: an insight into marker development in *Brachypodium*. PLoS One 6: e21298.

40. Holland D, Hatib K, Bar-Ya'akov I (2009) Pomegranate: botany, horticulture, breeding. Horticultural Reviews 35: 127–191.

41. Jbir R, Hasnaoui N, Mars M, Marrakchi M, Trifi M (2008) Characterization of Tunisian pomegranate (*Punica granatum* L.) cultivars using amplified fragment length polymorphism analysis. Scientia Horticulturae 115: 231–237.

42. Meisel L, Fonseca B, Gonzalez S, Baeza-Yates R, Cambiazo V, et al. (2005) A rapid and efficient method for purifying high quality total RNA from peaches (*Prunus persica*) for functional genomics analyses. Biol Res 38: 83–88.

43. Ben-Simhon Z, Judeinstein S, Nadler-Hassar T, Trainin T, Bar-Yaakov I, et al. (2011) A pomegranate (*Punica granatum* L.) WD40-repeat gene is a functional homologue of Arabidopsis *TTG1* and is involved in the regulation of anthocyanin biosynthesis during pomegranate fruit development. Planta 234: 865–881.

44. Porebski S, Bailey LG, Baum BR (1997) Modification of a CTAB DNA extraction protocol for plants containing high polysaccharide and polyphenol components. Plant Molecular Biology Reporter 15: 8–15.

45. Chevreux B, Pfisterer T, Drescher B, Driesel AJ, Muller WE, et al. (2004) Using the miraEST assembler for reliable and automated mRNA transcript assembly and SNP detection in sequenced ESTs. Genome Res 14: 1147–1159.

46. Rice P, Longden I, Bleasby A (2000) EMBOSS: the European Molecular Biology Open Software Suite. Trends Genet 16: 276–277.

47. Goodstein DM, Shu S, Howson R, Neupane R, Hayes RD, et al. (2012) Phytozome: a comparative platform for green plant genomics. Nucleic Acids Res 40: D1178–1186.

48. Conesa A, Gotz S, Garcia-Gomez JM, Terol J, Talon M, et al. (2005) Blast2GO: a universal tool for annotation, visualization and analysis in functional genomics research. Bioinformatics 21: 3674–3676.

49. Ewing B, Hillier L, Wendl MC, Green P (1998) Base-calling of automated sequencer traces using phred. I. Accuracy assessment. Genome Res 8: 175–185.

50. Weir BS (1990) Genetic data analysis. Methods for discrete population genetic data: Sinauer Associates, Inc. Publishers.

51. Bowcock AM, Ruiz-Linares A, Tomfohrde J, Minch E, Kidd JR, et al. (1994) High resolution of human evolutionary trees with polymorphic microsatellites. Nature 368: 455–457.

52. Odong TL, van Heerwaarden J, Jansen J, van Hintum TJ, van Eeuwijk FA (2011) Determination of genetic structure of germplasm collections: are traditional hierarchical clustering methods appropriate for molecular marker data? Theor Appl Genet 123: 195–205.

53. Paradis E, Claude J, Strimmer K (2004) APE: Analyses of Phylogenetics and Evolution in R language. Bioinformatics 20: 289–290.

54. Popescu AA, Huber KT, Paradis E (2012) ape 3.0: New tools for distance-based phylogenetics and evolutionary analysis in R. Bioinformatics 28: 1536–1537.

55. Pritchard JK, Stephens M, Donnelly P (2000) Inference of population structure using multilocus genotype data. Genetics 155: 945–959.

56. Evanno G, Regnaut S, Goudet J (2005) Detecting the number of clusters of individuals using the software STRUCTURE: a simulation study. Mol Ecol 14: 2611–2620.

57. Wright S (1950) Genetical structure of populations. Nature 166: 247–249.

58. Gotz S, Garcia-Gomez JM, Terol J, Williams TD, Nagaraj SH, et al. (2008) High-throughput functional annotation and data mining with the Blast2GO suite. Nucleic Acids Res 36: 3420–3435.

59. Blanca J, Canizares J, Roig C, Ziarsolo P, Nuez F, et al. (2011) Transcriptome characterization and high throughput SSRs and SNPs discovery in *Cucurbita pepo* (Cucurbitaceae). BMC Genomics 12: 104.

60. Der JP, Barker MS, Wickett NJ, dePamphilis CW, Wolf PG (2011) De novo characterization of the gametophyte transcriptome in bracken fern, *Pteridium aquilinum*. BMC Genomics 12: 99.

61. Wessling R, Schmidt SM, Micali CO, Knaust F, Reinhardt R, et al. (2012) Transcriptome analysis of enriched *Golovinomyces orontii* haustoria by deep 454 pyrosequencing. Fungal Genet Biol 49: 470–482.

62. Hamblin MT, Warburton ML, Buckler ES (2007) Empirical comparison of Simple Sequence Repeats and single nucleotide polymorphisms in assessment of maize diversity and relatedness. PLoS One 2: e1367.

63. Thiel T, Michalek W, Varshney RK, Graner A (2003) Exploiting EST databases for the development and characterization of gene-derived SSR-markers in barley (*Hordeum vulgare* L.). Theor Appl Genet 106: 411–422.

64. Kofler R, Schlotterer C, Lelley T (2007) SciRoKo: a new tool for whole genome microsatellite search and investigation. Bioinformatics 23: 1683–1685.

65. Xu Y, Ma RC, Xie H, Liu JT, Cao MQ (2004) Development of SSR markers for the phylogenetic analysis of almond trees from China and the Mediterranean region. Genome 47: 1091–1104.

66. Yasodha R, Sumathi R, Chezhian P, Kavitha S, Ghosh M (2008) Eucalyptus microsatellites mined in silico: survey and evaluation. J Genet 87: 21–25.

67. da Maia LC, de Souza VQ, Kopp MM, de Carvalho FI, de Oliveira AC (2009) Tandem repeat distribution of gene transcripts in three plant families. Genet Mol Biol 32: 822–833.

68. Ahmad R, Parfitt DE, Fass J, Ogundiwin E, Dhingra A, et al. (2011) Whole genome sequencing of peach (*Prunus persica* L.) for SNP identification and selection. BMC Genomics 12: 569.

69. Novaes E, Drost DR, Farmerie WG, Pappas GJ Jr, Grattapaglia D, et al. (2008) High-throughput gene and SNP discovery in Eucalyptus grandis, an uncharacterized genome. BMC Genomics 9: 312.

70. Bainbridge MN, Wang M, Wu Y, Newsham I, Muzny DM, et al. (2011) Targeted enrichment beyond the consensus coding DNA sequence exome reveals exons with higher variant densities. Genome Biol 12: R68.

71. Gojobori T, Li WH, Graur D (1982) Patterns of nucleotide substitution in pseudogenes and functional genes. J Mol Evol 18: 360–369.

72. Chakraborty R, Jin L (1993) Determination of relatedness between individuals using DNA fingerprinting. Hum Biol 65: 875–895.

73. Teixeira da Silva JA, Rana TS, Narzary D, Verma N, Meshram DT, et al. (2013) Pomegranate biology and biotechnology: A review. Scientia Horticulturae 160: 85–107.

74. Kaya HB, Cetin O, Kaya H, Sahin M, Sefer F, et al. (2013) SNP Discovery by Illumina-Based Transcriptome Sequencing of the Olive and the Genetic Characterization of Turkish Olive Genotypes Revealed by AFLP, SSR and SNP Markers. PLoS One 8: e73674.

75. Liu G, Li W, Zheng P, Xu T, Chen L, et al. (2012) Transcriptomic analysis of 'Suli' pear (Pyrus pyrifolia white pear group) buds during the dormancy by RNA-Seq. BMC Genomics 13: 700.

76. Wang L, Zhao S, Gu C, Zhou Y, Zhou H, et al. (2013) Deep RNA-Seq uncovers the peach transcriptome landscape. Plant Mol Biol 10.1007/s11103-013-0093-5.

77. Rajaram V, Nepolean T, Senthilvel S, Varshney RK, Vadez V, et al. (2013) Pearl millet [*Pennisetum glaucum* (L.) R. Br.] consensus linkage map constructed using four RIL mapping populations and newly developed EST-SSRs. BMC Genomics 14: 159.

78. Zhang T, Qian N, Zhu X, Chen H, Wang S, et al. (2013) Variations and transmission of QTL alleles for yield and fiber qualities in upland cotton cultivars developed in China. PLoS One 8: e57220.

79. Rauscher G, Simko I (2013) Development of genomic SSR markers for fingerprinting lettuce (*Lactuca sativa* L.) cultivars and mapping genes. BMC Plant Biol 13: 11.

80. Billot C, Ramu P, Bouchet S, Chantereau J, Deu M, et al. (2013) Massive Sorghum Collection Genotyped with SSR Markers to Enhance Use of Global Genetic Resources. PLoS One 8: e59714.

81. Emanuelli F, Lorenzi S, Grzeskowiak L, Catalano V, Stefanini M, et al. (2013) Genetic diversity and population structure assessed by SSR and SNP markers in a large germplasm collection of grape. BMC Plant Biol 13: 39.

82. Frascaroli E, Schrag TA, Melchinger AE (2013) Genetic diversity analysis of elite European maize (*Zea mays* L.) inbred lines using AFLP, SSR, and SNP markers reveals ascertainment bias for a subset of SNPs. Theor Appl Genet 126: 133–141.

83. Sharma PC, Grover A, Kahl G (2007) Mining microsatellites in eukaryotic genomes. Trends Biotechnol 25: 490–498.

84. Nielsen R, Paul JS, Albrechtsen A, Song YS (2011) Genotype and SNP calling from next-generation sequencing data. Nat Rev Genet 12: 443–451.

85. Wang Z, Gerstein M, Snyder M (2009) RNA-Seq: a revolutionary tool for transcriptomics. Nat Rev Genet 10: 57–63.

86. Bachlava E, Taylor CA, Tang S, Bowers JE, Mandel JR, et al. (2012) SNP discovery and development of a high-density genotyping array for sunflower. PLoS One 7: e29814.

87. Brachi B, Morris GP, Borevitz JO (2011) Genome-wide association studies in plants: the missing heritability is in the field. Genome Biol 12: 232.

88. Jalikop SH, Rawal RD, Kumar R (2005) Exploitation of Sub-temperate Pomegranate Daru in Breeding Tropical Varieties. Acta horticulturae: 696.

89. Jalikop SH (2007) Linked dominant alleles or inter-locus interaction results in a major shift in pomegranate fruit acidity of 'Ganesh' × 'Kabul Yellow'. Euphytica 158: 201–207.

90. Goor A, Liberman J (1956) The pomegranate. State of Israel, Ministry of Agriculture, AgrPubl Section. pp. 57.

Unraveling the Complex Trait of Harvest Index with Association Mapping in Rice (*Oryza sativa* L.)

Xiaobai Li[1,3], Wengui Yan[2]*, Hesham Agrama[4], Limeng Jia[1,2,4], Aaron Jackson[2], Karen Moldenhauer[4], Kathleen Yeater[5], Anna McClung[2], Dianxing Wu[1]*

1 State Key Lab of Rice Biology, International Atomic Energy Agency Collaborating Center, Zhejiang University, Hangzhou, People's Republic of China, 2 Agricultural Research Service, United States Department of Agriculture, Dale Bumpers National Rice Research Center, Stuttgart, Arkansas, United States of America, 3 Instiue of Horticulture, Zhejiang Academy of Agricultural Sciences, Hangzhou, People's Republic of China, 4 University of Arkansas, Rice Research and Extension Center, Stuttgart, Arkansas, United States of America, 5 Agricultural Research Service, United States Department of Agriculture, Southern Plains Area, College Station, Texas, United States of America

Abstract

Harvest index is a measure of success in partitioning assimilated photosynthate. An improvement of harvest index means an increase in the economic portion of the plant. Our objective was to identify genetic markers associated with harvest index traits using 203 *O. sativa* accessions. The phenotyping for 14 traits was conducted in both temperate (Arkansas) and subtropical (Texas) climates and the genotyping used 154 SSRs and an *indel* marker. Heading, plant height and weight, and panicle length had negative correlations, while seed set and grain weight/panicle had positive correlations with harvest index across both locations. Subsequent genetic diversity and population structure analyses identified five groups in this collection, which corresponded to their geographic origins. Model comparisons revealed that different dimensions of principal components analysis (PCA) affected harvest index traits for mapping accuracy, and kinship did not help. In total, 36 markers in Arkansas and 28 markers in Texas were identified to be significantly associated with harvest index traits. Seven and two markers were consistently associated with two or more harvest index correlated traits in Arkansas and Texas, respectively. Additionally, four markers were constitutively identified at both locations, while 32 and 24 markers were identified specifically in Arkansas and Texas, respectively. Allelic analysis of four constitutive markers demonstrated that allele 253 bp of RM431 had significantly greater effect on decreasing plant height, and 390 bp of RM24011 had the greatest effect on decreasing panicle length across both locations. Many of these identified markers are located either nearby or flanking the regions where the QTLs for harvest index have been reported. Thus, the results from this association mapping study complement and enrich the information from linkage-based QTL studies and will be the basis for improving harvest index directly and indirectly in rice.

Editor: Ivan Baxter, United States Department of Agriculture, Agricultural Research Service, United States of America

Funding: Funding was provided by the United States Department of Agriculture-Agricultural Research Service. The funders had no role in study design, data collection and analysis, decision to publish, or preparation of the manuscript.

Competing Interests: The authors have declared that no competing interests exist.

* E-mail: Wengui.Yan@ars.usda.gov (WY); dxwu@zju.edu.cn (DW)

Introduction

In food production, optimizing grain yield, reducing production costs, and minimizing risks to the environment have been the primary objectives since the beginning of the twentieth century [1]. Food crops grow by developing a vegetative canopy that transpires water and carries out photosynthesis, and a root system that takes up water and nutrition, which leads to the production of biomass. Following the reproductive stage, a portion of the plant biomass is partitioned to various yield components and determines harvest index [2] Harvest index is the ratio of grain yield to total biomass and is considered as a measure of biological success in partitioning assimilated photosynthate to the harvestable product [3,4,5]. In cereal crops, dramatic improvements in harvest index have made commercial cultivars greatly different from their wild ancestors [6]. Rice (*Oryza sativa* L.) is one of the most important staple foods [7]. It can be highly productive if high harvest index genotypes are grown with optimal management practices [2].

Harvest index of rice is the result of various integrated processes with an involvement of the number of panicles per unit area, the number of spikelets per panicle, the percentage of fully ripened grains, and the weight of 1,000 mature kernels [8]. Marri et al. [9] found that harvest index was negatively correlated with plant height, but positively correlated with grain number/panicle, grain number/plant, percentage spikelet fertility, test grain weight and yield/plant in rice. Sabouri et al. [10] verified the negative correlation of harvest index with plant height and positive correlation with spikelet number and grain weight per panicle, and reported the impact of some flag leaf characteristics on harvest index in rice. In maize, harvest index is negatively correlated with plant height, but positively correlated with grain yield both phenotypically and genotypically [11]. In sorghum, harvest index is negatively correlated with forage yield [12], but positively correlated with growth rate and grain filling rate [13]. Usually, the correlated traits are interrelated, so that increases in one component may lead to decreases or increases in others.

Therefore, scientists aim to identify genes/QTLs that directly improve a target trait without negatively affecting others, or improve the target trait indirectly through the improvement of its associated characteristics.

Crop harvest index is also highly influenced by environmental factors [14], such as soil condition [15,16] and temperature [17,18]. However, genetic control of harvest index plays important role in crop production. Large variation was observed for harvest index in rice: about 0.25 among wild species, 0.30 among tall cultivars and more than 0.40 for semi-dwarf cultivars [19]. The intrinsic regulation of harvest index is controlled by many genes. A few reports in the literature have examined QTLs in rice associated with harvest index. Mao et al. [20] reported four main-effect QTLs for harvest index on chromosome (Chr) 1, 4, 8 and 11 and other epistatic interaction between two QTLs respectively on Chr 1 and Chr 5. Sabouri et al. [10] identified three QTLs mapped on Chr 2, 3 and 5, and two QTLs close to each other on Chr 4. Lanceras et al. [21] described harvest index QTLs on Chr 1 and 3. However, a recurring complication of the QTL data showed that different parental combinations and/or experiments conducted in different environments often result in partly or wholly non-overlapping sets of QTLs [22]. Therefore, it is necessary to explore constitutive QTLs across different environments and adaptive QTLs specifically for a given environment [23].

Classical QTL mapping reveals only a portion of the genetic control of a trait because there are only two alleles that can differ at any locus between the two parental lines. More comprehensive analyses of genetic architecture require consideration of a larger sample of the genetic variation in the species. One approach is association mapping, which maps the QTLs either among extant breeding lines with known pedigree relationships or in a diverse germplasm collection. Given pedigree and marker information, the probability for different lines in complex populations to share identity by descent QTLs can be defined, permitting estimation of the effects of each QTL [24]. Association mapping provides an alternate route into identifying the QTLs that have effects across a broader spectrum of germplasm, if false-positives caused by population structure can be minimized [25]. Whole-genome association scans are expected to be effective when linkage disequilibrium (LD) and marker density are sufficiently high, so that the random markers could have a greater chance of being in disequilibrium with QTLs across diverse genetic materials [26]. Huang et al. [27] successfully performed genome-wide association study (GWAS) in a rice landrace collection of China for 14 agronomic traits and identified a substantial number of loci at close to gene resolution. Many other studies have minimized the large-scale population structure effects by analyzing associations separately for each heterotic group, and controlled the finer-scale population structure by explicitly incorporating pedigree relationships between lines in the analysis [25,26,28,29,30,31,32].

Recently, the USDA rice mini-core (URMC) subset was developed and serves as a genetically diversified panel for mining genes of interest to various users [33]. The URMC was derived from 1,794 accessions in the USDA rice core collection using PowerCore software based on 26 phenotypic traits and 70 molecular markers [34]. The core collection represents over 18,000 accessions in the USDA global genebank of rice [35]. The URMC contains 217 accessions originating from 76 countries and covering 14 geographic regions worldwide plus some of unknown origin. The URMC has a great genetic diversity and well represents the five sub-populations found in O. sativa [33]. As a result, it is an ideal population for exploring QTLs responsible for

harvest index traits with the powerful approach of association mapping.

We genotyped 203 O. sativa URMC accessions with 155 molecular markers and phenotyped 14 traits contributable to harvest index in both temperate (Stuttgart, Arkansas) and subtropical (Beaumont, Texas) locations. Our objectives were to identify the traits significantly correlated with harvest index *per se* and the markers significantly associated with component traits of harvest index. To control spurious associations, i.e., Type I error, we analyzed the genetic structure and familial relatedness in the collection. Different mapping models were tested for best fit of each trait. The chosen model was used to map markers associated with harvest index and associated traits phenotyped in two environments.

Results

Markers profile

The set of 154 SSRs and an *indel* with genome-wide distribution detected a total of 1993 alleles among 203 O. sativa accessions. The average number of alleles per locus was 12.86 ranging from 2 for RM338 to 57 for con673. Polymorphic Information Content (PIC) varied from 0.25 for AP5625-1 to 0.97 for con673 among the 155 markers with an average of 0.71. Nei's (1983) [36] genetic distances ranged from 0.0181 to 0.9667 with an average 0.7464 among each pair of 203 accessions in the URMC.

Population structure and geographic origin

Using *STRUCTURE* software with multi-loci genotype data, a five-group model was identified to sufficiently explain genetic structure among 203 accessions. Ancestry of each of these accessions was inferred for assignment into a genetic group (Figure 1A). A dendrogram tree created with *PowerMarker* had five main branches for the 203 accessions as well (Figure 1B). The principal components analysis (PCA) also displayed the pattern of genetic structure with five groups. The first three components of PCA for 45.07% of total variation were used to visualize the five groups derived from ancestry analyses (Figure 1C).

The resultant five groups of O. sativa categorized by the Q value (ancestry index) belong to *indica* (IND), *temperate japonica* (TEJ), *tropical japonica* (TRJ), *aus* (AUS) and *aromatic* (ARO) (Figure 1A), based on reference cultivars reported previously by Garris et al. [37], Agrama and Eizenga [38] and Agrama et al. [34]. Each accession with ancestry information was plotted on a world map using its latitude and longitude of geographic origin (Figure 2). TEJ accessions were mainly distributed between latitudes 30 and 50 degrees north and south of the equator (i.e. temperate zone) while the other four groups scattered between latitude N 30 and S 30 degrees (i.e. tropical and subtropical zone).

Morphological analysis

Statistical analysis using a mixed model demonstrated that the differences due to genotypes and genotype×location interactions were highly significant at the 0.001 level of probability for all of the 14 traits (Table 1). The differences due to location were also significant for 12 traits except for panicle branches and seed set. Heritability was very high for all of these 14 traits. Heading had the highest heritability which was close to 100%. Although seed set had the lowest heritability, it was still above 70%. Heritability ranged from 77 to 97% among the other 12 traits. Harvest index had a heritability of 83% at Stuttgart and 90% at Beaumont. Correlation coefficients for each pair of the 14 traits were calculated using Spearman rank for each location and presented in Table S1A and S1B, respectively. To visualize the complex

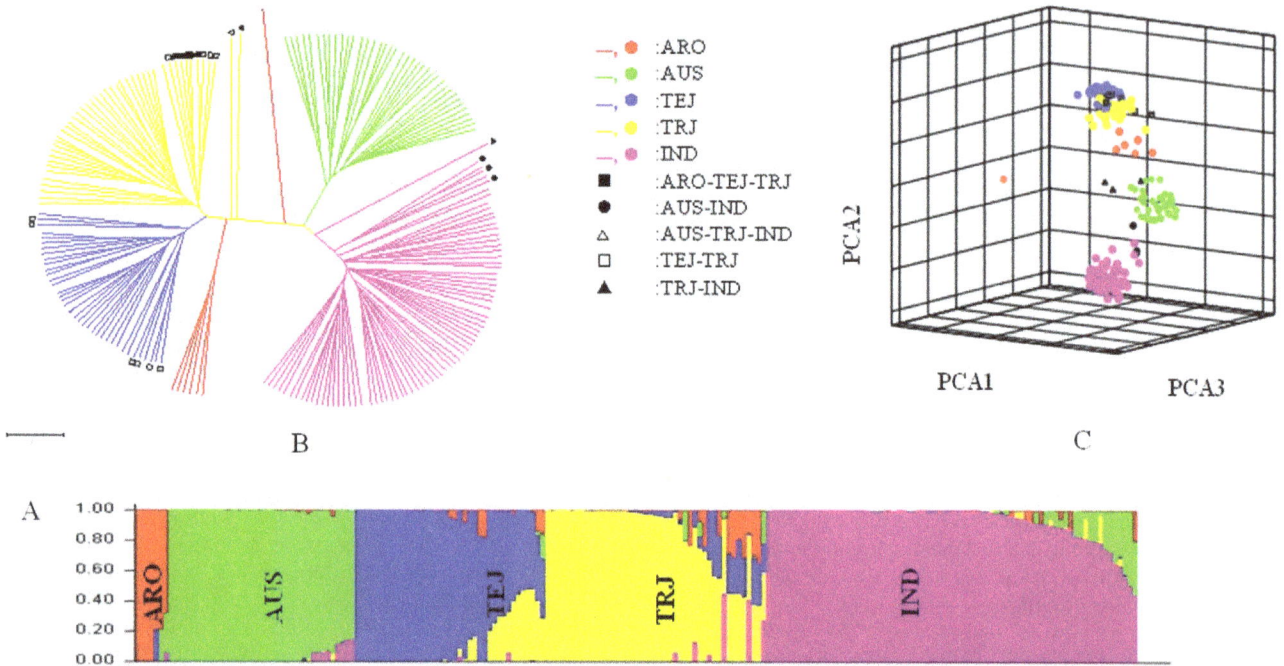

Figure 1. Structure analysis of USDA rice mini-core collection using A: *STRUCTURE*, B: Unrooted UPMGA and C: PCA. ARO: *aromatic* in red; AUS: *aus* in green; IND: *Indica* in purple; TRJ: *Tropical japonica* in yellow; TEJ: *Temperate japonica* in blue; ARO-TEJ-TRJ: admixture of ARO with TEJ and TRJ; AUS-IND: admixture of AUS with IND; AUS-TRJ-IND: admixture of AUS with TRJ and IND; TEJ-TRJ: admixture of TRJ with TEJ; TRJ-IND: admixture of TRJ with IND.

relationship among the 14 traits, PCA was used to construct plots with the first two axes accounting for more than 50% phenotypic variation (Figure 3A, B). At Stuttgart, 47 out of 91 correlations among the 14 traits were significant (<0.0001) (Table S1A, Figure 3A), and 40 correlations were significant at Beaumont (Table S1B, Figure 3B). Thirty four correlations were uniformly significant across two locations and their correlation directions (positive or negative) were also same across two locations (Table S1A, S1B).

Six traits were significantly correlated with harvest index and these correlation directions were the same across the two locations. The correlations with harvest index were negative for heading (-0.46 at Stuttgart and -0.61 at Beaumont), plant height (-0.50 and -0.50), plant weight (-0.36 and -0.30), panicle length (-0.45 and -0.32), while positive for seed set (0.52 and 0.61) and grain weight/panicle (0.32 and 0.40) (Figure 3A, B). In the PCA based on phenotypic traits of 203 mini-core accessions, four traits negatively correlated with harvest index were plotted on opposing

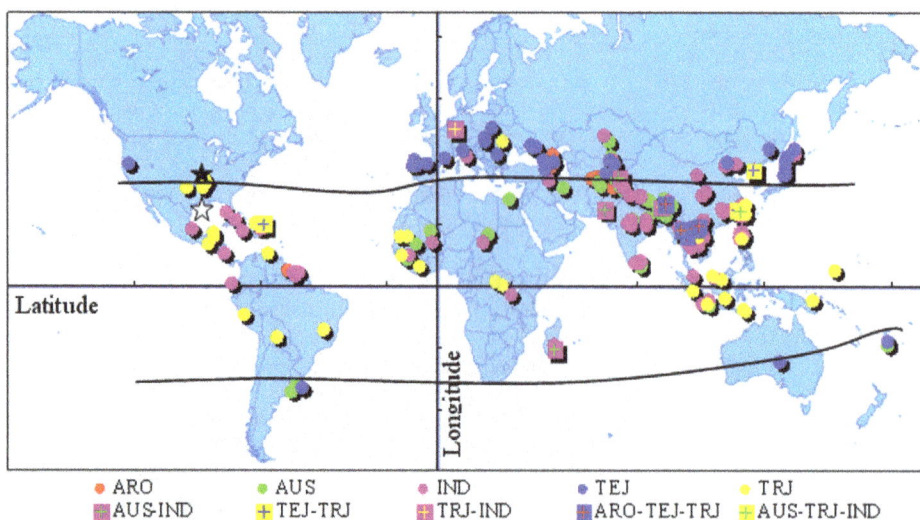

Figure 2. Geographic distribution of 203 accessions based on their latitude and longitude. ARO: *Aromatic*; AUS: *aus*; IND: *Indica*; TRJ: *Tropical japonica*; TEJ: *Temperate japonica*; ARO-TEJ-TRJ: admixture of ARO with TEJ and TRJ; AUS-IND: admixture of AUS with IND; AUS-TRJ-IND: admixture of AUS with TRJ and IND; TEJ-TRJ: admixture of TEJ with TRJ and TRJ-IND: admixture of TRJ with IND. ★: Stuttgart AR, ☆: Beaumont TX.

Table 1. Statistical analysis of 14 traits generated at Stuttgart, Arkansas and Beaumont, Texas in 2009 in the USDA rice mini-core collection.

Trait	Location	Mean ± SD	Range	Heritability (%)	Genotype		Location		Genotype*Location	
					F value	Pr>F	F value	Pr>F	F value	Pr>F
Heading (days)	Stuttgart	99.33±21.31	34.00~181.67	98.08	341.53	0.000000	2634.77	0.000001	12.45	0.000000
	Beaumont	87.55±22.63	38.00~182.00	98.64						
Plant height (cm)	Stuttgart	109.73±20.20	61.08~153.92	97.11	127.48	0.000000	1676.50	0.000002	45.63	0.000000
	Beaumont	124.74±22.45	67.00~178.78	95.73						
Plant weight (g)	Stuttgart	168.71±79.88	27.83~548.42	86.33	30.87	0.000000	122.48	0.000376	9.94	0.000000
	Beaumont	219.02±87.70	35.93~558.02	86.83						
Tillers	Stuttgart	23.95±11.20	6.42~67.75	86.53	35.27	0.000000	818.76	0.000009	7.10	0.000000
	Beaumont	41.13±15.83	13.00~85.89	87.16						
Grain yield (g)	Stuttgart	60.02±25.51	8.54~127.27	87.05	29.06	0.000000	98.37	0.000568	8.33	0.000000
	Beaumont	76.67±30.05	5.64~165.97	84.03						
Harvest index (%)	Stuttgart	30.44±7.02	3.40~45.06	82.75	35.79	0.000000	2174.76	0.000000	6.10	0.000000
	Beaumont	38.98±10.51	6.25~60.02	89.98						
Panicle length (cm)	Stuttgart	26.66±3.81	14.21~37.19	89.86	46.56	0.000000	293.26	0.000060	3.68	0.000000
	Beaumont	24.75±3.44	16.84~38.40	88.34						
Panicle branches	Stuttgart	10.97±2.15	5.44~17.78	85.65	29.97	0.000000	31.18	0.004559	2.40	0.000000
	Beaumont	10.64±2.06	5.56~16.33	81.68						
Kernels/panicle	Stuttgart	194.97±57.49	68.56~399.00	86.48	29.94	0.000000	367.90	0.000041	4.45	0.000000
	Beaumont	155.77±45.46	50.00~318.33	86.92						
Seed set (%)	Stuttgart	78.15±15.23	25.48~96.97	78.39	15.39	0.000000	14.26	0.019138	4.39	0.000000
	Beaumont	73.55±12.65	35.07~95.29	72.66						
1000 Seed weight (g)	Stuttgart	25.77±5.07	11.17~44.74	91.79	69.00	0.000000	75.18	0.000477	3.94	0.000000
	Beaumont	24.41±4.66	12.32~43.86	95.52						
Kernels/cm panicle	Stuttgart	7.30±1.80	3.25~14.61	84.71	28.72	0.000000	218.17	0.000104	3.60	0.000000
	Beaumont	6.31±1.63	2.80~12.27	87.02						
Kernels/branch panicle	Stuttgart	17.88±4.24	11.56~37.10	82.66	19.90	0.000000	353.27	0.000058	4.31	0.000000
	Beaumont	14.67±2.98	9.61~23.23	77.42						
Grain weight/panicle (g)	Stuttgart	3.79±1.18	0.68~8.62	82.29	21.86	0.000000	241.69	0.000075	3.94	0.000000
	Beaumont	2.75±0.95	0.63~6.27	80.72						

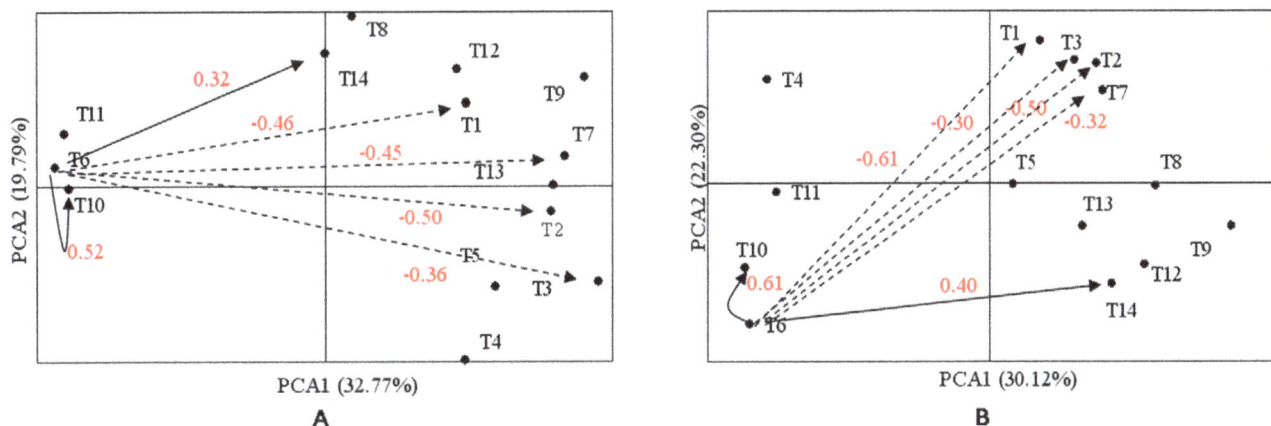

Figure 3. Relationship map constructed by PCA for 14 traits at A: Stuttgart, AR and B: Beaumont, TX. The distance between traits is inversely proportional to the size of the correlation coefficients. *Solid* and *dashed lines* indicate positive and negative correlations, respectively. Trait names are T1:Heading; T2:Plant height; T3:Plant weight; T4:Tillers; T5:Grain yield; T6:Harvest index; T7:Panicle length; T8:Panicle branches; T9:Kernels/panicle; T10:Seed set; T11:1000 Seed weight; T12:Kernels/cm panicle; T13:Kernels/branch panicle; T14: weight/panicle. The variation explained by the principal components is showed in the brackets.

axis from harvest index (Figure 3A,B). Conversely, two traits positively correlated with harvest index were plotted in the same axis relatively close to harvest index.

Model comparison and marker-trait associations

Dimension determination for PCA indicated that different dimensions should be included for testing associations for these traits. Further, relative performance of the association mapping models was also evaluated based on the criterion BIC (Table S2). The smaller BIC indicated the better model fit [25]. Among all possible models (naive, kinship, PCA, Q, PCA+kinship and Q+kinship), naive and kinship models showed the highest BIC value. The four other models (PCA, Q, P+kinship and Q+kinship) had a better performance, indicated by smaller BIC values. The model installed with kinship had a slightly higher BIC than the one without kinship. The PCA models containing different dimensions for different traits had the lowest BIC value. Thus, the PCA model was selected to conduct association mapping for harvest index traits.

At Stuttgart, a total of 36 markers were identified to be significantly associated with harvest index traits at the 6.45×10^{-3} level of probability (the Bonferroni corrected significance level) (Table S3). Among 36 markers, seven were associated with harvest index *per se*, five with heading, three with plant height, six with plant weight, five with panicle length, nine with seed set and one with grain weight/panicle. Eight of these trait-marker associations have been reported previously (Table S3). Additionally, seven markers were consistently associated with two or more harvest index traits [39]. Out of the seven consistent markers, RM600, RM5 and RM302 were co-associated with harvest index and seed

set, RM431 with heading and seed set, RM341 with plant height and panicle length, RM471 with heading and plant weight, and RM510 with three traits, plant height, harvest index and seed set.

At Beaumont, we identified 28 markers significantly associated with harvest index traits (Table S3). Among these, two were associated with harvest index, three with heading, nine with plant height, six with plant weight, four with panicle length, three with seed set and one with grain weight/panicle. At Stuttgart, eight of the trait-marker associations have been identified in previous QTL studies. Two consistent markers were RM208 co-associated with harvest index and seed set, and RM55 co-associated with plant height and plant weight.

Across two locations, the associations of RM431 with plant height, Rid12 and RM471 with plant weight, and RM24011 with panicle length were consistently true. The four markers that associated with the same trait across both locations are called "constitutive QTL" markers, while others that associated with a certain trait only at one location are called "adaptive QTL" markers [23].

Allelic effects

The allelic effects of the constitutive markers associated with their traits were estimated using the least square mean (LSMEAN) of phenotypic values and are presented in Figure 4 and Table S4. For RM431, allele 253 bp had a significantly larger effect than all other 6 alleles at Beaumont and than 4 others at Stuttgart to reduce plant height. For RM24011, allele 390 bp had the greatest effect on decreasing panicle length while allele 411 bp had the largest effect on increasing panicle length at both locations. However, for Rid12, the allelic effects were opposite between two

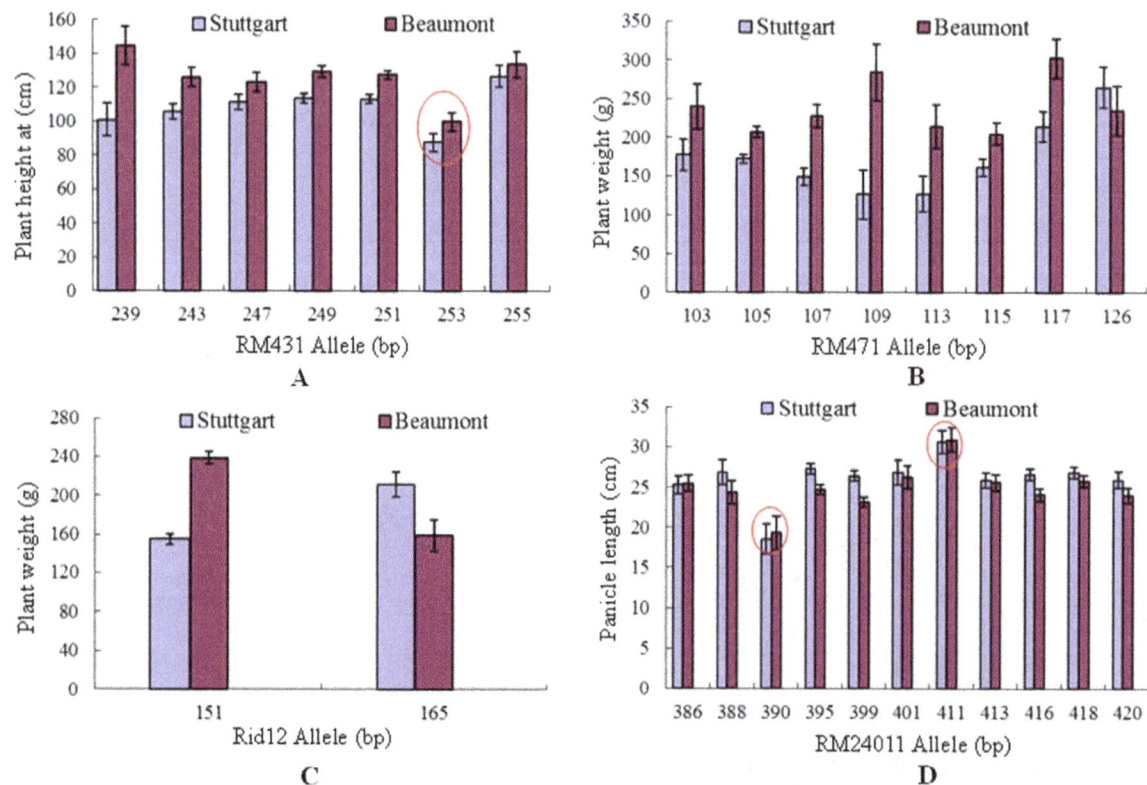

Figure 4. Comparisons of allelic effects of four constitutive marker loci. A: RM431 associated with plant height, B: RM471 and C: Rid12 associated with plant weight, D: RM24011 associated with panicle length constitutively at both Stuttgart, Arkansas and Beaumont, Texas.

locations. Allele 151 bp of Rid12 had a decreasing effect on plant weight at Stuttgart, but an increasing effect at Beaumont instead. The 165 allele of Rid12 had an opposite effect to 151 bp on plant weight. For RM471, the allelic effects on plant weight were not consistent from one location to another. The 109 bp allele was associated with one of the lowest means for plant weight at Stuttgart, but one of the largest means for plant weight at Beaumont.

Discussion

Genetic diversity and genetic structure

The average number of alleles per locus was 12.86 among 203 accessions in the URMC genotyped with 155 markers. The allele number per locus is the highest among the rice collections that have been reported to date [37,40], including an Indian germplasm collection [41], an Indonesian landrace collection [42] and a Brazilian rice core collection [43], with an exception of an Indonesian traditional and improved rice collection with 13 alleles per locus reported by Thomson et al. [44]. The average polymorphic information content (PIC) value in this study was 0.71, which is also the highest among previous studies for rice populations [37,40,41,42,44,45] with an exception of 0.75 PIC value in a study reported by Borba et al. [43]. The wide range of genetic diversity along with the manageable number of accessions in the URMC makes it one of the best collections for mining valuable genes in rice.

Population structure is an important component in association mapping analyses because it can be a source of Type I error in an autogamous species such as barley and rice [46,47,48]. In this study, the 203 O. sativa accessions in the URMC were divided into five model-based groups from ancestry analysis (Figure 1A). Both the dendrogram tree (Figure 1B) and the PCA analysis (Figure 1C) reached similar conclusions regarding population structure in this collection. The results obtained from these three separate analyses supported each other. The classification agreed with the previous study [33] except for the group of wild relatives of rice having a high rate of rare alleles. The high rate of rare alleles was suggested by its high percentage of private alleles and the small size of the group [33]. The wild rice accessions were not integrated into association mapping since low frequency alleles are known to inflate variance estimates of linkage disequilibrium and produce a greater chance of Type I error [46,47,49]. In addition, the population structure was observed to be tied with geographic origins, e.g. TEJ mainly distributed in the temperate zone (Figure 2) and wild rice relatives were from a relatively isolated area (data not shown). The distinctive geographic origins corresponding to the difference of ecological environments could be partially responsible for the genetic differentiation, which in turn contributes to the different responses to environmental factors and rare alleles in the germplasm accessions of wild relative species.

Morphological environment-sensitivity and trait-trait correlation

All 14 traits were significantly affected by environment and environment X genotype interaction, which suggested genotypic sensitivities to differences in environmental conditions at the two locations (Table 1). The sensitivity of panicle heading to temperature change and the variation of harvest index in response to photoperiod were previously observed in rice [50]. Others have reported that rice accessions derived from different geographic regions react to environmental signals differently as well [51,52]. Information on germplasm and environmental interaction is

helpful for parental selection for a specific or broad adaptation to environments.

The correlations among the 14 traits exhibited a complex relationship between pairs of traits. At both locations, the harvest index increased with an increase of seed set and grain weight/panicle, while decreased with an increase of heading, panicle length, plant height and plant weight. The negative and significant correlation between heading and harvest index was also reported in spring wheat [53], rice [54] and sorghum [11]. These studies concluded that harvest index could be easily influenced not only during the grain filling period [55,56], but also during the period from panicle initiation to heading [54] as affected by planting dates and temperature during the growing season [57]. Plant height is another important agronomic trait that is directly linked to harvest index [9,58]. Yoshida et al. [15] also reported a similar result to this study where harvest index was inversely correlated with plant height, which may be due to lodging in the tall varieties [54], or greater translocation of photosynthate from the vegetative tissues to grain in semi-dwarf varieties [59]). The positive correlation between harvest index and grain weight/panicle was also reported by Sabouri et al. [10]. However, panicle length was not found to be correlated with harvest index in Marri's study [9]. Similarly, plant weight was not correlated with harvest index in Sabouri's study [10]. These different results are understandable since different materials were used in those studies. In practice, highly correlated traits, such as heading, can be used to obtain indirect estimates of harvest index when direct estimates are difficult or impractical to obtain. Thus improvement of harvest index can be manipulated indirectly. In theory, the correlation of harvest index with its related traits determined in this study, indicates an interrelationship of physiological pathways controlling these traits.

Model comparison for association mapping of harvest index's traits

For harvest index traits, the number of dimensions in PCA was tested for each trait, and the appropriate number of dimensions was determined on the basis of BIC. Our simulated experiments showed that the dimension of PCA can exhibit phenotypic specificity. As an example with heading, the PCA model required a higher dimension number to capture the true population structure effects. Traditionally, the number of dimensions has been generally determined on the basis of random marker information without considering phenotypic information. However, the effects of population structure on different complex traits vary dramatically [60,61] and it is logical to hypothesize that the numbers of dimensions required for cofactors in detecting marker–trait association are not necessarily the same [62].

Comparing with other five models (naive, kinship, PCA+Kinship, Q and Q+Kinship model), the PCA showed the best fit with the smallest BIC value for harvest index traits. Interestingly, correction of the kinship model was not observed to be better than the naive model. Similarly, the models with Q+kinship or PCA+kinship did not perform better than the ones with only Q or PCA, either. Shao et al. [63] also found that Q+kinship model performed similarly to the Q model alone in a rice panel. The result did not agree with some other studies on cross-pollinated plants and humans [25,62], where the relatedness among accessions in a population is quite complex because of the mating style. The low complex relatedness in the URMC rice collection could be attributable to the restricted gene flow among these self-pollinated accessions and the diverse global origination of these accessions. Moreover, the low complex relatedness may be a result of the M strategy based on 26 phenotypic traits and 70 molecular markers [48] being used to develop this collection. This strategy is a powerful approach for selection of accessions with the most diverse alleles because it eliminates redundancies resulting from noninformative alleles that arise from co-ancestry [64]. The low-complexity relatedness was also confirmed by few secondary branches in the UPMGA tree (Figure 1B). In summary, different populations may have their own best fit model for a specific trait, which makes it necessary to compare different models.

Genetic dissection of harvest index

Harvest index is an integrative trait including the net effect of all physiological processes during the crop cycle and its phenotypic expression is generally affected by genes responsible for non-target traits, such as heading [20,65], plant height [20] and panicle architecture [66]. The magnitude and direction of these gene functions on different phenotypes would bear heavily on the utility of such genes for improvement of these traits. In the current study, the traits like heading, plant height, plant weight and panicle length had a strong negative correlation with harvest index, while seed set and grain weight/panicle were positively correlated with harvest index. These phenotypic correlations were consistently reflected in the identification of molecular markers associated with harvest index and related traits. For example, four consistent markers at Stuttgart, RM600, RM302, RM25, and RM431, were associated with not only harvest index itself, but also for one or more traits consistently correlated with harvest index. Another consistent marker, Rid12, associated with both heading and plant weight, was close to a reported QTL "qHID7-1" responsible for harvest index [67] and the gene "Ghd7" having major effects on grains per panicle, plant height and heading in rice [68]. At Beaumont, the consistent marker RM55, associated with both plant height and plant weight, was adjacent to a QTL "qHID3-2" for control of harvest index [67]. RM431 co-associated with plant height and harvest index in this study has been reported to be closely linked to gene "sd1" [69,70]. The sd1 that is involved in gibberellic acid biosynthesis decreases plant height, thus increases harvest index. The decreased height reduces lodging susceptiblity, is tolerant to heavy applications of nitrogen fertilizer, and can be planted at relatively high density, all contributing to improved grain yield that has resulted in the Green Revolution in cereal crops including rice [71].

Other markers were associated with the traits correlated with harvest index, but not with harvest index directly in this study. These markers have been reported either nearby or flanking the QTLs for harvest index. RM5, which was associated with plant height in the Stuttgart study, was close to a reported QTL for harvest index on Chr 1 [9]. RM471 associated with plant weight was close to the reported qHID4-1 and qHID4-2 for harvest index [67]. Furthermore, RM257 and RM22559 associated with seed set were co-localized with a known QTL on Chr 9 [9], and with qHID8-1 [67] for harvest index, respectively. Similarly, at Beaumont, RM44 associated with plant height was close to qHID8-1 [67], and RM263 associated with heading was adjacent to hi2.1 [9]. The chromosomal regions where numerous correlated traits are mapped indicate either pleiotropy of a single gene or tight linkage of multiple genes. Fine-mapping of such chromosomal regions would help discern the actual genetic control of these congruent traits. Development of markers for such traits in specific regions could lead to a highly effective strategy of marker-assisted selection for improving harvest index.

Environmental sensitivity and marker-assisted selection

Quantitative traits show a range of sensitivities to environmental changes [67]. In this study, 32 marker-trait associations were identified specifically adaptive to Stuttgart, whereas 24 marker-trait associations were adaptive to Beaumont. More importantly, we identified four constitutive markers associated with harvest index traits in both environments.

Environment-specific QTLs can be used for marker-assisted selection (MAS) at specific environments. For example, RM431 could be used to improve harvest index directly and indirectly through decreasing plant height and increasing seed set in Arkansas because it was co-associated with harvest index, plant height, and seed set. However, the constitutive marker-trait associations over multiple environments can be applied to MAS programs in a wide area. For example, results suggest that the constitutive markers Rid12 and RM471 could be used to improve harvest index indirectly through decreasing plant weight in the southern states of the USA.

Comparison of allelic effects of these constitutive markers can classify the alleles within a marker locus into superior or inferior ones, which helps decide which to use for MAS in the southern states. For example, allele 253 bp of RM431 and allele 390 bp of RM24011 had the largest effect on decreasing two traits, plant height and panicle length, negatively associated with harvest index. Thus, these superior alleles can be introduced for improvement of harvest index indirectly through decreasing the negative traits at both locations. Conversely, the allele 411 bp of RM24011 had the largest effect on increasing the panicle length and thus would not be useful for improving harvest index using MAS at either location. Interestingly, the two alleles of Rid12 associated with plant weight had opposite effects at the two locations. Allelic choice for this marker should be dependent on the particular environment targeted for breeding.

Results of the present study demonstrated that genome-wide association mapping in the URMC could complement and enrich the information derived from linkage-based QTL studies. After validation or fine mapping of these putative genomic regions, the information will help secure food production through either direct improvement of harvest index or indirect improvement via changes in seed set, grain weight per panicle, heading, plant height and weight, and panicle length using the MAS.

Materials and Methods

Rice association panel

Of 217 accessions in the URMC, 203 belong to O. *sativa* whereas the remaining belongs to other species in *Oryza*. Pure seed of these accessions were provided by the Genetic Stock *Oryza* Collection (GSOR) (www.ars.usda.gov/spa/dbnrrc/gsor) with cultivar name or designation, accession number, registration year, place of origin, longitude and latitude of origin, pedigree or genetic background (if available), morphological characteristics and references. The GSOR supplies seeds for research purposes to national and international users upon to request. In this study, only 203 O. *sativa* accessions were used for the following analysis because the wild relatives, O. *glaberrima*, *nivara*, *rufipogon*, *glumaepatula* and *latifolia*, contain many rare alleles. Rare alleles are one of the factors that increase the risk of Type I errors or spurious associations [47].

Location and field experiment

Evaluations were conducted for 14 traits in two field locations, USDA-ARS Dale Bumpers National Rice Research Center near Stuttgart, Arkansas and USDA-ARS Rice Research Unit near Beaumont, Texas during the 2009 growing season. The Stuttgart test site is located at N 34°27'44" and W 91°24'59", representing a temperate climate with a 243 d frost free period and average temperature of 23.9 C during the growing season. The Beaumont test site is located at N 30°03'47" and W 94°17'45", representing a subtropical climate with a 253 d frost free period and an average temperature of 26.1 C during the growing season. The experiments at both locations utilized a randomized complete block design having three replications with nine plants spaced 0.3×0.6 m in each plot. Three seeds were sown in each of nine hills in a plot using a Hege 1000 grain drill planter on April 23 and May 6 of 2009 at Stuttgart and Beaumont, respectively. Each hill was thinned to a single plant right after the permanent flood was applied at five leaf stage. Before flooding, fertilizer at 55 kg ha^{-1} of nitrogen as urea was applied. Weeds were controlled at both pre-planting and pre-flooding stages with locally recommended herbicides.

Phenotyping

Data collection followed procedures described by Yan et al. [72,73] with modifications. Heading was recorded as the number of days when 50% of the panicles in a plot had begun to emerge from the boot. Meanwhile, three plants were selected from the 9 in each plot and their main panicles were marked. Each plant was then bagged at the top to avoid panicle damage and supported by a bamboo pole to avoid lodging. Each plant was manually cut at ground level when mature and air-dried for two months before recording plant weight (g). Then, plant height (cm) was measured from the base to the panicle tip, the main panicle was removed at the panicle node and tillers of the plant were recorded before being threshed. Grain yield (g) was measured as total weight after the threshed grains were cleaned by an Almaco seed cleaner, plus seed weight of the removed main panicle. Harvest index (%) was calculated as the ratio of grain yield to plant weight. Each main panicle was measured for its length (cm), counted for its primary and secondary branches and manually threshed for kernels. All kernels from the panicle were placed in a cup half full of water and the cup was stirred with a spoon. Blank kernels floated to the top of the water and filled kernels sank to the bottom. The number of each was recorded after they were dried at 50°C for 12 hrs. Seed weight (mg) was determined by the filled kernel weight divided by its number, and seed set (%) was expressed by a ratio of the filled kernels to the total kernels including both filled and unfilled in the panicle. Panicle length and branch data were used to generate kernels/cm panicle and kernels/branch panicle using the total kernels.

Genotyping

Bulk tissue from five plants was collected from each accession as described by Brondani et al. [74] and total genomic DNA was extracted using a rapid alkali extraction procedure [75]. The bulked DNA allowed identification of the origin of heterogeneity, which can result from the presence of heterozygous individuals or from a mix of individuals with different homozygous alleles [76]. The 155 molecular markers covering the entire rice genome, approximately one marker per 10 cM on average, were used to genotype 203 accessions in the URMC. Among the markers, 149 SSRs were obtained from the Gramene database (http://www.gramene.org/), and five SSRs (AP5652-1, AP5652-2, AL606682-1, con673 and LJSSR1) were amplified in house [33]. The remaining was an *indel* at the *Rc* locus, named *Rid 12* and is responsible for rice pericarp color. Polymerase chain reaction (PCR) marker amplifications were performed as described in Agrama et al. [34]. The genetic positions and physical positions of these markers were estimated using the map of Cornell SSR 2001 and the map of Gramene Annotated Nipponbare Sequence 2009, respectively (http://www.gramene.org/). Markers labeled with

different colored fluorescence and that amplified products with size differences of 20 bp or more were multiplexed together post PCR.

Statistical analysis

Marker and phenotype profile. Genetic distance was calculated from the 155 molecular markers using Nei distance [36]. Phylogenetic reconstruction was based on the UPGMA method implemented in *PowerMarker* version 3.25 [77]. *PowerMarker* was also used to calculate the average number of alleles, gene diversity, and polymorphism information content (PIC) values. The tree to visualize the phylogenetic distribution of accessions and ancestry groups was constructed using MEGA version 4 [78].

Each of the 14 phenotypic traits was modeled independently with the MIXED procedure in SASv.9.2, where genotype, location and interaction of location with genotype were defined as fixed effects while replication within a location (block effect) was a random effect. Broad-sense heritability was calculated using formula $H^2 = \sigma_g^2/(\sigma_g^2+\sigma_e^2/n)$, where σ_g^2 as the genotypic variance, σ_e^2 as the environmental variance and n as the number of replications [79]. Spearman rank correlation coefficients between each pair of the 14 traits were calculated using the mean of 9 plants, 3 in each of three replications for an accession, using the CORR procedure in SASv.9.2. Correlation coefficients for the traits that significantly correlated with harvest index were displayed graphically using principal components analysis (PCA) performed with *NTSYSpc* software version 2.11 [80].

Population structure. The model-based program *STRUCTURE* [81] was used to infer population structure using a burn-in of 100,000, a run length of 100,000, and a model allowing for admixture and correlated allele frequencies. The number of groups (K) was set from 1 to 10, with ten independent runs each. The most probable structure number of (K) was calculated based on Evanno et al. [82] using an ad hoc statistic $D(K)$, assisted with $L(K)$, $L'(K)$ and $(L''K)$. The $D(K)$ perceives the rate of change in log probability of the data between successive (K) values rather than just the log probability of the data. Determination of mixed ancestry (an accession unable to be clearly assigned to only one group) was based on 60% (Q) as a threshold to consider an individual with its inferred ancestry from one single group. Principal component analysis (PCA), that summarizes the major patterns of variation in a multi-locus data set, was performed with *NTSYSpc* software version 2.11 [80]. The first three principal components were used to visualize the dispersion of the mini core accessions in a graph. Each accession was assigned into a group according to its maximum ancestry index assessed by *STRUCTURE* for the following linkage disequilibrium analysis.

Model comparison and association mapping. Following the procedures previously recommended [25,62] for various mixed models, we tested a subpopulation membership percentage (Q), PCA as fixed covariates and kinship (K) as a random effect. The kinship was calculated using SPAGeDi [83]. Phenotypic data were also incorporated into the process to determine the final number of dimensions for PCA based on Bayesian information criterion (BIC) [62]. The best fit model for each trait was determined based on the BIC among six models, naive, Kinship, PCA, PCA+Kinship, Q and Q+Kinship [25,84]. The selected model was then used to map the SSR markers significantly associated with harvest index's traits. The association analysis was conducted using the MIXED procedure in SASv.9.2. For multiple testing, P values were compared to the Bonferroni threshold $(1/155 = 6.45\times10^{-3})$ to identify statistically significant loci. Allelic effects at marker loci were compared using the LSMEANS and pdiff option in the MIXED procedure, using Saxton's PDMIX800 SAS macro [85].

Supporting Information

Table S1 A. Spearman correlation for each pair of 14 traits evaluated at Stuttgart, Arkansas in 2009. B. Spearman correlation for each pair of 14 traits evaluated at Beaumont, Texas in 2009.

Table S2 Fitness analysis of mapping model for harvest index traits using Bayesian information criterion (BIC) in both Arkansas and Texas.

Table S3 The marker loci associated with harvest index traits at Stuttgart, Arkansas and Beaumont, Texas in 2009.

Table S4 Comparison of allelic effect of four constitutive marker loci at two locations, Stuttgart, Arkansas and Beaumont, Texas. RM431 for Plant height, RM471 and Rid12 for Plant weight, and RM24011 for Panicle length.

Acknowledgments

The authors thank two anonymous reviewers and Ellen McWhirter for critical review, Tiffany Sookaserm, Tony Beaty, Yao Zhou, Biaolin Hu, Melissa Jia, LaDuska Simpson, Curtis Kerns, Sarah Hendrix, Bill Luebke, Jodie Cammack, Kip Landry, Carl Henry, Jason Bonnette, and Piper Roberts for technical assistance.

Author Contributions

Conceived and designed the experiments: WY AM DW KM. Performed the experiments: XL LJ AJ WY. Analyzed the data: XL LJ KY HA. Contributed reagents/materials/analysis tools: HA AJ. Wrote the paper: XL WY AM DW KM.

References

1. Koutroubas SD, Ntanos DA (2003) Genotype differences for grain yield and nitrogen utilization in indica and japonica rice under Mediterranean conditions. Field Crops Res 83: 251–260.
2. Raes D, Steduto P, Hsiao TC, Fereres E (2009) AquaCrop-The FAO Crop Model to Simulate Yield Response to Water: II. Main Algorithms and Software Description. Agron J 101: 438–447.
3. Donald CM, Hamblin J (1976) The biological yield and harvest index of cereals as agronomic and plant breeding criteria. Adv Agron 28: 361–405.
4. Hay RKM (1995) Harvest index: a review of its use in plant breeding and crop physiology. Annu Appl Biol 126: 197–216.
5. Sinclair TR (1998) Historical changes in harvest index and crop nitrogen accumulation. Crop Sci 38: 638–643.
6. Gepts P (2004) Crop domestication as a long-term selection experiment. Plant breeding reviews 24: 1–44.
7. Tyagi A, Khurana JP, Khurana P, Raghuvanshi S, Gaur A, et al. (2004) Structural and functional analysis of rice genome. J Genet 83: 79–99.
8. Terao T, Nagata K, Morino K, Hirose T (2010) A gene controlling the number of primary rachis branches also controls the vascular bundle formation and hence is responsible to increase the harvest index and grain yield in rice. Theor Appl Genet 120: 875–893.
9. Marri PR, Sarla N, Reddy LV, Siddiq EA (2005) Identification and mapping of yield and yield related QTLs from an Indian accession of *Oryza rufipogon*. BMC Genet 13: 33–39.
10. Sabouri H, Sabouri A, Reza DA (1999) Genetic dissection of biomass production, harvest index and panicle characteristics in indica-indica crosses of Iranian rice (*Oryza sativa* L.) cultivars. Aust J Crop Sci 3: 155–166.
11. Can ND, Yoshida T (1999) Genotypic and phenotypic variances and covariances in early maturing grain sorghum in a double cropping. Pl Prod Sci 2: 67–70.
12. Mohammad D, Cox PB, Posler GL, Kirkham MB, Hussain A, et al. (1993) Correlation of characters contributing to grain and forage yields and forage quality in sorghum (*Sorghum bicolor*). Indian J Agric Sci 63: 92–95.
13. Soltani A, Rezai AM, Pour MRK (2001) Genetic variability of some physiological and agronomic traits in grain sorghum (*Sorghumbicolor* L.). J Sci Tech Agric Nat Resources 5: 127–137.
14. Shrotria PK, Singh R (1988) Harvest index-A useful selection criteria in sorghum. Sorghum-Newsletter Utter Pardesh India 31: 4.
15. Yoshida S (1981) Fundamentals of rice crop science. Los BanosPhilippines: International Rice Research Institute. 109 p.
16. Dalling MJ (1985) The physiological basis of nitrogen redistribution during filling in cereals. In: Harper JE, Schrader LE, Howell HW, eds. Exploitation of physiological and genetic variability to enhance crop productivity. Rockville MD: American Society of Plant Physiology. pp 55–71.
17. Prasad PVV, Boote KJ, Allen JLH, Sheehy JE, Thomas JMG (2006) Species, ecotype and cultivar differences in spikelet fertility and harvest index of rice in response to high temperature stress. Field Crops Res 95: 398–411.
18. Peng S, Huang J, Sheehy JE, Laza RC, Visperas RM, et al. (2004) Rice yields decline with higher night temperature from global warming. Proc Natl Acad Sci USA 101: 9971–9975.
19. Jun F (1997) Formation of harvest index in rice and its improvement. Crop Res 2: 1–3.
20. Mao B-B, Cai W-j, Zhang Z-h, Hu Z-L, Li P, et al. (2003) Characterization of QTLs for Harvest Index and Source-sink Characters in a DH Population of Rice (*Oryza sativa* L.). Acta Genetica Sinica 30: 1118–1126.
21. Lanceras JC, Pantuwan GP, Jongdee B, Toojinda T (2004) Quantitative trait loci associated with drought tolerance at reproductive stage in rice. Plant Physiol 135: 384–399.
22. Rong J, Feltus FA, Waghmare VN, Pierce GJ, Chee PW, et al. (2007) Meta-analysis of polyploidy cotton QTL shows unequal contributions of subgenomes to a complex network of genes and gene clusters implicated in lint fiber development. Genetics 176: 2577–2588.
23. Hao Z, Li X, Liu X, Xie C, Li M, et al. (2010) Meta-analysis of constitutive and adaptive QTL for drought tolerance in maize. Euphytica 174: 165–177.
24. Zhang YM, Mao Y, Xie C, Smith H, Luo L, et al. (2005) Mapping quantitative trait loci using naturally occurring genetic variance among commercial inbred lines of maize (*Zea mays* L.). Genetics 169: 2267–2275.
25. Yu JM, Pressoir G, Briggs WH, Bi IV, Yamasaki M, et al. (2006) A unified mixed-model method for association mapping that accounts for multiple levels of relatedness. Nat Genet 38: 203–208.
26. Kim S, Zhao K, Jiang R, Molitor J, Borevitz JO, et al. (2006) Association mapping with single-feature polymorphisms. Genetics 173: 1125–1133.
27. Huang X, Wei X, Sang T, Zhao Q, Feng Q, et al. (2010) Genome-wide association studies of 14 agronomic traits in rice landraces. Nat Genetics 42: 961–969.
28. González-Martínez SC, Wheeler NC, Ersoz E, Nelson CD, Neale DB (2007) Association genetics in Pinus taeda L. I. Wood property traits. Genetics 175: 399–409.
29. Kang HM, Zaitlen NA, Wade CM, Kirby A, Heckerman D, et al. (2008) Efficient control of population structure in model organism association mapping. Genetics 178: 1709–1723.
30. Li X, Yan W, Agrama H, Jia L, Shen X, et al. (2011) Mapping QTLs for improving grain yield using the USDA rice mini-core collection. Planta 234: 347–361.
31. Parisseaux B, Bernardo R (2004) In silico mapping of quantitative trait loci in maize. Theor Appl Genet 109: 508–514.
32. Zhao K, Aranzana MJ, Kim S, Lister C, Shindo C, et al. (2007) An Arabidopsis example of association mapping in structured samples. PLoS Genet 19: 3(1): e4.
33. Li X, Yan W, Agrama H, Hu B, Jia L, et al. (2010) Genotypic and phenotypic characterization of genetic differentiation and diversity in the USDA rice mini-core collection. Genetica 138: 1221–1230.
34. Agrama HA, Yan WG, Lee FN, Fjellstrom R, Chen MH, et al. (2009) Genetic assessment of a mini-core developed from the USDA rice genebank. Crop Sci 49: 1336–1346.
35. Yan WG, Rutger JN, Bryant RJ, Bockelman HE, Fjellstrom RG, et al. (2007) Development and evaluation of a core subset of the USDA rice (*Oryza sativa* L.) germplasm collection. Crop Sci 47: 869–878.
36. Nei M, Takezaki N (1983) Estimation of genetic distances and phylogenetic trees from DNA analysis. Proc 5th World Cong Genet Appl Livstock Prod 21: 405–412.
37. Garris AJ, Tai TH, Coburn J, Kresovich S, McCouch SR (2005) Genetic structure and diversity in *Oryza sativa* L. Genetics 169: 1631–1638.
38. Agrama HA, Eizenga GC (2008) Molecular diversity and genome-wide linkage disequilibrium pattern in worldwide rice and its wild relatives. Euphytica 160: 339–355.
39. Pinto RS, Reynolds MP, Mathews KL, McIntyre CL, Olivares-Villegas JJ, et al. (2010) Heat and drought adaptive QTL in a wheat population to minimize confounding agronomic effects. Theor Appl Genet 121: 1001–1021.
40. Cho YG, Ishii T, Temnykh S, Chen X, Lipovich L, et al. (2000) Diversity of microsatellites derived from genomic libraries and genbank sequences in rice (*Oryza sativa* L.). Theor Appl Genet 100: 713–722.
41. Jain S, Jain RK, McCouch SR (2004) Genetic analysis of Indian aromatic and quality rice (*Oryza sativa* L.) germplasm using panels of fluorescently-labeled microsatellite markers. Theor Appl Genet 109: 965–977.

42. Thomson MJ, Polato NR, Prasetiyono J, Trijatmiko KR, Silitonga TS, et al. (2009) Genetic diversity of isolated populations of Indonesian landraces of rice (*Oryza sativa* L.) collected in east Kalimantan on the island of Borneo. Rice 2: 80–92.

43. Borba TCO, Brondani RPV, Rangel PHN, Brondani C (2009) Microsatellite marker-mediated analysis of the EMBRAPA Rice Core Collection genetic diversity. Genetica 137: 293–304.

44. Thomson MJ, Septiningsih EM, Suwardjo F, Santoso TJ, Silitonga TS, et al. (2007) Genetic diversity analysis of traditional and improved Indonesian rice (*Oryza sativa* L.) germplasm using microsatellite markers. Theor Appl Genet 114: 559–568.

45. Xu YB, Beachell H, McCouch SR (2004) A marker-based approach to broadening the genetic base of rice in the USA. Crop Sci 44: 1947–1959.

46. Breseghello F, Sorrells ME (2006) Association mapping of kernel size and milling quality in wheat (*Triticum aestivum* L.) cultivars. Genetics 172: 1165–1177.

47. Breseghello F, Sorrells ME (2006) Association analysis as a strategy for improvement of quantitative traits in plants. Crop Sci 46: 1323–1330.

48. Agrama HA, Eizenga GC, Yan W (2007) Association mapping of yield and its components in rice cultivars. Mol Breed 19: 341–356.

49. Remington DL, Thornsberry JM, Matsuoka Y, Wilson LM, Whitt SR, et al. (2001) Structure of linkage disequilibrium and phenotypic associations in the maize genome. Proc Natl Acad Sci USA 98: 11479–11484.

50. Matsumoto TK (2006) Gibberellic acid and benzyladenine promote early flowering and vegetative growth of miltoniopsis orchid hybrids hortscience. Hortscience 41: 131–135.

51. Tang T, Lu J, Huang J, He J, McCouch SR, et al. (2006) Genomic Variation in Rice: Genesis of Highly Polymorphic Linkage Blocks during Domestication. PLoS Genetics 2: 1824–1833.

52. Vaughan DA, Lu BR, Tomooka N (2008) The evolving story of rice evolution. Plant Sci 174: 394–408.

53. Din RU, Subhani GM, Ahmad N, Hussain M, Rehman AU (2010) Effect of temperature on development and grain formation in spring. Wheat Pak J Bot 42: 899–906.

54. Hommaa K, Horiea T, Shiraiwaa T, Sripodokb S, Supapoj N (2004) Delay of heading date as an index of water stress in rainfed rice in mini-watersheds in Northeast Thailand field. Crops Res 88: 11–19.

55. Shpiler L, Blum A (1991) Heat tolerance for yield and its components in different wheat cultivars. Euphytica 51: 257–263.

56. Din K, Singh RM (2005) Grain filling duration: An important trait in wheat improvement. SAIC Newsletter 15: 4–5.

57. Mahboob AS, Arain MA, Khanzada S, Naqvi MH, Dahot MU, et al. (2005) Yield and quality parameters of wheat genotypes as affected by sowing dates and high temperature stress. Pak J Bot 37: 575–584.

58. Yang X-C, Hwa CM (2008) Genetic modification of plant architecture and variety improvement in rice. Heredity 101: 396–404.

59. Zou JS, Yao KM, Lu CG, Hu XQ (2003) Study on individual plant type character of Liangyoupeijiu rice. Acta Agron Sin 29: 652–657.

60. Aranzana MJ, Kim S, Zhao K, Bakker E, Horton M, et al. (2005) Genome-wide association mapping in Arabidopsis identifies previously known flowering time and pathogen resistance genes. PLoS Genet 1: 531–539.

61. Flint-Garcia SA, Thuillet AC, Yu JM, Pressoir G, Romero SM, et al. (2005) Maize association population: a high-resolution platform for quantitative trait locus dissection. Plant J 44: 1054–1064.

62. Zhu C, Yu J (2009) Nonmetric multidimensional scaling corrects for population structure in whole genome association studies. Genetics 182: 875–888.

63. Shao Y, Jin L, Zhang G, Lu Y, Shen Y, et al. (2010) Association mapping of grain color, phenolic content, flavonoid content and antioxidant capacity in dehulled rice. Theor Appl Genet 122: 1005–1016.

64. Franco J, Crossa J, Warburton ML, Taba S (2006) Sampling strategies for conserving maize diversity when forming core subsets using genetic markers. Crop Sci 46: 854–864.

65. Hemamalini GS, Shashidhar HE, Hittalmani S (2000) Molecular marker assisted tagging of morphological and physiological traits under two contrasting moisture regimes at peak vegetative stage in rice (*Oryza sativa* L.). Euphytica 112: 69–78.

66. Ando T, Yamamoto T, Shimizu T, Ma XF, Shomura A, et al. (2008) Genetic dissection and pyramiding of quantitative traits for panicle architecture by using chromosomal segment substitution lines in rice. Theor Appl Genet 116: 881–890.

67. Hittalmani S, Huang N, Courtois B, Venuprasad R, Shashidhar HE, et al. (2003) Identification of QTL for growth- and grain yield-related traits in rice across nine locations of Asia. Theor Appl Genet 107: 679–690.

68. Xue W, Xing Y, Weng X, Zhao Y, Tang W, et al. (2008) Natural variation in *Ghd7* is an important regulator of heading date and yield potential in rice. Nat Genet 143: 1–7.

69. Peng J, Richards DE, Hartley NM, Murphy GP, Devos KM, et al. (1999) 'Green revolution' genes encode mutant gibberellin response modulators. Nature 400: 256–261.

70. Fu Q, Zhang P, Tan L, Zhu Z, Ma D, et al. (2010) Analysis of QTLs for yield-related traits in Yuanjiang common wild rice (*Oryza rufipogon Griff.*). J Genet Genomics 37: 147–157.

71. Hedden P (2003) The genes of the Green Revolution. Trends Genet 19: 5–9.

72. Yan WG, Rutger JN, Bockelman HE, Tai TH (2005) Agronomic evaluation and seed stock establishment of the USDA rice core collection. In: Norman RJ, Meullenet JF, Moldenhauer KAK, eds. BR Wells Rice Research Studies. Stuttgart: University of Arkansas, Agri Exp Sta Res Ser. pp 63–68.

73. Yan WG, Rutger JN, Bockelman HE, Tai TH (2005) Evaluation of kernel characteristics of the USDA rice core collection. In: Norman RJ, Meullenet JF, Moldenhauer KAK, eds. BR Wells Rice Research Studies. Stuttgart: University of Arkansas, Agricultural Experiment Station Research Serie. Agri Exp Sta Res Ser. pp 69–74.

74. Brondani C, Borba TCO, Rangel PHN, Brondani RPV (2006) Determination of traditional varieties of Brazilian rice using microsatellite markers. Genet Mol Biol 29: 676–684.

75. Xin Z, Velten JP, Oliver MJ, Burke JJ (2003) High throughput DNA extraction method suitable for PCR. Biotechniques 34: 820–826.

76. Borba TCO, Brondani RPV, Rangel PHN, Brondani C (2005) Evaluation of the number and information content of fluorescent-labeled SSR for rice germplasm characterization. Crop Breed Appl Biotechnol 2: 157–165.

77. Liu K, Muse SV (2005) Powermarker: an integrated analysis environment for genetic marker analysis. Bioinformatics 21: 2128–2129.

78. Tamura K, Dudley J, Nei M, Kumar S (2007) MEGA4: Molecular Evolutionary Genetics Analysis (MEGA) software version 4.0. Mol Bio Evol 24: 1596–1599.

79. Wang LQ, Liu WJ, Xu Y, He YQ, Luo LJ, et al. (2007) Genetic basis of 17 traits and viscosity parameters characterizing the eating and cooking quality of rice grain. Theor Appl Genet 115: 463–476.

80. Rohlf F (2000) FNTSYS-PC numerical taxonomy and multivariate analysis system ver 2.11L. Applied Biostatistics, NY.

81. Prichard JK, Stephens M, Rosenberg NA, Donnelly P (2000) Association mapping in structured populations. AM J Hum Genet 67: 170–181.

82. Evanno G, Regnaut S, Goudet J (2005) Detecting the number of clusters of individuals using the software STRUCTURE: a simulation study. Mol Ecol 14: 2611–2620.

83. Hardy QJ, Vekemans X (2002) SPAGeDi: a versatile computer program to analyse spatial genetic structure at the individual or population levels. Mol Ecol Notes 2: 618–620.

84. Wang ML, Hu C, Barkley NA, Chen Z, Erpelding JE, et al. (2009) Genetic diversity and population structure analysis of accessions in the US historic sweet sorghum collection. Theor Appl Genet 120: 13–23.

85. Saxton AM (1998) A macro for converting mean separation output to letter groupings in Proc Mixed. In: Proc 23rd SAS Users Group Cary NC: SAS Institute. pp 1243–1246.

Geographical Gradient of the *eIF4E* Alleles Conferring Resistance to Potyviruses in Pea (*Pisum*) Germplasm

Eva Konečná[1,3]**, Dana Šafářová**[2]**, Milan Navrátil**[2]**, Pavel Hanáček**[1,3]**, Clarice Coyne**[4]**, Andrew Flavell**[5]**, Margarita Vishnyakova**[6]**, Mike Ambrose**[7]**, Robert Redden**[8]**, Petr Smýkal**[9]*****

1 Department of Plant Biology, Mendel University in Brno, Brno, Czech Republic, **2** Department of Cell Biology and Genetics, Palacky University in Olomouc, Olomouc, Czech Republic, **3** CEITEC MENDELU, Mendel University in Brno, Brno, Czech Republic, **4** Western Regional Plant Introduction Station - USDA, Pullman, Washington, United States of America, **5** Division of Plant Sciences, University of Dundee at James Hutton Institute, Invergowrie, United Kingdom, **6** Vavilov Institute of Plant Industries, Saint Petersburg, Russian Federation, **7** John Innes Centre, Norwich, United Kingdom, **8** Australian Grains Genebank, Horsham, Victoria, Australia, **9** Department of Botany, Palacky University in Olomouc, Olomouc, Czech Republic

Abstract

Background: The eukaryotic translation initiation factor 4E was shown to be involved in resistance against several potyviruses in plants, including pea. We combined our knowledge of pea germplasm diversity with that of the *eIF4E* gene to identify novel genetic diversity.

Methodology/Principal findings: Germplasm of 2803 pea accessions was screened for *eIF4E* intron 3 length polymorphism, resulting in the detection of four $eIF4E^{A-B-C-S}$ variants, whose distribution was geographically structured. The $eIF4E^A$ variant conferring resistance to the P1 PSbMV pathotype was found in 53 accessions (1.9%), of which 15 were landraces from India, Afghanistan, Nepal, and 7 were from Ethiopia. A newly discovered variant, $eIF4E^B$, was present in 328 accessions (11.7%) from Ethiopia (29%), Afghanistan (23%), India (20%), Israel (25%) and China (39%). The $eIF4E^C$ variant was detected in 91 accessions (3.2% of total) from India (20%), Afghanistan (33%), the Iberian Peninsula (22%) and the Balkans (9.3%). The $eIF4E^S$ variant for susceptibility predominated as the wild type. Sequencing of 73 samples, identified 34 alleles at the whole gene, 26 at cDNA and 19 protein variants, respectively. Fifteen alleles were virologically tested and 9 alleles ($eIF4E^{A-1-2-3-4-5-6-7}$, $eIF4E^{B-1}$, $eIF4E^{C-2}$) conferred resistance to the P1 PSbMV pathotype.

Conclusions/Significance: This work identified novel *eIF4E* alleles within geographically structured pea germplasm and indicated their independent evolution from the susceptible $eIF4E^{S1}$ allele. Despite high variation present in wild *Pisum* accessions, none of them possessed resistance alleles, supporting a hypothesis of distinct mode of evolution of resistance in wild as opposed to crop species. The Highlands of Central Asia, the northern regions of the Indian subcontinent, Eastern Africa and China were identified as important centers of pea diversity that correspond with the diversity of the pathogen. The series of alleles identified in this study provides the basis to study the co-evolution of potyviruses and the pea host.

Editor: João Pinto, Instituto de Higiene e Medicina Tropical, Portugal

Funding: This work was supported by the Ministry of Agriculture IQ91A229 project. CC was supported by USDA (United States Department of Agriculture) project #5348-2100 = 017-00D. The funders had no role in study design, data collection and analysis, decision to publish, or preparation of the manuscript.

Competing Interests: The authors have declared that no competing interests exist.

* E-mail: petr.smykal@upol.cz

Introduction

Crop genetic diversity is an important pre-requisite for improving crop traits through breeding, particularly since the presence of closely related and genetically uniform varieties provides an ideal genetic environment for disease epidemics to occur, as evidenced by several historical and also recent events (the 1846 potato blight in Ireland, the 1970 corn blight in the USA, or the 1999 wheat rust in Africa). Available crop genetic resources function as reservoirs of often yet undiscovered allelic variants that provide an opportunity for genetic improvement of a given cultivated species [1,2]. However, the identification of specific, often rare traits requires specific screening and testing of entire, often very large collections. This is a time- and resource-intensive process. This situation improves, however, once a respective underlying gene is identified, especially in the case of monogenic

traits. Various available genomic technologies [3–7] can be applied to uncover such variation. Such screening of wild and cultivated germplasm for allelic variation of identified resistance genes has been receiving increased attention [2,8], as it can be efficiently substituted for phenotypic characterization.

Pea is an ancient legume crop, originating and domesticated in the Middle East and Mediterranean regions, from where its cultivation spread to today's Russia, and westwards through the Danube Valley and/or through ancient Greece and Rome into Europe. Pea likewise spread eastward into Persia, India and China [9]. Archaeological evidence indicates that the first instance of pea dates back to 8000 B.C. [10] in the Near East. Later, during the Stone and Bronze Ages, the existence of pea was documented in Europe and India. From there, it entered China by the first century B.C. [11]. Wild *Pisum sativum* subsp. *elatius* and subsp.

sativum are found naturally in Europe, Northwestern Asia and in temperate Africa, while *P. fulvum* range is restricted to the Middle East. *Pisum abyssinicum* is found in Ethiopia and Yemen [12,13]. There are 25 larger sized germplasm collections preserving pea diversity, which collectively hold around 72,000 accessions, while an additional 27,000 accessions are maintained in 146 smaller collections around the world [14,15]. The diversity of these collections has been studied using both by morphological descriptors and agronomical traits as well as molecular markers (reviewed in Smýkal et al. [14]), and diversity core collections were formed thereafter [16–19]. These studies showed that although *Pisum* is a comparably small genus containing two or three species, it has a wide and structured diversity, showing a range of degrees of relatedness that reflect taxonomic identifiers, eco-geography and breeding gene pools [13,17].

Along with abiotic stresses, plant pathogens are a major detriment to agriculture and threaten global food security. The use of genetic resistance is considered to be the most effective and sustainable strategy for controlling plant pathogens in agricultural practice, as it is environmentally friendly, targets specific pathogens, and provides reliable protection without additional labor or material costs [20]. Long before plants were domesticated and grown as monocultures, plant pathogens were co-evolving with wild plants growing in mixed-species communities. Evolution has continued to occur within domesticated plants growing as selected genotypes in denser populations than in the wild. Furthermore, the domestication of wild plants has distributed crops far from their places of origin [21,22], and the introduction of pathogens eventually accompanied this distribution. This co-evolutionary process shaped both plants and their pathogens, including viruses [23–25].

Pea seed borne mosaic virus (PSbMV), member of the genus *Potyvirus*, has been since identification in Czechoslovakia [26] reported worldwide and causes serious yield losses in a broad spectra of legumes including the most economically important like pea, lentil, faba bean, and chickpea [27]. The virus causes various symptoms, depending on the host and virus isolate/pathotype, such as the downward rolling of leaflets, the transient clearing and swelling of leaf veins, chlorotic mosaics, stunting, and delayed flowering. PSbMV is transmitted between plants in a non-persistent manner by aphids and then infects seeds [28]. Due to its seed-borne transmission, PSbMV presents a serious phytosanitary risk both for germplasm maintenance [29] and seed production.

The pea genome contains two virologicaly defined clusters of recessive resistance genes that are responsive to various potyviruses. One cluster (on pea linkage group II) includes *bcm*, *cyv-1*, *mo*, *sbm-2* and *sbm-3* loci, conferring resistance to *Bean common mosaic virus* (BCMV) *Clover yellow vein virus* (ClYVV), *Bean yellow mosaic virus* (BYMV-S s pathotype) and *Pea seed-borne mosaic virus* (PSbMV, pathotype P2) respectively. The second cluster (on pea linkage group VI) includes *cyv-2*, *wlv* and *sbm-1* loci, conferring resistance to *Clover yellow vein virus* (ClYVV), *Bean yellow mosaic virus* (BYMV-W pathotype) and the *Pea seed-borne mosaic virus* (PSbMV, pathotype P1), respectively [30,31]. With exception of *cyv-2* and *sbm-1* loci, shown to be identical [32,33], it is not clear if these clusters are closely linked separate genes or the same gene with alleles of different specificity.

The pea *eIF4E* gene was identified as a susceptibility factor corresponding to the recessive resistance gene, *sbm-1* locus, and a homologue *eIF(iso)4E* is presumed to be *sbm-2* locus [31–35]. PSbMV has been well-studied genomically [36–38] and viral P3-6K1 and VPg proteins have been identified as PSbMV determinants [30] responsible for physical interaction with host

eIF4E or eIF(iso)4E proteins and critical for viral infection. Studies on pepper and *Arabidopsis* suggest that potyviruses may selectively use either eIF4E and/or eIF(iso)4E proteins to achieve infection [39,40].

Recessive resistance of pea to PSbMV corresponds with the matching-allele model, explaining the interaction between potyviruses and plant hosts [41]. The mutant resistant allele of pea *eIF4E* (named *eIF4E^A* in this study) differs from its wild-type (sensitive) counterpart by five non-conservative amino acids [32,31,42]. Natural variation and functional analysis have revealed evidence of co-evolution between *eIF4E* and potyviral VPg [3,43,44]. The *eIF4E* allelic diversity has been systematically screened in various crop collections, such as pepper [5,45,46], melon [4], tomato [47] and barley [3]. A possible link between the spread of potyviruses and the origin of agriculture was demonstrated by Gibbs et al. [48], who clearly showed that the human-mediated spread of crop hosts was followed by the further diversification of viruses.

In this study, we have combined our knowledge of pea germplasm diversity with that of the *eIF4E* resistance gene and systematically screened 2803 accessions with known geographical origins including *Pisum* species with the aim of identifying allelic diversity present within the broader pea germplasm held in *ex situ* collections.

Results

Geographical distribution of eIF4E variants

A total of 2803 pea accessions were screened for *eIF4E* intron 3 length polymorphism using two sets of primer combinations, resulting in the detection of four respective *eIF4E$^{A-B-C-S}$* variants (Table 1, Fig. 1). The first corresponded to the already known *eIF4E^A* resistance [31] variant (amplified 243 bp fragment with primer combination A and 536 bp with primer combination B), the second to the susceptible *eIF4E^S* [31] (293 and 586 bp fragments), the third to the novel *eIF4E^B* (293 and 536 bp fragments) and the fourth to the novel *eIF4E^C* (293 and 592 bp fragments). The resistance *eIF4E^A* variant was found in seven accessions from Ethiopia-Sudan, fourteen accessions from India, five accessions from Pakistan-Nepal and one accession each from China, Russia and Afghanistan. In addition, this variant was detected in fifteen USA and Canadian, two South American and eleven European modern varieties or breeding lines, which have the Ethiopian line PI193835 in their pedigree. In total, the *eIF4E^A* variant was detected in 53 accessions, which represent 1.9% of those tested. The *eIF4E^B* variant was abundant in accessions of Ethiopian (29%), Afghan (23%), Nepal-Indian (20%), Caucasus (6.4%) and particularly Chinese (36.5%) origin (Table 2, Fig. 2). Moreover, it was found in five (e.g. 20%) accessions of South American origin. In contrast, this variant was under-represented in landraces of European origin, except in the Balkans (4 acc.), and absent from modern pea varieties. The *eIF4E^C* variant was the most frequent in Iberian peninsula (25%) and India-Nepal (16.6%), followed by Balkan (9%), Ethiopia (6%), and Afghan (6.2%) locations, and occurring in total of 91 (3.2%) accessions (Table 2). Due to bulking of 10 plants per sample in the case of Chinese origin accessions (ATFCC), we detected high proportion (15%) of sample heterogeneity. This heterogeneity was not possible to test in USDA, IPK, CGN and JIC samples, as they originated from single plants, while in CzNPC and VIR samples (originated again from bulks of 10 plants per samples) prevailed susceptible alleles. Finally, the susceptible *eIF4E^S* variant was found in 53.8% to 100% proportion within studied regions, in total 2331 accessions (83%). The lowest occurrence was in 145 studied accessions from India (53.8%), followed by China (63%) and Ethiopia (63%). In contrast, within the 1145 analyzed European origin accessions, including modern pea varieties, it was predominant (97%) (Fig. 2, Table 2). It is clear that there is bias towards European (1145 acc.) compare to other geographical regions, affecting likely allele distribution resulting in S allele over-representation.

Isolation of eIF4E sequences from the selected accessions

In order to assess the genetic diversity within the four classes (A, B, C and S) and to test whether intron 3 lengths correspond to DNA and amino acid haplotypes; we sequenced 32, 25, 9 and 7 accessions with respective $eIF4E^{A-B-C-S}$ variants from selected geographically diverse regions: Turkey, Nepal, Pakistan, Afghanistan, Ethiopia and China (Fig. 2, Table 2, 3, Table S2). The total length of genomic DNA clones of $eIF4E^{A-B-C-S}$ variants from ATG to stop codons were 2102, 2080, 2208, 1973–2892 bp long, respectively (Table S2), and length polymorphism was largely conferred by 50 and 56 bp insertions/deletions of minisatellite-like repeat sequences located at 102 and 421 bp respectively from the beginning of intron 3 (Fig. 1, Table 1). The extensive variation of $eIF4E^{S}$ was due to wild *Pisum* samples, especially of more distant subspecies (*P. abyssinicum*) and species (*P. fulvum*). Beside intron 3 polymorphism, extensive nucleotide variation was found; altogether, there were 156 SNPs identified in 73 sequences (Fig. 1, Table S2) resulting in 34 alleles at whole gene (exones-introns), 26 at cDNA and 19 alleles at protein levels, respectively. Of these 156 SNPs, 62 were unique, e.g. occurred only in once within the sequenced samples.

There were four different sequences of susceptible $eIF4E^{S}$ found within 25 sequenced samples of the cultivated *Pisum sativum* gene pool and three additional sequences within 7 samples of wild pea species (Table 3, Table S2). Translation of computationally spliced cDNA has resulted in four ($eIF4E^{S-1-2-3 \text{ and } 7}$) protein variants within the cultivated genepool and an additional three ($eIF4E^{S-4-5-6}$) in *P. fulvum* and *P. abyssinicum* (Fig. 3, Table S2). The typically resistant $eIF4E^{A-1}$ allele [31,39], as represented by the PI193835 accession from Ethiopia (genebank sequence number GU289735), differed from the $eIF4E^{S-1}$ allele (represented by JI194, genebank sequence number KF053441) by five nucleotide (G185T, C218A, C221A in exon 1, G410A in exon 2 and T687G in exon 3) exchanges, while the $eIF4E^{A-6}$ allele (PI269818 accession, genebank

sequence number KF053455) differed by an additional two nucleotide (G217C in exon 1, C993T in intron 3) exchanges, one AGC triplet (AGC229-231) deletion, and one single base (C768T) deletion, in addition to 1200 bp intron 3 length similar to susceptible 1201 bp (Table 2, Table S2). Within the 28 sequenced accessions with $eIF4E^{B}$ intron length variant, we detected seven alleles, both at nucleotide (Fig. 1) and amino acid (Fig. 3) levels. The three $eIF4E^{B-1-2-3}$ alleles found in nine sequenced samples differed from the $eIF4E^{S}$ alleles in 40 SNPs and one nucleotide insertion, resulting in four DNA and three protein variants. The deduced amino acid sequence of the $eIF4E^{B-1}$ allele differed from the susceptible $eIF4E^{S}$ allele at three amino acid (M207I and LD218-219QE) exchanges in exons 4 and 5. Two accessions from Turkey (JI1370, JI1090) differed by an additional three SNPs in exon 1, leading to two amino acid exchanges, V23D and V49A, resulting in two $eIF4E^{B-2-3}$ alleles (Table 3). We have detected two $eIF4E^{C}$ alleles among seven sequenced samples. The $eIF4E^{C-2}$ allele of the VIR1859 accession, besides having a 1257bp long intron 3 due to its extra copy of a second 56 bp repeat, differed by two (A56T in exon 1, G1872A in intron 3) nucleotides from the $eIF4E^{S}$ allele, leading to single amino acid (N19I) exchange at exon 1 (Fig. 3). The remaining six sequences of $eIF4E^{C}$ were identical to the $eIF4E^{S}$ allele, except for a 56 bp insertion in intron 3 and a single nucleotide exchange at intron 3 in PI505122 and PI639981, accessions from Albania and Bulgaria (Table 2, S2). We were interested in establishing which, if any, of the eIF4E A, B or C alleles can be found in wild pea. It turned out that all analyzed wild pea accesions displayed the $eIF4E^{S}$ variant as assessed by intron 3 length. Similarly, sequencing analysis of the seven selected accessions had shown that all *P. sativum* subsp. *elatius*, except for *P. fulvum* JI1007, JI1010 and *P. abyssinicum* JI1632 accessions with one or two amino acid exchanges, had the typically susceptible $eIF4E^{S-1}$ allele (Table 2, 3, S2). All had substantial sequence polymorphism compare to cultivated pea except for *P. sativum* subsp. *elatius* JI2630 from the Crimea.

Haplotype network analysis

Haplotype network analysis was used to reveal relationships between sequences. This showed the clear separation of *Pisum sativum* subsp. *elatius* (JI3157, JI1091, JI2630), *Pisum fulvum* (JI1007, JI1010) and *Pisum abyssinicum* (JI1632) accessions from cultivated germplasm (Fig. 4). The only exception was *Pisum sativum* subsp.

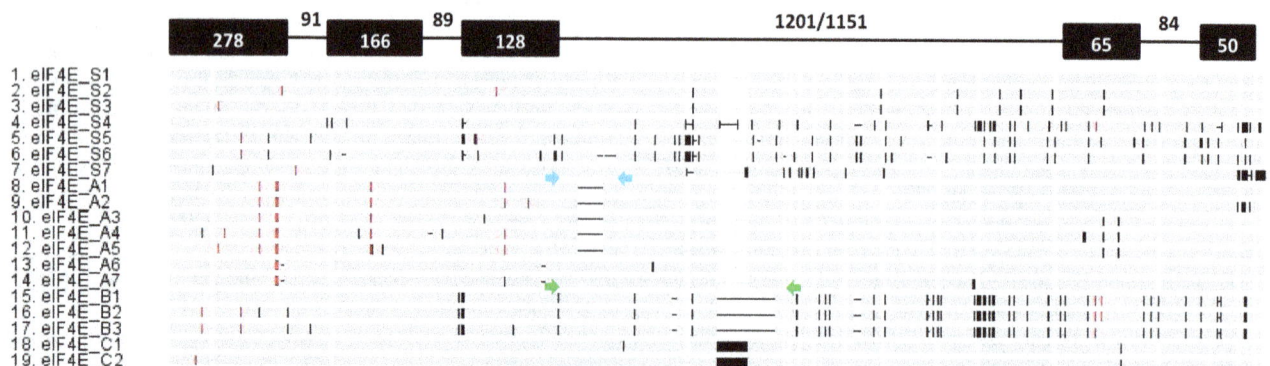

Figure 1. Schematic representation of sequence alignment of the all 19 identified protein *eIF4E* alleles. Four principal *eIF4E* variants identified by intron 3 polymorphism are designated A,B,C and S, while numbers indicate the respective allelic variant. Black bars indicate polymorphism nucleotides within both exons and introns, while red bars indicate polymorphism leading to amino acid exchanges. Horizontal lines indicate insertions/deletions. The heading line indicates nucleotide numbers and exon (solid black boxes)-intron (lines) positions and sizes. Blue arrows indicate primer A combination (Ps-eIF4E-750F and Ps-eIF4E-586gR) and green arrows indicate primer B combination (Ps-eIF4E-750F and Ps-eIF4E-1270R).

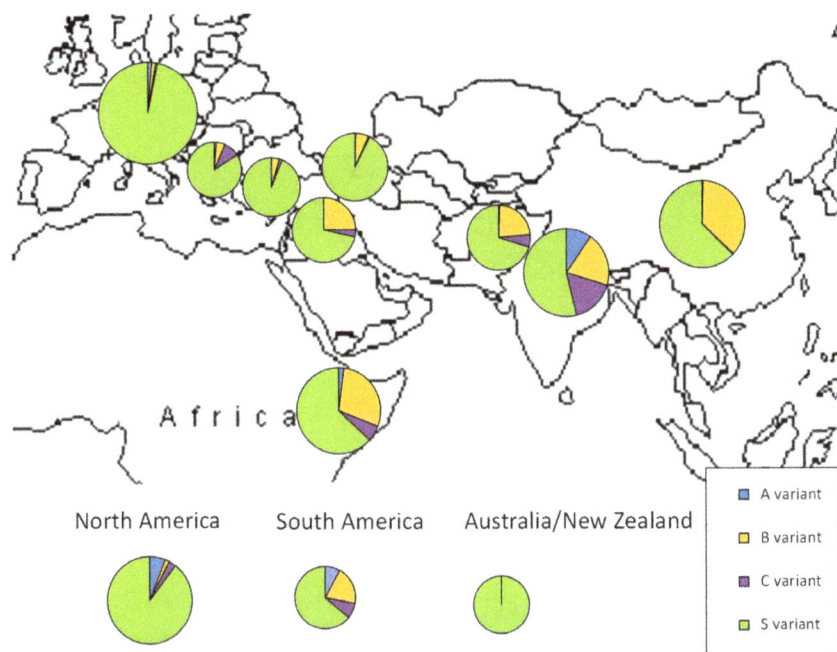

Figure 2. Geographical distribution of four *eIF4E* alleles expressed as percentage of total. Comparison of allele frequencies in 13 geographical regions, as detailed in Table 2.

elatius JI261 (genebank sequence number KF053440) from Turkey, which had a sensitive allele at the amino acid level, while exhibiting a transitory stage towards the *eIF4E*^{B} allele at the nucleotide level (Table S2). Moreover, intron 3 was 36 bp longer, with a length of 1187 bp. Interestingly, the JI261 sequence was identical to *eIF4E*^{S} until position 956 of intron 3, from which point forward it was similar to the *eIF4E*^{B} allele. An important difference was the absence of the exon 4 and 5 exchanges, where it was again identical to the susceptible *eIF4E*^{S} allele. The JI261 sequence occurred between the *eIF4E*^{S} and *eIF4E*^{B} alleles, with 23 mutational steps. A further 24 mutations separated the *eIF4E*^{B} alleles of JI1194 (genebank sequence number KF053439) of Afghanistan, ATC7134 (KF053433) of China, and JI1090 (KF053434) and JI1370 (KF053432) of Turkey (Fig. 4). The *eIF4E*^{A} alleles identified in Indian accessions (PI347464, PI347494, PI378158) proved to be derived from Ethiopian (PI193835, PIS479), Afghan (PIS468) and Indian (PI356991, ATC1044) accessions, and were identical. A further cluster contained Chinese (ATC6931, ATC6927) and Indian (PI347484, PI347422 and PI347328) origin S alleles, both at nucleotide and amino acid levels (Fig. 4).

Testing for response to PSbMV infection

The functional status of the 15 alleles of all four detected variants was verified through biological testing of their response to P1 PSbMV infection. All 19 of the pea accessions bearing *eIF4E*$^{A-1}$ tested showed resistance to viral infection i.e. the same phenotype as the non-infected or resistant controls (Table S2). DAS-ELISA testing proved negative, indicating the absence of viral coat protein. Six samples had *eIF4E*$^{A-2-3-4-5-6-7}$ alleles with single (PI116056, PI193584, JI1788) or multiple (PI269818, PI269774) amino acid exchanges, but none of these changed the expression of resistance in these accessions. The eight tested accessions bearing the *eIF4E*$^{B-1}$ allele split into two phenotypic groups. Three accessions (ATC7134 from China and CGN3311, PIS477 from Pakistan, all *eIF4E*$^{B-1}$) were resistant to the P1 pathotype of PSbMV, as the virus was not able to replicate and spread. The second group of accessions (ATC7140, ATC6928 from China, CGN3319 from Pakistan, all *eIF4E*$^{B-1}$; and JI1090, *eIF4E*$^{B-2}$, from Turkey), showed the typical susceptibility symptoms of infection, vein clearing and chlorotic leaf mosaics, leaf size decrease, internode shortening, and sporadically mild leaf-rolling, 14 days after inoculation by PSbMV. The symptoms were comparable

Table 1. Allele designation based on intron 3 length polymorphism, indicating size of obtained fragments with primer combinations A and B, number of repeat motives, intron 3 difference and total length.

eIF4E variant	Primer combination A product (bp)	Primer combination B product (bp)	copies of 50 bp repeat	copies of 56 bp repeat	intron 3 difference	intron 3 length (bp)
S	293	586	2	2	0	1201
A	243	586	1	2	−50 bp	1151
B	293	536	2	1	−70 bp	1131
C	293	592	2	3	+56 bp	1257

Table 2. Summary of germplasm screening of 2803 pea accessions, indicating four principal *elF4E* intron 3 length variants (A-B-C and S), distribution in total numbers and percentage over 13 geographical regions.

Regions						Percentage			
	Number of accesions	A variant	B variant	C variant	S variant	A variant	B variant	C variant	S variant
Ethiopia - East Africa	349	7	101	21	220	2.0	28.9	6.0	63.0
Afghanistan-Pakistan	178	1	41	11	125	0.6	23.0	6.2	70.2
India-Nepal	145	14	29	24	78	9.7	20.0	16.6	53.8
China-Mongolia	271	1	99	1	170	0.4	36.5	0.4	62.7
Turkey-Syria	64	0	3	1	60	0.0	4.7	1.6	93.8
Israel-Lebanon-Jordan	28	0	7	1	20	0.0	25.0	3.6	71.4
Balkan	75	1	4	7	63	1.3	5.3	9.3	84.0
Russia-Caucassus	295	1	19	2	273	0.3	6.4	0.7	92.5
Iberian Peninsula	36	0	2	9	25	0.0	5.6	25.0	69.4
Australia-New Zealand	11	0	0	0	11	0.0	0.0	0.0	100.0
South America	25	2	5	2	16	8.0	20.0	8.0	64.0
Europe	1145	15	15	7	1108	1.3	1.3	0.6	96.8
USA-Canada	181	11	3	5	162	6.1	1.7	2.8	89.5
Total	2803	53	328	91	2331	Average 1.9	11.7	3.2	83.2

Table 3. List of sequenced 73 accessions assigned to identified 19 protein alleles of four *elF4E* [A-B-C-S] intron 3 length variants with indicated country of origin.

elF4E allele	Number of accessions with given allele	Accession(s)	Origin
elF4E[S-1]	24	JI 182, JI 1785, JI 1030, JI 190, PI 357290, JI 193, JI 194, JI 205, JI 1107, JI 1085, JI 2065, JI 1845, JI 1532, JI 2607, JI 1121, JI 1756, JI 2571, JI 3001, JI 2630, JI 1091, JI 3157, JI 261, ATC_7173, ATC_6927	Nepal, Nepal, Iran, Sudan, Macedonia, Sudan, Sudan, Russia, Nepal, Turkey, Tanzania, Greece, Hindukush, Lybia, Nepal, Nepal, Gruzie, Iran, Crimea, Greece, Turkey, Turkey, China, China
elF4E[S-2]	1	PI 347328	India
elF4E[S-3]	3	PI 347422, PI 347484, ATC_6931	India, India, China
elF4E[S-4]	1	JI 1632	Ethiopia
elF4E[S-5]	1	JI 1007	Israel
elF4E[S-6]	1	JI 1010	Israel
elF4E[S-7]	1	JI 2646	Malawi
elF4E[A-1]	19	JI 967, JI 467, JI 1790, ATC_1044, PIS_468, PIS_479, CGN_3302, PI 116056, PI 193584, PI 249645, PI 356991, PI 356992, PI 347494, PI 347492, JI 1260, PI 193586, JI 2643, JI 1787, JI 1788	Ethiopia, Europe, Afghanistan, Afghanistan, Afghanistan, Ethiopia, Ethiopia, India, Ethiopia, India, India, India, India, India, India, Ethiopia, Malawi, Nepal, Nepal
elF4E[A-2]	1	JI 1546	Ethiopia
elF4E[A-3]	1	PI 193835	Ethiopia
elF4E[A-4]	1	PI 378158	India
elF4E[A-5]	1	PI 347464	India
elF4E[A-6]	1	PI 269818	England
elF4E[A-7]	1	PI 269774	England
elF4E[B-1]	7	ATC_7134, CGN_3311, CGN_3319, ATC_7140, PIS_477, JI 1194, ATC_3275	China, Pakistan, Pakistan, China, Hindukush-Pakistan, Afghanistan, China
elF4E[B-2]	1	JI 1370	Turkey
elF4E[B-3]	1	JI 1090	Turkey
elF4E[C-1]	6	JI 267, PI 505122, PI 639981, JI1109, JI1104, JI1108	Greece, Albania, Bulgaria, Nepal, Nepal, Nepal
elF4E[C-2]	1	VIR 1589	Russia, Sverdlovsk

	19	23	34	49	62*	73-74*	77	86	107*	108	115	148	163	169*	186	207	218-219
S allele																	
eIF4E^S-1	N	V	E	V	W	AA	S	S	G	A	H	L	I	N	N	M	LD
eIF4E^S-2	N	V	E	V	W	AA	R	S	G	A	H	L	T	N	N	M	LD
eIF4E^S-3	N	V	A	V	W	AA	S	S	G	A	H	L	I	N	N	M	LD
eIF4E^S-4	N	V	E	V	W	AA	S	S	G	A	H	L	I	N	N	I	LD
eIF4E^S-5	N	V	E	A	W	AA	S	S	G	A	H	W	I	N	N	M	LD
eIF4E^S-6	N	V	E	A	W	AA	S	S	G	A	H	L	I	N	N	M	LD
eIF4E^S-7	N	V	E	V	W	AA	S	C	G	A	H	L	I	N	N	M	LD
A allele																	
eIF4E^A-1	N	V	E	V	L	DD	S	S	R	A	H	L	I	K	N	M	LD
eIF4E^A-2	N	V	E	V	L	DD	S	S	R	A	H	L	I	K	H	M	LD
eIF4E^A-3	N	V	E	V	L	DD	S	S	R	A	H	L	I	N	N	M	LD
eIF4E^A-4	N	A	E	V	L	DD	S	S	R	A	H	L	I	K	N	M	LD
eIF4E^A-5	N	V	A	V	L	DD	S	S	R	P	R	L	I	K	N	M	LD
eIF4E^A-6	N	V	E	V	W	PD	_	S	G	A	H	L	I	N	N	M	LD
eIF4E^A-7	N	V	E	V	W	PD	_	S	G	A	H	L	I	K	N	M	LD
B allele																	
eIF4E^B-1	N	V	E	V	W	AA	S	S	G		H	L	I	N	N	I	QE
eIF4E^B-2	N	D	E	A	W	AA	S	S	G	A	H	L	I	N	N	I	QE
eIF4E^B-3	N	D	E	A	W	AA	S	S	G	A	H	L	I	N	N	I	LD
C allele																	
eIF4E^C-1	N	V	E	V	W	AA	S	S	G	A	H	L	I	N	N	M	LD
eIF4E^C-2	I	V	E	V	W	AA	S	S	G	A	H	L	I	N	N	M	LD

Figure 3. Summarized amino acid exchanges of identified 19 *eIF4E* alleles. Asterix indicates the position of exchanges previously identified as crucial for resistance.

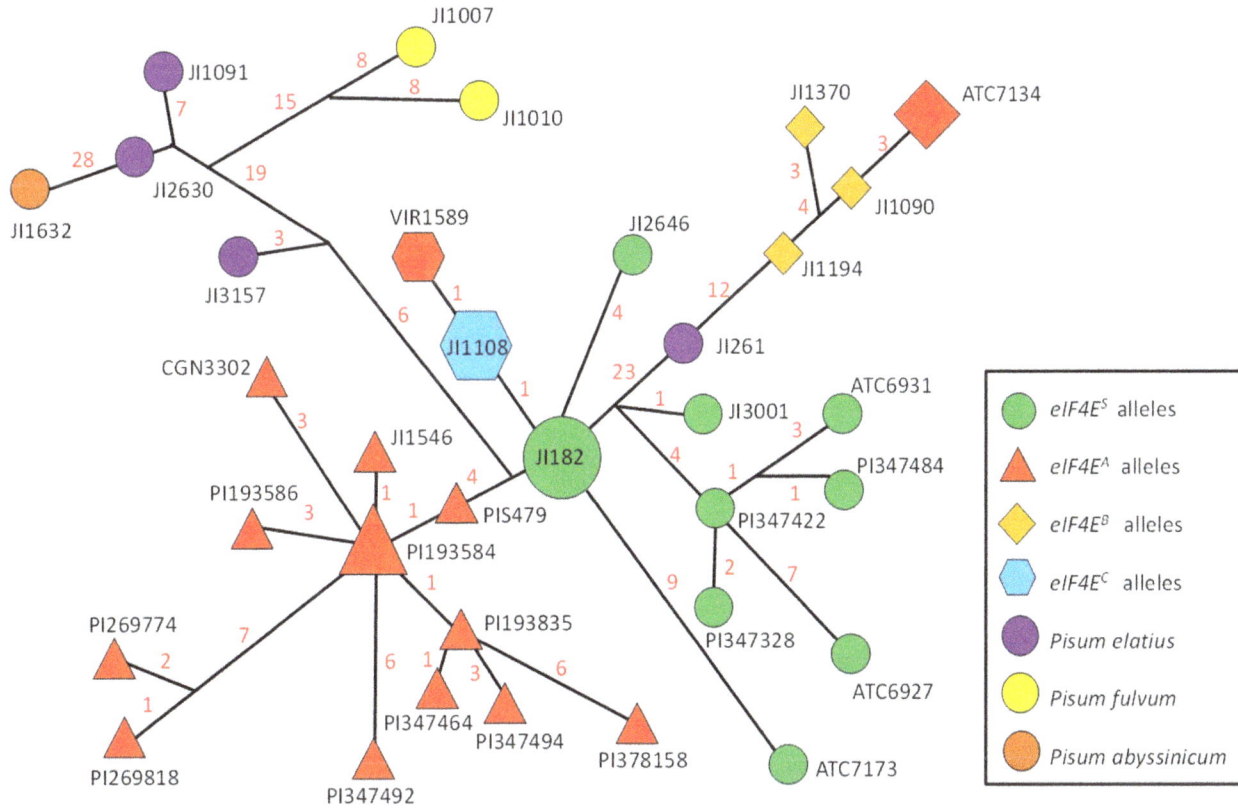

Figure 4. Haplotype network of 34 alleles identified at *eIF4E* whole gene level, using median-joining network algorithm, implemented in NETWORK. It is based on total of 156 SNP characters, and excludes 50, 56 bp indels in intron 3, as these would compromise analysis. Red numbers above lines indicate mutational steps. The shortest tree has 235 mutations. Triangles indicate A, squares B, hexagons C and circles S variants as detailed in Table 1. Red colour of symbols indicates accessions tested resistant to P1 PSbMV. Size of symbols is proportional to number of accessions with given haplotype.

with the response of the sensitive genotypes with the $eIF4E^{S-1}$ allele. All tested $eIF4E^{S-1}$ (18 accesions), $eIF4E^{S-2}$ (1 acc.) and $eIF4E^{S-3}$ (3 acc.) displayed sensitive response to P1 PSbMV pathotype (Table S2). We have not been able to test remaining $eIF4E^{S-4,5,6 \text{ and } 7}$ alleles, due to the shortage of homogenous plant material. Finally, of the three tested accessions with the $eIF4E^{C-1-2}$ alleles, only VIR1589 from the Sverdlovsk region of Russia was resistant, while JI1104 and JI1108 from Nepal proved susceptible (Table S2). This corresponded to an amino acid exchange at position N19I in exon 1 between the $eIF4E^{C-1}$ and $eIF4E^{C-2}$ alleles (Table S2).

Discussion

In this study, we combined our knowledge and access to a wide range of pea germplasm [13,17] with an interest in PSbMV resistance [31] and focused our analysis on the geographical distribution of selected, easily detectable alleles of the $eIF4E$ gene using the polymorphism of intron 3.

eIF4E gene structure and diversity

The pea $eIF4E$ gene is comprised of five exons, interrupted by four intron sequences, with a total length of 2.1 kb [31], which is spliced into a 687 nucleotide long (open reading frame) mRNA [32]. The size polymorphism of intron 3 is caused by 50 and 56 bp insertions of minisatellite-like repeat sequences [31], separated by 259 bp of common intron sequence. This polymorphism led to the initial mapping of the $sbm-1$ locus [35] and the development of a co-dominant marker for breeding [31]. Previous sequencing of 43 accessions of pea varieties showed four haplotypes of intron 1, eight haplotypes of the intron 2 and three haplotypes of intron 4, while 17 haplotypes were detected in intron 3 sequences ([31] and unpublished). Since all, except PI269774 (Sankia) and PI269818 (Aa134), previously tested susceptible and resistant accessions differed by 50 or 56 bp indels in intron 3 [31], we exploited this length polymorphism to screen a wider pea germplasm population. This led to the identification of novel polymorphisms in four principal variants ($eIF4E^S$, $eIF4E^{A-B-C}$). Another functional $eIF4E$ resistance allele found in the PI269818 and PI269774 accessions was not directly tested in this study, as its polymorphism is more difficult to analyze. It is clear that this type of analysis does not provide a comprehensive dataset, as more subtle, single-nucleotide mutations are missed, as documented by sequenced samples, whereas in 73 sequences, 19 $eIF4E$ alleles were found (Fig. 1, Table 3, S1).

The mutant $eIF4E^{A-1}$ allele of pea gene $eIF4E$ differs from its $eIF4E^{S-1}$ counterpart in five non-conservative amino acid exchanges in and around the cap-binding pocket of pea $eIF4E$, which impacts infection by PSbMV [32]. Previously reported mutations in pea $eIF4E$ were localized on the β1 (W62L), β1-β2 (AA73-74DD), β3-β4 (G107R) and β5 (N169K) loops [42]. Positions of the newly identified amino acid (V23D, V49A and LD218-219QE) changes in $eIF4E^{B-1-2-3}$ alleles do not correspond with the known mutations leading to resistance and are not localized on the variable β loops regions, with the exception of the M207I change located on the α3-β7 loop. The only resistance determinant mapped to the β7 loop is for BaYMV infection in barley [54]. In the *Capsicum* species, several $eIF4E$ haplotypes with amino acid changes were reported, some of which included amino acid changes in positions 218 and 219 similar to those in the pea $eIF4E^B$ allele. Systematic testing of individual mutations by *in planta* assay showed that only a specific combination of mutations leads to resistance [42]. This hypothesis confirms previously reported amino acid changes in pea eIF4E in PI269818 and PI269774

accessions [31] with $eIF4E^{A6-7}$ alleles, which have amino acid changes in the cap-binding pocket [31] and lack one (S78) amino acid (Fig. 3, Table 3 and Table S2, Supporting information). Amino acid changes in the loop near the cap-recognition pocket were shown to be directly associated with resistance to potyviruses in pepper, lettuce, and pea [35,55]. Moreover, not all $eIF4E$ proteins encoded by the resistance alleles are defective in their ability to bind the m7-GTP cap, suggesting that disrupted cap binding is not always required for potyvirus resistance [56]. Thus mutations conferring $sbm-1$ resistance to PSbMV act combinatorially [42]. The interactions between particular $eIF4E$ and particular viral VPgs are highly specific for host-virus interactions. Higher plants are unique in that they encode two distinct isoforms of eIF4F that have both overlapping and isoform-specific roles, eIF4F protein complex, which contains eIF4E and eIF4G, and eIF(iso)4F protein complex, which contains eIF(iso)4E and eIF(iso)4G [25]. Although these two complexes seem to be equivalent for the *in vitro* translation, they differ in their *in vivo* expression patterns and specificity for cellular mRNAs and likely also viral RNA. Potyviruses could selectively use either $eIF(iso)4E$ or $eIF4E$ to infect plants, while some are able to use both [39]. In pepper, the combination of $eIF4E$ ($pvr2$ locus on chromosome 4) and $eIF(iso)4E$ ($pvr6$ locus on chromosome 3) resistance alleles showed a complementary effect [45]. Also in pea, the $eIF(iso)4E$ homologous gene exists, mapped to LG II ([35] and close to the $sbm-2$ locus (not shown). However, its direct involvement in potyvirus resistance still needs to be demonstrated. Thus we cannot exclude the possibility that interaction between eIF4E and eIF(iso)4E proteins may also play a role in pea. This might partly explain the discrepancies in virological testing of $eIF4E^B$ accessions. Finally, the result of the interaction, the resistance or tolerance, depends not only on the genetic background but also on the virus pathotype [43,57]. It would be interesting to test region-specific potyvirus isolates [58] on the alleles identified in this study. The current phylogeny of potyviruses indicates that the progenitor probably infected plants growing in southwestern Eurasia and northern Africa, which evolved from a rymovirus by acquiring the ability to be transmitted by aphids [48]. It is notable that we did not find the resistant $eIF4E^A$ variant in any of the 146 tested wild pea samples, nor have we been able to detect any intermediate alleles between the susceptible $eIF4E^{S-1}$ and resistant $eIF4E^{A-1}$, except in JI261 accessions already separated by 23 mutations between $eIF4E^{S-1}$ and $eIF4E^{B-1}$. Such an intermediate allele was found for $Pm3$ powdery mildew resistance in wheat from the Himalayan Range, supporting the hypothesis that a recent evolution of these alleles took place in the hexaploid wheat gene pool [8]. This is a departure from the situation with other diseases for which resistance can often be found within the wild genepool, in contrast to the domestic gene pool. i.e. pathogens overcame resistance in the domestic gene pool. Haplotype analysis indicated the separate evolution of $eIF4E^A$ alleles, with Ethiopian and Indian accessions being the most distantly related, while Chinese and some Afghan and Nepalese accessions were closer to $eIF4E^{S-1}$ (Fig. 4). Thus the evolution of the $eIF4E$ gene is demonstrated in the available pea genepool.

Geographical distribution of eIF4E variants versus PSbMV diversity

Pea genotypes resistant to PSbMV were identified by Hagedorn and Gritton [59] in two Ethiopian lines (PI193586 and PI193835). Later, Hampton [60] found several resistant accessions from India. Since they are all identical in sequence [31], it raises the question of their origin. Hampton [60] proposed that India may have been the center of origin for the virus, and consequently also the center of PSbMV-immune germplasm. Despite this resistant germplasm, pea has not been systematically bred to include PSbMV resistance until recently; as a consequence, the vast majority of pea varieties are susceptible to this virus [31], especially when contrasted with landraces where the possibility of both natural and human selection for resistance exists. A relationship between pea host and PSbMV pathogen diversity in northern Pakistan and Afghanistan has been proposed [60,61]. More recently, phylogenetic analysis of PSbMV isolates from Australia and China grouped them on separate clades [55]. These results suggest the operation of co-evolutionary forces, which could be tested using the eIF4E alleles identified in this study. Interestingly, similar results were found in a study of barley eIF4E haplotype diversity, which was also found to be considerably higher in Central and East Asia, both regions with a long history of the bymovirus disease [3]. It is intriguing to see that in different crops, the resistance diversity center is outside the original place of domestication, especially in Central and Eastern Asia. It might be hypothesized that only some environments may be conducive for pathogen development in crops, hence co-evolution may be geographically limited. This agrees with results showing that the eIF4E gene is under positive selection pressure [3,44]. There is strong selection for the eIF4E^B allele in China and less elsewhere (Fig. 2, Table 2), but no A or C alleles in China, possibly consistent with a genetic bottleneck in the later migration of pea to China, compared with central/western Asia [18]. Moreover, assignment of Chinese accessions to provinces showed preferential occurrence of B alleles in the autumn/winter-sown provinces of Henan, Anhui, Hubei and Sichuan (Table S1) that experience high rainfall, long growing seasons and cold, frost-prone vegetative phases. Conversely, S alleles predominated in spring-sown types grown in northern provinces with low rainfall and high, variable temperatures that result in severe drought [62]. Analysis of pea germplasm diversity, using retrotransposon-based insertion polymorphism [13,14,63] showed that although there is substantial genetic diversity present, it is only partially geographically structured. Especially in case of the eIF4E^B alleles, there is substantial clustering of Indian and Ethiopian accessions, indicating a close relationship between materials from these areas (Fig. 5). The number of mutations makes independent origin unlikely, unless stringent positive selection is operating, as detected by Cavatorta [44]. We have also found evidence of positive selection in a set of 27 tested pea cDNA haplotypes (not shown). As proposed for barley eIF4E, the explanation for the unusually high overall degree of eIF4E sequence variation and haplotype diversity may be that these are important for adaptation to different local habitats [3]. The second, more plausible scenario is introgression and maintenance in populations by selective advantage. Based on the number of accesions with detected eIF4E^A and eIF4E^B alleles in the northern Indian subcontinent versus Ethiopia/eastern Africa, we speculate that these more likely originated in the Africa and were then brought by oversea trade to Afghanistan-Pakistan-Nepal-India region. The Silk Route, Amber Road and trans-Saharan trade routes were all instrumental in establishing links between Africa, India and beyond. This scenario, while not proven, is supported by a haplotype analysis network, where Indian eIF4E^A alleles were shown to be derived from Ethiopian accessions (Fig. 4). There is bias towards European origin accesions, comprising 40.8% of samples, followed by Ethiopia and East Africa (12.4%) and Russia (10.5%) which consequently affects proportion (57%) of S alleles (Table 2). However, these samples were included to provide more complete view on eIF4E^A diversity, including modern pea varieties. However this affects only

percentage in total dataset and not within studied regions (Fig. 2, Table 2).

Various approaches have been applied to identify variants of the eIF4E gene in several crops. These showed different levels of diversity [3,43,46,47]. In the most comprehensive study, 1090 barley samples from 84 countries screened using the high-resolution melting PCR method of cDNA, led to the identification of 47 eIF4E haplotypes [3]. In our study, we have identified 34 alleles on exon-intron level with 156 single-nucleotide polymorphic sites in total of 73 sequences. This suggests that substantial diversity within the eIF4E gene can be expected within available pea germplasm, which might be fully revealed only by sequencing. It is notable that none of the 146 tested wild pea samples had eIF4E^A or eIF4E^B alleles, but at the amino acid level, all corresponded to the susceptible eIF4E^{S-1} allele, despite substantial polymorphism both in introns and exons, indicating selection for functionality. The discovery of the intermediate allele in wild P. sativum subsp. elatius JI261 from Turkey is similar to data obtained in Arabidopsis thaliana [25], which suggests a distinct mode of evolution of resistance in wild species as opposed to crop species and supports the scenario in which potyviruses spread at the advent of agriculture, which brought plants into dense monocultures [48]. Very little is known about the selective forces that drive viral evolution in natural ecosystems, which contrasts with the more detailed population genetics studies in crop plants that have revealed the importance of mutation rates, recombination, genetic drift and migration in virus evolution [64]. Although we can not fully excluded the possibility, that JI261 allele could have resulted from domestic to wild introgression and not independent mutations, the analysis of 45 RBIP loci (Fig. 5) clustered it with typical P. s. subsp. elatius accessions [63]. The existence of resistance alleles only in the domesticated pea genepool leads us to speculate that the mutation originated during early cultivation. The series of alleles identified in this study provide an excellent basis for testing various potyviruses and pathotypes in which to study the co-evolution of potyviruses and their pea host.

Materials and Methods

Plant material

The 2803 pea accessions used in this study were drawn from the following collections: the Czech National Pea Collection (CzNPC, 1252 accessions); the USDA core collection (384); 119 accessions of Chinese origin from the Australian Temperate Field Crops Collection [(ATFCC) now reconstituted as Australian Grains Genebank, (AGG)], the Vavilov Institute (VIR) Russian Federation (69); the Centre for Genetic Resources (CGN), The Netherlands (14); Leibniz Institute of Plant Genetics and Crop Plant Research (IPK), Germany (35); and John Innes Centre (JIC), UK (836). The permission from the relevant institutions was obtained to access the collections, and the respective pea germplasm accessions were donated for this study under Standard Material Transfer Agreement of germplasm resources. Although we specifically targeted local landraces rather than globally grown, modern varieties, the analyses of 1252 accessions from the Czech National Pea collection represent the diversity present in the genepool of cultivated pea in the 19th to 21st centuries [16,49]. This collection is composed of 972 commercial varieties, 226 breeding lines and 54 landraces, which originate largely from Europe (925), the former Soviet Union (177), and the USA and Canada (77). In contrast, the John Innes Centre pea collection has, in addition to varieties (1071) and breeding lines (61), a large proportion of landraces (600), mutant stocks (585) and wild peas (445). Geographically, this collection is dominated by accessions of

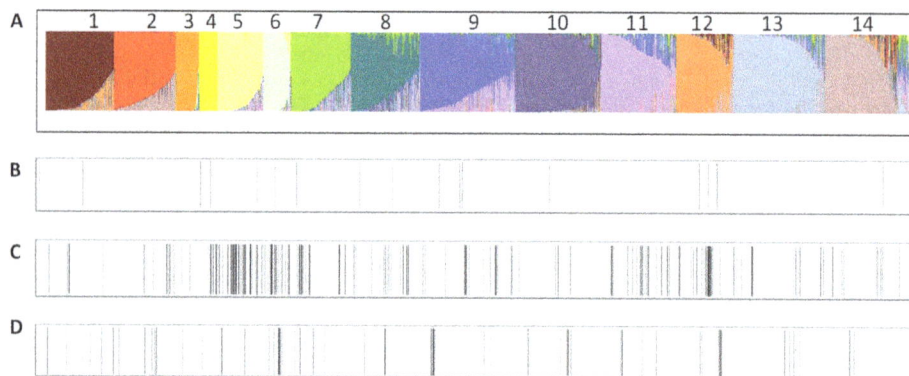

Figure 5. Visualization of genetic relationship of four identified *eIF4E* variants on the background of 14 BAPS identified clusters of 3,029 JIC accessions genotyped by 45 retrotransposon-insertion loci [13,63]. Notably, clusters 3 and 4 contain 145 acc. of wild *P. fulvum*, *P. s.* subsp. *elatius* and *P. abyssinicum*; cluster 7 contains 95 acc. from Afghanistan; cluster 8 has 225 acc. from Ethiopia; and cluster 9 contains 247 acc. from India and Ethiopia, while remaining clusters are more diverse (A). 18 accessions with the *eIF4E^A* (B), 241 accessions with the *eIF4E^B* (C), 81 accessions with the *eIF4E^C* variant (D).

Ethiopian (388), Mediterranean (199), Indian (53) and Chinese (37) origin. The Australian ATFCC and USDA pea collections are partially complementary. The former includes the diversity core set of Chinese origin [18], of which 119 were studied here, while the latter has over 6000 accessions, of which a core set of 384 [19] was analyzed in this study. Finally, 69 accessions selected from the VIR collection originated in the Caucasus and Central Asia (Turkmenistan, Kazakhstan, Georgia, and Armenia) as well as in North Africa (Morocco and Algeria). It is worth mentioning that although Ethiopia is known as the origin and occurrence for *P. abyssinicum*, the accessions used in this study were mainly *P. sativum* subsp. *sativum*. This set was complemented with wild *P. fulvum* (9), *P. sativum subsp. elatius* (86), *P. sativum subsp. sativum* (19) (formerly *P. humile/P.syriacum*) (see Smýkal et al. [13] for taxonomical classification) and *P. abyssinicum* (32) (Table S1). These represent both primary and secondary centers of pea diversity as well as primary and secondary gene pools.

DNA isolation and PCR analysis

Young leaves were harvested from ten (in the case of the Czech, ATFCC, CGN, IPK and VIR collections) or single (in the case of the JIC and USDA collections) randomly chosen plants per accession and stored at −80°C. Genomic DNA was isolated and PCR performed using standard protocols [31]. Products were resolved on 1.5% TBE agarose gel and visualized using ethidium bromide staining under UV-light.

eIF4E variants amplification

Intron 3 length polymorphism was used as screening criteria, as developed by Smýkal et al. [31]. In short, differences in intron 3 length were assayed using PCR with the following pairs of primers: *Ps*-eIF4E-750F (5′-GGACTAAGAATGCTTCAAAT-GAAGCTGC-3′) and *Ps*-eIF4E-586gR (5′-GAATCATTTAA-GAAGCTCGTGAAGTG-3′) primers (combination A, nested within the combination of B) that amplified 243 bp in the resistant and 293 bp in the susceptible accessions [31]; and (combination B) *Ps*-eIF4E-750F and *Ps*-eIF4E-1270R (5′-ATTCTCGATCACAC-TAGCCCCCTCC-3′) [35] primers that amplified 536 bp versus 586 bp fragment. The combination of these two assays resulted in the detection of respective *eIF4E^{A-B-C-S}* variants (Fig. 1, Table 1). PCR amplification of 1000 bp product from start codon to intron 3 was carried out using primers *Ps*-eIF4E-1F start codon (5′-ATGGTTGTAGAAGACACCCCCAAATC-3′) and *Ps*-eIF4E-

586gR. Primers *Ps*-eIF4E-794F (5′-GCTAGATGGTTGTTAT-GATGTTTATCAG-3′) and *Ps*-eIF4E-2188R stop codon (5′-TTGCTAGTTTGCTACCATGTAAGAACG-3′) were used to amplify a 1500 bp PCR product spanning the rest of *eIF4E* gene from intron 3 to exon 5. The PCR products were purified according to Werle et al. [50] protocol and sequenced using a BigDye Terminator kit (Applied Biosystems, UK) by Macrogene (Amsterdam, The Netherlands).

Bioinformatics

Primer design and restriction analyses were performed using FastPCR software version 5.1.83 (PrimerDigital Ltd., Finland). The DNA sequences were viewed and edited using Sequence Scanner version 2.0 (Applied Biosystems, UK). CLUSTALW alignment was performed using BioEdit version 7.09.0 [51]. The haplotype alignments were performed using a median-joining network algorithm, implemented in NETWORK 4.5.1.6 [52]. To reveal the genetic relationship of samples, we used previously made Bayesian clustering analysis of the genetic diversity of 3,029 JIC accessions genotyped by 45 retrotransposon-insertion loci [13]. Identified clusters were used as framework to visualize individual *eIF4E* alleles within the set of 836 JIC accessions analyzed in this study. Single nucleotide polymorphism mapping to reference sensitive allele was done using Geneious 6.1.6 analysis software (Biomatters, USA).

Virological testing

Plants were grown in a substrate Klassman no. 4 (Klasmann-Deilmann GmbH, Germany) in a growth chamber (Microclima 1000, Snijders Scientific, Holland) under a 16/8-h and 22/18°C day/night cycle. Evaluation of resistance/susceptibility to the PSbMV pathotype P-1 was conducted through mechanical inoculation using isolate PSB117CZ [31,53]. All together, 50 accessions were tested. Ten plants per accession were tested in same block, under the same conditions. Sensitive pea cultivars, Merkur and Raman, were used as a sensitive and B99 as a resistant controls [31]. Symptoms were observed at one-week intervals, and systemic infection was confirmed three weeks after infection with DAS-ELISA (Loewe Biochemica, Germany).

Supporting Information

Table S1 List of studied accessions, divided by 7 studied germplasm collection with indicated accession number, name, origin and *eIF4E* variant.

Table S2 Table of sequenced 73 accessions indicating single-nucleotide polymorphic (SNP) sites in exons and introns (SNP sheet) as well as resulting amino acid exchanges (protein sheet). Species name, germplasm, country of origin and *eIF4E* allele asignment are shown. Length of all introns and exons is shown in base pairs, with variable intron 3 being highlighted. Results of virological testing with P1 PSbMV are indicated as positive, susceptible (S) and resistant (R) reactions, respectively.

Acknowledgments

We thank Eva Fialová and Jana Veselá for technical assistance and numerous collegues for discussion over earlier versions of the manuscript.

Author Contributions

Conceived and designed the experiments: PS. Performed the experiments: EK PH CC DŠ PS. Analyzed the data: EK PS DŠ MN. Contributed reagents/materials/analysis tools: AF MA MV RR. Wrote the paper: EK MA RR CC PS.

References

1. Tanksley SD, McCouch SR (1997) Seed banks and molecular maps: Unlocking genetic potential from the wild. Science 277: 1063–1066.
2. Bhullar NK, Street K, Mackay M, Yahiaoui N, Keller B (2009) Unlocking wheat genetic resources for the molecular identification of previously undescribed functional alleles at the *Pm3* resistance locus. Proc Nat Academy of Sci USA 106: 9519–9524.
3. Hofinger BJ, Russell JR, Bass CHG, Baldwin T, dos Reis M, et al. (2011) An exceptionally high nucleotide and haplotype diversity and a signature of positive selection for the *eIF4E* resistance gene in barley are revealed by allele mining and phylogenetic analyses of natural populations. Mol Ecology 20: 3653–3668.
4. Nieto C, Piron F, Dalmais M, Marco CF, Moriones E, et al. (2007) EcoTILLING for the identification of allelic variants of melon eIF4E, a factor that controls virus susceptibility. BMC Plant Biology 7: 34.
5. Ibiza VP, Cañizares J, Nuez F (2010) EcoTILLING in *Capsicum* species: searching for new virus resistance. BMC Genomics 11: 631.
6. Reeves PA, Panella LW, Richards CHM (2012) Retention of agronomically important variation in germplasm core collections: implications for allele mining. Theor Appl Genet 124: 1155–1171.
7. Robaglia C, Caranta C (2006) Translation initiation factors: a weak link in plant RNA virus infection. Trends in Plant Sci 11: 40–45.
8. Bhullar NK, Zhang Z, Wicker T, Keller B (2010) Wheat gene bank accessions as a source of new alleles of the powdery mildew resistance gene *Pm3*: a large scale allele mining project. BMC Plant Biol 10: 88.
9. Chimwamurombe PM, Khulbe RK (2011) Domestication. In: Pratap A, Kumar J, editors. Biology and Breeding of Food Legumes. MA, USA: CABI, Cambridge. pp. 19–34.
10. Zohary D, Hopf M (1993) Domestication of Plants in the Old World—The Origin and Spread of Cultivated Plants in West Asia, Europe, and the Nile Valley. Oxford: Clarendon Press.
11. De Candolle A (1882) Origin of cultivated plants. Whitefish MT: Kesinger Publishing LCC.
12. Maxted N, Ambrose M (2001) *Peas (Pisum L.)*. In: Maxted N, Bennett SJ, editors. Plant Genetic Resources of Legumes in the Mediterranean. Dordrecht: Kluwer Academic Publishers. pp. 181–190.
13. Smýkal P, Kenicer G, Flavell AJ, Corander J, Kosterin O, et al. (2011) Phylogeny, phylogeography and genetic diversity of the *Pisum* genus. Plant Gen Res 9: 4–18.
14. Smýkal P, Aubert G, Burstin J, Coyne C, Ellis N, et al. (2012) Pea (*Pisum sativum* L.) in the genomic era. MDPI Agronomy 2: 74–115.
15. Smýkal P, Coyne C, Redden R, Maxted N (2013) Peas. In: Singh M, Bisht IS, editors. Genetic and Genomic Resources for Grain Legume Improvement. London: Elsevier Insights. pp. 41–80.
16. Smýkal P, Hybl M, Corander J, Jarkovsky J, Flavell AJ, et al. (2008) Genetic diversity and population structure of pea (*Pisum sativum* L.) varieties derived from combined retrotransposon, microsatellite and morphological marker analysis. Theor Appl Genet 117: 413–424.
17. Jing R, Ambrose MA, Knox MR, Smykal P, Hybl M, et al. (2012) Genetic diversity in European *Pisum* germplasm collections. Theor Appl Genet 125: 367–380.
18. Zong X, Redden R, Liu Q, Wang S, Guan J, et al. (2009) Analysis of a diverse global Pisum sp. collection and comparison to a Chinese local *P. sativum* collection with microsatellite markers. Theor Appl Gen 118: 193–204.
19. Kwon SJ, Brown AF, Hu J, McGee RJ, Watt CA, et al. (2012) Genetic diversity, population structure and genome-wide marker-trait association analysis emphasizing seed nutrients of the USDA pea (*Pisum sativum* L.) core collection. Genes & Genomics 34: 305–320.
20. Wang A, Krishnaswamy S (2012) Eukaryotic translation initiation factor 4E-mediated recessive resistance to plant viruses and its utility in crop improvement. Mol Plant Pat 13: 795–803.
21. Vavilov NI (1949-1950) The phytogeographic basis of plant breeding. In Chester KS trans. The origin, variation, immunity, and breeding of cultivated plants. Waltham MA: Chronica Botanica. pp. 13–54.
22. Abbo S, Lev-Yadun S, Gopher A (2012) Plant domestication and crop evolution in the Near East: on events and processes. Crit Rev in Plant Sci 31: 241–257.
23. Lovisolo O, Hull R, Rosler O (2003) Coevolution of viruses with hosts and vectors and possible paleontology. Adv Virus Res 62: 325–379.
24. Jones EI, Ferriere R, Bronstein JL (2009) Eco-evolutionary dynamics of mutualists and exploiters. American Naturalist 174: 780–794.
25. Le Gall O, Aranda MA, Caranta C (2011) Plant resistance to viruses mediated by translation initiation factors. In: Caranta C, Aranda MA, Tepfer M & López-Moya J, editors. Recent Advances in Plant Virology. Norfolk: Caister Academic Press. pp. 177–194.
26. Musil M (1966) Über das Vorkommen des Virus des Blattrollens der Erbse in der Slowakei (Vorlaufige Mitteilung). Biologia 21: 133–138.
27. Latham IJ, Jones RAC (2001) Alfalfa mosaic and pea seed-borne mosaic viruses in cool season crop, annual pasture, and forage legumes: susceptibility, sensitivity, and seed transmission. Austr J Agri Res 52: 771–790.
28. Hampton RO, Mink GI (1975) Pea seed-borne mosaic virus. CMI/AAB Descriptions of Plant Viruses. 146 n.
29. Alconero R, Weeden NF, Gonsalves D, Fox DT (1985) Loss of genetic diversity in pea germplasm by the elimination of individuals infected by pea seedborne mosaic virus. Ann of App Bio 106: 357–364.
30. Hjulsager CK, Olsen BS, Jensen DM, Cordea MI, Krath BN, et al. (2006) Multiple determinants in the coding region of pea seed-borne mosaic virus P3 are involved in virulence against *sbm-2* resistance. Virology 355: 52–61.
31. Smýkal P, Šafářová D, Navrátil M, Dostálová R (2010) Marker assisted pea breeding: *eIF4E* allele specific markers to pea seed-borne mosaic virus (PSbMV) resistance Mol Breeding 26: 425–438.
32. Bruun-Rasmussen M, Moller IS, Tulinius G, Hansen KR, Lund OS, et al. (2007) The same allele of translation initiation factor 4E mediates resistance against two Potyvirus spp. in *Pisum sativum*. Mol Plant Microbe Interact 20: 1075–1082.
33. Andrade C, Abe Y, Nakahara KS, Uyeda I (2009) The *cyv-2* resistance to Clover yellow vein virus in pea is controlled by the eukaryotic initiation factor 4E. J Gen Plant Pathol 75: 241–249.
34. Johansen EI, Lund OS, Hjulsager CK, Laursen J (2001) Recessive resistance in *Pisum sativum* and Potyvirus Pathotype resolved in a gene-for-cistron correspondence between host and virus. J Virology 75: 6609–6614.
35. Gao Z, Eyers S, Thomas C, Ellis N, Maule A (2004) Identification of markers tightly linked to sbm recessive genes for resistance to pea seed-borne mosaic virus. Theor Appl Genet 109: 488–494.
36. Olsen BS, Johansen IE (2001) Nucleotide sequence and infectious cDNA clone of the L1 isolate of Pea seed-borne mosaic potyvirus. Archives in Virology 146: 15–25.
37. Johansen IE, Rasmussen OF, Heide M, Borkhardt B (1991) The complete nucleotide sequence of pea seed-borne mosaic virus RNA. J Gen Virology 72: 2625–2632.
38. Johansen IE, Keller KE, Dougherty WG, Hampton RO (1996) Biological and molecular properties of a pathotype P-1 and a pathotype P-4 isolate of pea seed-borne mosaic virus. J Gen Virology 77: 1329–1333.
39. Ruffel S, Gallois J-L, Moury B, Robaglia C, Palloix A, et al. (2006) Simultaneous mutations in translation initiation factors eIF4E and eIF(iso)4E are required to prevent Pepper veinal mottle virus infection of pepper. J GenVirol 87: 2089–2098.
40. Sato M, Nakaharaa K, Yoshii M, Ishikawa M, Uyeda I (2005) Selective involvement of members of the eukaryotic initiation factor 4E family in the infection of *Arabidopsis thaliana* by potyviruses. FEBS Letters 579: 1167–1171.
41. Fraile A, Garcia-Arenal F (2010) The coevolution of plants and viruses: resistance and pathogenicity. Adv Virus Res 76: 1–32.
42. Ashby AJ, Stevenson CEM, Jarvis GE, Lawson DM, Maule AJ (2011) Structure-Based Mutational Analysis of eIF4E in Relation to sbm1 Resistance to Pea Seed-Borne Mosaic virus in pea. PLoS ONE 6: e15873.
43. Charron C, Nicolai M, Gallois JL, Robaglia C, Moury B, et al. (2008) Natural variation and functional analyses provide evidence for co-evolution between plant eIF4E and potyviral VPg. Plant J 54: 56–68.

44. Cavatorta JR, Savage AE, Yeam I, Gray SM, Jahn MM (2008) Positive Darwinian selection at single amino acid sites conferring plant virus resistance. J Mol Evol 67: 551–559.

45. Rubio M, Nicolaï M, Caranta C, Palloix A (2009) Allele mining in the pepper gene pool provided new complementation effects between pvr2-eIF4E and pvr6-eIF(iso)4E alleles for resistance to pepper veinal mottle virus. J GenVirol 90: 2808–2814.

46. Jeong HJ, Kwon JK, Pandeya D, Hwang J, Hoang NH, et al. (2012) A survey of natural and ethyl methane sulfonate-induced variations of eIF4E using high-resolution melting analysis in Capsicum. Mol Breeding 29: 349–360.

47. Rigola DJ, van Oeveren J, Janssen A, Bonne A, Schneiders H, et al. (2009) High throughput detection of induced mutations and natural variation using KeyPoint technology. PLoS ONE 4: e4761.

48. Gibbs AJ, Ohshima K, Phillips MJ, Gibbs MJ (2008) The Prehistory of Potyviruses: Their Initial Radiation Was during the Dawn of Agriculture. PLoS ONE 3: e2523.

49. Cieslarova J, Hýbl M, Griga M, Smýkal P (2012) Molecular Analysis of Temporal Genetic Structuring in Pea (Pisum sativum L.) Cultivars Bred in the Czech Republic and in Former Czechoslovakia Since the Mid-20th Century. Czech J Gen and Plant Breeding 48: 61–73.

50. Werle E, Schneider C, Renner M, Völker M, Fiehn W (1994) Convenient single-step, one tube purification of PCR products for direct sequencing. Nucleic Acids Research 22: 4354–4355.

51. Hall TA (1999) BioEdit: a user-friendly biologicalsequence alignment editor and analysis program for Windows 95/98/NT. Nucl Acids Symp Ser 41: 95–98.

52. Bandelt HJ, Forster P, Röhl A (1999) Median-joining networks for inferring intraspecific phylogenies. Mol Biol Evol 16: 37–48.

53. Šafářová D, Navrátil M, Petrusová J, Pokorný R, Piaková Z (2008) Genetic and biological diversity of the pea seed-borne mosaic virus isolates occurring in Czech Republic. Acta Virologica 52: 53–57.

54. Stein N, Perovic D, Kumlehn J, Pellio B, Stracke S, et al. (2005) The eukaryotic translation initiation factor 4E confers multiallelic recessive bymovirus resistance in Hordeum vulgare (L.). Plant J 42: 912–22.

55. Nicaise V, German-Retana S, Sanjuan R, Dubrana MP, Mazier M, et al. (2003) The eukaryotic translation initiation factor 4E controls lettuce susceptibility to the potyvirus Lettuce mosaic virus. Plant Physiol 132: 1272–1282.

56. Kang BC, Yeam I, Frantz DJ, Murphy JF, Jahn MM (2005) The pvr1 locus in pepper encodes a translation initiation factor eIF4E that interacts with tobacco etch virus VPg. Plant J 42: 392–405.

57. Hjulsager CK, Lund OS, Johansen E (2002) A new pathotype of pea seedborne mosaic virus explained by the properties of the p3-6k1 and viral genome-linked protein (VPg) coding regions. Mol Plant Microbe Interact 15: 169–171.

58. Wylie SJ, Coutts BA, Jones RAC (2011) Genetic variability of the coat protein sequence of pea seed-borne mosaic virus isolates and the current relationship between phylogenetic placement and resistance groups. Arch Virol 156: 1287–1290.

59. Hagedorn DJ, Gritton ET (1973) Inheritance of resistance to the pea seed-borne mosaic virus. Phytopathology 62: 1130–1133.

60. Hampton RO (1986) Geographic origin of pea seed-borne mosaic virus: a hypothesis. Pisum Newsletter 18: 22–26.

61. Ali A, Randles JW (2001) Genomic heterogeneity in Pea seed-borne mosaic virus isolates from Pakistan, the centre of diversity of the host species, Pisum sativum. Arch Virology 146: 1855–1870.

62. Ling Li, Redden RJ, Zong X, Berger JD, Bennett SJ (2013) Ecogeographic analysis of pea collection sites from China to determine potential sites with abiotic stresses. Genet Res Crop Evol 60: 1801–1815.

63. Jing R, Vershinin A, Grzebyta J, Shaw P, Smýkal P, et al. (2010) The genetic diversity and evolution of field pea (Pisum) studied by high throughput retrotransposon based insertion polymorphism (RBIP) marker analysis. BMC Evol Biol 10: 44.

64. Roossinck MJ (2011) The big unknown: plant virus biodiversity. Current Opinions in Virology 1: 63–67.

Allele Distributions at Hybrid Incompatibility Loci Facilitate the Potential for Gene Flow between Cultivated and Weedy Rice in the US

Stephanie M. Craig[1][¶], Michael Reagon[2][¶], Lauren E. Resnick[1], Ana L. Caicedo[1]*

1 Biology Department, University of Massachusetts, Amherst, Massachusetts, United States of America, 2 Department of Evolution, Ecology, and Organismal Biology, Ohio State University, Lima, Ohio, United States of America

Abstract

The accumulation of independent mutations over time in two populations often leads to reproductive isolation. Reproductive isolation between diverging populations may be reinforced by barriers that occur either pre- or postzygotically. Hybrid sterility is the most common form of postzygotic isolation in plants. Four postzygotic sterility loci, comprising three hybrid sterility systems (Sa, s5, DPL), have been recently identified in Oryza sativa. These loci explain, in part, the limited hybridization that occurs between the domesticated cultivated rice varieties, O. sativa spp. japonica and O. sativa spp. indica. In the United States, cultivated fields of japonica rice are often invaded by conspecific weeds that have been shown to be of indica origin. Crop-weed hybrids have been identified in crop fields, but at low frequencies. Here we examined the possible role of these hybrid incompatibility loci in the interaction between cultivated and weedy rice. We identified a novel allele at Sa that seemingly prevents loss of fertility in hybrids. Additionally, we found wide-compatibility type alleles at strikingly high frequencies at the Sa and s5 loci in weed groups, and a general lack of incompatible alleles between crops and weeds at the DPL loci. Our results suggest that weedy individuals, particularly those of the SH and BRH groups, should be able to freely hybridize with the local japonica crop, and that prezygotic factors, such as differences in flowering time, have been more important in limiting weed-crop gene flow in the past. As the selective landscape for weedy rice changes due to increased use of herbicide resistant strains of cultivated rice, the genetic barriers that hinder indica-japonica hybridization cannot be counted on to limit the flow of favorable crop genes into weeds.

Editor: Daniel Ortiz-Barrientos, The University of Queensland, St. Lucia, Australia

Funding: This project was funded by a grant from the U.S. National Science Foundation Plant Genome Research Program (IOS-1032023) to A.L.C., Kenneth M. Olsen and Yulin Jia (http://www.nsf.gov/awardsearch/showAward?AWD_ID=1032023). The funders had no role in study design, data collection and analysis, decision to publish, or preparation of the manuscript.

Competing Interests: The authors have declared that no competing interests exist.

* E-mail: caicedo@bio.umass.edu

¶ These authors are co-first authors on this work

Introduction

Population divergence, a critical step in the process of speciation, is often accompanied and reinforced by the evolution of reproductive isolating mechanisms, which can occur pre – or postzygotically. In plants, hybrid sterility is the most common form of postzygotic isolation [1]. Hybrid sterility is thought to evolve according to the Bateson-Dobzhansky-Muller (BDM) theory of speciation, which posits that independent mutations occurring in diverging populations become fixed and then interact negatively in the background of the hybrid [2].

The cultivated rice complex (Oryza sativa L.) affords a rare opportunity to investigate the evolutionary history and underlying genetics of traits influencing postzygotic barriers to hybridization. The sterility observed in crosses of two subspecies of Asian cultivated rice, O. sativa indica and japonica, is one of the most extensively studied of all hybrid incompatibilities in plants [3–6]. Indica and japonica cultivars were domesticated ~10,000 years ago and differ in various morphological characteristics and in their responses to a multitude of biotic and abiotic stresses [5]. Gene exchange between these rice subspecies would be highly beneficial

to rice breeding practices, however, full exploitation of hybrid rice is limited by the tendency of hybrids to exhibit some degree of sterility [7], which can vary from 5 to 95% depending on the cross [8]. The identification of sterility-causing loci between these cultivars and subsequent determination of their contributions to sterility has been a topic of much research. Despite the ~57 hybrid-incompatibility quantitative trait loci (QTL) detected so far in rice [9], only a few have been cloned and subjected to experimental testing. Four such loci, encompassing three hybrid sterility systems, are Sa, s5, and DOPPELGANGER1 (DPL1) and DOPPELGANGER2 (DPL2). Sa and DPL1/2 affect pollen viability and s5 sterility results in embryo-sac abortion [10–12].

All three of these hybrid sterility systems have been shown to cause semi-sterility in hybrids between indica and japonica cultivars, and could be partial contributors to the low levels of gene flow observed between these two rice subspecies. However, the possible roles of these loci in influencing gene flow between these main rice varieties and other Oryza groups have not been explored. There is potential for gene flow between indica and japonica cultivars and other Asian rice cultivars (e.g. aus, aromatic), the wild ancestor of cultivated rice (*O. rufipogon*), and the conspecific weed of cultivated rice known as weedy or red rice (*O. sativa*). Red rice is a troublesome weed that invades cultivated rice fields worldwide and displays competitive traits such as dormancy, high shattering, and rapid growth [13]. Weedy rice infestations can lead to a reduction of rice yields and considerable financial losses [14]. Gene flow between cultivated and weedy rice can have very negative agricultural consequences, such as the potential for crop traits to escape into weedy rice populations and unfavorable weedy traits contaminating seed stocks.

While weedy red rice is a worldwide problem of rice agriculture [13], the evolutionary origins of weedy rice and its relationship with the local rice crop differs throughout the world [15–17], affecting expectations of the potential for gene flow. In the US, local cultivated rice belongs to the *tropical japonica* variety of the *japonica* subspecies, while weedy rice is related to the *indica-aus* lineage. Two main genetically differentiated populations of red rice are known to co-occur in rice fields in the US: the most common straw hull (SH) group, which is characterized by straw colored grains, and the black hull awned (BHA) types, which typically have black colored grains with awns [16]. Studies of polymorphism have shown that the SH and BHA weedy groups are most closely related to, and likely descendant from, the *indica* and *aus* cultivated varieties, respectively [16], [18]. *Indica* and *aus* are closely related crop varieties, typically grown in lowland tropical regions of Asia, and distinct from the *tropical japonica* cultivars widely cultivated in the US. Further population structure is observed in the BHA group, which can be partitioned into two subpopulations (BHA1 and BHA2). Additionally, hybridization between SH and BHA groups has given rise to a group of weeds known as the BRH (brown hulled) group that occurs at lower frequency [16].

Although weedy rice is classified as the same species, is interfertile with, and co-occurs in fields of cultivated rice, weedy rice in the US shows limited hybridization with the local *japonica* crop [19]. Some hybrids between the local US *japonica* crop and SH or BHA weeds have been identified [16], as well as some evidence for past introgression [20] however there is little genetic evidence for extensive crop-weed hybridization. This may be due to self-pollination tendencies of cultivated and weedy rice [21] or differences in flowering time between crops and weeds in the field [13] (e.g. prezygotic mating barriers). However, since *indica* and *japonica* cultivars have been shown to experience limited hybrid compatibility due to various deleterious genetic interactions, it is possible that similar hybrid barriers limit the amount of outcrossing between weedy rice and the local *japonica* cultivar.

In this study we examine the allelic diversity of the characterized rice hybrid sterility loci (*Sa*, *s5*, and *DPLs*) in US weedy rice populations. We find very few barriers to intercrossing between weedy and cultivated rice at these loci, including the near fixation of a rare allele at the *Sa* locus, which seemingly confers wide-compatibility to some populations of weedy rice. Despite current low frequencies of hybrids in US rice fields, our results suggest that no postzygotic barriers should prevent widespread gene flow between weedy and cultivated rice.

Materials and Methods

Plant Material

Diverse *Oryza* seeds obtained from the United States Department of Agriculture (USDA), the International Rice Research Institute (IRRI), collections contributed by Dr. David Gealy of the Dale Bumpers National Rice Research Center, and Susan McCouch of Cornell University (Table S1) were grown at the University of Massachusetts Amherst. Our panel consisted of 107 individuals from multiple *Oryza* species, including weedy rice (51), *O. rufipogon* (25), *O. nivara* (2) and various *O. sativa* cultivars including *aus* (7), *indica* (10), and *japonica* (10); this latter group contained both US and Asian cultivars. Other AA genome *Oryza* species, *O. meridionalis* (1) and *O. glaberrima* (1), were included as outgroups (Table S1). The weedy groups used in our panel were previously defined by Reagon *et al.* [16] based on 48 sequence tagged site (STS) markers, and consisted of the main weedy groups in the US that have putative *indica* and *aus* ancestry (SH, BHA1 and BHA2), the BRH group believed to be a hybrid between SH and BHA weeds, and rarer weedy individuals classified as MX, which are likely early generation hybrids between the main weedy groups and the local *japonica* crop. Twenty *aus* and thirty *indica* individuals obtained from USDA Genetic Stocks – *Oryza* Collection (GSOR) and IRRI were later added to our survey for further genotyping at the *s5* and *SaF* loci (Table S2).

Sequencing and Genotyping

DNA was extracted from leaf material of all accessions using a CTAB method. Primers were designed using Primer3 [22] to amplify portions of each hybrid sterility locus, taking into account previously described causal polymorphisms at each locus [10–12] (Table S3). *Indica* or *japonica* type alleles were genotyped for all individuals at each locus using either DNA sequencing, when alleles were differentiated by a single nucleotide polymorphism (SNP), or differential gel migration, when alleles could be visualized by a size difference due to insertion-deletions (indels). DNA sequences were aligned and edited using BioLign Version 2.09.1 (Tom Hall, NC State University). DNA sequences obtained were deposited into GenBank as population datasets under accession numbers KF892880–KF893259.

Genetic Diversity and Phylogenetic Analysis

Summary statistics for each sequenced locus were obtained with DnaSP version 5.0 [23]. Statistics included Watterson's estimator of nucleotide variation (θ_W), the average pairwise nucleotide diversity (π) [24] and Tajima's D (TD) [25]. Summary statistics for each locus were compared against 48 genome-representative STS loci [16] for outlier behavior. Heterozygotes were phased using the haplotype subtraction method [26]. Genealogical relationships among sequenced haplotypes at each locus were determined with Neighbor-Joining analyses using a Kimura-2-parameter model in MEGA5 [27]. For all loci, the Nipponbare sequence, the *temperate japonica* accession with a sequenced genome, was included as haplotype 1. Weedy accessions were examined for novel alleles, clade membership expectations based on known ancestry, and the likelihood of introgression with US cultivars based on genotypes at each locus. For simplicity, in the remainder of the manuscript the term "haplotype" is used to refer to the DNA sequence content of a given allele, while "allele-type" refers to the functional classification of an allele as *indica*-type, *japonica*-type, or (in some cases) wide compatibility-type (Table S4 and Table S5).

Crosses and Quantifying Pollen Viability

Crosses were performed between *Oryza* accessions to compare pollen production among individuals with different *Sa* genotypes. Parents were planted in January 2012 and were grown in a walk-in Conviron PGW36 growth chamber at the University of Massachusetts Amherst under 11 hour days at 25°C. Panicles newly emerged from the boot, but not dehisced were chosen as the female. The top of each floret was cut off and the anthers removed with forceps, leaving the stigma intact; 20–30 florets were cut per panicle. A panicle on the verge of dehiscence was chosen to be the male parent. Both male and female panicles were placed in a single glycine bag and secured with a paperclip. Bags were collected after one month to check for hybrid seed. F_1 seeds were heat-treated overnight at 34°C and then either plated on a petri dish or planted in soil. Heterozygosity of the F_1 was confirmed *via* PCR.

Both homozygous parents of various *Sa* genotypes, as well as hybrid offspring were examined for pollen quality. For each sample, pollen grains of six anthers, from six different florets were suspended in 100 μl of Lugol's Solution (LS), and then serially diluted in 90 μl of LS. Pollen viability was quantified by obtaining a nonviable-to-viable ratio of pollen observed under a Leica MZ 16FA microscope using LS, a potassium iodide stain that reacts with starch in viable cells dying them black; non-viable cells do not stain and appear clear. Three ratios were calculated using ImageJ (http://rsb.info.nih.gov/ij) within a fixed area from three fields of view, and averaged for percent viable pollen.

The quantity of pollen produced was measured using a 0.5 mm deep Nageotte Bright Line Hemacytometer, resulting in a pollen grain/μl concentration. Three measurements were obtained per individual and averaged to calculate mean pollen quantity.

Results

The Genealogy of *s5*

S5 encodes an aspartic protease on chromosome six that is expressed in ovule tissue. Three allele types have been described: an *indica*-type (*s5-i*), a *japonica*-type (*s5-j*) and a wide compatibility allele (*s5-n*) [10]. SNPs C282A and C877T differentiate *s5-i* and *s5-j* alleles, while a 136 bp deletion in the N-terminus of *s5-n* renders it non-functional and confers wide-compatibility [10]. The dimerization of *s5-i* and *s5-j* causes embryo sac abortion and has been reported to reduce spikelet fertility by 46% [10].

We sequenced a total of 1897 bp for *s5*, beginning 949 bp downstream of the start codon in the second exon to 1016 bp after the stop codon, in samples of weedy, wild and cultivated rice. Since sequencing of outgroups failed repeatedly for this locus, we included the published sequence of an *O. barthii* (Gen Bank Accession # JF298922). Thirty-five SNPs and five indels were identified within this region encompassing thirty-four haplotypes (Figure 1A; Table S4, Table S5A). We classified alleles as either *indica*- or *japonica*-type based on the SNPs reported by Chen *et. al* [10]. *Indica*-type alleles were further grouped based on whether they contained the wide-compatibility deletion (Table S4).

Previous analysis of the *s5* locus in wild groups found the majority of wild accessions to carry *indica*-type alleles [28]. Likewise, 97% of wild alleles in our study (29/30) were *indica*-type (Table 1). Du *et al.* [28] also found *indica* and *japonica*-type alleles fixed within their respective groups, although the *aus* group was not explicitly characterized. We found that all *indica* and *aus* individuals in our panel carried *indica*-type alleles, while only 62.5% (5/8) of *japonica* individuals carried the expected *japonica* allele types (Table 1). The wide-compatibility deletion was detected in 50% of cultivars possessing *indica*-type alleles, including one grown in the US (sus02), and in 14% of wild *indica*-type alleles (Figure 1A, Table 1, and Table S4). Interestingly, no *indica* individuals surveyed possessed the wide-compatibility deletion, but all *aus* individuals did (Figure 1A, Table 1).

As expected based on their putative ancestry, all weedy BHA1, SH, and BRH individuals examined possessed *indica*-type alleles (Table 1, Table S4). All MX weeds also possessed *indica*-type alleles, despite their mixed weed x *japonica* ancestry. The only weedy group not fixed for *indica*-type alleles was BHA2, in which

37% (3/8) of the individuals carried *japonica*-type alleles (Table 1 and Table S4).

Despite the wide occurrence of *indica*-type alleles, not all sequenced weedy haplotypes were identical to those detected in cultivars. Nine novel haplotypes were detected in the weedy populations, (Haplotypes 2–8, 14 and 15) (Figure 1A). These haplotypes are not shared with wild or cultivated individuals in our panel, and, except for haplotype 2 [28], have not been previously reported in the literature. Remarkably, not a single SH weed sequenced carried an identical haplotype to any of the *indica*, its putative progenitor group, in our panel (Figure 1A, Table S4).

Regardless of population, most weeds (87%) contained the wide compatibility deletion (*s5-n* allele; Figure 1A, Table 1). This deletion occurred in only six of our cultivated individuals, all from the *aus* and *japonica* cultivar groups, and, unlike their putative *indica* progenitors, SH weeds often have the deletion. Based on this finding, we further explored the possible origins of *s5-n* by genotyping an additional 20 *aus* and 30 *indica* accessions from south Asia (Table S2). The deletion was very common in *aus*, detected in 19/20 individuals. The deletion was rare in *indica* occurring in only six individuals, mostly from Nepal. The widespread presence of the wide-compatibility allele in US weedy rice groups suggests that this locus poses no postzygotic barrier to hybridization with US crops or hybridization between weedy groups.

Genealogy and Allelic Distribution at the *DPL* loci

DPL1 (chromosome 1) and *DPL2* (chromosome 6) are paralogous hybrid incompatibility genes that encode small plant proteins and are highly expressed in mature anthers [12]. *Japonica* cultivars have been described as containing a functional copy of *DPL1* (*DPL1-N+*), and a non-functional allele of *DPL2* (*DPL2-N−*) due to a SNP at A434G [12]. *Indica* and *aus* cultivars have been described as carrying a non-functional allele of *DPL1* (*DPL1*-K−) due to a 517 bp insertion 204 bp downstream of the start codon, and a functional copy of *DPL2* (*DPL2*-K+) [12]. Pollen carrying non-functional alleles at both loci (e.g. *DPL1*-K− *DPL1*-K−// *DPL2-N− DPL2-N−*) is non-viable [12].

Allelic Distribution of *DPL1*. *DPL1* was genotyped for functionality based on the presence/absence of the 517 bp insertion (Mizuta *et. al* 2010). Individuals without the insertion were categorized as functional (*N+*), and individuals with the insertion were categorized as non-functional (*K−*) (Table S4). Previously, the presumably ancestral functional *DPL1* alleles were found to be more prevalent in wild rice populations [12]. Similarly, we found *DPL1-N+* alleles were present at higher frequencies in our *O. rufipogon/nivara* sample (56%) (Table 1, Table S4). Within cultivated populations, Mizuta *et al.* [12] found both alleles at equal frequencies in *indica*; however we found the *DPL1-K−* allele at higher frequency than the *DPL1-N+* allele in *indica* (4/6), and fixed in all *aus* individuals surveyed (Table 1). Mizuta *et al.* [12] described the *japonica* group as fixed for *DPL1-N+* alleles. Unexpectedly, we found 21% of our *japonica* individuals to carry nonfunctional *DPL1-K−* alleles; however, among surveyed US *japonica* cultivars, all but one carried *DPL1-N+* alleles (Table 1, Table S4).

All BHA2 weedy individuals possessed the non-functional *indica*-type allele (*DPL1-K−*), which is consistent with their *aus* ancestry. However, three BHA1 individuals, a group that also has *aus* ancestry, carried *DPL1-N+*, which we did not find in any *aus* individual. MX and SH populations only had *DPL1-N+* alleles, even though we found this allele at lower frequency within *indica*, and the BRH group also carried predominantly *DPL1-N+* alleles (Table 1, Table S4).

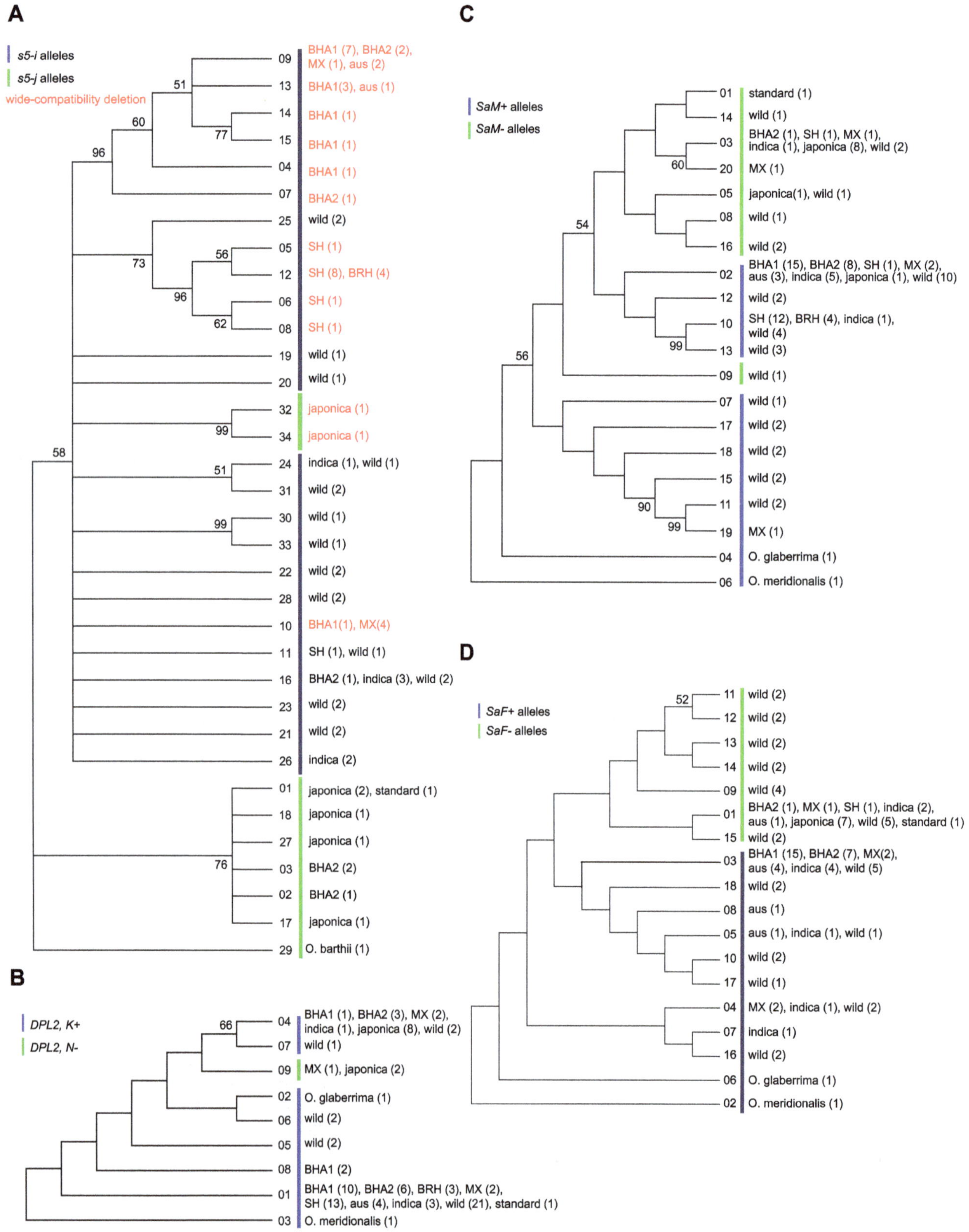

Figure 1. Neighbor joining trees of hybrid-incompatibility loci sequenced haplotypes. Numbers on branches correspond to bootstrap percentages from 500 replicates. Bootstrap values below 50 are not shown. Tip labels correspond to haplotype numbers as in Table S4, and to *Oryza* groups in which the haplotype was found. Numbers in parentheses correspond to the number of alleles found for each haplotype. Green bars designate haplotypes reported as typically from the *japonica* group, and blue bars designate haplotypes typical of the *indica* group. A. Haplotype tree for *s5*. B. Haplotype tree for *DPL2*. C. Haplotype tree for *SaM*. D. Haplotype tree for *SaF*.

Table 1. Frequency (in percentage) of allele types found at each population and each hybrid incompatibility locus in our core set of accessions.

Locus	Allele Type	Oryza group			SH	BHA1	BHA2	MX	BRH	Wild	Total samples
		indica	aus	japonica							
s5	s5-i	100	100	37	100	100	63	100	100	97	
	wc#	0	100	100	86	100	80	100	100	14	
	s5-j	0	0	63	0	0	37	0	0	3	
	alleles sampled*	6	3	8	14	14	8	5	4	30	92
DPL1	DPL1-K−	67	100	21	0	77	100	0	25	44	
	DPL1-N+	33	0	79	100	23	0	100	75	56	
	alleles sampled	6	5	14	11	13	7	5	4	32	97
DPL2	DPL2-K+	100	100	85	100	100	100	80	100	100	
	DPL2-N−	0	0	15	0	0	0	20	0	0	
	alleles sampled	4	4	13	13	13	9	5	3	28	92
SaM	SaM+	89	80	10	13	100	89	80	20	66	
	SaM−	11	20	90	6	0	11	20	0	18	
	SaM+X	0	0	0	75	0	0	0	60	16	
	SaM-X	0	0	0	6	0	0	0	20	0	
	alleles sampled	9	5	10	16	15	9	5	5	38	112
SaF	SaF+	78	86	0	0	100	88	80	0	39	
	SaF−	22	14	100	7	0	12	20	0	43	
	SaFX	0	0	0	93	0	0	0	100	18	
	alleles sampled	9	7	8	14	15	8	5	4	44	114

#wc = wide compatibility; wide compatibility percentages are out of total indica-type alleles.
*Due to differences in mating system, O. rufipogon genotypes are considered diploid; all other samples are considered as haploid, except in cases of rare heterozygotes.

The Genealogy of DPL2. We sequenced 499 bp of DPL2, encompassing from 16 bp into the 1st exon through 19 bp into the second exon of the gene. Four indels and 22 SNPs were found in our sample, encompassing a total of nine haplotypes (Figure 1B; Table S5B). DPL2-K+ has previously been found to be more frequent in wild populations [12]. Accordingly, this allele was fixed within our wild accessions (Table 1; Table S4). In cultivated populations, the non-functional DPL2-N− and functional DPL2-K+ alleles have previously been reported as fixed within the japonica and indica populations, respectively. The DPL2-K+ allele was also fixed in our indica samples, but, strikingly, the DPL2-N− allele was rare in our japonica individuals, with only 15% (2/13) carrying japonica-like alleles (Table 1; Table S4). Moreover, only one of our US japonica individuals carried DPL2-N− (Table 1).

Similar to wild and domesticated groups, the indica-type allele (DPL2-K+) was nearly fixed in weedy groups (98%, 42/43). The only japonica-type allele was carried by one MX weed thought to be a putative hybrid between an SH weed and a japonica cultivar [16] (Table 1; Table S4). The haplotype carried by this weed, haplotype 9, is the only japonica-type haplotype we found in all our sequences, indicating low sequence diversity for DPL2-N− in contrast with the ancestral DPL2-K+ alleles (Figure 1B). The remaining weedy groups fell within three haplotypes, 1, 4, and 8, with weedy individuals grouping with their domesticated progenitors in haplotypes 1 and 4 (Figure 1B). Haplotype 8, a novel haplotype that occurs in two BHA1 individuals, is characterized by a 1 bp deletion in the coding region of DPL2, which could lead to a non-functional gene.

The Distribution of DPL Genotypes. For simplicity, for hybrid incompatibility loci consisting of two genes, we will list genotypes in haploid format in the remainder of this manuscript; only in the case of hybrid progeny/heterozygotes will we write out the complete diploid genotype. Only two genotypes across both DPL loci were detected in our wild populations: DPL1-K−/DPL2-K+ (42%) and DPL1-N+/DPL2-K+ (58%) (Table S4). Typical japonica cultivars are reported to carry DPL1-N+/DPL2-N− [12], but the predominant genotype within our japonica panel was DPL1-N+/DPL2-K+. Only two of our japonica individuals (one from the US) possessed the genotype expected of typical japonicas, suggesting that previous characterizations of DPL loci may have been too narrow (Table S4). Mizuta et al. [12] found that half of the indica population had DPL1-K−/DPL2-K+ genotypes, similar to what we found in our indica panel (Table S4). This genotype was also fixed among our aus cultivars and was found in three japonica individuals within our panel (Table S4).

In weedy populations, the DPL1-N+/DPL2-K+ genotype, with functional alleles at both loci, was fixed in SH, nearly fixed in MX and BRH individuals, and present in three individuals from the BHA1 group (Table S4). The DPL1-K−/DPL2-K+ genotype was fixed in BHA2 and present at high frequency (73%) in the BHA1 group (Table S4). Given the high frequency of the DPL1-N+/DPL2-K+ genotype in US cultivars as well as several of the weed groups, the DPL loci do not seem to present a barrier to gene flow between cultivated and weedy rice in the US. Only crosses between DPL1-K− carrying weeds (primarily in the BHA groups) and the very rare DPL2-N− carrying cultivars (only 1 out of 8 surveyed cultivars possessed this allele type), would be expected to result in the DPL1-K−/DPL2-N− sterility-causing genotype.

Genealogical Relationships at the *Sa* Locus

The *Sa* locus comprises two adjacent genes on chromosome 1, *SaM* and *SaF*. *SaM* encodes a small ubiquitin-like modifier E3 ligase-like protein and *SaF* encodes an F-box protein [11]. *Indica* cultivars typically possess a *SaM+/SaF+* genotype, while *japonica* cultivars have a derived *SaM−/SaF−* genotype. A *SaM* heterozygote and a *SaF+* allele are required to cause male semi-sterility (usually about 50%; [11]). *SaM+* and *SaM−* are differentiated by a G-to-T polymorphism in the fifth intron, resulting in a truncated protein [11]. *SaF+* and *SaF−* are differentiated by a C-to-T transition at position 287 that leads to an amino acid change. Mechanistically, during the uninucleate stage of gamete development, direct interaction of SaF+ with SaM− causes selective abortion of SaM− bearing microspores [11].

The Genealogy of *SaM*. We amplified a 634 bp portion of *SaM* starting four bp into the 4th exon through 44 bp downstream of 5th exon. Within this portion of the gene, we found seven indels and 20 SNPs, distributed among 20 haplotypes (Table S5C). Alleles were further classified as either *indica* (*SaM+*) or *japonica* (*SaM−*)-like based on differentiating polymorphisms assigned by [11] (Table S4). Consistent with two previous studies [11], [29], we found 82% of wild alleles in our panel were *SaM+* (Table 1; Table S4). The *SaM+* allele has also been documented as the most common in *indica* cultivars [11], [29]. Likewise, 89% of our *indica* individuals carried the *SaM+* allele (Figure 1C, Table 1). The *SaM−* allele is nearly fixed in all *japonica* populations reported to date [11], [29]. Consistently, we found only one *japonica* individual (str02) with a *SaM+* allele in our sample. Consistent with its close relationship with *indica*, we found that four of five *aus* individuals had the *SaM+* allele (Table 1; Table S4).

As expected based on US weed ancestry, most of our weed alleles (90% or 45/50) were *SaM+*. Only four individuals from several weedy groups (BHA2, SH, and MX) possessed the *SaM−* allele (Table 1; Table S4). Weeds largely possessed haplotypes identical to those in progenitor groups or other cultivated groups (Figure 1C). Two novel haplotypes were observed in the MX groups (19 and 20) (Figure 1C).

The Genealogy of *SaF*. We amplified a 1.3 kb portion of *SaF* starting 674 bp into the first exon through 357 bp into the 3rd exon. Repeated amplification failures were observed mostly in SH (15/16) and BRH (5/5) groups, so initial genotyping was carried out in the remaining *Oryza* groups (Table S4). Within this region, we found one indel and 24 SNPs and 18 different haplotypes (Figure 1D; Table S5D). Alleles were further classified as either *indica* (*SaF+*) or *japonica* (*SaF−*)-like based on differentiating polymorphisms assigned by [11] (Table S4).

Previous reports found wild populations to carry both allele types at relatively equal frequencies, which was consistent with our findings among individuals that amplified (Table 1). Twenty-two percent of our *indica* individuals had a *SaF−* allele, while previous reports found 10% of *indica* individuals carrying this *japonica*-type allele [11]. *SaF+* was nearly fixed within our *aus* panel (Figure 1D, Table 1; Table S4). *SaF−* was fixed in our *japonica* samples (Table 1; Table S4), consistent with previous studies [11], [29].

Among samples that amplified, we found that 90% of weedy individuals (26/29) carried *SaF+* alleles (Table S4). Consistent with their *aus* ancestry, *SaF+* is fixed in BHA1, and is found in 87.5% of BHA2 individuals (Table 1; Table S4). The majority of MX weeds also carry *SaF+* alleles (Table 1; Table S4). The three weedy individuals detected with *japonica*-type alleles fall into haplotype 1, the most frequent *japonica*-type haplotype, and also the only *japonica*-type haplotype we detected in *indica* cultivars (Figure 1D).

A Novel *SaF* Allele: *SaFX*. Early attempts to sequence the *SaF* gene were complicated by amplification failures for the

majority of samples belonging to the SH and BRH weedy groups and a few other *Oryza* samples. We obtained the whole-genome sequence of a single SH individual (Young and Caicedo, unpublished information), and noted a large deletion spanning the entire *SaF* gene. To determine the exact genomic boundaries of the *SaF* deletion, we designed primers on both sides of the inferred deletion breakpoints (Table S3) and amplified and sequenced four SH samples. We found that the deletion spans 8,628 bp (from coordinates 22,371,187–22,379,815 of chromosome 1 in the MSU 6.0 rice genome) and begins 2,902 bp upstream from the start of *SaF*. This deletion partially knocks out the gene *Os01g39660*, a putative transposon protein, as well as the first four exons (1,240 bp) of *SaM*, the gene located immediately downstream of *SaF* (Figure 2).

We found no evidence of this deletion having been reported before, and therefore named the resulting *SaF* deletion allele *SaFX*. To genotype for the *SaFX* allele, a complementary set of primers that amplified specifically only in the presence (SaFdel primers) or absence (SaF primers) of the deletion was used (Table S3). We found that 61% of individuals that failed amplification for *SaF* (20/33) amplified with our *SaFX* primers (Table 1; Table S4).

The boundary of the *SaFX* deletion extends four exons into *SaM*, but prior to the *SaM+/SaM−* differentiating SNP in the fifth intron (Figure 2). Taking this into account, we re-assigned *SaM* alleles in individuals containing *SaFX* as either *SaM+X* or *SaM-X*, to indicate which type of allele it carries (*indica* (+) or *japonica* (−)) and that a large portion of the gene is missing (Table 1; Table S4). Only two individuals with *SaM-X* alleles were found (Table S4), and the deletion breakpoint seemed identical as in the *SaM+X* alleles.

Given the frequency of this deletion in the SH groups, we expected to find this deletion in both cultivated and wild groups, particularly in *indica*, which is believed to be ancestral to the SH weed group. While the deletion was found in six wild individuals (Table 1; Table S4), we found no individuals with the deletion among the *indica* included in our initial panel. We thus expanded our sample set to genotype the same 50 individuals belonging to the *aus* and *indica* cultivar groups that were genotyped for *s5*. The *SaFX* deletion was detected only in four *indica* individuals from Nepal (3) and India (1) and one *aus* cultivar from India, suggesting a low frequency of this allele in wild and cultivated *Oryza* groups (Table S2).

The Distribution of *Sa* Genotypes. Wild populations have previously been reported to carry primarily *SaM+/SaF+* genotypes [29]. Of the 36 wild samples with genotypes at both loci, 47% carried *SaM+/SaF+* and 14% carried *SaM−/SaF−* (Table 1; Table S4). However, we also detected novel allelic combinations of *SaM−/SaF+* (5%), *SaM+/SaF−* (17%) and *SaM+X/SaFX* (17%) (Table S4). We believe that *SaFX* arose in a *SaM+* background, since we found it only in combination with *SaM+X* alleles in wild individuals; however we did find the *SaFX/SaM-X* combination in two weeds (Table S4). Consistent with previous research [11] the *SaM−/SaF−* genotype was nearly fixed in our *japonica* individuals, including the US cultivars surveyed (Table S4). Likewise, the typical *indica* genotype, *SaM+/SaF+*, was found to be nearly fixed in our *indica* and *aus* panels as well (Table S4).

Consistent with *aus* ancestry, *SaM+/SaF+* is fixed and nearly fixed in BHA1 and BHA2, respectively. Given the common occurrence of *SaM−/SaF−* genotypes in US cultivars, interactions at this locus could limit hybridization with BHA weeds. *SaM+X/SaFX* was nearly fixed in the BRH and SH groups. However, we were not able to predict how *SaFX* affects hybrid sterility levels in crop-weed hybrids since this is the first report of this allele. We

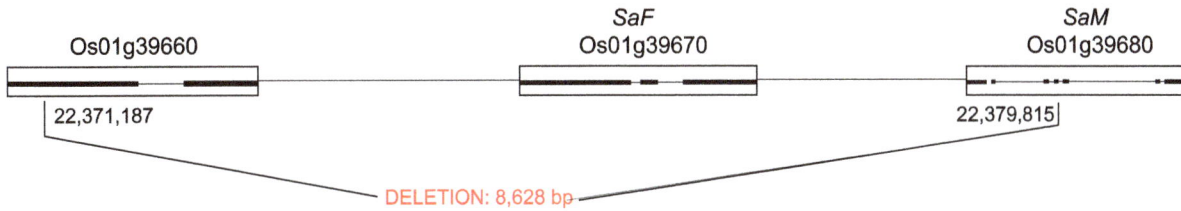

Figure 2. Diagrammatic representation of the 8,628 bp *SaFX* deletion. The deletion begins within the gene upstream of *SaF*, a putative transposon protein, and knocks out the first four exons of *SaM*.

speculate that *SaFX* is a non-functional allele since it knocks out the entire *SaF* gene and four exons of *SaM* (see below).

The Consequences of *SaFX* on Pollen Viability

Because the *SaFX* gene has not been reported in the literature, we attempted to assess the phenotypic consequences of carrying this allele on pollen production. We designed crosses to test the phenotypic consequences of *SaFX* in various genetic backgrounds (Table 2). We quantified both the quantity and quality of pollen produced in parental individuals and in crosses performed between these parentals.

We assessed five common parental genotypes: *SaM+/SaF+*, *SaM+/SaF−*, *SaM−/SaF−*, *SaM+X/SaFX*, and *SaM-X/SaFX*, which were present within our panel. Ideally, our goal was to have each genotype be both the male and female parent, but varied flowering dates prevented us from making all intended crosses (Table 2). The resulting, verified hybrid genotypes we obtained are shown in Table 2.

Pollen Quality. Non-viable pollen ratios obtained for each individual are shown in Table 3. The parental individual with the highest non-viable ratio possessed a *SaM-X/SaFX* genotype with an average of 44% non-viable pollen (Table 3). However, another individual with the same genotype only had 6.8% non-viable pollen. The homozygous genotype that produced the most viable pollen was *SaM+/SaF−* with an average of 5.6% non-viability, but variance was high among individuals within the same genotype (Table 3). We performed a t-test comparing the parental individuals with a *SaFX* allele to those without *SaFX* and found no effect of *SaFX* on pollen quality ($P = 0.9$).

SaFX does not seem to affect pollen quality in the resulting F_1 hybrids, though our small sample size limits our statistical power. The typical sterility-causing genotype at *Sa*, a *SaM* heterozygote and at least one *SaF+* allele, usually leads to a 50% decrease in total pollen viability [11]. Such drastic decreases in pollen viability were not seen in any of our F_1 (Table 3). Non-viable ratios among the progeny do not differ significantly from parental values ($P = 0.35$). Our hypothesis of *SaM* being non-functional due to the *SaFX* deletion is supported by the observation of no decrease in pollen quality in our *SaM+X/SaM−//SaF+/SaFX* hybrid, as a

functional *SaM* heterozygote and *SaF+* allele should have resulted in a substantial decrease in total pollen viability.

Pollen Quantity. We also determined if the amount of pollen produced was affected by the presence of a *SaFX* allele (Table 4). There was considerable variation in the amount of pollen produced among parents regardless of allele type (Table 4). The highest pollen-producing individual had a *SaM−/SaF−* genotype (97.5 pollen grains/µl) and the lowest was *SaM−/SaFX* (14.1 pollen grains/µl) (Table 4). We found no statistically significant differences in pollen production between parents with and without the *SaFX* deletion (t-test, $P = 0.54$).

There was no obvious effect on pollen production in the hybrid progeny carrying the *SaFX* allele. Pollen production ranged from close to the highest observed in parents (*SaM+/SaM-X//SaF−/SaFX*), to among the lower values (*SaM−/SaM+X//SaF−/SaFX*) (Table 4). Hybrid pollen production did not differ significantly from the parents ($P = 0.44$). As before, the *SaM+X/SaM−//SaF+/SaFX* hybrid did not display any evidence of reduced pollen production, suggesting that a functional *SaM+* protein is not being produced due to the deletion.

Summary Statistics

Genetic diversity statistics were calculated for all members of the main weedy populations, as well as *indica*, *japonica*, *aus* and wild populations at each locus (Table 5). We attempted to look for any common patterns in polymorphism trends across the three hybrid incompatibility loci.

A previous study based on STS loci among *Oryza* and weedy groups and found wild populations to harbor the most genetic diversity, followed by intermediate levels in the cultivars and low levels in US weedy groups, due to a genetic bottleneck upon US colonization [16]. In our study, wild populations were the most diverse, as expected, except at *s5* and *DPL2*, where BHA2 and *japonica* have the highest levels of nucleotide diversity (π), respectively (Table 5). Levels of diversity for hybrid incompatibility genes in weed groups varied among loci, but were sometimes an order of magnitude larger than the genome-wide averages. Particularly noticeable were the high levels of diversity observed in the BHA groups at *s5*. At this locus, all three weed groups surpassed the levels of diversity seen in their putative cultivated

Table 2. Successful crosses carried out to assess the effects of the *SaFX* allele.

Cross	Parental 1 genotype	Parental 1 accession	Parental 2 genotype	Parental 2 accession	F1 genotype
1	*SaM+X/SaFX*	rr07	*SaM−/SaF−*	rr06	*SaM+X/SaM−//SaFX/SaF−*
2	*SaM-X/SaFX*	rr15	*SaM+/SaF−*	or18	*SaM-X/SaM+//SaFX/SaF−*
3	*SaM-X/SaFX*	rr33	*SaM+/SaF+*	rr03	*SaM-X/SaM+//SaFX/SaF+*

Table 3. Pollen viability in individuals with different *Sa* genotypes.

Genotype[#]	Accession	Average % Non-Viable pollen*	Genotype Average (sd)^
SaM+/SaF+	rr03	8.4	
	sin02	6.1	
	rr39	8.8	
			7.7 (6.5)
SaM−/SaF−	rr22	12.3	
	rr22	29.5	
	rr06	19	
	rr53	6	
			16.7 (14)
SaM+/SaF−	or18	8	
	sin08	3.2	
			5.6 (2.5)
SaM+X/SaFX	rr07	5.4	
	rr08	6.6	
	rr12	15	
	rr11	6	
			8.25 (5.4)
SaM-X/SaFX	rr33	6.8	
	rr15	44	
			25.4 (24.4)
SaM+X/SaM−//SaF−/SaFX	F1	13.1	
			13.1 (4.4)
SaM+/SaM-X//SaF−/SaFX	F1	15	
			15 (2)
SaM+/SaM-X//SaF+/SaFX	F1	5.3	
			5.3 (3)

[#]parental genotypes listed as haploid for simplicity. All parents are homozygous.
*Values are averages of three measurements per individual.
^Averages and standard deviations calculated from all original raw measurements; standard deviations are in parentheses.

Table 4. Pollen quantity in individuals with different *Sa* genotypes.

Genotype[#]	Accession	Average Pollen Quantity (grains/μl)*	Genotype Average (sd)^
SaM+/SaF+	rr39	21.2	
	rr03	59.6	
			40.4 (22.2)
SaM−/SaF−	sus02	27.47	
	rr56	46	
	sus01	97.5	
			57 (37.6)
SaM+/SaF−	sin08	42.4	
			42.4 (11.4)
SaM+X/SaFX	rr12	26.5	
	rr07	28.5	
	rr11	37.5	
			30.8 (14.6)
SaM-X/SaFX	rr33	32.8	
	rr15	14.1	
			23.45 (11.2)
SaM+/SaM-X//SaF−/SaFX	F1	97.2	
			97.2 (20.3)
SaM−/SaM+X//SaF−/SaFX	F1	27.6	
			27.6 (13)
SaM+/SaM-X//SaF+/SaFX	F1	51.7	
			51.7 (1.2)

[#]parental genotypes listed as haploid for simplicity. All parents are homozygous.
*Values are averages of three measurements per individual.
^Averages and standard deviations calculated from all original raw measurements; standard deviations are in parentheses.

ancestors (Table 5); moreover, all weed groups possessed novel haplotypes not seen in cultivated or weed groups.

When polymorphism was present, TD values for many weed groups across all hybrid incompatibility loci tended to be more negative than genomic averages, indicating a tendency towards excess of rare mutations. At the *s5* locus, SH and BHA1 TD values are consistent with the occurrence of novel alleles. Also, noticeably, BHA2 at *s5* is associated with a very positive TD, which is consistent with the moderate frequencies of *indica* and *japonica*-type alleles in this group (Figure 1, Table 5).

Discussion

The Origin and Evolution of Hybrid Sterility Loci in US Weedy Groups

The demographic history of US weed groups based on random [16] and "weediness" candidate loci [20], [30] has been previously described, providing a framework for our expectations at hybrid sterility loci. Evidence to date suggests that the most common US

weed groups, SH and BHA, which co-occur in crop fields, are directly descended from the *indica* and *aus* cultivated groups respectively. Additional low-frequency weedy groups are products of weed-weed hybridization (BRH) or hybridization with the local *japonica* crop (MX). US weeds tend to harbor very low levels of genetic diversity compared to wild and cultivated *Oryza* groups, due to a genetic bottleneck upon US colonization. The expectations based on genome-wide surveys, however, were not always borne out at hybrid incompatibility loci.

Levels of diversity among our weedy groups varied across hybrid incompatibility loci, but were sometimes an order of magnitude higher than the genome-wide averages (Table 5). Particularly striking were the high levels of variation observed at *s5* (Table 5). At this locus, all weed groups had greater levels of diversity than those seen in their putative cultivated ancestors and possessed novel haplotypes. It should be noted that we sequenced more noncoding sequence at this locus than the other hybrid incompatibility and STS loci. However, this cannot account for greater diversity within weed groups compared to their cultivated ancestors. A possible explanation is that having the wide-compatibility deletion, which makes the *s5* gene non-functional, has removed selection at this locus, allowing for higher levels of polymorphism in weedy groups. In the case of the BHA2 group,

Table 5. Summary statistics for sequenced loci.

| Locus | Statistic | Oryza group | | | | | | |
		BHA1	BHA2	SH	indica	Aus	japonica	wild
s5	π	0.0013	0.004	0.0007	0.0005	0.0004	0.003	0.0021
	θ	0.002	0.003	0.0011	0.0005	0.0004	0.0022	0.0035
	TD	−1.396	1.846	−1.269	−0.05	n/a[#]	0.826	−1.385
DPL2	π	0.0003	0.0011	0	0.0011	0	0.0031	0.0033
	θ	0.0007	0.0008	0	0.0012	0	0.0031	0.0104
	TD	−1.149	0.986	n/a	−0.612	n/a	0.022	**−2.374***
SaM	π	0	0.0014	0.0019	0.0035	0.002	0.0021	0.0072
	θ	0	0.002	0.003	0.0044	0.0023	0.0028	0.0074
	TD	n/a	−1.609	−1.183	−0.901	−1.048	−1.035	−0.096
SaF	π	0	0.0024	0	0.0012	0.0008	0	0.003
	θ	0	0.0004	0	0.0014	0.0018	0	0.0042
	TD	n/a	−1.055	n/a	−0.689	−1.358	n/a	−1.037
STS^	π	0.0007	0.0005	0.0006	0.0016	0.0012	0.0011	0.0044
	θ	0.0008	0.0006	0.0004	0.0017	0.0011	0.0014	0.0056
	TD	−0.177	0.042	−0.441	−0.026	0.092	−0.773	−0.729

[#]TD values only calculated when more then four sequences available.
*Bolded values indicate significant TD.
^based on averages from Reagon et. al 2010.

the high levels of diversity and positive TD could be due to hybridization, given the presence of several *japonica*-type alleles (Table S4, Table 1). No overall evolutionary trend was observed among hybrid sterility loci in the weedy groups, suggesting that each locus has been subjected to independent evolutionary forces.

As expected, the allele types found within weedy group were also largely found within their ancestral cultivated populations, albeit at varying frequencies (Table 1). However, the occurrence of occasional allele types within the main weedy populations that are not found in their putative ancestors, suggest possible hybridization. For example, several members of BHA1 carry *DPL1-N+* and *DPL2-K+* alleles, a genotype not observed in any of their putative *aus* ancestors. This could indicate hybridization with the local *japonica* crop, or with SH, BRH or MX individuals. As mentioned above, some BHA2 individuals carry *s5-j* alleles, suggesting possible hybridization with the local *japonica* crop.

For weed groups of known hybrid origin, contributions of each parental group vary at each locus. BRH, a weedy hybrid of SH and BHA, carries alleles common in both parental groups; however, at *DPL2* and *Sa*, the genotype found in the SH population is more common. The MX individuals, comprising hybrids of *japonica* cultivars and either SH or BHA weeds, carry alleles found in all three parental groups at *s5*, *DPL2*, and *Sa*. Curiously, however, *SaFX* is nearly fixed in SH but is not found in any MX individuals.

Origins and Implications of the *SaFX* Allele

This is the first report of the *SaFX* allele, a deletion knocking out the *SaF* gene as well as portions of the *SaM* gene. We believe *SaFX* arose before domestication, as it was found in three individuals in our wild rice samples originating from Laos, Cambodia and Papua New Guinea (Table S4). However, it has remained at low frequency in both wild and cultivated populations, where it also seems to be geographically restricted. The *indica* and *aus* individuals we detected with the deletion are from Nepal and

India (Table S4). The low frequency of this allele in wild and cultivated rice groups suggests lack of selection for the *SaFX* allele in these populations.

The low frequency of *SaFX* in wild and cultivated groups is in contrast to its near-fixation in the SH and BRH weedy groups. While further studies need to be done to evaluate how *SaFX* behaves between inter-subspecific crosses, our analyses suggest that *SaFX* has no significant effects on pollen quality or quantity. This deletion may counteract the sterility-causing interaction between a *SaM* heterozygote and *SaF+* allele, making *SaFX* comparable to the wide-compatibility allele at the *s5* locus in enabling gene flow between *indica* and *japonica* populations. The prominence of *SaFX* in some weedy groups may be a consequence of founder effects – perhaps SH weeds descend from *indica* cultivars from Nepal. This view is supported by the high frequency of *s5-n* alleles in the SH group, which was also only found in *indica* cultivars from Nepal (Table S2). However, it is also possible that selection may have favored the *SaFX* allele in weeds, either as a way to decrease hybridization barriers with the local crop or as a way to circumvent other possible fitness effects of the incompatibility interaction.

SaFX may create a new version of a tri-allelic system involved in the evolution of speciation genes. Another wide-compatibility allele has been reported at *Sa*, characterized by two polymorphisms, a 6 bp insertion in *SaM* and a SNP in *SaF* [29]. The occurrence of wide-compatibility alleles in hybrid sterility systems is not uncommon, as these alleles have also been reported at other hybrid sterility loci [31–33].

All known wide-compatibility alleles act to restore fertility. Relaxation of sterility barriers can be favored if hybrids with wide-compatibility alleles have higher or equal fitness to their parents. Additionally, as is seen in the killer-protective system at *s5*, some sterility genes do not function directly in pollen or seed production and deleterious interactions between alleles at these loci, while promoting sterility, can cause other problems within the organism unrelated to gamete development (endoplasmic reticulum stress in the case of s5 [34]. However, intra-cellular complications caused by incompatible interactions at *Sa*, besides hybrid semi-sterility, have not been reported, though the fitness effects of the *SaF+−SaM* heterozygote interaction have not been fully investigated. Interestingly, in our study, only the hybrid sterility gene(s) that function directly in gamete development (the *DPLs*), which are implicated in pollen development; [12] did not display a putative wide-compatibility allele.

The Potential for Weed-Crop Gene Flow

Because US weedy rice groups descend from closely related *indica* and *aus* cultivars, while the local US rice crop is of *japonica* descent, we expected that typical *indica-japonica* postzygotic hybrid sterility barriers would occur between these two groups. Surprisingly, examination of three cloned hybrid sterility systems suggests that fewer postzygotic barriers exist between US weeds and the local crop than what is typically observed between *indica* and *japonica* cultivated rice subspecies. Given the prominence of the wide compatibility deletion at the *s5* locus in all weedy groups, there do not seem to be postzygotic barriers decreasing the possibility of gene flow with the local *japonica* crop at this locus (Table 1). Functional alleles at both *DPL* loci predominated in the SH, MX and BRH weed groups, implying that no crosses involving these weed groups can give rise to the sterility causing genotype *DPL1-K−/DPL2-N−* (Table 1). While BHA groups did have a high incidence of nonfunctional *DPL1* alleles, the dearth of nonfunctional alleles at either *DPL* locus in the local *japonica* crop also suggests few barriers to gene flow (Table 1). Few barriers for crop-weed gene flow are also apparent at the *Sa* locus for BRH and SH groups, as *SaFX*, a seemingly wide-compatibility allele, is nearly fixed. *SaFX* is absent in both BHA groups, indicating these groups are less likely to able to hybridize without issues of sterility with the local crop. However, the overall lack of postzygotic barriers to gene flow given the ancestry of US weeds is remarkable, and could suggest that *japonica*-compatible weed types have been favored during weed evolution in the US.

That weed-crop gene flow occasionally occurs in US rice fields is known. As mentioned above, MX weeds are known hybrids of SH or BHA weeds with the local *japonica* crop [16], and we have also found evidence of more localized genomic introgression of *japonica* alleles in some members of the BHA1 group [20]. However, the rate of crop-weed hybridization historically in the US has been low, <1% [13], [19]. Most hybrid incompatibility systems described in rice lead to only partial sterility in hybrids, and are expected to decrease the rate of hybridization, not eliminate it, which could account for low rates of gene flow between crops and weeds. Additional possible postzygotic fitness consequences in F1, such as overly late flowering that could compromise seed set [35], have not been sufficiently explored to draw firm conclusions about their overall contribution to reproductive isolation. In general, our study suggests that genetically based postzygotic barriers to hybridization are almost nonexistent, implying that prezygotic barriers [36] are more likely the main barriers to extensive gene flow between weedy and cultivated rice grown in the US. Possible prezygotic barriers to weed-crop outcrossing include the high self fertilizing levels of both weedy and cultivated rice due to short pollen longevity [37], and differences in flowering time among groups. For example, BHA weeds tend to flower later than SH weeds or the *japonica* crop, although variability in flowering time exists in all groups [20], [38]. Our study also suggests that under the right environmental circumstances or selective pressure, gene flow levels between crops and weeds have the potential to increase.

The use of herbicide resistance (HR) rice cultivars is increasing in the US, and in 2011 60–65% of Southern US rice was reported to be HR [39]. If the hybrid sterility loci used in this study adequately represent how other hybrid sterility loci have evolved within weedy groups, then the capacity of weedy groups to freely cross with these HR cultivars and produce fertile offspring lends itself to the creation of HR red rice, undermining many weed prevention strategies. In this respect, use of HR rice and herbicide applications changes the selective environment for weeds, such that HR hybrid weeds will be selectively favored. Alternatively, gene flow of unfavorable weedy traits, including shattering and dormancy, could be passed into native cultivated fields, which could interfere with uniform harvesting conditions. Likewise, the escape of an HR gene from red rice could also contaminate fields dedicated to non-HR rice [19].

Hybrid sterility is the most common form of postzygotic isolation in plants, and has long been of interest to evolutionary biologists, due to its importance in the speciation process, and to breeders, due to its impact on crop-improvement strategies. The importance of hybrid sterility barriers to crop-weed gene flow has not explicitly been considered, and here we have shown that there is greater potential than expected for crop-weed hybridization in US cultivated rice fields. As planting of HR rice has increased in the southern rice belt over the last 10 years, and reports of herbicide resistant weedy rice begin to surface [40], it is apparent that neither prezygotic nor postzygotic mating barriers are likely to impede weedy rice from acquiring the crop alleles that will enable them to survive in an herbicide-rich environment. Development of effective weed management strategies must take into account the greater than expected capacity for gene flow between weedy rice and its conspecific crop.

Acknowledgments

We are grateful to Lauren Bishop, Sara Weil, and Sherin Perera at UMass Amherst for help with genotyping, and to Caleb Rounds and Peter Hepler at UMass Amherst for all their help with the pollen microscopy. Additional thanks to the Bezanilla lab at UMass Amherst for use of their equipment.

Author Contributions

Conceived and designed the experiments: ALC MR. Performed the experiments: SMC MR LER. Analyzed the data: SMC MR LER. Wrote the paper: ALC SMC MR.

References

1. Ouyang Y, Liu Y-G, Zhang Q (2010) Hybrid sterility in plant: stories from rice. Curr Opin Plant Biol 13: 186–192. doi:10.1016/j.pbi.2010.01.002.

2. Orr HA (1996) Dobzhansky, Bateson, and the genetics of speciation. Genetics 144: 1331–1335.

3. Harushima Y, Nakagahara M, Yano M, Sasaki T, Kurata N (2002) Diverse variation of reproductive barriers in three intraspecific rice crosses. Genetics 160: 313–322.

4. Kubo T, Yamagata Y, Eguchi M, Yoshimura A (2008) A novel epistatic interaction at two loci causing hybrid male sterility in an inter-subspecific cross of rice (Oryza sativa L.). Genes Genet Syst 83: 443–453.

5. Oka H-I (1957) Genic analysis for the sterility of hybrids between distantly related varieties of cultivated rice. J Genet 55: 397–409. doi:10.1007/BF02984059.

6. Yang CY, Chen ZZ, Zhuang CX, Mei MT, Liu YG (2004) Genetic and physical fine-mapping of the Sc locus conferring indica-japonica hybrid sterility in rice (Oryza sativa L.). Chin Sci Bull 49: 1718–1721. doi:10.1360/04wc0197.

7. Reflinur, Chin JH, Jang SM, Kim B, Lee J, et al. (2012) QTLs for hybrid fertility and their association with female and male sterility in rice. Genes Genom 34: 355–365. doi:10.1007/s13258-011-0209-8.

8. Oka H-I (1974) Analysis of Genes Controlling F1 Sterility in Rice by the Use of Isogenic Lines. Genetics 77: 521–534.

9. Ouyang YD, Chen JJ, Ding JH, Zhang QF (2009) Advances in the understanding of inter-subspecific hybrid sterility and wide-compatibility in rice. Chin Sci Bull 54: 2332–2341. doi:10.1007/s11434-009-0371-4.

10. Chen J, Ding J, Ouyang Y, Du H, Yang J, et al. (2008) A triallelic system of S5 is a major regulator of the reproductive barrier and compatibility of indica-japonica hybrids in rice. Proc Natl Acad Sci USA 105: 11436–11441. doi:10.1073/pnas.0804761105.

11. Long Y, Zhao L, Niu B, Su J, Wu H, et al. (2008) Hybrid male sterility in rice controlled by interaction between divergent alleles of two adjacent genes. Proc Natl Acad Sci USA 105: 18871–18876. doi:10.1073/pnas.0810108105.

12. Mizuta Y, Harushima Y, Kurata N (2010) Rice pollen hybrid incompatibility caused by reciprocal gene loss of duplicated genes. Proc Natl Acad Sci USA 107: 20417–20422. doi:10.1073/pnas.1003124107.

13. Delouche JC, Burgos NR, Gealy DR, de San Martin GZ, Labrada R, et al. (2007) Weedy Rices - Origin, Biology, Ecology, and Control. FAO Plant Production and Protection Paper 188.

14. Sha XY, Linscombe SD, Groth DE (2007) Field evaluation of imidazolinone-tolerant Clearfield rice (Oryza sativa L.) at nine Louisiana locations. Crop Sci 47: 1177–1185. doi:10.2135/cropsoc2006.09.0592.

15. Bres-Patry C, Lorieux M, Clement G, Bangratz M, Ghesquiere A (2001) Heredity and genetic mapping of domestication-related traits in a temperate japonica weedy rice. Theor Appl Genet 102: 118–126. doi:10.1007/s001220051626.

16. Reagon M, Thurber CS, Gross BL, Olsen KM, Jia Y, et al. (2010) Genomic patterns of nucleotide diversity in divergent populations of US weedy rice. BMC Evol Biol 10. doi:10.1186/1471-2148-10-180.

17. Sun J, Qian Q, Ma D-R, Xu Z-J, Liu D, et al. (2013) Introgression and selection shaping the genome and adaptive loci of weedy rice in northern China. New Phytol 197: 290–299. doi:10.1111/nph.12012.

18. Londo JP, Schaal BA (2007) Origins and population genetics of weedy red rice in the USA. Mol Ecol 16: 4523–4535. doi:10.1111/j.1365-294X.2007.03489.x.

19. Shivrain VK, Burgos NR, Gealy DR, Sales MA, Smith KL (2009) Gene flow from weedy red rice (Oryza sativa L.) to cultivated rice and fitness of hybrids. Pest Management Science 65: 1124–1129. doi:10.1002/ps.1802.

20. Reagon M, Thurber CS, Olsen KM, Jia Y, Caicedo AL (2011) The long and the short of it: SD1 polymorphism and the evolution of growth trait divergence in U.S. weedy rice. Mol Ecol 20: 3743–3756. doi:10.1111/j.1365-294X.2011.05216.x.

21. Gealy DR, Gressel J (2005) Gene movement between rice (Oryza sativa) and weedy rice (Oryza sativa) - a U.S. temperate rice perspective. Crop Fertility and Volunteerism. CRC Press. 323–354.

22. Rozen S, Skaletsky H (2000) Primer3 on the WWW for general users and for biologist programmers. Methods Mol Biol 132: 365–386.

23. Librado P, Rozas J (2009) DnaSP v5: a software for comprehensive analysis of DNA polymorphism data. Bioinformatics 25: 1451–1452. doi:10.1093/bioinformatics/btp187.

24. Nei M, Li W (1979) Mathematical-Model for Studying Genetic-Variation in Terms of Restriction Endonucleases. Proc Natl Acad Sci USA 76: 5269–5273. doi:10.1073/pnas.76.10.5269.

25. Tajima F (1989) Statistical method for testing the neutral mutation hypothesis by DNA polymorphism. Genetics 123: 585–595.

26. Clark AG (1990) Inference of haplotypes from PCR-amplified samples of diploid populations. Mol Biol Evol 7: 111–122.

27. Tamura K, Peterson D, Peterson N, Stecher G, Nei M, et al. (2011) MEGA5: molecular evolutionary genetics analysis using maximum likelihood, evolutionary distance, and maximum parsimony methods. Mol Biol Evol 28: 2731–2739. doi:10.1093/molbev/msr121.

28. Du H, Ouyang Y, Zhang C, Zhang Q (2011) Complex evolution of S5, a major reproductive barrier regulator, in the cultivated rice Oryza sativa and its wild relatives. New Phytol 191: 275–287. doi:10.1111/j.1469-8137.2011.03691.x.

29. Wang Y, Zhong ZZ, Zhao ZG, Jiang L, Bian XF, et al. (2010) Fine mapping of a gene causing hybrid pollen sterility between Yunnan weedy rice and cultivated rice (Oryza sativa L.) and phylogenetic analysis of Yunnan weedy rice. Planta 231: 559–570. doi:10.1007/s00425-009-1063-7.

30. Thurber CS, Reagon M, Gross BL, Olsen KM, Jia Y, et al. (2010) Molecular evolution of shattering loci in US weedy rice. Mol Ecol 19: 3271–3284. doi:10.1111/j.1365-294X.2010.04708.x.

31. LeiGang S, XiangDong L, Bo L, XingJuan Z, Lan W, et al. (2009) Identifying neutral allele Sb at pollen-sterility loci in cultivated rice with Oryza rufipogon origin. Chin Sci Bull 54: 3813–3821. doi:10.1007/s11434-009-0571-y.

32. Wang GW, Cai HY, Xu Y, Yang SH, He YQ (2009) Analysis on the spectra and level of wide-compatibility conferred by three neutral alleles in inter-subspecific hybrids of rice (Oryza sativa L.) using near isogenic lines. Plant Breed 128: 451–457. doi:10.1111/j.1439-0523.2008.01514.x.

33. Li JQ, Shahid MQ, Feng JH, Liu XD, Zhao XJ, et al. (2012) Identification of neutral alleles at pollen sterility gene loci of cultivated rice (Oryza sativa L.) from wild rice (O. rufipogon Griff.). Plant Syst Evol 298: 33–42. doi:10.1007/s00606-011-0520-5.

34. Yang J, Zhao X, Cheng K, Du H, Ouyang Y, et al. (2012) A Killer-Protector System Regulates Both Hybrid Sterility and Segregation Distortion in Rice. Science 337: 1336–1340. doi:10.1126/science.1223702.

35. Rajguru SN, Burgos NR, Shivrain VK, McD Stewart J (2005) Mutations in the red rice ALS gene associated with resistance to imazethapyr. Weed Sci 53: 946–946. doi:10.1614/WS-04-111R1.1.

36. Rieseberg LH, Willis JH (2007) Plant Speciation. Science 317: 910–914. doi:10.1126/science.1137729.

37. Song ZP, Lu BR, Chen JK (2001) A study of pollen viability and longevity in Oryza rufipogon, O. sativa, and their hybrids. International Rice Research Notes 26: 31–32.

38. Shivrain VK, Burgos NR, Sales MA, Kuk YI (2010) Polymorphisms in the ALS gene of weedy rice (Oryza sativa L.) accessions with differential tolerance to imazethapyr. Crop Prot 29: 336–341. doi:10.1016/j.cropro.2009.10.002.

39. Salassi ME, Wilson Jr CE, Walker TW (2012) Proceedings of the Thirty-Fourth Rice Technical Working Group Hot Springs Convention Center, Hot Springs, Arkansas. 17–26.

40. Burgos NR, Norsworthy JK, Scott RC, Smith KL (2008) Red Rice (Oryza sativa) Status after 5 Years of Imidazolinone-Resistant Rice Technology in Arkansas. Weed Technology 22: 200–208. doi:10.1614/WT-07-075.1.

Variation in Broccoli Cultivar Phytochemical Content under Organic and Conventional Management Systems: Implications in Breeding for Nutrition

Erica N. C. Renaud[1]*, Edith T. Lammerts van Bueren[1], James R. Myers[2], Maria João Paulo[3], Fred A. van Eeuwijk[3], Ning Zhu[4], John A. Juvik[4]

1 Wageningen UR Plant Breeding, Plant Sciences Group, Wageningen University, Wageningen, The Netherlands, **2** Department of Horticulture, Oregon State University, Corvallis, Oregon, United States of America, **3** Biometris, Plant Sciences Group, Wageningen University, Wageningen, The Netherlands, **4** Department of Crop Sciences, University of Illinois, Urbana, Illinois, United States of America

Abstract

Organic agriculture requires cultivars that can adapt to organic crop management systems without the use of synthetic pesticides as well as genotypes with improved nutritional value. The aim of this study encompassing 16 experiments was to compare 23 broccoli cultivars for the content of phytochemicals associated with health promotion grown under organic and conventional management in spring and fall plantings in two broccoli growing regions in the US (Oregon and Maine). The phytochemicals quantified included: glucosinolates (glucoraphanin, glucobrassicin, neoglucobrassin), tocopherols (δ-, γ-, α-tocopherol) and carotenoids (lutein, zeaxanthin, β-carotene). For glucoraphanin (17.5%) and lutein (13%), genotype was the major source of total variation; for glucobrassicin, region (36%) and the interaction of location and season (27.5%); and for neoglucobrassicin, both genotype (36.8%) and its interactions (34.4%) with season were important. For δ- and γ-tocopherols, season played the largest role in the total variation followed by location and genotype; for total carotenoids, genotype (8.41–13.03%) was the largest source of variation and its interactions with location and season. Overall, phytochemicals were not significantly influenced by management system. We observed that the cultivars with the highest concentrations of glucoraphanin had the lowest for glucobrassicin and neoglucobrassicin. The genotypes with high concentrations of glucobrassicin and neoglucobrassicin were the same cultivars and were early maturing F_1 hybrids. Cultivars highest in tocopherols and carotenoids were open pollinated or early maturing F_1 hybrids. We identified distinct locations and seasons where phytochemical performance was higher for each compound. Correlations among horticulture traits and phytochemicals demonstrated that glucoraphanin was negatively correlated with the carotenoids and the carotenoids were correlated with one another. Little or no association between phytochemical concentration and date of cultivar release was observed, suggesting that modern breeding has not negatively influenced the level of tested compounds. We found no significant differences among cultivars from different seed companies.

Editor: Hany A. El-Shemy, Cairo University, Egypt

Funding: Funding was provided by Seeds of Change (www.seedsofchange.com) and monetary funding from Wageningen University, University of Illinois and Oregon State University. The funder had no role in the study design, data collection and analysis, decision to publish, or preparation of the manuscript.

Competing Interests: The principle researcher, Erica Renaud was employed by Seeds of Change and later Vitalis Organic Seeds, the organic division of Enza Zaden, a vegetable seed company.

* Email: E.Renaud@enzazaden.com

Introduction

Organic food consumption is in part driven by consumer perception that organic foods are more nutritious and simultaneously less potentially harmful to human health [1–2]. Studies, such as Smith-Sprangler et al. [3], have concluded that there is little evidence for differences in health benefits between organic and conventional products, but other studies have indicated that organic vegetables and fruits contain higher concentrations of certain plant phytochemicals associated with health promotion than those produced conventionally [4–8]. A number of these compounds are produced by plants in response to environmental stress or pathogen infection, providing a potential explanation of why concentrations of these compounds might be higher in plants grown in organic systems without application of pesticides [9]. In addition, higher phytochemical levels may be due to the effects

that different fertilization practices have on plant metabolism. Synthetic fertilizers used in conventional agriculture are more readily available to plants than organic fertilizers [10]. Nutrients derived from organic fertilizers need to be mineralized, and the availability of these nutrients depends on soil moisture, temperature and level of activity of soil organisms [11]. Conventional systems seek to maximize yields, resulting in a relative decrease of plant phytochemicals and secondary metabolites [12–15]. Correspondingly, compounds such as phenolics, flavonoids, and indolyl glucosinolates may be induced by biotic or abiotic stress [16–17].

Broccoli is an abundant source of nutrients, including provitamin A (β-carotene), vitamin C (ascorbate), and vitamin E (tocopherol) [18]. It is also a source of phytochemicals associated with health benefits and these include glucosinolates, carotenoids, tocopherols, and flavonoids [19–21]. Verhoeven et al. [22], Keck and Finley [23] and Here and Büchler [24], reported that diets rich in broccoli reduce cancer incidence in humans. Strong associations between consumption level and disease risk reduction exists for glucosinolates (anti-cancer), tocopherols (cardiovascular), and the carotenoids (eye-health) [25].

Sulfur containing glucosinolates are found in the tissues of many species of the *Brassicaceae* family. When glucosinolates are consumed, they are hydrolyzed into isothiocyanates (ITC) and other products that up-regulate genes associated with carcinogen detoxification and elimination. Aliphatic glucoraphanin (up to 50% of total glucosinolates) and the indolylic glucosinolates, glucobrassicin and neoglucobrassicin are abundant in broccoli florets [20,19,26]. Glucoraphanin is hydrolyzed either by the endogenous plant enzyme myrosinase [27–28] or by gut microbes to produce sulforaphane, an ITC. The indole glucosinolates are tryptophan-derived in a similar but alternate biosynthetic pathway [29]. The health promoting effects of the indolyl glucosinolates are attributed to indole-3-carbinol, a hydrolysis product of gluco-brassicin, N-methoxyindole-3-carbinol and neoascorbigen, hydro-lysis products from neoglucobrassicin, and the catabolic products derived from alkyl glucosinolates. Clinical studies have shown that the glucosinolate hydrolysis products reduce the incidence of certain forms of cancer (e.g., prostate, intestinal, liver, lung, breast, bladder) [30–35]. The lipophilic phytonutrients found in broccoli include the carotenoids lutein, zeaxanthin, β-carotene, and tocopherols (forms of vitamin E) [36–37]. In addition to their role as vitamins, these compounds are powerful antioxidants [38–39]. Consumption of vegetables high in tocopherols and caroten-oids has decreased the incidence of certain forms of cancer [40]. Lutein and zeaxanthin protect against development of cataracts and age-related macular degeneration [41]. Tocopherols have also been associated with reduced risk of cardiovascular disease by preventing oxidative modification of low-density lipoproteins in blood vessels [42].

The genetic potential for high nutrient content has long been a concern of the organic industry in order to meet the expectations of organic consumers. This has often been manifested by questioning whether modern elite cultivars may have lower levels of nutritional content than older open pollinated cultivars. Indirect evidence supporting this argument comes from Davis et al. [43], who compared USDA nutrient content data for 43 garden crops released between 1950 and 1999. Statistically significant decreases were noted for six nutrients (protein, calcium, potassium, iron, riboflavin, and ascorbic acid), with declines ranging from 6% for protein to 38% for riboflavin. Crop varieties in 1950 had been bred to be adapted to specific regions and a relatively low input agriculture system, but contemporary cultivars are selected for yield, disease resistance, broad adaptation to high input agriculture systems, and for increased 'shipability' and shelf life. Traka et al. [44] recommend breeding with greater genetic diversity when the goal is enhanced phytochemical content by exploiting wild crop relatives. The genotype is important in determining the level of nutrients in a crop cultivar [45–47]. What is unclear, however, is whether the nutritional content of a cultivar is associated with certain genotypic categorization, e.g. old versus modern, open pollinated versus F$_1$ hybrid cultivars. In addition, there is no clear differentiation as to what extent nutritional content in a crop is determined by genotypic or by field management factors or by the interaction of both. Some studies comparing performance of genotypes in organic and conventional production systems have shown that for certain agronomic traits, cultivars perform differently between the two production systems (e.g. for winter wheat: Murphy et al. [48], Baresel et al. [49]; for lentils: Vlachostergios et al. [50]; for maize: Goldstein et al.[51]), while others have shown no differences in ranking performance (for

maize: Lorenzana and Bernardo [52]; for onions: Osman et al. [53]; for cereals: Przystalski et al.[54]). The results of these studies have profound implications for organic cultivar selection and breeding strategies and raise questions as to the need for cultivars to be bred with broad adaptability or specific adaptation for the requirements of regional organic production and for designing breeding programs that optimize phytochemicals in an adapted management system.

Previous studies comparing organically versus conventionally grown broccoli for nutritional quality have been 'market basket' (off-the-shelf) studies [55–56]. Harker [57] explained that the limitation of market basket studies is that they either have purchased the products from the store shelf and cannot relate differences to specific growing conditions or that the number of cultivars is too small to generalize the results. While other studies have compared cultivars from one production season time period to another, knowledge of the actual cultivar and production system (soil quality, temperature, rainfall) was not available [58,43]. The concentrations and form of health-promoting nutrients in *Brassica* vegetables have been reported to vary significantly due to (1) genotype (cultivar and genotypic class) [59,20,26,60,21,37,61,44], (2) environmental conditions such as season [62–67], light [19], max/min temperature, irrigation [68–69], (3) genotype by environment interactions [19,70–71]; (4) management system including soil fertility [72–73], organic versus conventional [13,74–75], days to harvest [63–64], and (5) post-harvest management [76–77]. Identifying specific growing conditions and genotypes that produce cultivars with varying phytochemical content and putative disease-prevention activity could offer value-added commercial opportunities to the seed and food industry.

In addition to research conducted on how broccoli genotypes, management system and environment interact for horticultural traits [78], we address in this paper the question of how do genotypes, management system and environment interact to determine the nutritional contributions of broccoli to the human diet. We studied the relative importance and interaction among genotypes (cultivars, genotypic classes) and environment {management system [M: organic (O) or conventional (C)], season (S, a combination of year and season within year, i.e., fall 2006, spring 2007, fall 2007, spring 2008), location (E)} in a set of 23 broccoli cultivars for floret glucosinolate, tocopherol and carotenoid concentrations grown under organic and conventional production systems in two contrasting broccoli production regions of the US: Oregon and Maine. Specifically we addressed the following questions: (1) what is the impact of organic management system compared to the environmental factors including climatic region, season and their interactions [Genotype (G) x Environment (E) x Management System (M)]? (2) is there a significant difference in phytochemical content between different genotypes and genotypic classes (old and modern cultivars; open pollinated and F$_1$ hybrid cultivars; early and late maturing cultivars; and between different commercial seed sources)? (3) what is the best selection environment for a broccoli breeding program for enhanced phytochemical content?

Materials & Methods

Plant Material and Field Trial Locations

Twenty-three broccoli cultivars including open pollinated (OP) cultivars, inbred lines, and F$_1$ hybrids were included in field trials (**Table 1**). Cultivars were grown in a randomized complete block design with three replicates in Maine (ME)-Monmouth (Latitude 44.2386°N, Longitude 70.0356°W); and Oregon (OR)-Corvallis (Latitude 44.5647°N, Longitude123.2608°W)] with each location including organically (O) and conventionally (C) managed treatments. Plots contained 36 plants, planted in three rows of 12 plants at 46 cm equidistant spacing within and between rows. The 2006 trials had only 18 of the 23 entries, and the Oregon 2006 trial had only two replicates at the organic location. Field trials were conducted for three consecutive years with one production cycle in Fall 2006, two production cycles in Spring and Fall 2007 and one production cycle in Spring 2008. The primary management differences between the organic and conventional field trial sites are outlined in **Table S1 in File S1**, which describes the production system, soils, fertility applications, the applied supplemental irrigation, and weather conditions for the area of study. Further details of the field design are reported in Renaud et al. [78].

Field Data Collection

As plots approached maturity they were evaluated three times a week for field quality and broccoli heads that had reached commercial market maturity (approximately 10 to 12 cm in diameter for most of the cultivars while retaining firmness). Field quality traits evaluated on a 1 to 9 ordinal scale included head color, bead size, and bead uniformity. Average head weight was determined by taking the mean of the five individual heads per plot. Head diameter averaged for five heads at harvest maturity from each plot. Maturity was based on days to harvest from transplanting date. Detailed procedures and horticulture trait performance data are reported in Renaud et al. [78].

Broccoli Floret Samples and glucosinolate, tocopherol, and carotenoid analysis

In order to analyse nutritional compounds of the broccoli heads, the following procedure was followed: As plots approached maturity, five broccoli head tissue samples were harvested fresh from each subplot at each trial location and were composited into a single sample per replication. The samples were frozen at $-20°C$ and shipped in a frozen state to the University of Illinois, Urbana-Champaign where they were freeze-dried and assessed for nutritional phytochemicals. Each sample was analyzed for the glucosinolates (glucoraphanin, glucobrassicin and neoglucobrassicin), carotenoids (β-carotene, lutein, and zeaxanthin), and tocopherols (δ-, γ-, α- tocopherol) by high-performance liquid chromatography (HPLC) analysis using analytical protocols described in Brown et al. [19] for glucosinolates, and Ibrahim and Juvik [37] for tocopherols and carotenoids. Glucosinolates in lyophilized floret tissue samples were extracted and analysed by HPLC using a reverse phase C18 column. Three hundred mg samples of broccoli floret tissue were weighed out for extraction and the HPLC quantification of the tocopherols and carotenoids.

Statistical Analysis

Various linear mixed models were used for the analysis of trait variation. We followed the same methodology as described in Renaud et al. [78], which was comparable to the approach followed by Lorenzana and Bernardo [52]. For fitting the linear

Table 1. Overview of commercially available broccoli cultivars, showing origin, main characteristics, included in paired organic - conventional field trials 2006–2008.

Cultivar	Abbreviation	Origin	Cultivar Type[a]	Date of Market Entry	Maturity Classification[b]
Arcadia	ARC	Sakata	F_1	1985	L
B1 10	B11	Rogers	F_1	1988	M
Batavia	BAT	Bejo	F_1	2001	M
Beaumont	BEA	Bejo	F_1	2003	L
Belstar	BEL	Bejo	F_1	1997	L
Diplomat	DIP	Sakata	F_1	2004	L
Early Green	EGR	Seeds of Change	OP	1985	E
Everest	EVE	Rogers	F_1	1988	E
Fiesta	FIE	Bejo	F_1	1992	L
Green Goliath	GRG	Burpee	F_1	1981	M
Green Magic	GRM	Sakata	F_1	2003	M
Gypsy	GYP	Sakata	F_1	2004	M
Imperial	IMP	Sakata	F_1	2005	L
Marathon	MAR	Sakata	F_1	1985	L
Maximo	MAX	Sakata	F_1	2004	L
Nutribud	NUT	Seeds of Change	OP	1990	E
OSU OP	OSU	Jim Myers, OSU	OP	2005	E
Packman	PAC	Petoseed	F_1	1983	E
Patriot	PAT	Sakata	F_1	1991	M
Patron	PAN	Sakata	F_1	2000	M
Premium Crop	PRC	Takii	F_1	1975	E
USVL 048	U48	Mark Farnham, USVL	Inbred	not released	L
USVL 093	U93	Mark Farnham, USVL	Inbred	not released	M

[a]Cultivar Type: F_1: hybrid; OP: Open Pollinated; Inbred.
[b]Maturity Classification: E: Early; M: Mid; L: Late.

mixed models, GenStat 15 (VSNi, 2012) was used. The models followed the set-up:

$$y = E + R(E) + G + G \times E + e.$$

Here y is the phytochemical response. Term E represents the environment in a very general sense, it includes all main effects and interactions of Season (S), Location (L) and Management (M). For analyses per location, the terms involving L were dropped. Similarly, for analyses regarding a specific management regime, the terms involving M were dropped. Term $R(E)$ is the effect of replicate within environment, and there were two or three replicates in individual trials. G and $G \times E$ are genotype and genotype by environment interaction effects, respectively. Finally e is a residual.

Variance components were reported as coefficients of variation, i.e., $CV = 100\sqrt{V}/x$ with V the variance corresponding to specific effects and x the trait mean. Repeatability was calculated from the variance components in its most general form as $H^2 = V_G/(V_G$ $V_{GL}/nL + + V_{GS}/nS + V_{GM}/nM + V_{GLS}/(nL.nS) + V_{GLM}/(nL.nM)$ $+ V_{GSM}/(nS.nM) + V_{GLSM}/(nL.nS.nM) + V_e/(nL.nS.nM.nR))$, where the variance components correspond to the terms in the mixed model above. The terms nL, nS, nM and nR stand for the number of locations (2: Maine and Oregon), number of 'seasons' (4: Fall 2006, Spring 2007, Fall 2007, Spring 2008), management (2; organic and conventional), and replicates (2 or 3).

Genotypic means were calculated by taking genotypic main effects fixed instead of random in the mixed models above. Pairwise comparisons between genotypic means were performed using GenStat procedure *VMCOMPARISON*. Correlations on the basis of genotypic means were referred to as genetic correlations. Genotypic stabilities under organic and conventional conditions were calculated as the variance for individual genotypes across all trials in the system.

To assess the feasibility of selection for organic conditions (the target environment) under conventional conditions, we calculated the ratio of correlated response (for organic conditions using conventional conditions), CR, to direct response (for organic conditions in organic conditions), DR, as the product of the genetic correlation between organic and conventional systems (r_G) and the ratio of the roots of conventional and organic repeatabilities (H_C and H_O respectively): $CR/DR = r_G H_C/H_O$. A ratio smaller than 1 indicates that selection is better done directly under organic conditions when the aim is indeed to improve the performance in organic conditions.

Results

Comparison of phytochemicals means over the environments

Glucosinolates. Across all trials, glucoraphanin levels were comparable between locations and seasons but were more variable at the individual location and season trial analysis level (**Table 2**). Glucoraphanin, glucobrassicin and neoglucobrassicin levels were comparable between organic and conventional treatments. Comparisons of organic versus conventional by location and season for the glucosinolate phytochemicals are presented in **Figure S1**. Comparable levels of glucosinolates were observed in the organic - conventional comparisons within locations and seasons.

Tocopherols. Across trials compared regionally, Oregon had higher levels of all three tocopherols compared to Maine (**Table 2, Figure S2**). The tocopherols δ- and γ- were higher in Fall compared to Spring, but not so for α-tocopherol (**Figure S2**).

Organic and conventional levels for all tocopherol concentrations were in the same range and not significantly different. When the three tocopherols were analysed by organic versus conventional within location and season, there were no clear significant differences in management system across the season and location combinations (**Table 2, Figure S2**).

Carotenoids. Overall, Oregon had higher levels of lutein and β-carotene compared to Maine (**Table 2, Figure S3**) and comparative levels of zeaxanthin (**Table 2, Figure S3**). Spring produced higher levels of all carotenoids compared to Fall levels in contrast to the glucosinolates and the δ- & γ- tocopherol concentrations. There were no significant differences between organic and conventional for any carotenoid measured. When carotenoids were analysed by management system within location and season, β-carotene showed significantly lower levels in Maine in the Fall compared to other location and season combinations (**Figure S3**).

Partitioning of variance components

Glucosinolates. For glucoraphanin across all trials in both regions, Genotype (G) main effect accounted for the largest proportion of variance, followed by G×L×S interaction (**Table 3**). There was no Management (M) main effect, but M contributed to the three (L×S×M and G×S×M) and four-way interactions (G×L×S×M). In contrast to glucoraphanin, Location (L) had the largest effect for glucobrassicin and neoglucobrassicin across all trials in both regions, followed by the L×S interactions. For neoglucobrassicin the S and G main effect was more important than for glucobrassicin. When trials were further partitioned by location, a G and S main effect was apparent for neoglucobrassicin in both locations; for glucobrassicin the S main effects was only apparent in Oregon and not in Maine (**Table S2 in File S1**). There was M main effect for glucobrassicin and neoglucobrassicin, but not for glucoraphanin, and no G×M interaction for all glucosinolates.

Tocopherols. For δ- and γ-tocopherol across all trials in both regions, the Season (S) main effect accounted for the largest proportion of variance (**Table 3**). In contrast the proportion of the variation associated with S for α-tocopherol across all trials was minor. For all three tocopherols there was minor to no M effect, but a large L main effect, being the greatest for γ-tocopherol. The G main effect showed a similar pattern to L.

Carotenoids. For all three carotenoids across all trials in both regions, the G main effect described a significant component of total variance and was of largest influence for lutein (**Table 3**). The S main effect played an important role for zeaxanthin, and to a lesser extent for lutein but not for β-carotene. For all three carotenoids the L effect was minor, but the L×S interaction for β-carotene was relatively large and mostly associated with Maine (**Table S2 in File S1**). There was no M main effect; only for β-carotene was there a small effect of the G×M interaction (mainly driven by Maine).

Repeatability, genetic correlation and ratio of correlated response to direct response

Organic versus conventional. In the present study, we were able to estimate the proportion of the genotypic variance relative to phenotypic variance, but because we did not have a genetically structured breeding population, we apply the term repeatability rather than broad sense heritability. Of the phytochemicals studied, repeatabilities for concentrations of seven of the nine were comparable or higher in organic compared to conventional systems (**Table 4**). Only for glucobrassicin and δ-tocopherol was repeatability under organic conditions lower than under conven-

Table 2. Trait means[1] of phytochemicals of 23 broccoli cultivars grown across four pair combinations of location (Maine/Oregon), season (Fall/Spring) two-years combined and management system (Conventional/Organic), 2006–2008.

	Maine					Oregon				
	Fall 2006–2007 Combined		Spring 2007–2008 Combined		Mean	Fall 2006–2007 Combined		Spring 2007–2008 Combined		Mean
	C	O	C	O		C	O	C	O	
Glucoraphanin	5.31 e	3.77 bc	3.56 b	4.06 c	4.18	3.46 b	3.03 a	4.64 d	4.51 d	3.91
Glucobrassicin	1.06 b	0.90 a	1.45 c	1.33 c	1.19	5.14 f	5.51 g	2.24 d	2.70 e	3.90
Neoglucobrassicin	0.46 a	0.40 a	2.16 c	1.85 b	1.22	2.34 c	3.20 d	4.32 e	5.10 f	3.74
δ-Tocopherol	2.34 c	2.77 d	1.91 b	1.70 a	2.18	3.53 e	3.66 e	1.91 b	2.24 c	2.83
γ-Tocopherol	4.67 c	4.40 c	2.63 a	2.98 b	3.67	8.48 d	8.73 d	3.31 b	3.22 b	5.94
α-Tocopherol	25.83 a	27.33 a	38.61 b	40.51 bc	33.07	43.04 c	43.20 c	40.52 bc	42.25 c	42.25
Lutein	11.49 a	12.47 a	15.53 b	15.93 b	13.85	15.91 b	16.04 b	16.48 b	17.81 c	16.56
Zeaxanthin	0.81 a	0.83 ab	0.87 ab	0.88 b	0.85	0.83 ab	0.84 ab	1.02 c	1.02 c	0.93
β-Carotene	12.98 a	13.25 a	28.73 c	29.71 c	21.16	29.10 c	30.10 c	25.16 b	25.80 b	27.54

[1]Values in the table are means. Means of the same letter in the same row are not significantly different at the P<0.05 level.

Table 3. Partitioning of variance components (%) presented as coefficients of variation for phytochemicals of 23 broccoli cultivars grown across eight pair combinations of location (Maine/Oregon), season (Fall/Spring) and management system (Conventional/Organic), 2006–2008.

	Location (L)	Season (S)	Management (M)	L×S	L×M	S×M	L×S×M	L×S×M×R Rep (R)	Genotype (G)	G×L	G×S	G×M	G×L×S	G×L×M	G×S×M	G×L×S×M	Residual
Glucoraphanin	0.01	5.45	0.00	7.20	0.01	0.00	11.86	1.56	17.45	0.01	0.01	0.00	15.97	0.01	7.49	12.62	11.94
Glucobrassicin	36.00	0.00	0.00	27.51	5.58	4.18	1.34	1.86	9.42	7.77	0.01	0.00	13.84	0.00	0.00	10.63	10.91
Neoglucobrassicin	36.81	13.51	0.00	34.36	8.47	6.84	4.50	4.76	15.16	6.24	0.01	0.00	16.40	0.00	0.01	13.40	15.80
δ-Tocopherol	6.83	35.22	0.43	7.87	0.01	0.01	3.39	0.01	5.57	5.65	6.01	0.00	13.65	0.01	0.00	12.21	12.74
γ-Tocopherol	12.02	19.09	0.01	12.11	0.01	3.64	4.47	2.12	13.79	4.85	15.82	0.00	12.95	0.00	0.01	11.08	12.03
α-Tocopherol	6.73	0.01	0.28	10.20	0.01	0.01	0.01	1.29	2.79	3.42	0.01	0.97	10.07	0.00	0.00	8.18	8.92
Lutein	3.71	4.91	0.00	7.52	0.01	2.70	1.85	1.40	13.03	4.14	0.01	0.01	10.76	0.73	0.01	9.21	9.95
Zeaxanthin	1.97	11.55	0.01	3.99	0.01	0.00	0.01	0.00	8.44	3.18	0.01	1.05	6.91	0.01	0.83	8.52	11.36
β-Carotene	4.61	0.00	0.00	17.84	0.01	0.70	0.01	0.71	8.41	4.45	0.00	2.31	11.32	4.63	0.65	12.83	10.99

tional. In the analyses δ- and α-tocopherol had relatively low repeatabilities. The highest repeatabilities were for glucoraphanin (0.82–0.84), neoglucobrassicin (0.75–0.76), γ-tocopherol (0.72–0.75), lutein (0.83–0.85) and zeaxanthin (0.76–0.77). Genetic correlations were high between organic and conventional for the glucosinolates, γ-tocopherol and lutein (0.84–0.95), while δ-tocopherol, α-tocopherol, zeaxanthin and β-carotene were lower (0.63–0.77). The ratio of the correlated response to direct response for selection in the organic system was less than 1.0 for all traits.

By location and season. For the glucosinolates, glucoraphanin and glucobrassicin repeatability at each location, season and treatment trial were comparable and generally high (0.83–0.97) between organic and conventional trials, while no clear trend for neoglucobrassicin repeatabilities was observed between organic and conventional aside from being much lower than glucoraphanin and glucobrassicin (**Table S3 in File S1**). For γ- and α-tocopherol, repeatabilities were comparable between organic and conventional, while for δ-tocopherol repeatabilities were comparable between systems or higher in conventional except for one paired trial. For the carotenoids, repeatabilities were comparable or higher in organic for all paired trials, while for lutein in seven of the eight paired trials organic was comparable or greater than conventional. Repeatabilities for zeaxanthin concentrations were comparable for six of the eight paired trials.

Comparison of cultivar ranking for phytochemical concentration and stability across trials

To determine trends in cultivars with both the highest concentration of phytochemical groups most stable across locations, seasons and production systems, phytochemical concentrations were plotted against stability per genotype across trials. A group of cultivars were identified as both highest in concentration and most stable and are indicated in the highlighted 'red circle' per phytochemical (**Figure 1A–I**). For glucoraphanin, the same group of cultivars had both the highest concentrations and were the most stable across production systems (**Figure 1A; Table S4 in File S1**). While for glucobrassicin, a different set of cultivars had the highest concentrations across production systems (**Figure 1B; Table S5 in File S1**). Overall stability of all cultivars across production system was less related to cultivar mean concentrations for glucobrassicin than for glucoraphanin. None of the cultivars with the highest concentration for neoglucobrassicin were in the top quartile for stability across trials; all cultivars with the highest neoglucobrassicin content were in the bottom half for stability (**Figure 1C; Table S6 in File S1**). Some but not all cultivars that had the highest concentrations of α-tocopherol were among the top group for δ- and/or γ-tocopherol. There was no relationship between δ-tocopherol concentrations and stability, but both γ- and α- tocopherols had higher concentrations associated with greater stability (**Figure 1D–F; Tables S7–9 in File S1**). Open pollinated and early maturing cultivars had the highest and most stable concentrations for all carotenoids. (**Figure 1G–I; Tables S10–12 in File S1**).

Comparison of phytochemical concentration by genotype classification

The open pollinated and F₁ hybrid cultivars were compared across trials for each phytochemical analysed (**Figure 2A**). The levels of glucoraphanin in F_1 hybrids tended to be higher than the open pollinated cultivars. But the inverse trend was observed for glucobrassicin, which was supported by the ranking and stability analysis where the F_1 hybrids showed higher levels and more stability across trials than the open pollinated cultivars for

Table 4. Repeatabilities, genetic correlation and ratio of correlated response to direct response for broccoli phytochemicals comparing organic versus conventional management systems over all trial season/location combinations, 2006–2008.

| | Repeatability (H) | | $r_A{}^a$ | $CR_{org}/R_{org}{}^b$ |
	C	O		
Glucoraphanin	0.84	0.82	0.84	0.83
Glucobrassicin	0.70	0.64	0.88	0.84
Neoglucobrassicin	0.75	0.76	0.94	0.94
δ-Tocopherol	0.50	0.42	0.73	0.66
γ-Tocopherol	0.75	0.72	0.95	0.93
α-Tocopherol	0.23	0.35	0.61	0.76
Lutein	0.83	0.85	0.93	0.94
Zeaxanthin	0.76	0.77	0.77	0.78
β-Carotene	0.62	0.72	0.63	0.68

[a] Average genetic correlation between conventional and organic production systems across locations.
[b] Ratio of correlated response to direct response.

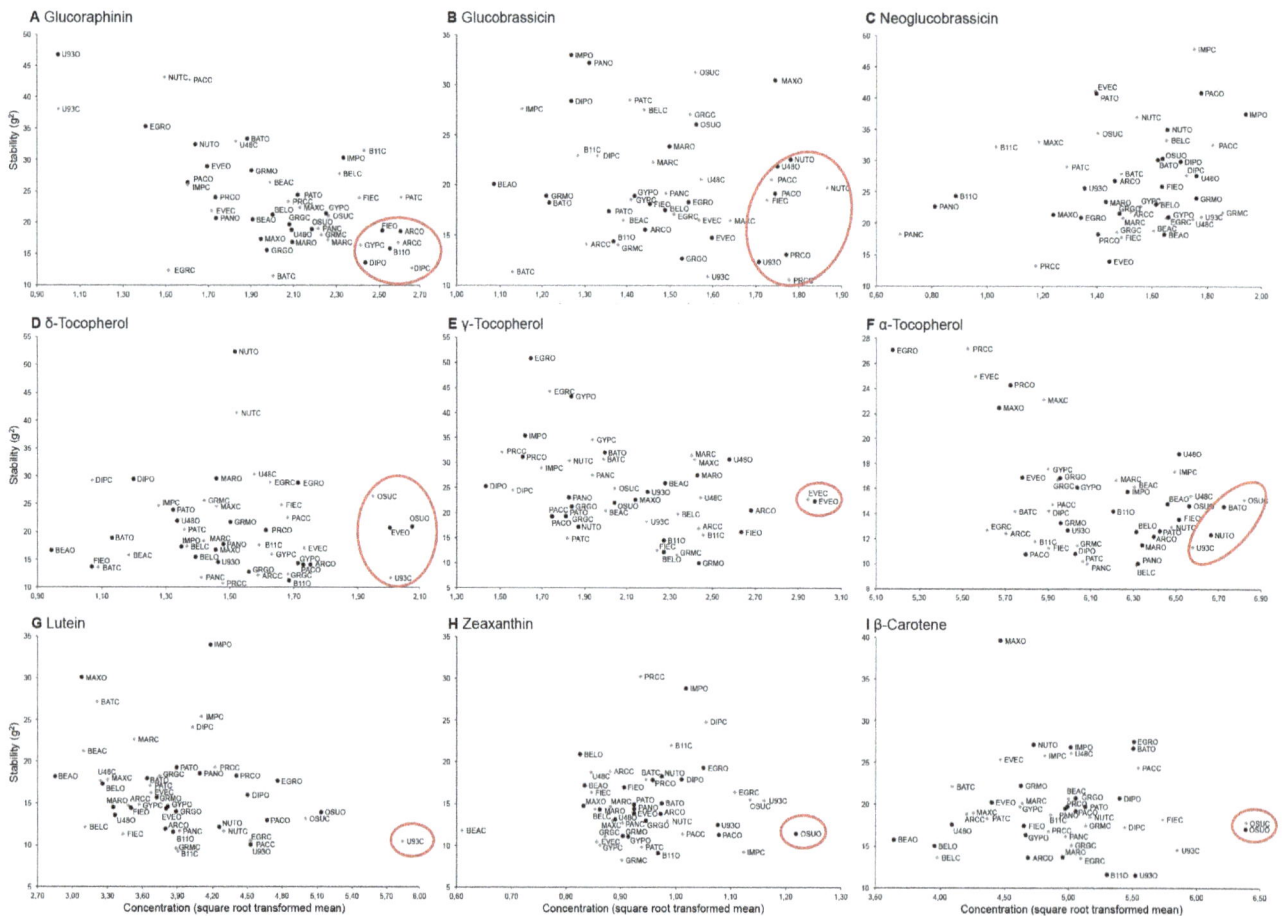

Figure 1. Broccoli cultivar stability from trials conducted in two locations over four seasons with two management systems plotted against phytochemical content. A. Glucoraphanin, B. Glucobrassicin, C. Neoglucobrassicin, D. δ-tocopherol, E. γ-tocopherol, F. α-tocopherol, G. Lutein, H. Zeaxanthin, I. β-carotene. See Table 1 for cultivar name abbreviations. The C or O at the end of the cultivar abbreviation indicates conventional or organic management system, respectively.

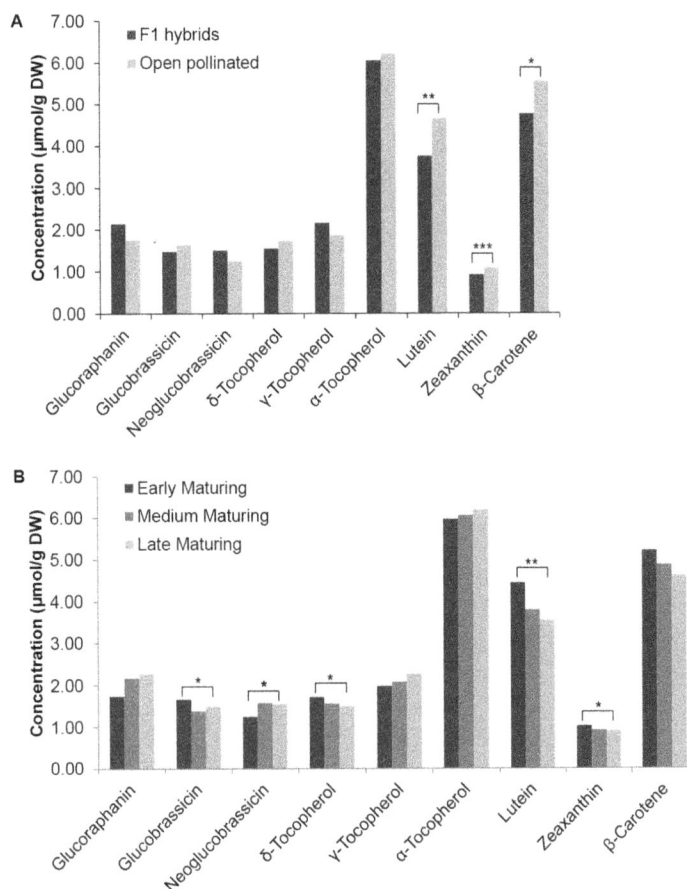

Figure 2. Mean phytochemical content of broccoli genotypic classes. A. Mean phytochemical content of broccoli F_1 hybrids versus open pollinated cultivars, and B. Mean phytochemical content of early, mid- and late-maturing cultivars grown across all trials at two locations (Maine and Oregon), in two seasons (Fall and Spring) and in two management systems (Conventional and Organic) and conventional management systems. See Table 1 for key to cultivar F1 hybrid versus open pollinated classification and maturity classification. Significance (* $= P<0.05$, ** $= P<0.01$, *** $= P< 0.001$).

glucoraphanin. The reverse was observed for glucobrassicin. For the carotenoids, the open pollinated cultivars had a significantly higher mean value of lutein and zeaxanthin and tended to be higher for β-carotene compared to the F_1 hybrids.

Based on the results of our field trials, the 23 cultivars of broccoli were grouped into three distinct maturity classes: Early (55–63 days); Mid (64–71 days); and Late (72–80 days) and analysed for the effect of the maturity class on phytochemical content (**Figure 2B**). For glucoraphanin, late maturing cultivars had significantly higher content levels, while for the carotenoids, early maturing cultivars tended to have higher concentrations and were significantly higher for lutein.

When cultivar performance between genetic material originating from two primary broccoli breeding companies was compared for phytochemical content there were no significant differences with the exception of lutein, where company 1's cultivars had significantly higher concentrations than those of company 2 (**data not shown**).

A negative correlation between the date of release and levels of glucobrassicin ($R^2 = 0.21$; p $= 0.03$) (**Figure 3**) was observed, but no significant correlations for any other phytochemical were seen when 21 cultivars (the total set minus the two inbred lines) were analysed by their date of commercial release (1975–2005).

Correlation analysis among phytochemicals and horticulture traits

Phytochemical correlation across trials. Correlation among phytochemicals indicated that glucoraphanin was significantly negatively correlated to glucobrassicin (**Table 5**). Correlations between the glucosinolates and the tocopherols were not significant. Glucoraphanin and neoglucobrassicin were negatively correlated to all carotenoids but only lutein and glucoraphanin were statistically significant. Glucobrassicin demonstrated a positive trend with all carotenoids. No statistically significant correlations were observed within tocopherols. Δ-tocopherol was positively correlated, while γ-tocopherol was negatively correlated to all carotenoids. There were no significant correlations for α-tocopherol with carotenoids. All carotenoids were highly positively correlated with one another.

Phytochemical correlation to horticulture traits across trials. A correlation analysis was conducted for six horticulture traits, derived from the field study component of this research, Renaud et al. [78], and the nine phytochemicals across trials. The results indicated that greater head weight and head diameter were significantly positively correlated with glucoraphanin and negatively correlated with glucobrassicin, δ-tocopherol and the carotenoids. Increasing days to maturity was positively correlated with glucoraphanin, and negatively correlated to carotenoids.

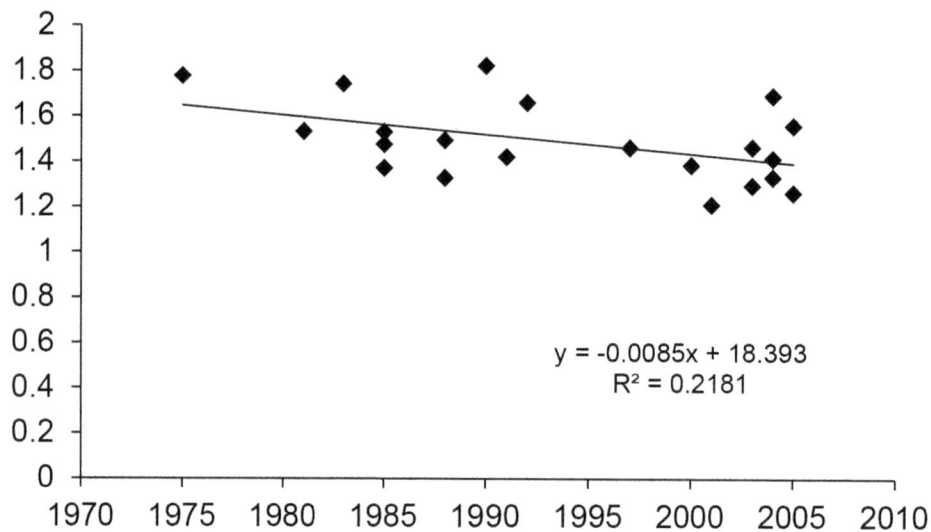

$$y = -0.0085x + 18.393$$
$$R^2 = 0.2181$$

Figure 3. Regression of broccoli floret glucobrassicin concentrations on date of cultivar release for 23 cultivars grown across all trials in two locations (Maine and Oregon), in two seasons (Fall and Spring), in two management systems (Conventional and Organic), 2006–2008.

Head color was significantly correlated with δ-tocopherol and the carotenoids, but not with glucosinolates or γ- and α-tocopherol. Bead size and bead uniformity were positively correlated with glucoraphanin, neoglucobrassicin and γ-tocopherol and negatively correlated with glucobrassicin and the carotenoids.

Principal component biplot analysis: correlation between phytochemicals and cultivars by production system

In the principal component analysis the first PC axis accounted for similar amounts of the total variation in both conventional and organic production systems (43.5% vs. 39.6%). The second PC axis showed a similar trend with 17.02% for conventional and 16.93% for organic (**Figures 4A and 4B**). The first two PC axes together accounted for 60.53% and 56.57% of total variation for conventional and organic, respectively. The PCA biplot analysis supported our findings that carotenoids were highly associated across systems, while tocopherols were highly associated in conventional, but not in organic (tocopherols demonstrated the largest shift between production systems). Glucoraphanin and neoglucobrassicin were associated with one another, but not with glucobrassicin across production systems. Glucoraphanin was associated with α-tocopherol in organic, but not in conventional treatments. Glucobrassicin was associated with δ- and α-tocopherol in conventional, but not in organic treatments. δ-tocopherol had a higher association with the carotenoids in organic than conventional. The biplots show response of both cultivars and phytochemical traits to environment. Those cultivars close to the origin reveal little about the relationship of cultivars and trait vectors, whereas those located near the extremes of trait vectors are those with the highest (or lowest) values for those traits.

Discussion

Impact of organic management system compared to environmental factors on phytochemical content

Few studies have specifically compared the levels of health promoting compounds in *Brassica* vegetable species grown under organic and conventional production systems [13,74–75]. To our knowledge, this investigation is the most comprehensive study with the broadest range of phytochemical compounds (9) and a diverse set of broccoli cultivars (23) over regions (2), and management systems (2), with Fall and Spring season trials (2 each). In this study organic versus conventional management systems contributed the smallest source of variation compared to genotype, region and season. Within the phytochemicals studied individual compound concentrations responded differently. All compounds showed genetic variation, but also a substantial proportion of variance components were accounted for by high level interactions (**Table 3; Table S2 in File S1**). While M main effect was generally small, it had a substantial contribution in three- and four-way interactions. In particular, many G×L×S×M interactions were large relative to other variance components. This indicates that for the phytochemicals, M did have an influence on G, but that there were no consistent patterns across locations and seasons that would have shown up as significant G×M. Rather in each season and location, the paired organic and conventional environments differed significantly from one another but each situation was unique. In contrast to many comparisons between organic and conventional production systems [79], it should be noted that in our trials, yields averaged over the years did not differ significantly between the organic and conventional management systems [78].

Among the nine compounds, glucoraphanin was the most strongly influenced by genotype followed by lutein: supporting the findings of several other broccoli studies where variation in concentrations for glucoraphanin [19,70,65–66] and lutein [21,37] was primarily due to genotype. For γ-tocopherol, genotype was a large source of variation, but this compound was equally influenced by location and season (also found by Ibrahim and Juvik [37]). For glucobrassicin and neoglucobrassicin the location was the largest source of variation, but also L×S interaction was very influential, particularly for neoglucobrassicin, which is supported by Kushad et al. [20] and Schonhof et al. [26]. Jasmonic acid, a signal transduction compound in plants, is up-regulated under conditions of plant stress, wounding, and herbivory. Increased endogenous levels or exogenous application of this compound (or methyl jasmonate) increases biosynthesis and transport of neoglucobrassicin to broccoli florets. This up-

Table 5. Correlations coefficients (r) for six horticultural traits and nine phytochemicals, calculated using data standardized across trials.

	Head Weight	Head Diameter	Maturity	Head Color	Bead Size	Bead Uniformity	Glucoraphanin	Glucobrassicin	Neoglucobrassicin	δ-Tocopherol	γ-Tocopherol	α-tocopherol	Lutein	Zeaxanthin	β-Carotene
Head Weight															
Head Diameter	0.81														
Maturity															
Head Color															
Bead Size	0.63		0.69												
Bead Uniformity	0.49	0.48													
Glucoraphanin	0.47	0.44	0.43		0.63	0.51									
Glucobrassicin	-0.54	-0.50			-0.56	-0.64	-0.51								
Neoglucobrassicin			0.58		0.48										
δ-Tocopherol	-0.55														
γ-Tocopherol				0.49	0.43										
α-Tocopherol															
Lutein	-0.65		-0.70	0.56	-0.69		-0.41			0.55	-0.54				
Zeaxanthin	-0.68	-0.43	-0.62	0.49	-0.64					0.60	-0.42		0.95		
β-Carotene	-0.53		-0.54	0.59	-0.48					0.50	-0.43		0.90	0.90	

Correlation results include means from 23 cultivars, across eight pair combinations of location (Maine/Oregon), season (Fall/Spring) and management system (Conventional/Organic), 2006–2008[a].
[a]For empty cells, r is not significantly different from zero (P<0.05).

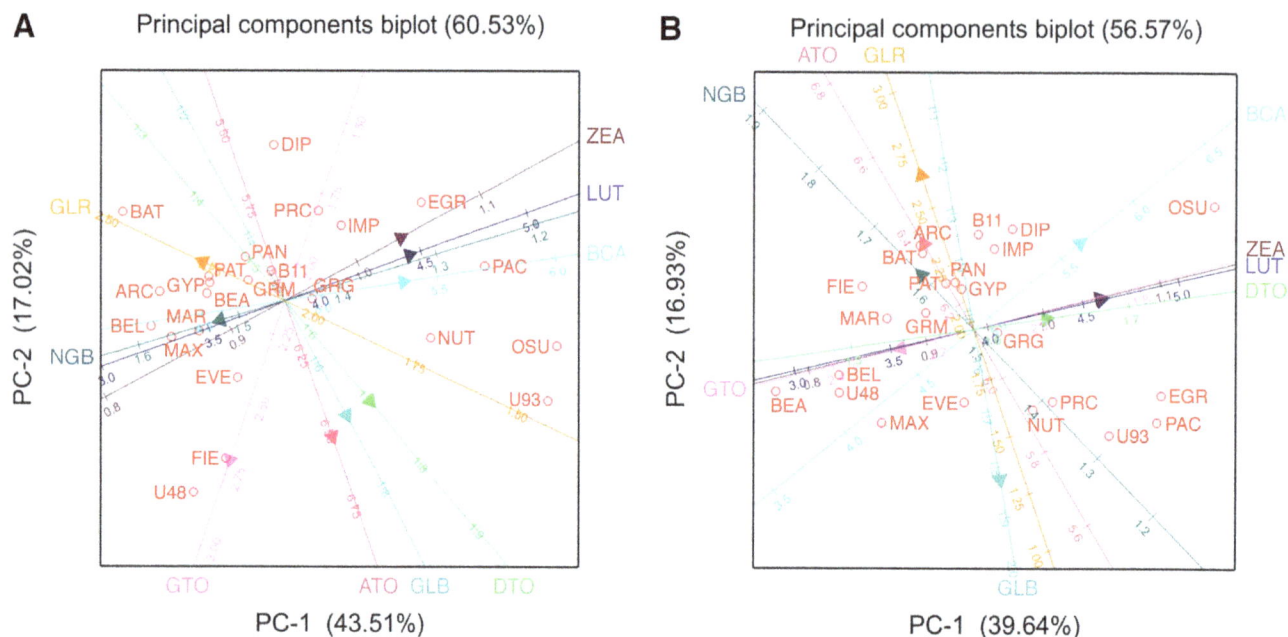

Figure 4. Principal components biplot of phytochemicals (vectors) and 23 cultivars (circles) grown in four seasons in Oregon and Maine. A. Biplot for conventional production, B. Biplot for organic production. See Table 1 for cultivar name abbreviations. Trait abbreviations: GLR: Glucoraphanin; GLB: Glucobrassicin; NGB: Neoglucobrassicin; DTO: δ-tocopherol; GTO: γ-tocopherol; ATO: α-tocopherol; LUT: Lutein; ZEA: Zeaxanthin; BCA: β-Carotene.

regulation was not observed for glucobrassicin biosynthesis [17]. This could explain why neoglucobrassicin was primarily under the control of Location and L×S interaction in our study. Season was the largest variance component for δ-tocopherol and zeaxanthin, which contrasts with the work of Ibrahim and Juvik [37] who found genotype had the largest influence on these compounds, followed by genotype by environment interaction although this study was constrained by the fact that the experiment was conducted in only one location over two growing seasons. For the other compounds such as α-tocopherol and β-carotene, L×S and the G×L×S interactions were most important.

Overall we found high genetic correlations between glucosinolates in organic and conventional trials. When trial locations were analysed separately, M main effect was present for glucobrassicin and neoglucobrassicin. The mean concentrations of glucobrassicin and neoglucobrassicin in broccoli from Oregon organic trials had higher concentrations compared to Oregon conventional trials, while Maine trials were comparable between management systems (**Table 2, Figure S1**). These results can be explained by the larger environment effect on glucobrassicin and genotype by environment effect on neoglucobrassicin found in the variance component analysis indicating sensitivity of these compounds to abiotic and/or biotic stresses. Our location specific findings are supported by those of Meyer and Adam [13] who performed a comparative study of the glucosinolate content of store bought organic and conventional broccoli and determined that the indolyl glucosinolates, glucobrassicin and neoglucobrassicin were significantly higher in the organically grown versus the conventionally grown. Evaluation of 10 broccoli genotypes over two years by Brown et al. [19] further supports our findings and those of Rosa and Rodrigues [62], Vallejo et al. [64], and Farnham et al. [70], that variation in concentration for glucoraphanin was primarily due to genetic variation, while differences in glucobrassicin was due to environmental variation (e.g. season, temperature) and genotype by environment interaction. The significantly higher

levels of glucobrassicin in Oregon in the Fall harvested trials compared to Maine could be attributed to the higher maximum temperatures and GDD in Oregon compared to Maine (**Table S1 in File S1**).

Compared to glucosinolates, there is substantially less research on the genotype by environment interaction of tocopherol and carotenoid phytochemical groups in broccoli, and no specific studies exploring the influence of organic production system. In our study, minor management system effect at the overall trial analysis level was observed for the tocopherols and for carotenoids, there was management system effect only for lutein in Oregon Spring trials. Picchi et al. [75] also did not find differences in levels of carotenoids in cauliflower in organic versus conventional systems. In the tocopherols, there were no significant differences in location, but for δ- and γ- tocopherol concentration levels were higher in the fall compared to the spring, while for α-tocopherol, concentration levels were higher in the spring compared to the fall. For the carotenoids, there were no significant location differences, however there was a seasonal trend that all carotenoids were higher in spring compared to fall. Ibrahim and Juvik [39] found significant environmental variation among 24 broccoli cultivars for carotenoids and tocopherols which they attributed to the stressful production environments. Factors explaining the genotype and genotype by environment interaction components of variation in the carotenoids and tocopherols could be clarified by the fact that environmental stimuli are both up- and down-regulating genes associated with carotenoid and tocopherol biosynthesis. There is evidence in the literature that there are coordinated responses of the carotenoid and tocopherol antioxidants *in vivo*. There was a reduction in rape seed (*Brassica napus*) tocopherol content in response to increased carotenoid levels due to over expression of the enzyme phytoene synthase [80]. This response could explain the negative correlation between γ-tocopherol concentration and the carotenoids observed in our trials.

Differences in phytochemical content between different genotypes and genotypic classes

The partitioning of variance indicated that genotype was an important source of variation for all glucosinolates. The cultivar ranking and rank correlation analysis demonstrated that there was a pattern in genotype content of glucosinolates where cultivars with the highest concentrations of glucoraphanin had the lowest levels for glucobrassicin (**Figures S1**). In our trials, the range in glucoraphanin concentrations across cultivars was (1.15–7.02 µmol/g DW, **Table S4 in File S1**), while glucobrassicin was 1.46–3.89 µmol/g DW, **Table S5 in File S1**). Several of the cultivars with the highest concentrations of neoglucobrassicin were those that had the highest concentrations of glucobrassicin. Range in neoglucobrassicin concentrations across cultivars was 0.68–4.54 µmol/g DW, **Table S6 in File S1**). In earlier studies, glucosinolate concentrations in broccoli have shown dramatic variation among different genotypes. Rosa and Rodriguez [62] studied total glucosinolate levels in eleven cultivars of broccoli and found ranges from 15.2–59.3 µmol/g DW. Among 50 accessions of broccoli Kushad et al. [20] found glucoraphanin content ranges from 0.8–22 µmol/g DW with a mean concentration of 7.1 µmol/g DW, while Wang et al. [61] found glucoraphanin content of five commercial hybrids and 143 parent materials ranging from 1.57–5.95 µmol/g for the hybrids and 0.06–24.17 µmol/g in inbred lines and Charron et al. [65] found ranges from 6.4–14.9 µmol/g DW. While the means in our study are somewhat lower, they are within the range of other studies.

A genotype effect was observed for tocopherols, but predominantly for γ-tocopherol. The PCA biplots (**Figure 4AB**) and the correlation analysis (**Table 5**) demonstrated the high positive correlations between δ-tocopherol, α-tocopherol and the carotenoids (α-tocopherol and β-carotene were also highly correlated in the Kushad et al. [20] study). The cultivar relationship to different phytochemicals was represented in the biplots as well as in the cultivar content and stability analysis (**Figure 1**). Many cultivars with the highest concentrations in the tocopherols and carotenoids were open pollinated cultivars, inbreds and early maturing, older F1 hybrids. Many of this same group were also relatively high in glucobrassicin concentrations. Kurilich et al. [38] found that carotenoid and tocopherol concentrations among 50 broccoli lines were highly variable and primarily genotype dependent. Specifically, levels of β-carotene ranged from 0.4–2.4 mg/100 g FW. Ibrahim and Juvik [37] also found broad ranges for total carotenoid and tocopherol concentrations among 24 genotypes ranging from 55–154 µg/g DW and 35–99 µg/g DW, respectively. Farnham and Kopsell [21] studied the carotenoid levels of nine double haploid lines of broccoli. Similar to our findings, lutein was the most abundant carotenoid in broccoli ranging from 65.3–139.6 µg/g DM. The sources of variation for lutein were predominantly genotype, followed by environment and G×E interaction, which also supports our findings. No genotypic differences were found for β-carotene in Farnham and Kopsell [21], which is in contrast to our findings. Overall, they found that most of the carotenoids measured were positively and highly correlated to one another as was observed in our study (**Table 5**). Kopsell et al. [81] found lutein levels in kale of 4.8–13.4 mg/100 g FW where the primary variance components for both lutein and β-carotene were also genotype and season.

Our research aimed also to address the question whether the phytochemical content of broccoli cultivars is associated with certain genotypic classes, e.g. open pollinated vs. F1 hybrids; older vs. newer cultivar releases; and between commercial sources. Broccoli is typically a cross-pollinated, self-incompatible crop species and cultivars are either open pollinated and composed of heterogeneous genetically segregating individuals, or F1 hybrids produced by crossing of two homozygous inbred lines, resulting in homogeneous populations of heterozygous individuals. In the 1960's virtually all broccoli grown was derived from OPs. By the 1990's almost all commercial cultivars were hybrids [82].

In our trials with 18 F1 hybrids (released between 1975–2005) and 3 open pollinated cultivars (released from 1985–2005), we found several interesting trends related to genotype and genotypic class performance as it related to the three groups of phytochemicals. When analysing F1 hybrid and open pollinated cultivars, they also demonstrated different performance patterns depending upon the individual phytochemical or group of compounds analysed. When cultivars were ranked for content and stability per phytochemical, there were distinct trends for certain compounds such as late maturing, F1 hybrids outperforming early maturing F1 hybrids and open pollinated cultivars for glucoraphanin, while the inverse was found for glucobrassicin and all carotenoids studied. This analysis was further supported by the PCA biplots that showed a strong relationship for select cultivars to certain phytochemicals or groups of phytochemicals such as 'OSU OP' to the carotenoids. When the full set of cultivars was divided into F1 hybrid and open pollinated groups and the means compared by phytochemical, the results further supported the individual cultivar analysis where F1 hybrids had higher mean values for glucoraphanin than the open pollinated cultivars (**Figure 2A**). Clear cultivar performance differences were identified where early maturing versus late maturing cultivars performed differently depending upon the phytochemical (**Figure 2B**). We also found that late maturing cultivars had higher concentrations for glucoraphanin than early maturing lines (and the inverse for glucobrassicin and the carotenoids). Picchi et al. [75] studied the quantity of glucosinolates of an early and late maturing cultivar of cauliflower grown in one conventional and three organic production systems, and found a significantly higher level of glucoraphanin in the later maturing cultivar compared to the early maturing cultivar in the organic production system. Another interesting trend was that cultivars with higher concentration levels for those phytochemicals whose expression is heavily influenced by environmental factors were not necessarily the most stable across trial environments; as was the case with neoglucobrassicin, δ- and γ-tocopherol in our study. For traits where genotype played a more significant role in contributing to variation, cultivars with a higher concentration level tended to also be those that were most stable across environments as was seen for lutein and glucoraphanin concentrations.

No significant differences were found for cultivar performance in phytochemical concentrations between genetic materials originating from two distinct commercial sources, with the exception of lutein (data not shown). When the full set of broccoli cultivars were analyzed for a correlation between date of release and mean level of phytochemical content across trials, no significant correlation was found with the exception of a negative trend for glucobrassicin (**Figure 3**). Our data does not support the idea that modern breeding for high yield performance and disease resistance necessarily leads to a trade-off in level of phytochemicals. Previous reports examining the relationship between year of release and performance had focussed on wheat vitamin and mineral content [83–85], and mineral content in broccoli [86–87]. However these authors did not study phytochemical content and their results were equivocal on the question on an innate biological trade-off between increased yield and nutritional content.

Not many studies have included two or more groups of phytochemicals. In our study with three phytochemical groups we found that phytochemicals demonstrating a negative correlation with one another (e.g. glucoraphanin with the carotenoids), showed an inverse cultivar response: e.g. cultivars with highest concentrations of glucoraphanin were the lowest in the carotenoids and vice versa. When both horticultural traits and phytochemicals were analysed for their phenotypic correlation, head weight was significantly and positively correlated with glucoraphanin and negatively correlated with δ- and α-tocopherol and the carotenoids. Farnham and Kopsell [21] explained that negative correlations may occur as a result of increased biomass accumulation in a certain genotype that is not accompanied by increased carotenoid production, effectively lowering the carotenoid concentration in the immature broccoli florets when pigments are expressed. Comparatively, head color was highly correlated to the carotenoids and negatively correlated to the glucosinolates overall. The cultivar 'OSU OP' was explicitly bred for a dark green stem and head color, not only for a darker green dome surface but also for a dark green interior color between the florets of the dome and in the stem (personal communication, Jim Myers 2013). 'OSU OP' was the highest in overall carotenoid concentrations across trials as it is known that carotenoids are correlated with chlorophyll concentrations and the intensity of green pigmentation [88].

Perspectives on breeding broccoli for enhanced phytochemical content specifically for organic agriculture

Our study included predominantly broccoli cultivars selected for broad adaptability in conventional production systems and not purposely bred for high phytochemical content nor for adaptation to organic agriculture. What we can conclude from our data is that there has been little change in levels of several phytochemicals over three decades of breeding. This may indicate genetic variation for phytochemicals is limited in elite germplasm, or it may be the result of the lack of selection tools for these traits. This may be changing with recent efforts to introgress high glucoraphanin from *B. villosa* to produce the high-glucoraphanin F$_1$ cultivar 'Beneforté' [89–90,44]. The seed industry needs to exploit known sources of variation in the genus *Brassica* to enhance levels of other health-promoting phytochemicals and to broaden the genetic diversity of commercial broccoli germplasm. Our finding of a strong correlation between dark green color and high carotenoid levels provides breeders with a simple and efficient means of increasing carotenoids. The three groups of phytochemicals studied contribute to health promotion in different ways. As these groups are related to different metabolic pathways selecting for one compound does not necessarily inadvertently improve the other compounds, and may even result in negative correlation as we have seen in our data between glucoraphanin and the carotenoids. Although these compounds belong to different metabolic pathways, their production may be coordinated through regulatory feedback loops, or the structural and/or regulatory genes controlling these pathways may be genetically linked.

Designing a breeding program for broccoli high in glucosinolates would require the following considerations generated from our research: (1) Glucoraphanin is a highly genetically determined compound with minor location and season main effects but with substantial G×L×S interaction. (2) Comparatively, glucobrassicin and neoglucobrassicin are more impacted by location and season and L×S interaction with highest glucobrassicin concentrations and largest range in our Oregon Fall trials and neoglucobrassicin highest in Oregon Spring trials. (3) Cultivar performance for glucoraphanin and glucobrassicin and neoglucobrassicin was negatively correlated indicating that there may be a trade-off between glucoraphanin on the one hand, and glucobrassicin and neoglucobrassicin on the other hand. (4) Selection for glucor-

aphanin without consideration of horticultural traits would probably result in larger headed and later maturing cultivars. Conversely, selection for smaller headed, early maturing cultivars would favor glucobrassicin and neoglucobrassicin at the expense of glucoraphanin.

A breeding program for broccoli for high tocopherol content would require: (1) Overall the tocopherols were more season, location and L×S dependent and had lower overall repeatabilities compared to the glucosinolates. In a structured genetic population where additive genetic variance could be partitioned, narrow sense heritability would likely be low, and increasing tocopherol content would best be conducted with breeding methods suited to low heritability traits; (2) δ- and γ-tocopherols were both season dependent and fall grown broccoli had higher concentrations of these compounds across trials and a wider range of content levels, whereas levels of α-tocopherol were higher in spring but the range was comparable under both seasons. Thus, fall would be the preferred environment for breeding for these compounds; (3) there were no significant differences for location for δ- or γ-tocopherol, but the average levels of α-tocopherol levels were significantly higher in Oregon than Maine, suggesting greater potential for genetic gain in the Oregon environment.

If the goal is to design a breeding program for broccoli enhancing the levels of carotenoids it would require the following considerations: (1) For all three carotenoids studied, genotypic variation, particularly for lutein, was relatively more important than location and season; (2) however, zeaxanthin exhibited a large S (spring) and L×S interaction. For both β-carotene and lutein, spring grown broccoli had significantly higher levels than fall produced. Thus, selection for carotenoids would probably be more effective in spring than in fall; (3) early maturing and small headed cultivars had higher levels of carotenoids. Since most of the carotenoids are associated with the outer surfaces of the inflorescence, smaller broccoli heads with a greater surface area to volume ratio should show higher concentrations of these compounds; (4) because carotenoids have high G main effect good germplasm sources as indicated in **Figure 1** have high concentrations of carotenoids and demonstrated stability across environments. As all three carotenoids are highly correlated with one another, selecting for one should effectively select for all; (5) selection for darker green colour more widely distributed throughout the tissues of the head should allow the breeder to relatively efficiently increase carotenoid content in broccoli.

In closing, we want to address the question of selecting in an organic or a conventional environment. The argument commonly used to support selecting in productive environments is that heritabilities are higher compared to resource poor environments [91–92]. Organic is often considered a low-external input environment, resulting on average in 20% less yield compared to conventional production [79]. Nevertheless, in our trials repeatabilities for some phytochemicals were higher or comparable to conventional (**Table 3**). Narrow sense heritabilities would be expected to be significantly lower. For those traits where repeatabilities were higher or comparable, direct selection under organic systems could enhance selection gain. In all cases, the ratio of correlated response to direct response was less than one suggesting that direct selection would allow more rapid progress than correlated selection. Our data on phytochemicals did not show a wider range of levels under organic conditions as we found for horticultural traits in the same trials [78], however, in several cases, repeatabilities in organic production were higher than in conventional.

To maximize efficiency in a breeding program, commercial breeders may seek to combine breeding for both conventional and organic markets, and a combination of strategies can be proposed. Some studies that utilized highly heritable (agronomic) traits, where cultivar yield performance ranked similarly between organic and conventional management systems and which had high genetic correlations, suggested that early breeding be conducted under conventional conditions, with the caveat that advanced breeding lines be tested under organic conditions for less heritable traits (e.g. Löschenberger et al. [93]; Lorenzano and Bernardo, 2008) [52]. In studies where cultivar yield performance differed between management systems and there were significant differences in cultivar ranking, and in some cases low genetic correlations for lower heritability traits (e.g. Kirk et al. [94]; Murphy et al. [48]), these studies recommended that cultivars intended for organic agriculture be selected only under organic conditions. In our study of phytochemicals, we would recommend for organic purposes selection under organic conditions for the compounds where genetic correlations between organic and conventional were moderate.

Supporting Information

Figure S1 Comparison of broccoli cultivars for glucosinolates (μmol/g DW) grown across all trials in two locations (Maine and Oregon), in two seasons (Fall and Spring), in two management systems (Conventional and Organic), and at the individual trial level, 2006–2008. A. Glucoraphanin, B. Glucobrassicin, C. Neoglucobrassicin.

Figure S2 Comparison of broccoli cultivars for tocopherols (μmol/g DW) grown across all trials in two locations (Maine and Oregon), in two seasons (Fall and Spring), in two management systems (Conventional and Organic), and at the individual trial level, 2006–2008. A. δ-tocopherol, B. γ-tocopherol, C. α-tocopherol.

Figure S3 Comparison of broccoli cultivars for carotenoids (μmol/g DW) grown across all trials in two locations (Maine and Oregon), in two seasons (Fall and Spring), in two management systems (Conventional and Organic), and at the individual trial level, 2006–2008. A. Lutein, B. Zeaxanthin, and C. β-carotene.

File S1 Supporting tables. Table S1. Description of agronomic and environmental factors of the trial locations with paired organically and conventionally managed fields, 2006–2008. **Table S2.** Partitioning (%) of variance components for various traits of 23 broccoli cultivars grown across four pair combinations in Maine, season (Fall/Spring) and management system (Conventional/Organic), 2006–2008. Variance components reported as coefficients of variation. **Table S3.** Repeatability for broccoli for phytochemicals and per trial of 23 broccoli cultivars grown across eight pair combinations of location (Maine/Oregon), season (Fall/Spring) and management system (Conventional/Organic), 2006–2008. **Table S4.** Glucoraphanin level (μmol/g DW) of 23

cultivars grown under conventional (C) and organic (O) conditions in two locations (Maine and Oregon) in two seasons (Fall and Spring) from 2006–2008. **Table S5.** Glucobrassicin level (μmol/g DW) of 23 cultivars grown under conventional (C) and organic (O) conditions in two locations (Maine and Oregon) in two seasons (Fall and Spring) from 2006–2008. **Table S6.** Neoglucobrassicin level (μmol/g DW) of 23 cultivars grown under conventional (C) and organic (O) conditions in two locations (Maine and Oregon) in two seasons (Fall and Spring) from 2006–2008. **Table S7.** δ-tocopherol level (μmol/g DW) of 23 cultivars grown under organic (O) and conventional (C) conditions in two locations (Maine and Oregon) in two seasons (Fall and Spring) from 2006–2008. **Table S8.** γ-tocopherol level (μmol/g DW) of 23 cultivars grown under conventional (C) and organic (O) conditions in two locations (Maine and Oregon) in two seasons (Fall and Spring) from 2006–2008. **Table S9.** α-tocopherol level (μmol/g DW) of 23 cultivars grown under conventional (C) and organic (O) conditions in two locations (Maine and Oregon) in two seasons (Fall and Spring) from 2006–2008. **Table S10.** Lutein level (μmol/g DW) of 23 cultivars grown under conventional (C) and organic (O) conditions in two locations (Maine and Oregon) in two seasons (Fall and Spring) from 2006–2008. **Table S11.** Zeaxanthin level (μmol/g DW) of 23 cultivars grown under conventional (C) and organic (O) and conditions in two locations (Maine and Oregon) in two seasons (Fall and Spring) from 2006–2008. **Table S12.** β-carotene level (μmol/g DW) of 23 cultivars grown under conventional (C) and organic (O) conditions in two locations (Maine and Oregon) in two seasons (Fall and Spring) from 2006–2008.

Acknowledgments

For support in understanding the genotype selection and providing select elite inbred lines for the study, the authors thank Dr. Mark Farnham from the USDA, Charleston, South Carolina. For the Oregon trials, the authors wish to thank the organic growers Jolene Jebbia and John Eveland at Gathering Together Farm for providing the location and support for the organic broccoli trials. Deborah Kean, Faculty Research Assistant at the Oregon State University Research Station, and the students Hank Keogh, Shawna Zimmerman, Miles Barrett, and Jennifer Fielder provided support in data collection of the field trials. For the Maine trials, the authors wish to thank the students Heather Bryant, Chris Hillard and Greg Koller for support in data collection of the field trials. We also thank the University of Maine Highmoor Farm Superintendent, Dr. David Handley University of Maine Cooperative Extension. For the soil analysis, we thank Dr. Michelle Wander from the University of Illinois, Urbana and her students. At Wageningen University, we thank Paul Keizers and Dr. Chris Maliepaard for support with the statistical analysis. We thank Carl Jones for his valuable input on iterations of this research paper and Ric Gaudet for support in data organization.

Author Contributions

Conceived and designed the experiments: ER ELvB JM MP FvE NZ JJ. Performed the experiments: ER JM NZ JJ. Analyzed the data: ER ELvB JM MP FE JJ. Contributed reagents/materials/analysis tools: ER JM MP FvE NZ JJ. Wrote the paper: ER ELvB JM JJ FvE.

References

1. Saba A, Messina F (2003) Attitudes towards organic foods and risk/benefit perception associated with pesticides. Food Quality and Preferences 14: 637–645.
2. Stolz H, Stolze M, Hamm U, Janssen M, Ruto E (2011) Consumer attitudes towards organic versus conventional food with specific quality attributes. Netherlands Journal of Agricultural Science 58: 67–72.
3. Smith-Spangler C, Brandeau ML, Hunter GE, Bavinger JC, Pearson M, et al (2012) Are organic foods safer or healthier than conventional alternatives? A systematic review. Annals of Internal Medicine 157: 348–366.
4. Asami DK, Hong YJ, Barrett DM, Mitchell AE (2003) Comparison of the total phenolic and ascorbic acid content of freeze-dried and air-dried marionberry, strawberry, and corn grown using conventional, organic and sustainable agricultural practices. Journal of Agriculture & Food Chemistry 51(5): 1237–1271.

5. Chassy AW, Bui L, Renaud ENC, Mitchell AE (2006) Three-year comparison of the content of antioxidant microconstituents and several quality characteristics in organic and conventionally managed tomatoes and bell peppers. Journal of Agriculture and Food Chemistry 54: 8244–8252.

6. Brandt K, Leifert C, Sanderson R, Seal CJ (2011) Agroecosystem management and nutritional quality of plant foods: The case of organic fruits and vegetables. Critical Reviews in Plant Sciences 30: 177–197.

7. Hunter D, Foster M, McArthur JO, Ojha R, Petocz P, et al (2011) Evaluation of the micronutrient composition of plant foods produced by organic and conventional methods. Critical Reviews in Food Science and Nutrition 51: 571–582.

8. Koh E, Charoenpraset S, Mitchell AE (2012) Effect of organic and conventional cropping systems on ascorbic acid, vitamin C, flavonoids, nitrate, and oxalate in 27 varieties of spinach (*Spinacia oleracea* L.) Journal of Agriculture and Food Chemistry. 60(12): 3144–3150.

9. Crozier A, Clifford MN, Ashihara H (eds) (2006) Plant secondary metabolites: occurrence, structure and role in the human diet. Blackwell Publishing Ltd.

10. Bourn D, Prescott J (2002) A comparison of the nutritional value, sensory qualities and food safety of organically and conventionally produced foods. Critical Review Food Science Nutrition 42(1): 1–34.

11. Mäder P, Fliessbach A, Dubois D, Gunst L, Fried P, et al (2002) Soil fertility and biodiversity in organic farming. Science 296: 1694–1697.

12. Martinez-Bellesta MC, Lopez-Perez L, Hernandez M, Lopez-Berenguer C, Fernandez-Garcia N, et al (2008) Agricultural practices for enhanced human health. Phytochemical Review 7: 251–260.

13. Meyer M, Adam ST (2008) Comparison of glucosinolate levels in commercial broccoli and red cabbage from conventional and ecological farming. European Food Research Technology 226: 1429–1437.

14. Mozafar A (1993) Nitrogen fertilizers and the amount of vitamins in plants: a review. Journal of Plant Nutrition 16: 2479–2506.

15. Zhao X, Carey EE, Wang W, Rajashekar CB (2006) Does organic production enhance phytochemical content of fruit and vegetables? Current knowledge and prospects for research. HortTechnology 16(3): 449–456.

16. Dixon RA, Paiva NL (1995) Stress-induced phenylpropanoid metabolism. The Plant Cell 7: 1085–1097.

17. Kim HS, Juvik JA (2011) Effect of selenium fertilization and methyl jasmonate treatment on glucosinolate accumulation in broccoli floret. Journal of the American Society for Horticulture Science 136(4): 239–246.

18. USDA National Nutrient Database, 2011. http://ndb.nal.usda.gov/(last visited 14 April 2014).

19. Brown AF, Yousef GG, Jeffrey EH, Klein BP, Wallig MA, et al (2002) Glucosinolate profiles in broccoli: variation in levels and implications in breeding for cancer chemoprotection. Journal of American Society Horticulture Science 127(5): 807–813.

20. Kushad MM, Brown AF, Kurilich AC, Juvik JA, Klein BP, et al (1999) Variations of glucosinolates in vegetable crops of *Brassica oleracea*. Journal of Agriculture and Food Chemistry 47: 1541–1548.

21. Farnham MW, Kopsell DA (2009) Importance of genotype on carotenoid and chlorophyll levels in broccoli heads. HortScience 44(5): 1248–1253.

22. Verhoeven DT, Goldbohm PA, van Poppel GA, Verhagen H, van den Brandt PA (1996) Epidemiological studies on *Brassica* vegetables and cancer risk. Biomarkers Prevention 5: 733–748.

23. Keck A-S, Finley JW (2004) Cruciferous vegetables: cancer protective mechanisms of glucosinolate hydrolysis products and selenium. Integrative Cancer Therapies 3(1): 5–12.

24. Here I, Büchler MW (2010) Dietary constituents of broccoli and other cruciferous vegetables: implications for prevention and therapy of cancer. Cancer Treatment Review 36: 377–383.

25. Higdon JV, Delage B, Williams DE, Dashwood RH (2007) Cruciferous vegetables and human cancer risk: epidemiologic evidence and mechanistic basis. Pharmacology Research 55(3): 224–236.

26. Schonhof I, Krumbein A, Bruchner B (2004) Genotypic effects on glucosinolates and sensory properties of broccoli and cauliflower. Nahrung/Food 48(1): 25–33.

27. Fenwick GR, Heaney RK, Mulllin WJ (1983) Glucosinolates and their breakdown products in food and food plants. Critical Reviews in Food Science and Nutrition 18(2): 123–201.

28. Juge N, Mithen RF, Traka M (2007) Molecular basis for chemoprevention by sulforaphane: a comprehensive review. Cellular and Molecular Life Sciences 64: 1105–1127.

29. Mithen RF, Dekker M, Verkerk R, Rabot S, Johnson IT (2000) The nutritional significance, biosynthesis and bioavailability of glucosinolates in human foods. Journal of the Science of Food and Agriculture 80: 967–984.

30. Wang LI, Giovannucci EL, Hunter D, Neuberg D, Su L, et al (2004) Dietary intake of cruciferous vegetables, glutathione S-transferase (GST) polymorphisms and lung cancer risk in a Caucasian population. Cancer Causes Control 15: 977–985.

31. Hsu CC, Chow WH, Boffetta P, Moore L, Zaridze D, et al (2007) Dietary risk factors for kidney cancer in eastern and central Europe. American Journal of Epidemiology 166: 62–70.

32. Kirsh VA, Peters U, Mayne ST, Subar AF, Chatterjee N, et al (2007) Prospective study of fruit and vegetable intake and risk of prostate cancer. Journal of the National Cancer Institute 99: 1200–1209.

33. Lam TK, Ruczinski I, Helzlsouer KJ, Shugart YY, Caulfield LE, et al (2010) Cruciferous vegetable intake and lung cancer risk: a nested case-control study matched on cigarette smoking. Cancer Epidemiology, Biomarkers and Prevention 19: 2534–2540.

34. Bosetti C, Filomeno M, Riso P, Polesel J, Levi F, et al (2012) Cruciferous vegetables and cancer risk in a network of case-control studies. Annals of Oncology 23: 2198–2203.

35. Wu QJ, Yang Y, Vogtmann E, Wang J, Han LH, et al (2012) Cruciferous vegetables intake and the risk of colorectal cancer: a meta-analysis of observational studies. Annals of Oncology 00: 1–9. doi:10.1093/annonc/mds601.

36. Kopsell DA, Kopsell DE (2006) Accumulation and bioavailability of dietary carotenoids in vegetable crops. Trends in Plant Science 11(10): 499–507.

37. Ibrahim KE, Juvik JA (2009) Feasibility for improving phytonutrient content in vegetable crops using conventional breeding strategies: case study with carotenoids and tocopherols in sweet corn and broccoli. Journal of Agriculture and Food Chemistry 57: 4636–4644.

38. Kurilich AC, Tsau GJ, Brown A, Howard L, Klein BP, et al (1999) Carotene, tocopherol, and ascorbate contents in subspecies of *Brassica oleracea*. Journal of Agriculture and Food Chemistry 47: 1576–1581.

39. Kurilich AC, Juvik JA (1999) Quantification of carotenoid and tocopherol antioxidants in *Zea mays*. Journal of Agriculture and Food Chemistry 47: 1948–1955.

40. Mayne ST (1996) βeta carotene, carotenoids, and disease prevention in humans. The FASEB Journal. 10(7): 690–701.

41. Krinsky NI, Landdrum JT, Bone RA (2003) Biologic mechanisms of the protective role of lutein and zeazanthin in the eye. Annual Review Nutrition. 23: 171–201.

42. Kritchevsky SB (1999) β-Carotene, carotenoids and the prevention of coronary heart disease. Journal of Nutrition 129: 5–8.

43. Davis D, Epp MD, Riordan HD (2004) Changes in USDA food composition data for 43 garden crops, 1950 to 1999. Journal of the American College of Nutrition 23(6): 1–14.

44. Traka MH, Saha S, Huseby S, Kopriva S, Walley PG, et al (2013). Genetic regulation of glucoraphanin accumulation in Beneforte broccoli. New Phytologist, doi: 10.1111/nph.12232.

45. Munger HM (1979) The potential of breeding fruits and vegetables for human nutrition. HortScience 14(3): 247–250.

46. Welch RM, Graham RD (2004) Breeding for micronutrients in staple food crops from a human nutrition perspective. Journal of Experimental Botany 55(396): 353–364.

47. Troxell Alrich H, Salandanan K, Kendall P, Bunning M, Stonaker F, et al (2010) Cultivar choice provides options for local production of organic and conventionally produced tomatoes with higher quality and antioxidant content. Journal of Science Food and Agriculture 90: 2548–2555.

48. Murphy KM, Campbell KG, Lyon SR, Jones SS (2007) Evidence of varietal adaptation to organic farming systems. Field Crops Research 102: 172–177.

49. Baresel JP, Zimmermann G, Reents HJ (2008) Effects of genotype and environment on N uptake and N partition in organically grown winter wheat (*Triticum aestivum* L.) in Germany. Euphytica 163: 347–354.

50. Vlachostergios DN, Roupakias DG (2008) Response to conventional and organic environment of thirty-six lentil (*Lens culinaris* Medik.) varieties. Euphytica 163: 449–457.

51. Goldstein W, Schmidt W, Burger H, Messmer M, Pollak LM, et al (2012) Maize: Breeding and field testing for organic farmers. In: Lammerts van Bueren ET, Myers, JR editors. Organic crop breeding. West Sussex, UK, Wiley-Blackwell, John Wiley & Sons, Inc.

52. Lorenzana RE, Bernardo R (2008) Genetic correlation between corn performance in organic and conventional production systems. Crop Science 48(3): 903–910.

53. Osman AM, Almekinders CJM, Struik PC, Lammerts van Bueren, ET (2008) Can conventional breeding programmes provide onion varieties that are suitable for organic farming in the Netherlands? Euphytica 163: 511–522.

54. Przystalski M, Osman AM, Thiemt EM, Rolland B, Ericson L, et al (2008) Do cereal varieties rank differently in organic and non-organic cropping systems? Euphytica 163: 417–435.

55. Wunderlich SM, Feldmand C, Kane S, Hazhin T (2008) Nutritional quality of organic, conventional, and seasonally grown broccoli using vitamin C as a marker. International Journal of Food Sciences and Nutrition 59(1): 34–45.

56. Koh E, Wimalasin KMS, Chassy AW, Mitchell AE (2009) Content of ascorbic acid, quercetin, kaempferol and total phenolics in commercial broccoli. Journal of Food Composition and Analysis. 22(7–8): 637–643.

57. Harker FR (2004) Organic food claims cannot be substantiated through testing of samples intercepted in the marketplace: a horticulturist's opinion. Food Quality and Preference 15: 91–99.

58. Benbrook C (2009) The impact of yield on nutritional quality: lessons from organic farming. HortScience 44(1): 12–14.

59. Carlson DG, Daxenbichler ME, VanEtten CH, Kwolek WF, Williams PH (1987) Glucosinolates in crucifer vegetables: broccoli, brussels sprouts, cauliflower, collards, kales, mustard greens, and kohlrabi. Journal of the American Society for Horticulture Science 112(1): 173–178.

60. Farnham MW, Stephenson KK, Fahey JW (2005) Glucoraphanin level in broccoli seed is largely determined by genotype. HortScience 40(1): 50–53.

61. Wang J, Gu H, Yu H, Zhao Z, Sheng X, Zhang X (2012) Genotypic variation of glucosinolates in broccoli (*Brassica oleracea* var. italica) florets from China. Food Chemistry 133: 735–741.

62. Rosa EAS, Rodrigues AS (2001) Total and individual glucosinolate content in 11 broccoli cultivars grown in early and late seasons. HortScience 36(1): 56–59.

63. Vallejo F, Tomas-Barberán FA, García-Viguera C (2003a) Effect of climatic and sulphur fertilisation conditions, on phenolic compounds and vitamin C, in the inflorescences of eight broccoli cultivars. European Food Research Technology 216: 395–401.

64. Vallejo F, Tomas-Barberán FA, Gonzalez Benavente-Garcia A, García-Viguera C (2003b) Total and individual glucosinolate contents in inflorescences of eight broccoli cultivars grown under various climatic and fertilisation conditions. Journal of the Science of Food and Agriculture 83: 307–313.

65. Charron CS, Saxton A, Sams CE (2005a) Relationship of climate and genotype to seasonal variation in the glucosinolate – myrosinase system. I. Glucosinolate content in ten cultivars of *Brassica oleracea* grown in fall and spring seasons. Journal of the Science of Food and Agriculture 85: 671–681.

66. Charron CS, Saxton A, Sams CE (2005b) Relationship of climate and genotype to seasonal variation in the glucosinolate – myrosinase system. II. Myrosinase activity in ten cultivars of *Brassica oleracea* grown in fall and spring seasons. Journal of the Science of Food and Agriculture 85: 682–690.

67. Aires A, Fernandes C, Carvalho R, Bennett RN, Saavedra MJ, et al (2011) Seasonal effects on bioactive compounds and antioxidant capacity of six economically important brassica vegetables. Molecules 16: 6816–6832.

68. Pek Z, Daood H, Nagne MG, Berki M, Tothne MM, et al (2012) Yield and phytochemical compounds of broccoli as affected by temperature, irrigation, and foliar sulphur supplementation. HortScience 47(11): 1646–1652.

69. Schonhof I, Blandenburg D, Müller S, Krumbein A (2007) Sulfur and nitrogen supply influence growth, product appearance, and glucosinolate concentration of broccoli. Journal of Plant Nutrition and Soil Science 170: 65–72.

70. Farnham MW, Wilson PE, Stephenson KK, Fahey JW (2004) Genetic and environmental effects on glucosinolate content and chemoprotective potency of broccoli. Plant Breeding 123: 60–65.

71. Björkman M, Klingen I, Birch ANE, Bones AM, Bruce TJA, et al (2011) Phytochemicals of Brassicaceae in plant protection and human health – influences of climate, environment and agronomic practice. Phytochemistry 72: 538–556.

72. Robbins RJ, Keck AS, Banuelos G, Finley JW (2005) Cultivation conditions and selenium fertilization alter the phenolic profile, glucosinolate, and sulforaphane content of broccoli. Journal of Medicinal Food 8(2): 204–214.

73. Xu C, Guo R, Yan H, Yuan J, Sun B, et al (2010) Effect of nitrogen fertililization on ascorbic acid, glucoraphanin content and quinone reductase activity in broccoli floret and stem. Journal of Food, Agriculture and Environment 8(1): 179–184.

74. Naguib AM, El-Baz FK, Salama ZA, Hanna HAEB, Ali HF, et al (2012) Enhancement of phenolics, flavonoids and glucosinolates of broccoli (*Brassica oleracea*, var. Italica) as antioxidants in response to organic and bio-organic fertilizers. Journal of the Saudi Society of Agricultural Sciences 11: 135–142

75. Picchi V, Migliori C, Scalzo RL, Campanelli G, Ferrari V, et al (2012) Phytochemical content in organic and conventionally grown Italian cauliflower. Food Chemistry 130: 501–509.

76. Hansen M, Moller P, Sorensen H, Cantwell de Trejo M (1995) Glucosinolates in broccoli stored under controlled atmosphere. Journal of the American Society for Horticulture Science 120(6): 1069–1074.

77. Tiwari U, Cummins E (2013) Factors influencing levels of phytochemicals in selected fruit and vegetables during pre- and post-harvest food processing operations. Food Research International 50: 497–506.

78. Renaud ENC, Lammerts van Bueren ET, Paulo MJ, van Eeuwijk FA, Juvik JA, et al(2014) Broccoli cultivar performance under organic and conventional management systems and implications for crop improvement. Crop Science DOI: 10.2135/cropsci201.

79. De Ponti T, Rijk B, van Ittersum MK, (2012) The crop yield gap between organic and conventional agriculture. Agricultural Systems 108: 1–9.

80. Shewmaker CK, Sheehy JA, Daley M, Colburn S, Ke DY (1999) Seed-specific overexpression of phytoene synthase: increase in carotenoids and other metabolic effects. Plant Journal 20: 401–412.

81. Kopsell DA, Kopsell DE, Lefsrud MG, Curran-Celentano J, Dukach LE (2004) Variation in lutein, β-carotene, and chlorophyll concentrations among *Brassica oleracea* cultigens and seasons. HortScience 39(2): 361–364.

82. Hale AL, Farnham MW, Ndambe Nzaramba M, Kimbeng CA (2007) Heterosis for horticultural traits in broccoli. Theoretical Applied Genetics 115(3): 351–60.

83. Murphy KM, Reeves PG, Jones SS (2008) Relationship between yield and mineral nutrient concentrations in historical and modern spring wheat cultivars. Euphytica 163: 381–390.

84. Hussain A, Larsson H, Kuktaite R, Johansson E (2010) Mineral composition of organically grown wheat genotypes: contribution to daily minerals intake. International Journal of Environmental Research and Public Health 7: 3443–3456.

85. Jones H, Clarke S, Haigh Z, Pearce H, Wolfe M (2010) The effect of the year of wheat variety release on productivity and stability of performance on two organic and two non-organic farms. Journal of Agricultural Science 148: 303–317.

86. Farnham MW, Keinath AP, Grusak MA (2011) Mineral concentration of broccoli florets in relation to year of cultivar release. Crop Science 51: 2721–2727.

87. Troxell-Alrich H, Kendall P, Bunning M, Stonaker F, Kulen O, et al (2011) Environmental temperatures influence antioxidant properties and mineral content in broccoli cultivars grown organically and conventionally. Journal of AgroCrop Science 2(2): 1–10.

88. Khoo H-E, Prasad KN, Kong K-W, Jiang Y, Ismail A (2011) Carotenoids and their isomers: color pigments in fruits and vegetables. Molecules 16(2): 1710–1738.

89. Faulkner K, Mithen R, Williamson G (1998) Selective increase of the potential anticarcinogen 4-methylsulphinylbutyl glucosinolate in broccoli. Carcinogenis 19(4): 605–609.

90. Mithen R, Faulkner K, Magrath R, Rose P, Williamson G, Marquez J (2003) Development of isothiocyanate-enriched broccoli, and its enhanced ability to induce phase 2 detoxification enzymes in mammalian cells. Theoretical Applied Genetics 106: 727–734.

91. Ceccarelli S (1996) Adaptation to low/high input cultivation. Euphytica 92: 203–214.

92. Ceccarelli S (1994) Specific adaptation and breeding for marginal conditions. Euphytica 77: 205–219.

93. Löschenberger F, Fleck A, Grausgruber H, Hetzendorfer H, Hof G, et al (2008) Breeding for organic agriculture: the example of winter wheat in Austria. Euphytica 163: 469–480.

94. Kirk AP, Fox SL, Entz MH (2012) Comparison of organic and conventional selection environments for spring wheat. Plant Breeding 131: 687–694.

Identification of Suitable Reference Genes for Gene Expression Normalization in qRT-PCR Analysis in Watermelon

Qiusheng Kong[9], Jingxian Yuan[9], Lingyun Gao, Shuang Zhao, Wei Jiang, Yuan Huang, Zhilong Bie*

Key Laboratory of Horticultural Plant Biology, Ministry of Education/College of Horticulture and Forestry, Huazhong Agricultural University, Wuhan, China

Abstract

Watermelon is one of the major Cucurbitaceae crops and the recent availability of genome sequence greatly facilitates the fundamental researches on it. Quantitative real-time reverse transcriptase PCR (qRT–PCR) is the preferred method for gene expression analyses, and using validated reference genes for normalization is crucial to ensure the accuracy of this method. However, a systematic validation of reference genes has not been conducted on watermelon. In this study, transcripts of 15 candidate reference genes were quantified in watermelon using qRT–PCR, and the stability of these genes was compared using geNorm and NormFinder. geNorm identified ClTUA and ClACT, ClEF1α and ClACT, and ClCAC and ClTUA as the best pairs of reference genes in watermelon organs and tissues under normal growth conditions, abiotic stress, and biotic stress, respectively. NormFinder identified ClYLS8, ClUBCP, and ClCAC as the best single reference genes under the above experimental conditions, respectively. ClYLS8 and ClPP2A were identified as the best reference genes across all samples. Two to nine reference genes were required for more reliable normalization depending on the experimental conditions. The widely used watermelon reference gene 18SrRNA was less stable than the other reference genes under the experimental conditions. Catalase family genes were identified in watermelon genome, and used to validate the reliability of the identified reference genes. ClCAT1and ClCAT2 were induced and upregulated in the first 24 h, whereas ClCAT3 was downregulated in the leaves under low temperature stress. However, the expression levels of these genes were significantly overestimated and misinterpreted when 18SrRNA was used as a reference gene. These results provide a good starting point for reference gene selection in qRT-PCR analyses involving watermelon.

Editor: Meng-xiang Sun, Wuhan University, China

Funding: This work was founded by the earmarked fund for Modern Agro-industry Technology Research System (CARS-26-16), the National Science & Technology Pillar Program during the 12th Five-year Plan Period (2012BAD02B03-16), and Special Fund for Agro-scientific Research in the Public Interest (201003066-6). The funders had no role in study design, data collection and analysis, decision to publish, or preparation of the manuscript.

Competing Interests: The authors have declared that no competing interests exist.

* E-mail: biezhilong@hotmail.com

9 These authors contributed equally to this work.

Introduction

Watermelon (*Citrullus lanatus*), a major Cucurbitaceous crop and the fifth most consumed fresh fruit globally, is an important horticultural crop, and its planting area accounts for 6% of the worldwide area devoted to vegetable production in 2011 (FAOSTAT 2013, http://faostat3.fao.org). Watermelon is the third crop in the Cucurbitaceae family in which the genome has been sequenced, after cucumber and melon. An available genome sequence and large-scale transcriptome data will greatly facilitate molecular biology studies in watermelon [1–3].

Gene expression analysis is an effective and widely used approach to elucidate the regulatory networks and identify novel genes in molecular biology. Quantitative real-time reverse transcriptase PCR (qRT–PCR) has become the preferred method for gene expression studies because of its rapidity, sensitivity, and specificity [4]. However, the accuracy of the results obtained by this method depends on accurate transcript normalization using stably expressed reference genes, which allows the regulation of possible non-biological variations when the reference genes are exposed to the same preparation processes as the genes of interest

[5,6]. Therefore, appropriate reference genes should be validated with minimal variability in expression relative to the test samples before qRT–PCR analysis.

However, no validated reference genes have been reported for normalization of gene expression in watermelon as of this writing. *18SrRNA* is frequently used in watermelon as a reference gene for normalization in the fruit [2,3], root under *Fusarium* wilt infection [7], and leaf under cold stress [8], in qRT-PCR analyses without prior validation. In addition, *actin* was also used in watermelon leaf under water deficit stress [9]. The choice of such traditional genes as references may be inappropriate because their status as housekeeping genes is generally based on methods that are mainly qualitative (e.g., Northern blot) and is inconsistent with the high accuracy associated with qRT-PCR [10]. The disadvantages of using *18SrRNA* as a reference gene include its absence in purified mRNA samples and high abundance compared with target mRNA transcripts, which complicates the accurate subtraction of the baseline value in qRT-PCR data analysis [11].

The use of unstable references can dramatically change the interpretation of an expression pattern of a given target gene, and introduce flaws in the understanding of the function of the gene [12,13]. Systematic validation of reference genes is essential for producing accurate and reliable data in qRT-PCR analyses, and should be included as an integral component of these analyses [6]. A proper normalization strategy is also among the essential key elements on the Minimum Information for Publication of Quantitative Real-Time PCR Experiments (MIQE) guidelines [14]. The advent of watermelon genome sequence will greatly expedite the completion of studies related to gene expression, whereas the absence of these validated reference genes on watermelon will significantly impede the accurate quantification of gene expression. Validation of suitable reference genes for watermelon can guarantee the accurate quantification of the target genes in qRT-PCR analysis.

In this study, 15 candidate reference genes used in watermelon or validated in other crops were selected, and their transcripts were quantified in the organs and tissues of watermelon under various experimental conditions by qRT-PCR. NormFinder [15] and geNorm [11] were used to identify the suitable reference genes for normalization of gene expression in watermelon.

Catalase family genes are considered as the peroxisomal redox guardians in plants, and the proteins encoded by these genes have relatively specific functions in determining the accumulation of H_2O_2, which is an important signal molecule involved in plant development and environmental responses [16,17]. In *Arabidopsis*, catalase family genes are differentially expressed under different stresses to control reactive oxygen species (ROS) homeostasis [18]. Watermelon is susceptible to many biotic and abiotic stresses during production. However, the functions of catalase family genes in response to environmental stresses on watermelon remain unclear. The reliability of the identified reference genes was further validated by analyzing the expression patterns of catalase family genes under low temperature stress in watermelon leaves using the stable and unstable genes for normalization. The results provide valuable information for suitable reference gene selection in gene expression studies in watermelon.

Materials and Methods

Plant Materials and Treatments

The sequenced watermelon inbred line 97103 (*C. lanatus* (Thunb.) Matsum. & Nakai var. *lanatus*) was used as plant material [1]. The seeds were first sterilized with 1.5% sodium hypochlorite, soaked in distilled water for 4 h, and maintained at 30°C for germination. The germinated seedlings were planted in sterilized peat-perlite substrate (2:1, v/v) and cultured in the greenhouse. Seedlings with two true leaves were used for the following treatments.

The devastating diseases of *Fusarium* wilt and bacterial fruit blotch, which frequently occur on watermelon, were considered as the biotic stresses. Artificial inoculation of bacterial fruit blotch was conducted with 10^8 cfu·mL^{-1} suspension of *pslbtw20* strain (*Acidovorax avenae* subsp. *citrulli*) in accordance with the methods in a previous report [19]. The inoculation of *Fusarium* wilt pathogen was performed as described by Lu et al. [7]. Seedlings were infected with *Fusarium oxysporum* f. sp. *niveum* isolate FON1 by dipping their roots in a conidial suspension of 5×10^6 spore·mL^{-1} for 15 min. The inoculated seedlings were replanted in sterilized substrate and cultured under a 12 h diurnal light cycle at 26°C with 80% to 85% relative humidity inside a controlled environment chamber. Plants were watered every 2 d with 1/2 Hoagland nutrient solution. Root and leaf samples were obtained at 3 d post-inoculation. The remaining seedlings were maintained until

typical symptoms of *Fusarium* wilt or bacterial fruit blotch were visible to confirm the success of artificial inoculation.

Low temperature, salinity, and drought, which are the major environmental stresses in watermelon production, were adopted as the abiotic stresses. For low temperature treatment, the seedlings were stored at 10±1°C for 24 h in a controlled environment chamber with a 12 h diurnal light cycle. Seedlings used for salt and drought treatments were transplanted and cultivated hydroponically in 1/2 Hoagland nutrient solution for 5 d in the greenhouse so the seedlings could adapt to the growth environment. For salt treatment, NaCl was gradually added into the nutrition solution until a final concentration of 100 mM was reached, after which the seedlings were cultured for 48 h. For drought treatment, polyethylene glycol 6000 was gradually added into the nutrient solution until a final concentration of 10% was reached, after which the seedlings were cultured for 24 h. Roots and leaves were sampled for each abiotic stress treatment.

Under normal growth conditions, the organs of root, stem, and leaf were collected from the seedlings at the stage of two true leaves. Tendrils were collected from the flowering plants. The day before anthesis, the female flowers were covered by paper bags to prevent natural pollination. On the day of anthesis, flower tissues, including stamen, stigma, petal, and unfertilized ovary, were sampled in the morning. Fruit flesh tissues were collected at 15 and 31 d after pollination.

The biotic and abiotic stress treatments were performed simultaneously, and the seedlings at the stage of two true leaves under normal growth conditions served as controls. Three biological replicates were adopted for the aforementioned treatments, and each replicate contained 15 plants. Samples were randomly collected from five plants for each replicate, immediately frozen in liquid nitrogen, and maintained at −80°C for subsequent RNA extraction.

Candidate Reference Selection and Primer Design

A total of 15 candidate reference genes were evaluated. These genes were chosen based on their previous use in watermelon or their validation as best reference genes in other crops, including 18S ribosomal RNA (*18SrRNA*), β-actin (*ACT*), clathrin adaptor complex subunit (*CAC*), elongation factor 1-α (*EF1α*), glyceraldehy-3-phosphate-dehydrogenase (*GAPDH*), NADP-isocitrate dehydrogenase (*IDH*), leunig (*LUG*), protein phosphatase 2A regulatory subunit A (*PP2A*), polypyrimidine tract-binding protein 1 (*PTB*), ribosomal protein S (*RPS2*), SAND family protein (*SAND*), α-tubulin (*TUA*), ubiquitin-conjugating enzyme E2 (*UBC2*), ubiquitin carrier protein (*UBCP*), and yellow-leaf-specific procein8 (*YLS8*).

For each candidate reference gene, blastn was carried out in the Cucurbit Genomics Database (http://www.icugi.org) against watermelon coding DNA sequences (CDS) (v1) using *Arabidopsis* homolog as a query. The CDS of the best hit was retrieved and uploaded to Primer3Plus (http://primer3plus.com/cgi-bin/dev/primer3plus.cgi) for primer design. The product size was set at the range of 80 bp to 150 bp. The forward and reverse primers were intentionally targeted on the adjoining exons, which were separated by an intron. The generated primer pair for each gene was then aligned against all watermelon CDS to confirm its specificity *in silico*. The specificity of the PCR amplification product for each primer pair was further determined by electrophoresis in 2% agarose gel and melting curve analysis. Finally, the watermelon species name abbreviation of '*Cl*' was added as a prefix to the specificity-validated gene to specify the watermelon orthologous gene. For more comparable results, the primer pair of *18SrRNA*, which was previously published, was used

Table 1. Information on the selected reference genes.

Gene name	Gene description	Gene ID[a]	Arabidopsis homolog locus[b]	E-value	Forward primer seqence 5'-3'	Reverse primer seqence 5'-3'	Product size (bp)	Amplification efficiency (%)	R²
ClACT	β-Actin	Cla007792	AT3G18780	0	F:CCATGTATGTTGCCATCCAG	R:GGATAGCATGGGGTAGAGCA	140	103.9	0.989
ClCAC	Clathrin adaptor complex subunit	Cla020794	AT4G24550	5e-50	F:AATTGTGGTTGATGCTGCAC	R:TGACAGCTGTACCTGGCATC	94	97.2	0.992
ClEF1α	Elongation factor 1-α	Cla010539	AT5G60390	0	F: AGCACGCTCTTCTTGCTTTC	R:ACGATTTCGTCGTACCTTGC	115	96.4	0.996
ClGAPDH	Glyceraldehy-3-phosphate-dehydrogenase	Cla013454	AT1G16300	2e-80	F:CTGGCAGTACTTTGCCAACA	R:AGGATTGGGAGAGAGGTCGT	87	92.3	0.991
ClIDH	NADP-isocitrate dehydrogenase	Cla009135	AT1G65930	2e-71	F:TGGCCTCTTTACCTAAGCACA	R:ATATGCCAGCAGCCTCAAAC	124	94.0	0.991
ClUG	Leunig	Cla022288	AT4G32551	3e-62	F:TTGCTGGTCATTGGATGCTA	R:GCCGAAGCAACTAGACCTGA	138	103.3	0.995
ClPP2A	Protein phosphatase 2A regulatory subunit A	Cla021905	AT1G69960	e-125	F:AAGAGCCCACCAGCTTGTAA	R:TGTTCTCCCCAATCTCAAGG	136	101.0	0.996
ClPTB	Polypyrimidine tract-binding protein 1	Cla004906	AT3G01150	2e-15	F:GGAGCAAACAGAAATCAAGC	R:AGCAGGCTCAGAGAAGATG	133	107.9	0.992
ClRPS2	Ribosomal protein S	Cla021565	AT1G04270	3e-67	F:TGGCACTGATCAAGAAGCTG	R:TGATCATATTGCGGAGGTGA	97	98.4	0.996
ClSAND	SAND family protein	Cla001870	AT2G28390	4e-42	F:CAATTAGCAGCCGTCAACAA	R:GTTTTGTGAGGGCCAATTTC	100	109.6	0.996
ClTUA	α-Tubulin	Cla003129	AT1G64740	e-120	F:CTTGCTGGGAGTCTATTGC	R:AACGGATTAAAAGCGTCGTG	105	94.6	0.992
ClUBC2	Ubiquitin-conjugating enzyme E2	Cla010164	AT2G02760	2e-81	F:CCAAATAAGCCACCGACAGT	R:TCATAGATTGGGCTCCATTG	118	93.7	0.991
ClUBCP	Ubiquitin carrier protein	Cla010163	AT1G14400	6e-81	F:ACCAACAGTCCGCTTTGTGT	R:ATTGGGCTCCACTGATTTTG	101	96.6	0.995
ClYLS8	Yellow-leaf-specific proein8	Cla020175	AT5G08290	e-78	F:AGAACGGCTTGGTGGTCATTC	R:GAGGCCAACACTTCATCCAT	83	94.4	0.995
Cl18SrRNA[c]	18S ribosomal RNA				F:AGCCTGAGAAACGGCTACCACATC	R:ACCAGACTCGAAGAGCCCGGTAT		92.1	0.995

[a]Watermelon gene ID in Cucurbit Genomics Database (http://www.icugi.org);

[b]Arabidopsis gene ID in TAIR database (http://www.arabidopsis.org/);

[c]The primer pair of *18SrRNA* previously published was used here (Guo et al. 2011).

in this study [2]. Data on the selected reference genes and their amplification characters are listed in Table 1.

Total RNA Extraction, cDNA Synthesis, DNA Isolation, and PCR Amplification

The eleven golden rules of qRT–PCR were adopted as guidelines for RNA isolation, cDNA synthesis, and subsequent qRT–PCR analysis [20]. Total RNA was isolated using TransZol (TransGen) according to the manufacturer's instructions. The integrity of RNA was determined by electrophoresis in 2% agarose gel. The quantity and purity of RNA were determined using a NanoDrop^{TM} 2000 spectrophotometer (Thermo Scientific). Only high-quality samples in which $A_{260}/A_{280}>1.8$ and $A_{260}/A_{230}>2.0$ were used for subsequent cDNA synthesis. Genomic DNA elimination and cDNA synthesis were conducted using Prime-Script RT Reagent Kit with genomic DNA (gDNA) Eraser (Perfect Real Time, TaKaRa) according to the manual. For each sample, 1 µg of total RNA was used for each 20 µL reverse transcription reaction system. Genomic DNA was isolated from the leaves using Plant Genomic DNA Kit (Tiangen), and PCR amplifications were conducted using 2× PCR Reagent (Tiangen) according to the manual. The amplification products were resolved on 2% agarose gel.

qRT–PCR Analysis

qRT–PCR was carried out on a LightCycler480 System (Roche) using TransStart Top Green qPCR SuperMix (Trans-Gen). Reactions were performed using a total volume of 10 µL, which contained 100 ng of cDNA template, 0.2 µM each primer, and 1×Top Green qPCR SuperMix. The PCR cycling conditions were as follows: 94°C for 30 s, followed by 40 cycles of 95°C for 5 s, 55°C for 15 s, and 72°C for 10 s. The melting curve was recorded after 40 cycles to verify primer specificity by heating from 65°C to 95°C. Two technical replicates were adopted for each sample. Controls that were obtained without a template were included. For each gene, the full sample set in a replication was run on the same plate to exclude any technical variation. Amplification efficiencies for all primer pairs were evaluated using the serial fivefold dilutions of the pooled cDNA (500, 100, 20, 4, and 0.8 ng).

Data Analysis

Expression levels of the tested reference genes are determined by crossing point (Cp) values. The amplification efficiency (E) for each reference gene was calculated according to the following equation: E (%) = $(10^{-1/slope}-1)\times100$, where the slope is the standard curve slope calculated by the LightCycle 480 system. geNorm [11] and NormFinder [15] were used to assess the expression stability. Based on the principle that the expression ratio between two ideal reference genes should be invariable if the genes are stably expressed across the investigated sample set, geNorm calculates the gene expression stability value M, which is the average pairwise variation of a given gene with all other candidate reference genes, for each tested reference gene. The candidate reference genes are then ranked based on M value, and the gene with lower M value is considered to have higher expression stability. The least stable genes with the highest M are excluded stepwise until only the two most stable genes remain. The best two genes are ranked without distinguishing between them [11]. NormFinder calculates the expression stability for each single reference gene using a model-based approach with consideration of variations across groups. Lower stability value means higher expression stability of the gene [15]. For both

algorithms, the input data should be on a linear scale. Therefore, the raw Cp values were transformed to relative quantities Q using $Q = 2^{(minCp-sampleCp)}$ equation. R package (http://www.r-project.org/) was used to draw the plots.

To identify more stable genes under specific experimental conditions, all 20 samples were subdivided into three subsets based on their origins. The organ and tissue subset comprised organs and tissues from root, leaf, stem, tendril, pistil, stamen, petal, unfertilized ovary, and fruit flesh tissues at 15 and 31 d after pollination under normal growth conditions (10 samples); the biotic stress subset comprised roots and leaves collected from plants infected with *Fusarium* wilt and bacterial fruit blotch and the control (six samples); and the abiotic stress subset comprised roots and leaves collected from plants subjected to drought, salt, and low temperature stresses and the control (eight samples).

Normalization of Catalase Family Genes

To validate the reliability of the identified reference genes, the relative expression levels of watermelon catalase family genes under low temperature stress were analyzed. To identify the catalase family genes in the watermelon genome, the Hidden Markov Model of the catalase family (PF00199) was downloaded from the Pfam database (http://pfam.sanger.ac.uk/) and used as a query against the watermelon proteins (watermelon_v1.pep, ftp://www.icugi.org) using HMMER [21]. The primers were designed using Primique [22], which can design specific PCR primer pairs for each gene in a family. Specificity of the primers was further checked by electrophoresis in agarose gel and melting curve analysis. Low temperature treatment was conducted as mentioned above, and samples from leaves were obtained at 0, 12, 24, 36, and 48 h. Three biological replicates were adopted for each treatment. The transcripts of catalase genes were quantified by qRT–PCR, and two technical replicates were adopted for each sample. The relative expression levels were calculated using the $2^{-\Delta\Delta Cp}$ method. The single, pair of, and multiple best reference genes identified in this study, as well as the widely used reference gene *18SrRNA* in watermelon, were used as reference genes for normalization.

Results

Amplification Specificity and Efficiency for Each Candidate Reference Gene

To identify suitable reference genes for watermelon, 15 candidate reference genes were selected, including 14 reference genes previously identified as the most stable genes in other crops and the most frequently used reference gene *18SrRNA* in watermelon in qRT–PCR analyses. The watermelon orthologous genes were obtained by searching for watermelon CDS using *Arabidopsis* genes as queries. The best hit for each query was selected, and the same annotation as that in the *Arabidopsis* query was found in the watermelon genome database for each watermelon orthologous gene. Information on the selected reference genes is listed in Table 1.

To prevent the interference of gDNA contamination and pseudogene on qRT–PCR results, the forward and reverse primers were specifically located on the neighboring exons during primer design. The generated primer pairs were aligned with all watermelon CDS using blastn to confirm their specificity on a genomic scale *in silico*. When the target reference gene was the only output result of blastn, the primer pair was selected. For more comparable results, a primer pair of *18SrRNA* published in previous qRT–PCR analyses on watermelon was used in the present study.

PCR amplification specificity for each candidate reference gene was checked by electrophoresis in agarose gel using cDNA and gDNA as templates. As shown in Fig. 1, the primer pair for each reference gene amplified a specific product on both cDNA and gDNA templates. Amplicons with different sizes were observed between the cDNA and gDNA templates for the tested candidate genes, except for *18SrRNA*, which demonstrates the success of primer design and confirms that gDNA contamination did not occur in the cDNA samples. Melting curve analyses were also conducted for all the primer pairs. As shown in Fig. S1, the presence of a single peak with no visible primer-dimer formation further confirmed the specific amplification for each reference gene. Meanwhile, no signals were detected in the no-template controls. The specificity-validated primer sequences and amplification characters for the candidate reference genes are summarized in Table 1.

The qRT–PCR efficiency was determined for each primer pair by standard curve analysis, ranging from 92.1% (*Cl18SrRNA*) to 109.6% (*ClSAND*). The determination coefficients (R^2) of the standard curve regression equation varied from 0.989 for *ClACT* to 0.996 for *ClEF1α*, *ClPP2A*, *ClRPS2*, and *ClSAND* (Table 1). The aforementioned results prove that specific and high-efficiency qRT–PCR systems were established for the selected reference genes.

Expression Profiles of the Candidate Reference Genes

Expression levels of the 15 candidate reference genes were measured in the 20 samples collected from watermelon organs and tissues under normal growth conditions and biotic and abiotic

stresses by qRT–PCR, and presented as Cp values. Expression variations for these genes across the 20 samples are shown in Fig. 2. Different levels of transcription abundance were observed among these genes. *ClCAC*, which had the highest mean Cp value of 24.8, was expressed at the lowest level among the candidate reference genes. By contrast, *Cl18SrRNA* with the lowest average Cp value of 8.4 exhibited the highest transcription abundance. However, the expression levels for most of the genes were comparable and ranged from 18 to 24 cycles. None of the tested reference genes exhibited a constant expression level among the samples. The variability of Cp values in the 20 samples was highest for *Cl18SrRNA* and *ClRPS2*, whereas *ClYLS8* and *ClPTB* showed the lowest gene expression variations (Fig. 2). Five to six cycles of expression variations were observed for most candidate reference genes. However, given the variations in the amount of starting materials between the samples and subsequent operations of qRT–PCR, a direct comparison of the raw Cp values did not result in an accurate estimate of the expression stability of each reference gene. Therefore, the expression variation must be evaluated by more powerful methods.

Expression Stability Analyses

geNorm and NormFinder were used to evaluate the stability of the candidate reference genes. The ranks of the selected reference genes were determined by geNorm and are listed in Table 2. When all 20 samples were considered, *ClYLS8* and *ClPP2A* showed the lowest average expression stability value (M = 0.763), whereas *ClUBC2* showed the highest M value of 1.743. These results suggest that *ClYLS8* and *ClPP2A* had the most stable expression,

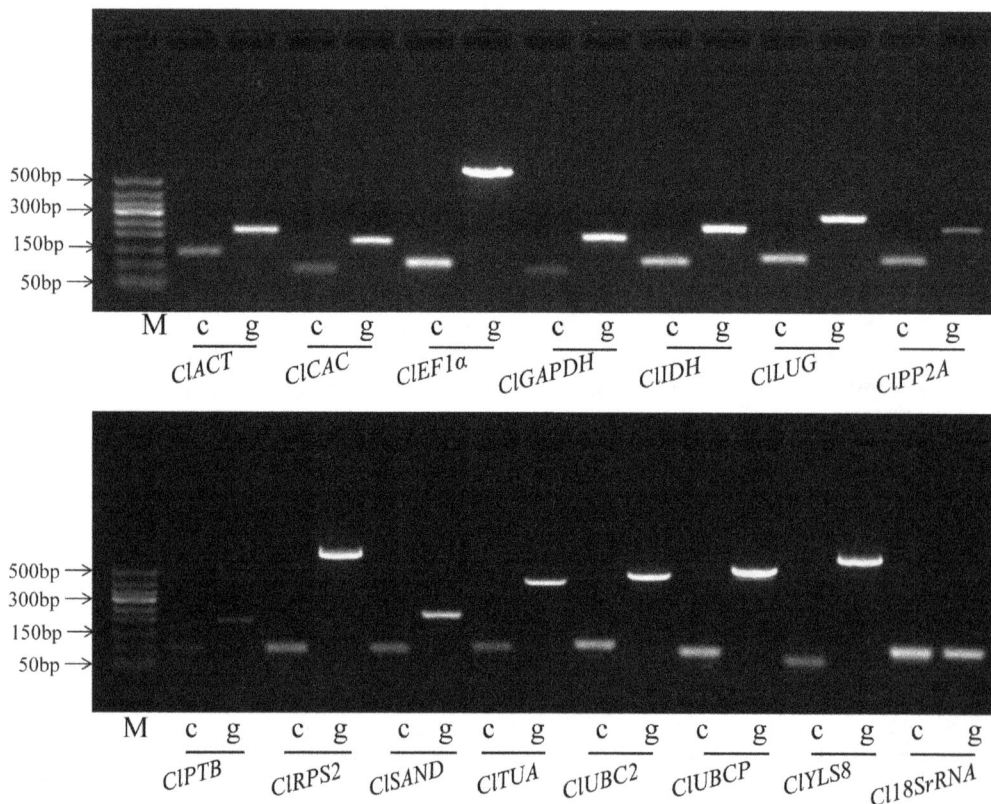

Figure 1. PCR amplification patterns of the candidate reference genes using cDNA and genomic DNA as templates, respectively, as visualized on 2% agarose gel. 'c' represents the use of cDNA as template. 'g' indicates the use of genomic DNA as template. 'M' represents the marker of 50 bp ladder.

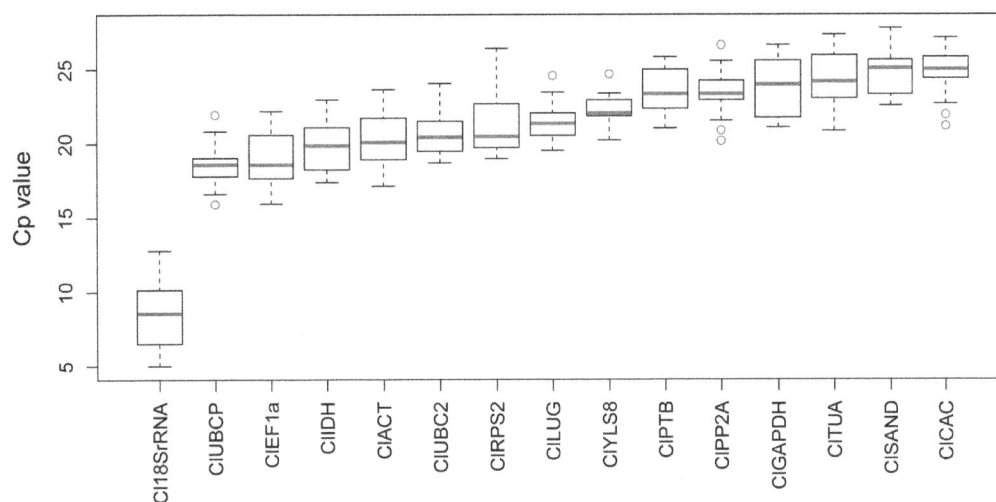

Figure 2. Expression profiles of the tested reference genes in raw Cp values in all 20 samples. Expression data are displayed as Cp values for each reference gene in all samples. The line across the box is the median. The boxes indicate the 25/75 percentiles. Whisker caps indicate the minimum and maximum values. The circles represent the outliers.

whereas $ClUBC2$ had the highest level of expression variation. Furthermore, pairwise variation between two sequential normalization factors (NFs) containing an increasing number of genes was also calculated using geNorm to determine the optimal number of genes required for normalization. A high pairwise variation indicates that the added gene had a significant effect and should preferably be included in the calculation of a reliable NF. The cut-off value of 0.15 was proposed, below which the inclusion of an additional reference gene is not required [11]. The results are illustrated in Fig. 3. The analysis showed that the pairwise variation V8/9 was higher than 0.15 (V = 0.154), whereas V9/10 was 0.134, which indicates that nine genes were required for more reliable normalization of target genes across the 20 samples. The results of expression stability, as evaluated by NormFinder, are summarized in Table 3. Compared with the geNorm results, $ClYLS8$ with the lowest stability value of 0.386 was also ranked as the most stable reference gene by NormFinder, and $ClUBC2$ with the highest stability value of 1.467 was identified as the least stable reference gene.

To determine more stable reference genes under specific experimental conditions, the 20 samples were further divided into three different subsets, as described in the data analysis section, and the experimental condition-specific reference genes were identified for each subset.

In the organ and tissue subset, which included different vegetative and reproductive organs or tissues of watermelon under normal growth conditions, $ClTUA$ and $ClACT$ were identified as the best pair of reference genes by geNorm (Table 2), and as many as nine genes were satisfactory for normalization because the pairwise variation value V9/10 was less than 0.15 (Fig. 3). NormFinder identified $ClYLS8$ as the best reference gene for this subset (Table 3). Both algorithms identified $ClRPS2$ as the gene with unstable expression.

In the abiotic stress subset, $ClEF1\alpha$ and $ClACT$ were identified as the best pair of reference genes by geNorm, and five genes, namely, $ClEF1\alpha$, $ClACT$, $ClUBCP$, $ClPTB$, and $ClIDH$, comprised the optimal reference genes for more accurate normalization. $ClUBCP$ was also identified as the most suitable reference gene by NormFinder. Similarly, $ClUBC2$ was identified as the least stable gene by both programs.

In the biotic stress subset, geNorm ranked $ClCAC$ and $ClTUA$ as the best reference genes, which were sufficient for more reliable normalization because the pairwise variation value V2/3 (0.129) was below the cut-off value of 0.15. $ClCAC$ was also ranked as the best reference gene by NormFinder. Both geNorm and Norm-Finder identified $ClGAPDH$ as the least stable reference gene.

$Cl18SrRNA$, the most frequently used reference gene in watermelon in qRT–PCR analyses, ranked from second to seventh from the bottom in all 20 samples and different subsets both by geNorm and NormFinder, which indicates that this gene was unsuitable for normalization in qRT–PCR analyses in watermelon.

Expression Profiles of Catalase Family Genes

Using the Hidden Markov Model of the catalase family, four genes ($Cla023447$, $Cla023448$, $Cla021932$, and $Cla003205$) were identified in watermelon genome using HMMER, with the threshold E-value <0.01. These genes were further confirmed by searching $Arabidopsis$ proteins in the TAIR database (http://www.arabidopsis.org/) using blastp. The genes $Cla023448$, $Cla023447$, and $Cla021932$ significantly matched three members of $Arabidopsis$ catalase family genes, and were designated as $ClCAT1$, $ClCAT2$, and $ClCAT3$, respectively, according to the best match with their $Arabidopsis$ counterparts. However, the best hit for $Cla003205$ was $AT5G38120$, with the annotation of AMP-dependent syntheses (E-value = 8e−17). Consequently, $Cla003205$ was dropped from subsequent analyses. The primer pair for each watermelon catalase gene was specifically designed to avoid amplifying multiple transcripts in qRT–PCR analysis. Information regarding watermelon catalase family genes and their primers is summarized in Table S1. Amplification specificity of the primers was further confirmed by electrophoresis in agarose gel and melting curve analysis (Fig. S2).

The single ($ClUBCP$ determined by NormFinder), pair ($ClEF1\alpha$ and $ClACT$ identified by geNorm), and multiple best reference genes (determined by geNorm) identified in the abiotic stress subset were used for normalizing the expression of catalase genes under low temperature stress. geNorm analysis showed that five reference genes were required for more reliable normalization in the abiotic stress subset. The use of five reference genes to

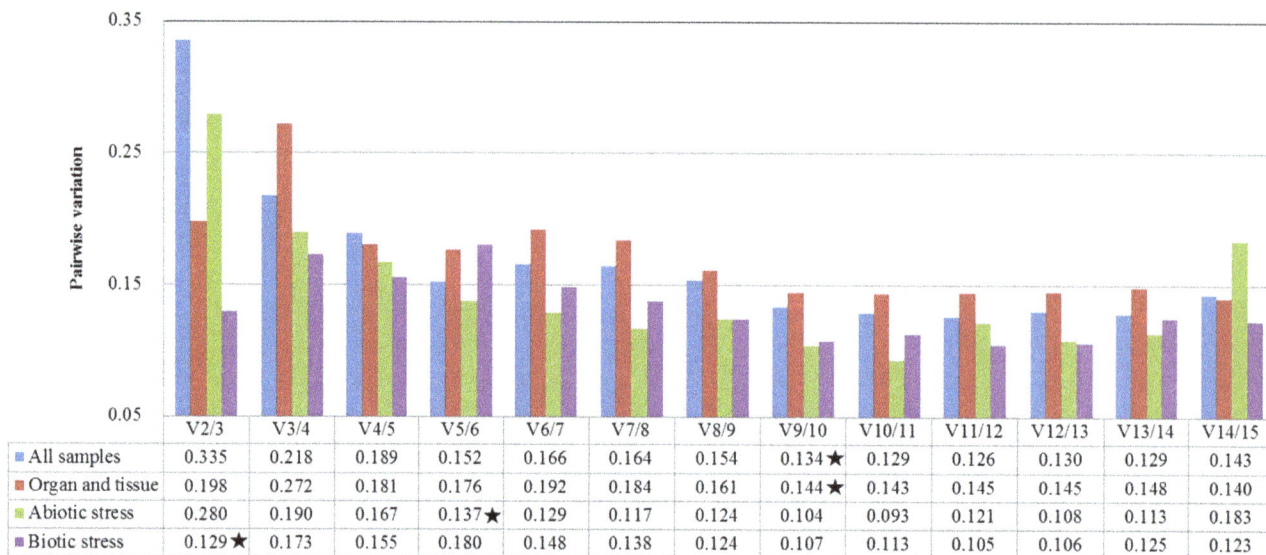

	V2/3	V3/4	V4/5	V5/6	V6/7	V7/8	V8/9	V9/10	V10/11	V11/12	V12/13	V13/14	V14/15
■ All samples	0.335	0.218	0.189	0.152	0.166	0.164	0.154	0.134 ★	0.129	0.126	0.130	0.129	0.143
■ Organ and tissue	0.198	0.272	0.181	0.176	0.192	0.184	0.161	0.144 ★	0.143	0.145	0.145	0.148	0.140
■ Abiotic stress	0.280	0.190	0.167	0.137 ★	0.129	0.117	0.124	0.104	0.093	0.121	0.108	0.113	0.183
■ Biotic stress	0.129 ★	0.173	0.155	0.180	0.148	0.138	0.124	0.107	0.113	0.105	0.106	0.125	0.123

Figure 3. Pairwise variation analyses of candidate reference genes in different sample sets. Pairwise variation (V) was calculated by geNorm to determine the minimum number of reference genes required for accurate normalization in different sample sets. "★" indicates the value of pairwise variation less than the recommended 0.15 for each sample set.

normalize three target genes is not feasible in practice. Therefore, Vandesompele et al. proposed that at least three best genes are required for more reliable normalization [11]. Accordingly, *ClEF1α*, *ClACT*, and *ClUBCP* were used to compose the multiple reference gene set for normalization. NF was calculated as the geometric mean of the used reference genes. The widely used reference gene *Cl18SrRNA* was ranked second from the bottom by both geNorm and NormFinder in the abiotic stress subset, and used as the control. The results are presented in Fig. 4.

All watermelon catalase family genes were induced by low temperature stress in leaves and showed different expression patterns. Transcripts of *ClCAT1* and *ClCAT2* were upregulated in the first 24 h, and subsequently declined. *ClCAT1* showed higher relative expression levels than *ClCAT2*. By contrast, the expression of *ClCAT3* was obviously downregulated when the best reference genes were used for normalization. The expression levels of catalase genes showed similar change patterns with slight differences when the identified single best reference gene of *ClUBCP*, pair of *ClEF1α* and *ClACT*, and multiple best reference

Table 2. Watermelon reference genes ranked according to their expression stability as determined by geNorm in different sample sets.

All samples	Stability value (M)	Organ and tissue	Stability value (M)	Abiotic stress	Stability value (M)	Biotic stress	Stability value (M)
*ClYLS8**	0.763	*ClTUA**	0.348	*ClEF1α**	0.481	*ClCAC**	0.320
*ClPP2A**	0.763	*ClACT**	0.348	*ClACT**	0.481	*ClTUA**	0.320
*ClACT**	0.970	*ClEF1α**	0.529	*ClUBCP**	0.742	*ClEF1α*	0.386
*ClEF1α**	1.000	*ClPP2A**	0.831	*ClPTB**	0.809	*ClRPS2*	0.550
*ClTUA**	1.049	*ClYLS8**	0.913	*ClIDH**	0.859	*ClYLS8*	0.665
*ClCAC**	1.075	*ClCAC**	1.014	*ClRPS2*	0.895	*ClUBCP*	0.826
*ClUBCP**	1.152	*ClUBCP**	1.146	*ClLUG*	0.934	*ClSAND*	0.918
*ClPTB**	1.239	*ClPTB**	1.271	*ClTUA*	0.977	*ClPTB*	0.989
*ClLUG**	1.318	*ClSAND**	1.362	*ClCAC*	1.042	*ClIDH*	1.044
ClSAND	1.375	*ClGAPDH*	1.430	*ClYLS8*	1.083	*ClLUG*	1.079
ClIDH	1.430	*ClLUG*	1.503	*ClPP2A*	1.113	*ClUBC2*	1.131
ClGAPDH	1.491	*ClIDH*	1.585	*ClSAND*	1.194	*Cl18SrRNA*	1.175
ClRPS2	1.565	*Cl18SrRNA*	1.675	*ClGAPDH*	1.259	*ClPP2A*	1.234
Cl18SrRNA	1.640	*ClUBC2*	1.775	*Cl18SrRNA*	1.330	*ClACT*	1.328
ClUBC2	1.743	*ClRPS2*	1.862	*ClUBC2*	1.536	*ClGAPDH*	1.419

*represents the multiple reference genes determined by pairwise variation analysis which are presented in Fig. 3. The recommended threshold of 0.15 is adopted.

Table 3. Watermelon reference genes ranked according to their expression stability as determined by NormFinder in different sample sets.

All samples	Stability value	Organ and tissue	Stability value	Abiotic stress	Stability value	Biotic stress	Stability value
ClYLS8	0.386	ClYLS8	0.353	ClUBCP	0.277	ClCAC	0.064
ClUBCP	0.492	ClPP2A	0.479	ClRPS2	0.340	ClTUA	0.271
ClPP2A	0.572	ClUBCP	0.535	ClTUA	0.432	ClRPS2	0.308
ClEF1α	0.652	ClPTB	0.713	ClLUG	0.473	ClYLS8	0.368
ClCAC	0.659	ClACT	0.759	ClYLS8	0.515	ClEF1α	0.403
ClPTB	0.706	ClEF1α	0.766	ClACT	0.569	ClUBC2	0.649
ClTUA	0.731	ClTUA	0.829	ClCAC	0.574	ClUBCP	0.660
ClACT	0.770	ClCAC	0.834	ClPTB	0.650	ClSAND	0.684
ClLUG	0.874	ClSAND	0.951	ClIDH	0.664	Cl18SrRNA	0.767
ClSAND	0.898	ClLUG	1.020	ClPP2A	0.682	ClPP2A	0.802
ClIDH	0.967	ClGAPDH	1.062	ClEF1α	0.692	ClPTB	0.849
ClGAPDH	1.050	ClIDH	1.179	ClSAND	0.949	ClIDH	0.958
ClRPS2	1.103	Cl18SrRNA	1.218	ClGAPDH	0.970	ClLUG	0.999
Cl18SrRNA	1.247	ClUBC2	1.364	Cl18SrRNA	1.196	ClACT	1.164
ClUBC2	1.467	ClRPS2	1.438	ClUBC2	1.898	ClGAPDH	1.265

gene set of ClEF1α, ClACT, and ClUBCP were used for normalization. Relative expression levels of the catalase genes that were normalized by the multiple reference gene set showed moderate changes compared with those of the single and pair of best reference genes. However, significantly different normalization results were observed between the best reference genes and unstable Cl18SrRNA. The relative expression levels of ClCAT1 and ClCAT2 were obviously overestimated when Cl18SrRNA was used for normalization. Moreover, an expression pattern of upregulation in the first 12 h and a subsequent downregulation was observed for ClCAT3 when normalized by Cl18SrRNA, which completely differed from that for ClCAT3 normalized by the best reference genes. The function of ClCAT3 in response to low temperature stress was misinterpreted.

Discussion

In previous gene expression studies on watermelon, only 18SrRNA or ACT was used as a reference gene in qRT–PCR analyses. However, 18SrRNA and ACT have not been systematically evaluated for their expression stability in watermelon, which is not in compliance with the MIQE guidelines [14,23]. The accuracy of the qRT–PCR results is highly dependent on the robust normalization strategy of employing an invariant reference gene. No reference gene has been validated in watermelon for qRT–PCR analyses as of this writing, which has resulted in the misinterpretation of quantification results. Thus, validation of suitable reference genes for watermelon is required to ensure the accuracy of gene expression studies by qRT–PCR.

The selection of potential reference genes from the genes that had been validated in other crops is a good starting point and an efficient approach for identifying suitable reference genes for crops that lack validated reference genes [10]. The previously published reference genes in other crops, together with the frequently used reference gene 18SrRNA in watermelon, were selected in this study to compare their expression stability under various experimental conditions to select the optimal reference genes for normalization in qRT–PCR analyses in watermelon.

Primer specificity is vital for reference gene validation [24]. Most of the commonly used reference genes belong to different gene families. Genes in the same family share conserved sequences. Designing specific primer pairs for a family of genes is difficult, particularly for crops without full genome sequences. When multiple family members are amplified, an increase in the expression abundance and variation in the tested gene may occur. To overcome this limitation in the present study, the generated primer pairs were aligned with all watermelon CDS to ensure their specificity on a genomic scale. Moreover, the possible gDNA contaminations in the samples can also introduce errors in the results. The strategy of at least one primer of a pair covering an exon-exon junction was adopted in *Arabidopsis* and tomato to overcome this problem [24,25]. However, finding a suitable primer binding site in the limited region of exon-exon junction for many candidate genes is difficult. Therefore, the forward and reserve primer sequences were intentionally targeted on the neighboring exons separated by an intron to control possible genomic DNA contamination in the present study. This strategy was also successfully used in peanut [26]. All the designed primer pairs amplified a larger product on gDNA templates than on cDNA templates, which could be used to check for gDNA contamination in the cDNA samples. However, the most frequently used reference gene 18SrRNA in watermelon was not powerful, and amplified the same product on cDNA and gDNA. The specific primer pairs developed in the study ensured the specificity and efficiency of qRT–PCR analysis for the selected reference genes, and offered a good starting point for other reference validation studies in watermelon.

An ideal reference gene should have stable expression in the developmental stages and under various experimental conditions. However, considering the obtained Cp values of the tested reference genes, an invariant expression level was not found among the 20 samples (Fig. 2), which highlights the importance of seeking appropriate reference genes via statistical approaches. The widely used tools geNorm and NormFinder were used to evaluate the expression stability of the candidate reference genes. Differences between the results of geNorm and NormFinder were found

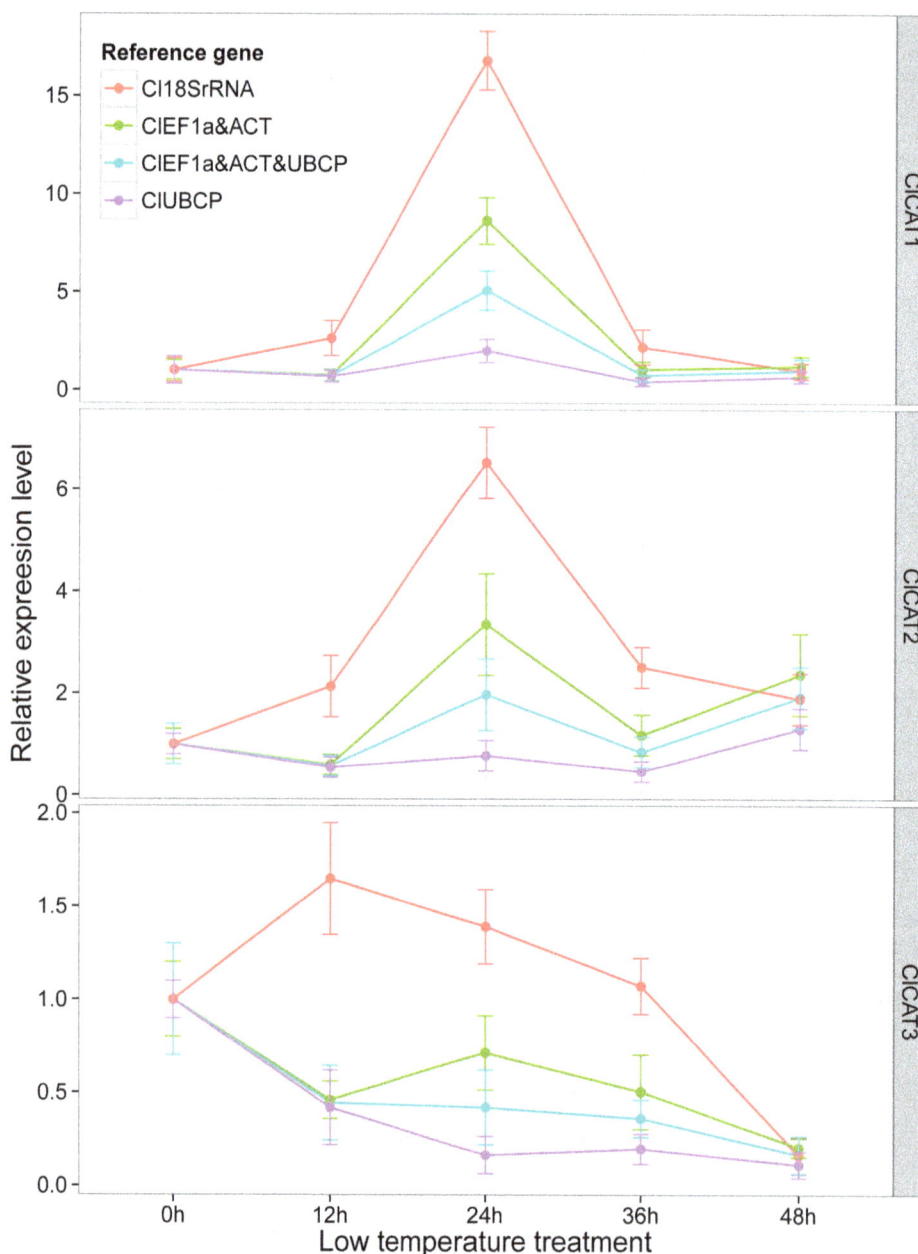

Figure 4. Expression profiles of catalase family genes in watermelon leaves under low temperature stress. Geometric mean was calculated for the pair of best reference genes and multiple reference gene set, and used for normalization. The relative expression levels are depicted as the mean ± SD, which was calculated from three biological replicates.

(Tables 2 and 3). The top seven stable genes (half of the total) were almost similar in the two algorithms for each sample set, but differences were found in the ranking order. Regardless of the changes in ranking order, the most unstable gene was the same for geNorm and NormFinder in each sample set.

In the present study, *ClYLS8* and *ClPP2A* were ranked as the best reference genes by geNorm in all 20 samples. However, when the samples were subdivided into different subsets based on experimental conditions, the best reference genes changed accordingly (Table 2). *ClTUA* and *ClACT*, *ClCAC* and *ClTUA*, *ClEF1α* and *ClACT* were identified as the best reference gene pairs by geNorm in the organs and tissues under normal growth conditions, biotic stress, and abiotic stress, respectively. A similar

trend was also observed by NormFinder (Table 3). These changes illustrate the impossibility of compiling a list of suitable genes that can be used as references across a wide range of experimental conditions [6], which highlights the necessity of systematic validation of reference genes under every set of specific experimental conditions [27,28]. However, data obtained by wide-scale gene expression analyses can be used as a starting point to choose candidates for the subsequent systematic validation of reference genes [10].

Reference gene validations have been performed in cucurbit crops, such as cucumber, melon, and zucchini. In cucumber, *EF1α*, *Fbox*, *CAC*, and *TIP41* are stable genes under different abiotic stresses, growth regulator treatments, and nitrogen nutrition treatments [29,30]. *EF1α*, *UBIep*, and *TUA* are the suitable reference genes in another study on cucumber [31]. In zucchini, the combination of *UFP*, *EF1α*, *RPL36aA*, *PP2A*, and *CAC* genes is the best strategy for reliable normalization [32]. *RPL2* is stably expressed in melon stem infected with *Fusarium* wilt [33]. Similarly, in the present study, the individual genes or combinations of *ClEF1α*, *ClTUA*, *ClCAC*, *ClUBCP*, *ClYLS8*, and *ClPP2A* were identified as the best reference genes by geNorm or NormFinder under different experimental conditions. However, considerable differences in the ranking of the candidate reference genes were also observed among the cucurbit crops. For example, *TUA* was found to be less stable in zucchini and cucumber [29,32]. *EF1α* was found to be unsuitable in melon [33]. *YLS8* was the most variant gene in cucumber under different nitrogen nutrition treatments [30]. These discrepancies on the most and least stable genes highlight the species-specific and experimental condition-specific characteristics of reference genes [34]. To guarantee the accurate quantification of target gene expression, validating the expression stability of reference genes prior to their use for normalization in qRT–PCR data analysis should be performed.

Only a single reference gene was used for normalization in previous studies on watermelon gene expression. However, an increasing number of studies shows that using multiple reference genes for normalization can improve the reliability of results because a certain level of variation always exists for any reference gene [11,35]. geNorm analysis in the present study indicated that two to nine reference genes were required for reliable normalization depending on the experimental conditions when the recommended cut-off value of 0.15 was used, which is infeasible particularly when a small number of target genes is tested. Vandesompele et al. recommended that the cut-off value of 0.15 should not be considered as a strict cut-off, and using three of the best reference genes is a valid normalization strategy in most cases [11]. Thus, at least three reference genes were needed for more reliable normalization of gene expression in watermelon.

18SrRNA has been widely used as a reference gene in different organs and tissues of watermelon under various experimental conditions. In this study, *18SrRNA* was ranked from second to seventh from the bottom in all samples and different subsets both by geNorm and NormFinder, which demonstrates that *18SrRNA* was not a suitable reference gene under specific experimental conditions and across all the tested experimental conditions. Moreover, *18SrRNA* exhibited a significantly higher expression level than other genes, which was unsuitable for normalization of target genes with middle or low expression levels. Therefore, the use of *18SrRNA* as a reference gene should be avoided in qRT–PCR analysis in watermelon in the future. *18SrRNA* is also unstable in cucurbit crops of melon [33], zucchini [32], and cucumber [31].

NormFinder usually generates a single best reference gene, whereas geNorm generates a pair of best reference genes and a set of multiple reference genes. To test the reliability of the identified reference genes, the expression levels of watermelon catalase family genes were quantified and normalized using the best reference genes, including the single, pair, and multiple best reference genes. The unstable and widely used *18SrRNA* in watermelon was also used for normalization. The expression of watermelon catalase family genes showed similar change patterns when the best reference genes were used for normalization. Compared with the single and pair of best reference genes, moderate changes in the expression levels were observed for the catalase genes when the multiple reference gene set was used for normalization, which indicates that multiple reference genes

resulted in more reliable normalization results than single reference genes. However, significantly higher expression levels and different expression patterns were observed for these genes when *18SrRNA* was used for normalization, which resulted in overestimation and misinterpretation of the transcripts of catalase family genes. Similar results were also observed in other crops [36–39]. Thus, the selection of reference genes greatly affected the normalization results, and an inappropriate reference gene may introduce bias in the analysis and lead to misleading results.

Catalase genes encode a small family of proteins, which can catalyze the decomposition of H_2O_2 and have important functions in controlling ROS homeostasis. Watermelon originated in Africa and is susceptible to low temperature stress. Analysis of the expression profiles of catalase genes in leaves under low temperature stress is crucial to reveal their regulatory functions. Differential expression patterns in response to low temperature were observed among catalase genes in watermelon. Upregulations of *ClCAT1* and *ClCAT2* in the first 24 h indicate that they had major functions in the removal of H_2O_2 generated under low temperature stress, whereas transcriptional downregulation of *ClCAT3* could sustain increased H_2O_2 availability, which was necessary for ROS homeostasis. Similar results were also observed in *Arabidopsis* [18].

Thus, the identified reference genes in the present study outperformed the currently used reference gene *18SrRNA* in watermelon in terms of expression stability during plant development and under different environmental conditions. Suitable reference genes should be selected depending on the experimental conditions. Multiple reference genes are recommended for more reliable normalization of gene expression in watermelon. In addition, the identified reference genes were expressed at much lower levels than *18SrRNA*, making them highly suitable for normalization of gene expression over a wide range of transcript levels. The identified reference genes with their specific primers will lead to better normalization and quantification of transcript levels in watermelon in the future.

Acknowledgments

The authors wish to thank Dr. Yong Xu from National Engineering Research Center for Vegetables for kindly offering the 97103 seeds and *Fusarium oxysporum* f. sp. *niveum* isolate FON1. We also extend our thanks to Dr. Tingchang Zhao from Chines Academy of Agricultural Sciences for offering *pslbtw20* strain (*Acidovorax avenae* subsp. *Citrulli*) for the study.

Author Contributions

Conceived and designed the experiments: QK JY ZB. Performed the experiments: JY LG. Analyzed the data: QK JY. Contributed reagents/materials/analysis tools: SZ WJ YH. Wrote the paper: QK.

References

1. Guo S, Zhang J, Sun H, Salse J, Lucas WJ, et al. (2013) The draft genome of watermelon (*Citrullus lanatus*) and resequencing of 20 diverse accessions. Nature Genetics 45: 51–58.

2. Guo S, Liu J, Zheng Y, Huang M, Zhang H, et al. (2011) Characterization of transcriptome dynamics during watermelon fruit development: sequencing, assembly, annotation and gene expression profiles. BMC Genomics 12: 454.

3. Wechter WP, Levi A, Harris KR, Davis AR, Fei Z, et al. (2008) Gene expression in developing watermelon fruit. BMC Genomics 9: 275.

4. Gachon C, Mingam A, Charrier B (2004) Real-time PCR: what relevance to plant studies? Journal of Experimental Botany 55: 1445–1454.

5. Dheda K, Huggett JF, Chang JS, Kim LU, Bustin SA, et al. (2005) The implications of using an inappropriate reference gene for real-time reverse transcription PCR data normalization. Analytical Biochemistry 344: 141–143.

6. Guénin S, Mauriat M, Pelloux J, Van Wuytswinkel O, Bellini C, et al. (2009) Normalization of qRT-PCR data: the necessity of adopting a systematic, experimental conditions-specific, validation of references. Journal of Experimental Botany 60: 487–493.

7. Lu GY, Guo SG, Zhang HY, Geng LH, Song FM, et al. (2011) Transcriptional profiling of watermelon during its incompatible interaction with *Fusarium oxysporum* f. sp *niveum*. European Journal of Plant Pathology 131: 585–601.

8. Sun Y, Li Y, Luo D, Liao DJ (2012) Pseudogenes as Weaknesses of ACTB (Actb) and GAPDH (Gapdh) Used as Reference Genes in Reverse Transcription and Polymerase Chain Reactions. PLoS ONE 7: e41659.

9. Akashi K, Yoshida K, Kuwano M, Kajikawa M, Yoshimura K, et al. (2011) Dynamic changes in the leaf proteome of a C(3) xerophyte, *Citrullus lanatus* (wild watermelon), in response to water deficit. Planta 233: 947–960.

10. Gutierrez L, Mauriat M, Pelloux J, Bellini C, van Wuytswinkel O (2008) Towards a systematic validation of references in real-time RT-PCR. Plant Cell 20: 1734–1735.

11. Vandesompele J, De Preter K, Pattyn F, Poppe B, Van Roy N, et al. (2002) Accurate normalization of real-time quantitative RT-PCR data by geometric averaging of multiple internal control genes. Genome Biology 3(7): research0034.1–0034.11.

12. Gutierrez L, Mauriat M, Guénin S, Pelloux J, Lefebvre J–F, et al. (2008) The lack of a systematic validation of reference genes: a serious pitfall undervalued in reverse transcription-polymerase chain reaction (RT-PCR) analysis in plants. Plant Biotechnology Journal 6: 609–618.

13. Ferguson BS, Nam H, Hopkins RG, Morrison RF (2010) Impact of Reference Gene Selection for Target Gene Normalization on Experimental Outcome Using Real-Time qRT-PCR in Adipocytes. PLoS ONE 5: e15208.

14. Bustin SA, Benes V, Garson JA, Hellemans J, Huggett J, et al. (2009) The MIQE Guidelines: Minimum Information for Publication of Quantitative Real-Time PCR Experiments. Clinical Chemistry 55: 611–622.

15. Andersen CL, Jensen JL, Orntoft TF (2004) Normalization of real-time quantitative reverse transcription-PCR data: A model-based variance estimation approach to identify genes suited for normalization, applied to bladder and colon cancer data sets. Cancer Research 64: 5245–5250.

16. Mhamdi A, Noctor G, Baker A (2012) Plant catalases: Peroxisomal redox guardians. Archives of Biochemistry and Biophysics 525: 181–194.

17. Mhamdi A, Queval G, Chaouch S, Vanderauwera S, Van Breusegem F, et al. (2010) Catalase function in plants: a focus on Arabidopsis mutants as stress-mimic models. Journal of Experimental Botany 61: 4197–4220.

18. Du Y-Y, Wang P-C, Chen J, Song C-P (2008) Comprehensive Functional Analysis of the Catalase Gene Family in Arabidopsis thaliana. Journal of Integrative Plant Biology 50: 1318–1326.

19. Carvalho FCQ, Santos LA, Dias RCS, Mariano RLR, Souza EB (2013) Selection of watermelon genotypes for resistance to bacterial fruit blotch. Euphytica 190: 169–180.

20. Udvardi MK, Czechowski T, Scheible WR (2008) Eleven golden rules of quantitative RT-PCR. Plant Cell 20: 1736–1737.

21. Finn RD, Clements J, Eddy SR (2011) HMMER web server: interactive sequence similarity searching. Nucleic Acids Research 39: W29–W37.

22. Fredslund J, Lange M (2007) Primique: automatic design of specific PCR primers for each sequence in a family. BMC Bioinformatics 8: 369.

23. Bustin SA (2010) Why the need for qPCR publication guidelines?-The case for MIQE. Methods 50: 217–226.

24. Czechowski T, Stitt M, Altmann T, Udvardi MK, Scheible W-R (2005) Genome-Wide Identification and Testing of Superior Reference Genes for Transcript Normalization in Arabidopsis. Plant Physiology 139: 5–17.

25. Exposito-Rodriguez M, Borges A, Borges-Perez A, Perez J (2008) Selection of internal control genes for quantitative real-time RT-PCR studies during tomato development process. BMC Plant Biology 8: 131.

26. Reddy DS, Bhatnagar-Mathur P, Cindhuri KS, Sharma KK (2013) Evaluation and Validation of Reference Genes for Normalization of Quantitative Real-Time PCR Based Gene Expression Studies in Peanut. PLoS ONE 8: e78555.

27. Marum L, Miguel A, Ricardo CP, Miguel C (2012) Reference Gene Selection for Quantitative Real-time PCR Normalization in *Quercus suber*. PLoS ONE 7: e35113.

28. Chandna R, Augustine R, Bisht NC (2012) Evaluation of Candidate Reference Genes for Gene Expression Normalization in *Brassica juncea* Using Real Time Quantitative RT-PCR. PLoS ONE 7: e36918.

29. Migocka M, Papierniak A (2011) Identification of suitable reference genes for studying gene expression in cucumber plants subjected to abiotic stress and growth regulators. Molecular Breeding 28: 343–357.

30. Warzybok A, Migocka M (2013) Reliable Reference Genes for Normalization of Gene Expression in Cucumber Grown under Different Nitrogen Nutrition. PLoS ONE 8: e72887.

31. Wan HJ, Zhao ZG, Qian CT, Sui YH, Malik AA, et al. (2010) Selection of appropriate reference genes for gene expression studies by quantitative real-time polymerase chain reaction in cucumber. Analytical Biochemistry 399: 257–261.

32. Obrero A, Die JV, Roman B, Gomez P, Nadal S, et al. (2011) Selection of Reference Genes for Gene Expression Studies in Zucchini (*Cucurbita pepo*) Using qPCR. Journal of Agricultural and Food Chemistry 59: 5402–5411.

33. Sestili S, Sebastiani M, Belisario A, Ficcadenti N (2013) Reference gene selection for gene expression analysis in melon infected by *Fusarium oxysporum* f.sp. *melonis*. Journal of Plant Biochemistry and Biotechnology 3: 1–11.

34. Le DT, Aldrich DL, Valliyodan B, Watanabe Y, Ha CV, et al. (2012) Evaluation of Candidate Reference Genes for Normalization of Quantitative RT-PCR in Soybean Tissues under Various Abiotic Stress Conditions. PLoS ONE 7: e46487.

35. Die JV, Rowland LJ (2013) Superior Cross-Species Reference Genes: A Blueberry Case Study. PLoS ONE 8: e73354.

36. Zhu X, Li X, Chen W, Chen J, Lu W, et al. (2012) Evaluation of New Reference Genes in Papaya for Accurate Transcript Normalization under Different Experimental Conditions. PLoS ONE 7: e44405.

37. Park SC, Kim YH, Ji CY, Park S, Jeong JC, et al. (2012) Stable Internal Reference Genes for the Normalization of Real-Time PCR in Different Sweetpotato Cultivars Subjected to Abiotic Stress Conditions. Plos One 7: e51502.

38. Mafra V, Kubo KS, Alves-Ferreira M, Ribeiro-Alves M, Stuart RM, et al. (2012) Reference Genes for Accurate Transcript Normalization in Citrus Genotypes under Different Experimental Conditions. PLoS ONE 7: e31263.

39. Fan C, Ma J, Guo Q, Li X, Wang H, et al. (2013) Selection of Reference Genes for Quantitative Real-Time PCR in Bamboo (*Phyllostachys edulis*). PLoS ONE 8: e56573.

13

Whole-Genome Quantitative Trait Locus Mapping Reveals Major Role of Epistasis on Yield of Rice

Anhui Huang[1], Shizhong Xu[2], Xiaodong Cai[1]*

1 Department of Electrical and Computer Engineering, University of Miami, Coral Gables, Florida, United State of America, **2** Department of Botany and Plant Sciences, University of California Riverside, Riverside, California, United State of America

Abstract

Although rice yield has been doubled in most parts of the world since 1960s, thanks to the advancements in breeding technologies, the biological mechanisms controlling yield are largely unknown. To understand the genetic basis of rice yield, a number of quantitative trait locus (QTL) mapping studies have been carried out, but whole-genome QTL mapping incorporating all interaction effects is still lacking. In this paper, we exploited whole-genome markers of an immortalized F_2 population derived from an elite rice hybrid to perform QTL mapping for rice yield characterized by yield per plant and three yield component traits. Our QTL model includes additive and dominance main effects of 1,619 markers and all pairwise interactions, with a total of more than 5 million possible effects. The QTL mapping identified 54, 5, 28 and 4 significant effects involving 103, 9, 52 and 7 QTLs for the four traits, namely the number of panicles per plant, the number of grains per panicle, grain weight, and yield per plant. Most identified QTLs are involved in digenic interactions. An extensive literature survey of experimentally characterized genes related to crop yield shows that 19 of 54 effects, 4 of 5 effects, 12 of 28 effects and 2 of 4 effects for the four traits, respectively, involve at least one QTL that locates within 2 cM distance to at least one yield-related gene. This study not only reveals the major role of epistasis influencing rice yield, but also provides a set of candidate genetic loci for further experimental investigation.

Editor: Tongming Yin, Nanjing Forestry University, China

Funding: This work was supported by the National Science Foundation (NSF) under NSF CAREER Award no. 0746882 to XC and by the Agriculture and Food Research Initiative (AFRI) of the USDA National Institute of Food and Agriculture under the Plant Genome, Genetics and Breeding Program 2007-35300-18285 to SX. The funders had no role in study design, data collection and analysis, decision to publish, or preparation of the manuscript.

Competing Interests: The authors have declared that no competing interests exist.

* E-mail: x.cai@miami.edu

Introduction

Given the paramount importance in sustaining food demanding, great efforts have been made in large scale genetic research and extensive breeding programs in almost all rice (*Oryza sativa* L.) producing countries [1,2]. Gains in rice yield in recent decades are mainly owed to advancements in breeding technologies including selection of cultivars with higher productivity and significant increase of agricultural inputs such as fertilizers and insecticides [3]. While global environmental degradation has limited further yield increase through more agricultural inputs, studying the underlying biological processes of rice yield, and transferring the knowledge gains into improvement in breeding and agronomic productivity have become the key for further increase of food production [4].

Rice yield is determined by several factors including the number of panicles per plant, the number of grains per panicle and grain weight. These component traits and the overall yield per plant exhibit continuous variation since they are influenced by multiple genetic factors named quantitative trait loci (QTLs) and other environmental factors. Genetic markers such as restriction fragment length polymorphisms (RFLPs) [5] and simple sequence repeats (SSRs) [6] have been utilized to identify QTLs for understanding genetic basis controlling rice yield [7–11]. A recent study on QTL mapping for rice yield derived a high density single nucleotide polymorphism (SNP) bin map from genomic sequences

obtained using deep sequencing technology, and demonstrated that such high density SNP bin map enabled to identify more QTLs with higher location precision than the traditional approach based on RFLP and SSR markers [12]. However, these studies attempted to identify QTLs individually via single interval mapping [5] or composite interval mapping with a small scan window [13], which had limited power of detection, given that many agronomic traits are controlled simultaneously by multiple QTLs and influenced by environmental factors [14,15].

Whole-genome marker QTL mapping employs a multiple QTL model that includes all available markers and evaluates effects of these markers simultaneously [16–18]. Such approach overcomes the limitations of the traditional single marker-based QTL mapping methods [16]. However, when genetic interactions are considered, a multiple QTL model can have a huge number of variables, which makes model inference very challenging. Early methods for multiple QTL mapping usually rely on Markov chain Monte Carlo (MCMC) simulation to fit a Bayesian model [16–20], which is computationally intensive and unpractical when a large number of markers are considered. Recently, more efficient and accurate methods have been developed [21,22], which make whole-genome marker QTL mapping feasible. With whole-genome marker QTL mapping considering main effects and interactions of all additive and dominance effects simultaneously, contributions of numerous genetic effects to rice yield can be assessed.

Table 1. Estimated QTL effects from the full model for the number of panicles per plant.

Loci(i, j)[a]	$\hat{\beta}(s_{\hat{\beta}})$[b]	p-value[c]	$\hat{h}_j^2$[d]	Loci(i, j)	$\hat{\beta}(s_{\hat{\beta}})$	p-value	$\hat{h}_j^2$
(757_add, 757_add)	−0.12(0.04)	2.16×10^{-3}	0.0030	(104_dom, 732_add)	−0.11(0.05)	9.54×10^{-3}	0.0012
7_add, 220_dom)	0.20(0.06)	2.39×10^{-4}	0.0032	(186_dom, 735_add)	−0.20(0.06)	1.72×10^{-4}	0.0039
(10_add, 887_dom)	−0.25(0.05)	3.78×10^{-6}	0.0061	(518_dom, 759_add)	−0.27(0.06)	1.64×10^{-6}	0.0075
(18_add, 1407_dom)	0.28(0.06)	1.45×10^{-6}	0.0080	(220_dom, 784_add)	0.23(0.06)	2.38×10^{-5}	0.0045
(20_add, 1026_dom)	0.27(0.06)	1.82×10^{-6}	0.0060	(561_dom, 828_add)	−0.36(0.05)	4.88×10^{-12}	0.0140
(44_add, 532_dom)	−0.27(0.05)	1.16×10^{-7}	0.0071	(861_add, 918_add)	0.41(0.05)	1.11×10^{-15}	0.0182
(69_add, 913_dom)	0.21(0.05)	2.39×10^{-5}	0.0040	(904_add, 1113_dom)	0.33(0.05)	3.50×10^{-10}	0.0098
(123_add, 1132_add)	0.23(0.05)	7.78×10^{-7}	0.0053	(213_dom, 929_add)	0.27(0.05)	4.85×10^{-8}	0.0076
(166_add, (684_add)	0.46(0.04)	$<10^{-15}$	0.0279	(967_add, 1515_add)	0.20(0.05)	2.91×10^{-5}	0.0040
(186_add, 1372_add)	−0.11(0.04)	2.86×10^{-3}	0.0013	(908_dom, 994_add)	0.90(0.05)	$<10^{-15}$	0.0782
(192_add, 580_add)	−0.11(0.04)	8.35×10^{-3}	0.0013	(1026_add, 1173_add)	0.21(0.05)	9.05×10^{-6}	0.0044
(199_add, 782_dom)	−0.08(0.03)	2.53×10^{-3}	0.0006	(1037_add, 1510_add)	0.14(0.05)	1.13×10^{-3}	0.0018
(208_add, 309_add)	−0.31(0.05)	1.50×10^{-9}	0.0092	(1089_dom, 1096_add)	0.34(0.12)	2.17×10^{-3}	0.0016
(227_add, 364_dom)	−0.36(0.05)	3.18×10^{-12}	0.0145	(1119_add, 1471_add)	−0.19(0.04)	1.29×10^{-5}	0.0045
(244_add, 1303_dom)	−0.11(0.04)	5.10×10^{-3}	0.0012	(229_dom, 1160_add)	−0.11(0.04)	5.70×10^{-3}	0.0012
(249_add, 417_dom)	0.12(0.04)	3.14×10^{-3}	0.0015	(1208_add, 1583_dom)	−0.14(0.05)	3.20×10^{-3}	0.0015
(333_add, 991_add)	0.24(0.05)	3.92×10^{-7}	0.0057	(64_dom, 1223_add)	−0.43(0.05)	2.22×10^{-15}	0.0191
(335_add, 372_add)	0.20(0.05)	2.05×10^{-5}	0.0041	(1237_add, 1370_add)	0.54(0.05)	$<10^{-15}$	0.0279
(349_add, 1425_dom)	−0.23(0.05)	3.20×10^{-6}	0.0060	(1334_add, 1576_add)	0.22(0.05)	2.94×10^{-6}	0.0049
(354_add, 358_dom)	−0.50(0.05)	$<10^{-15}$	0.0233	(408_dom, 1356_add)	−0.73(0.05)	$<10^{-15}$	0.0488
(371_dom, 381_add)	0.37(0.05)	6.01×10^{-11}	0.0113	(1065_dom, 1394_add)	−0.17(0.05)	1.55×10^{-4}	0.0026
(421_add, 1079_add)	−0.15(0.04)	1.04×10^{-4}	0.0023	(981_dom, 1558_add)	−0.79(0.05)	$<10^{-15}$	0.0735
(456_add, 1282_add)	0.38(0.05)	9.99×10^{-15}	0.0167	(1094_dom, 1558_add)	0.21(0.05)	4.05×10^{-5}	0.0046
(517_add, 1346_add)	−0.10(0.04)	8.46×10^{-3}	0.0009	(1217_dom, 1615_add)	0.37(0.05)	5.88×10^{-13}	0.0144
(520_add, 595_dom)	−0.37(0.05)	4.03×10^{-11}	0.0109	(54_dom, 1117_dom)	0.27(0.05)	3.96×10^{-7}	0.0052
(15_dom, 534_add)	0.47(0.05)	$<10^{-15}$	0.0246	(627_dom, 681_dom)	−0.25(0.05)	1.05×10^{-6}	0.0048
(649_add, 1364_add)	−0.20(0.04)	3.80×10^{-6}	0.0045	(786_dom, 810_dom)	0.15(0.05)	1.11×10^{-3}	0.0021
Parameter(s)				$a = 0.5$, $b = 0.5$			
μ				0.0035			
σ_0^2				0.1444			
$\hat{h}^2$				0.9405			

[a]add: additive effect; dom: dominance effect. If i equals j, then it is a main effect, otherwise, it is an interaction between locus i and locus j. Total number of effects is 112, only 54 effects with a p-value ≤ 0.01 are listed in this table.
[b]The estimated marker effect is denoted by $\hat{\beta}$ and the standard deviation is denoted by $s_{\hat{\beta}}$.
[c]P-value is obtained via t-test.
[d]Phenotypic variation explained.

In this study, we applied our empirical Bayesian least absolute shrinkage and selection operator (EBlasso) method [21,22] to whole-genome QTL mapping for an elite *indica* rice hybrid, Shanyou 63 [7,23]. Our EBlasso model includes additive and dominance main effects of 1,619 markers, all additive × additive interactions, additive × dominance interactions, dominance × additive interactions, and dominance × dominance interactions, with a total of more than 5 million possible effects. The quantitative traits considered in this study include yield per plant and three yield component traits, namely the number of panicles per plant, the number of grains per panicle and grain weight. We will demonstrate that our EBlasso identifies a number of QTLs, most of which are involved in digenic interactions, and coincide with or are close to experimentally investigated genes related to yield.

Results

Four quantitative traits including three rice yield component traits (the number of panicles per plant, the number of grains per panicle and grain weight) and overall yield per plant were analyzed using the EBlasso method. The full QTL model includes main additive and dominance effects of 1,619 markers and all their pair-wise interactions, with a total of $k = 5,242,322$ variables (see the Materials and Methods section for the genetic map). To understand the performance gain of the full model, we also performed QTL mapping for the four traits with a QTL model including $k = 3,238$ main effects, which is referred to as the main effect model.

We estimated the phenotypic variance explained by a particular QTL j as $h_j^2 = \dfrac{\hat{\beta}_j^2 \text{var}(x_j)}{\sigma_y^2}$, $j = 1, 2, \ldots, k'$, where $\text{var}(x_j)$ is the

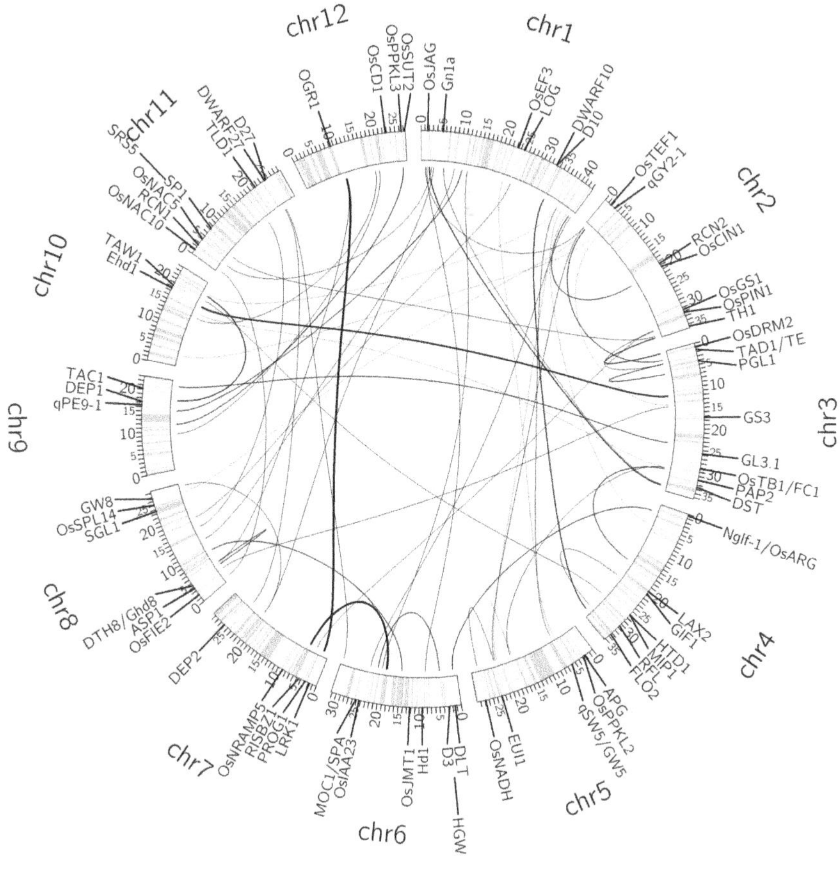

Figure 1. Interaction network of 103 QTLs for the number of panicles per plant. The circle shows the bin map and columns indicate position of the makers (ticks in million base pairs). The thickness of a link is proportional to the strength of the interaction effect. A short straight line indicates a main effect. Molecularly characterized genes related to yield are also labeled in the appropriate positions of the genome.

variance of the coefficient of QTL j and the total phenotypic variance σ_y^2 was estimated from the data. To estimate the total variance explained by all identified QTLs, we refitted the data to an ordinary linear regression model that includes variables corresponding to the identified QTLs. The phenotypic values were predicted from the linear regression model as $\hat{y}$, and the total phenotypic variance explained by all identified QTLs was calculated as

$$h^2 = \frac{\text{var}(\hat{y})}{\text{var}(y)} = \frac{\text{var}(\hat{y})}{\sigma_y^2} \quad (1)$$

QTL mapping for the number of panicles per plant

The three-step cross validation (CV) procedure (detailed in the Materials and Methods section) for the full model identified the optimal pair of parameters as $(a, b) = (0.5, 0.5)$ (Table S1 in File S1). Using the optimal values of (a, b), the EBlasso algorithm shrunk most of k variables to zero and yielded a QTL model with 111 nonzero effects. The statistical test, described in the Materials and Methods section, for each nonzero effect identified 54 significant effects at a p-value ≤ 0.01 (Table 1). Among them, one was main additive effect, 18 were additive × additive interaction, 32 were additive ×dominance interaction, and three were dominance × dominance interaction. The 54 effects involved 103 QTLs and explained 94.05% of the total phenotypic variance.

Table 2. Estimated QTL effects from the main effect model for the number of panicles per plant.

locus[a]	$\hat{\beta}(s_{\hat{\beta}})$[b]	p-value[c]	$\hat{h}_j^2$[d]
3_add	0.22(0.09)	6.86×10^{-3}	0.0088
228_add	−0.24(0.09)	2.98×10^{-3}	0.0125
353_add	−0.24(0.09)	2.68×10^{-3}	0.0126
757_add	−0.54(0.10)	4.18×10^{-8}	0.0625
818_add	0.40(0.10)	4.82×10^{-5}	0.0300
908_add	0.31(0.09)	4.78×10^{-4}	0.0206
994_add	0.54(0.11)	3.53×10^{-7}	0.0524
1363_add	−0.26(0.09)	2.29×10^{-3}	0.0135
461_dom	0.30(0.10)	1.92×10^{-3}	0.0091
861_dom	−0.45(0.10)	1.17×10^{-5}	0.0209
Parameter(s)		$a = -0.01$, $b = 0.5$	
μ		−0.0600	
σ_0^2		1.5260	
$\hat{h}^2$		0.3976	

[a]add: additive effect; dom: dominance effect. Total number of effects is 10, all with a p-value ≤ 0.01.
[b]The estimated marker effect is denoted by $\hat{\beta}$ and the standard deviation is denoted by $s_{\hat{\beta}}$.
[c]P-value is obtained via t-test.
[d]Phenotypic variation explained.

Table 3. Estimated QTL effects from the full model for the number of grains per panicle.

Loci(i, j)[a]	$\hat{\beta}(s_{\hat{\beta}})$[b]	p-value[c]	$\hat{h}_j^2$[d]
(436_{add}, 436_{add})	6.79(0.98)	1.58×10^{-11}	0.0846
(10_{dom}, 50_{add})	$-8.74(1.41)$	1.05×10^{-9}	0.0695
(875_{add}, 1156_{dom})	7.15(1.44)	6.37×10^{-7}	0.0427
(595_{dom}, 1004_{add})	$-12.78(1.88)$	3.34×10^{-11}	0.0853
(381_{dom}, 1057_{add})	$-8.82(1.33)$	9.40×10^{-11}	0.0801
Parameter(s)		$a = 0.05$, $b = 0.1$	
μ		-0.3228	
σ_0^2		156.7998	
$\hat{h}^2$		0.4651	

[a]add: additive effect; dom: dominance effect. If i equals j, then it is a main effect, otherwise, it is an interaction between locus i and locus j. Total number of effects is 5, all with a p-value ≤ 0.01.
[b]The estimated marker effect is denoted by $\hat{\beta}$ and the standard deviation is denoted by $s_{\hat{\beta}}$.
[c]p-value is obtained via t-test.
[d]Phenotypic variation explained.

We did a literature survey and found 99 genes with known genomic locations that had experimental evidence showing that they were related to rice yield and yield component traits. For each of the 103 QTLs, we identified genes from 99 experimentally investigated genes that were within 20 centi-Morgan (cM) distance and associated such genes with the QTL. In total, we found 58 genes for 103 QTLs. For the ease of presentation, we organized QTLs within 20 cM distance into a group, which resulted in 51 groups for 103 QTLs. These 51 QTL groups and associated genes are listed in Table S2 in File S1. It is seen that 36 groups of QTLs have at least one associated gene and the distances between QTLs and their associated genes are relatively small (median distance 5.37 cM). Moreover, 21 QTLs involved in 19 of 54 effects locate within 2 cM distance to at least one gene influencing rice yield. The interaction network of the 103 QTLs and their associated genes are visualized in Figure 1.

The three-step CV for the main effect model identified the optimal pair of parameters as $(a, b) = (-0.01, 0.5)$ (Table S1 in File S1), with which eight additive and two dominance effects involving ten QTLs were identified with a p-value ≤ 0.01 (Table 2). The ten effects totally explained 39.76% of the phenotypic variance, and nine of them had genes related to yield within 20 cM distance (median distance 9.29 cM) (Table S3 in File S1). Seven QTLs were identical to QTLs or within the QTL group identified from the full model (Bins 228, 353, 757, 861, 908, 994, 1363), and the other three (Bins 3, 461 and 818) were close to QTLs identified by

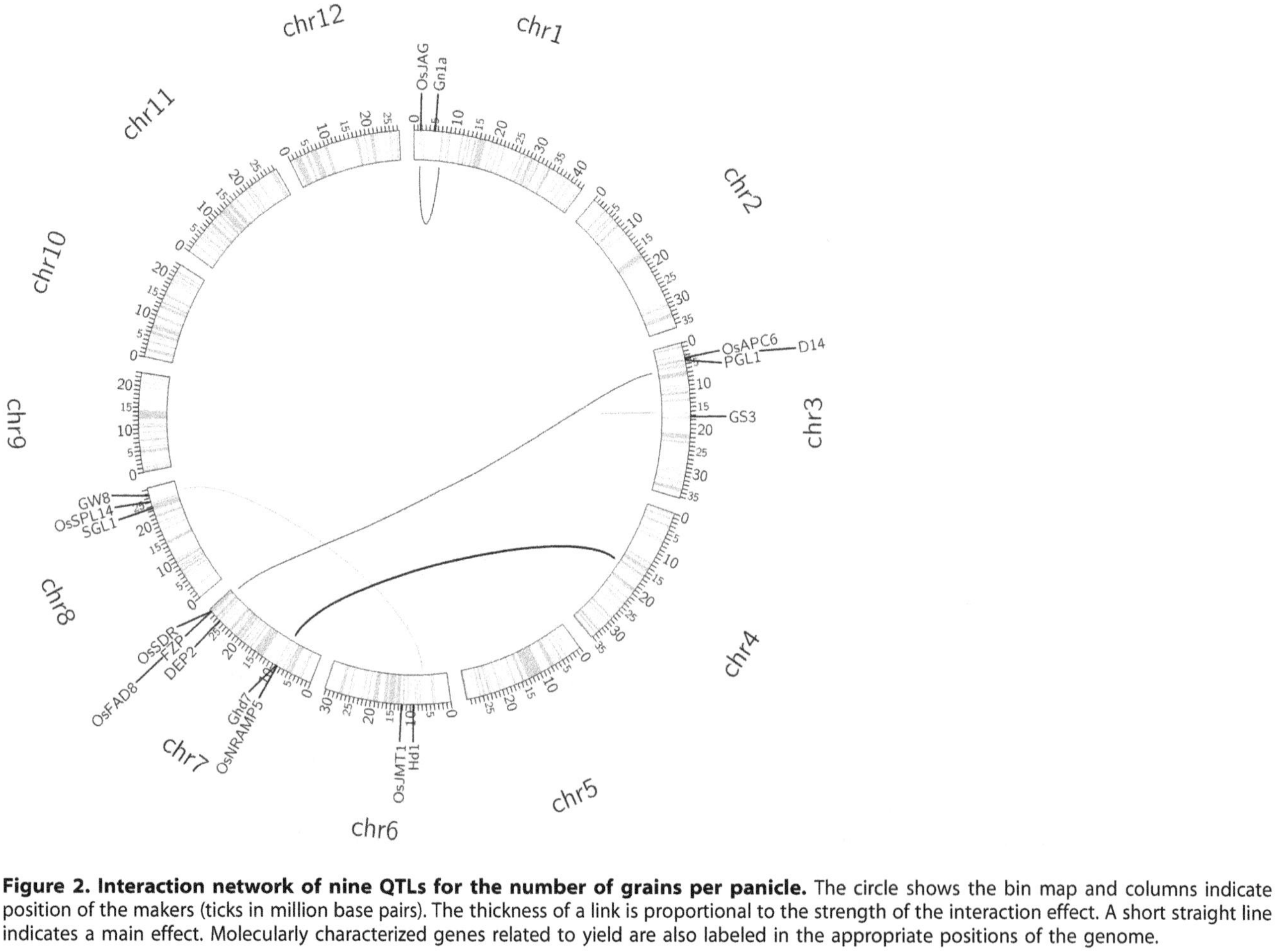

Figure 2. Interaction network of nine QTLs for the number of grains per panicle. The circle shows the bin map and columns indicate position of the makers (ticks in million base pairs). The thickness of a link is proportional to the strength of the interaction effect. A short straight line indicates a main effect. Molecularly characterized genes related to yield are also labeled in the appropriate positions of the genome.

the full model. Specifically, Bin3 was 3.97 cM away from Bin7 identified from the full model; Bin461 was 8.29 cM away from Bin456 identified from the full model; and Bin818 was 6.15 cM away from Bin810 identified from the full model. Comparing the results obtained from the two models, we see that the full model identified more QTLs, which included all those identified by the main effect model, and explained a much larger percentage of the phenotypic variance.

QTL mapping for the number of grains per panicle

The CV analysis identified the optimal pair of parameters $(a, b) = (0.05, 0.1)$ for the full QTL model for the number of grains per panicle (Table S4 in File S1), with which EBlasso identified five nonzero effects. All of these nonzero effects were significant at a p-value ≤ 0.01 (Table 3), including one main additive effect and four additive $\times$ dominance interactions. The five effects involved nine QTLs, and explained 46.51% of the overall phenotypic variance. Eight of the nine QTLs have experimentally verified genes related to rice yield within 20 cM distance (median distance 4.86 cM) (Table S5 in File S1). Moreover, four of these QTLs involved in four effects locate within 2 cM distance to at least one yield-related gene. The interaction network of the nine QTLs and their associated genes are depicted in Figure 2.

The same three-step CV for the main effect model identified the optimal pair of parameters $(a, b) = (-0.4, 0.5)$ (Table S4 in File S1), with which five additive effects were identified, all having a p-value ≤ 0.01 (Table 4). The five QTLs (Bins 43, 436, 877, 1006, 1057) totally explained 41.48% of the phenotypic variance, and all had molecularly characterized genes related to rice yield within 19 cM distance (median distance 1.59 cM) (Table S6 in File S1). All five QTLs were identical or very close to the QTLs identified from the full model. Specifically, Bin436 and Bin1057 were identified in both models; Bin43 is 3.40 cM away from Bin50 identified from the full model; Bin877 is 0.47 cM away from Bin875 identified from the full model; and Bin1006 is 0.72 cM away from Bin1004 identified from the full model. Comparing the results obtained from the two models, we observed that although both models identified five effects, the full model identified four more QTLs and explained a slightly larger percentage of phenotypic variance. Moreover, the main effect model identified five additive effects, but the full model identified QTLs with both additive and dominance effects.

QTL mapping for grain weight

The CV analysis determined the optimal $(a, b) = (1, 1)$ (Table S7 in File S1) for the full QTL model for grain weights. Using the optimal a and b, EBlasso yields a QTL model including 89 nonzero effects, among which 28 effects were identified as significant at a p-value ≤ 0.01 (Table 5). Among them, one was a main additive effect, 10 were additive $\times$ additive, 15 were additive $\times$ dominance, and two were dominance $\times$ dominance interactions. The 28 effects involved 52 QTLs, and explained 93.79% of the phenotypic variance. QTLs with a distance $\leq$ 20 cM were placed into a group, resulting in 32 groups, and 26 of the 32 QTL groups had at least one gene within 20 cM distance (median distance 5.06 cM) (Table S8 in File S1). Moreover, 15 QTLs involved in 12 of 28 effects locate within 2 cM distance to at least one yield-related gene. The interaction network of the 52 QTLs and their associated genes are shown in Figure 3.

The CV analysis for the main effect model identified the optimal pair of parameters $(a, b) = (1, 1)$ (Table S7 in File S1), with which 26 QTLs (19 additive and 7 dominance effects) were identified with a p-value ≤ 0.01 (Table 6). The 26 QTLs totally explained 84.24% of the overall phenotypic variance, and 23 of

Table 4. Estimated QTL effects from the main effect model for the number of grains per panicle.

locus[a]	$\hat{\beta}(s_{\hat{\beta}})$[b]	p-value[c]	$\hat{h}_j^2$[d]
43 _add	−5.15(1.08)	1.38×10^{-6}	0.0454
436 _add	8.06(1.03)	6.02×10^{-14}	0.1190
877 _add	3.64(1.05)	3.21×10^{-4}	0.0225
1006 _add	−8.00(1.15)	1.48×10^{-11}	0.1036
1057 _add	−7.51(1.11)	3.55×10^{-11}	0.0973
Parameter(s)		$a = -0.4, b = 0.5$	
μ		−0.6700	
σ_0^2		171.21	
$\hat{h}^2$			0.4148

[a]add: additive effect; dom: dominance effect. Total number of effects is five, all with a p-value ≤ 0.01.
[b]The estimated marker effect is denoted by $\hat{\beta}$ and the standard deviation is denoted by $s_{\hat{\beta}}$.
[c]P-value is obtained via t-test.
[d]Phenotypic variation explained.

them had molecularly characterized genes related to rice yield within 16 cM distance (median distance 4.08 cM) (Table S9 in File S1). Twenty three of the 26 QTLs were identical to or within a QTL group identified from the full model, but three QTLs (Bins 228, 843, and 894) do not correspond to any QTLs identified from the full model within 20 cM distance. Again, the full model identified more QTLs than the main effect model and the QTLs detected by the full model explained more phenotypic variance than those detected by the main effect model.

QTL mapping for yield per plant

The CV analysis determined the optimal pair of parameters $(a, b) = (1, 1)$ for the full QTL model for rice yield (Table S10 in File S1). Using the optimal values of (a, b), EBlasso yielded four nonzero effects, all were significant at a p-value ≤ 0.01: one main additive effect, one additive $\times$ additive interaction, one additive $\times$dominance interaction, and one dominance $\times$ dominance interaction (see Table 7). The four effects involved seven QTLs and explained 34.01% of the overall phenotypic variance. Five out of the seven QTLs have an experimentally verified gene within 15 cM distance (median distance 2.21 cM) (Table S11 in File S1). Moreover, two QTLs involved in two of four effects locate within 2 cM distance to at least one yield-related gene. The interaction network of the seven QTLs and their associated genes are described in Figure 4.

The optimal pair of parameters determined by the CV analysis for the main effect model was $(a, b) = (-0.5, 0.1)$ (Table S10 in File S1), with which four QTLs with a p-value ≤ 0.01 were identified (Table 8). The four QTL effects explained 23.79% of the phenotypic variance, and all had at least one gene within 17 cM distance (median distance 7.82 cM) (Table S12 in File S1). Two of the four QTLs (Bin1014, Bin1057) were identical to the QTLs identified from the full model, but the other two QTLs do not correspond to any QTL identified from the full model within 20 cM distance. Overall, although the full model did not detect all QTLs identified by the main effect model, it still detected more QTLs and explained more phenotypic variance.

Table 5. Estimated QTL effects from the full model for grain weight.

Loci(i, j)[a]	$\hat{\beta}(s_{\hat{\beta}})$[b]	p-value[c]	$\hat{h}_j^2$[d]
(729_add, 729_add)	1.02(0.07)	$<10^{-15}$	0.1548
(37_add, 547_dom)	0.71(0.09)	1.29×10^{-14}	0.0428
(67_add, 772_add)	$-0.25(0.08)$	6.56×10^{-4}	0.0047
(96_add, 1117_dom)	0.21(0.08)	2.95×10^{-3}	0.0035
(119_add, 987_add)	0.18(0.07)	6.73×10^{-3}	0.0024
(151_add, 1262_add)	$-0.15(0.07)$	9.79×10^{-3}	0.0018
(71_dom, 184_add)	$-0.67(0.09)$	1.55×10^{-13}	0.0374
(210_add, 1400_add)	0.19(0.08)	7.79×10^{-3}	0.0025
(329_add, 727_dom)	0.22(0.09)	4.63×10^{-3}	0.0040
(310_dom, 419_add)	$-0.21(0.05)$	1.86×10^{-5}	0.0043
(431_add, 1111_add)	0.35(0.08)	1.01×10^{-5}	0.0107
(71_dom, 500_add)	$-0.76(0.08)$	$<10^{-15}$	0.0493
(583_add, 1578_dom)	0.35(0.08)	9.50×10^{-6}	0.0092
(107_dom, 700_add)	0.19(0.07)	4.80×10^{-3}	0.0035
(708_dom, 714_add)	$-1.15(0.32)$	1.79×10^{-4}	0.0076
(818_add, 1100_add)	0.26(0.08)	3.93×10^{-4}	0.0053
(916_add, 1026_add)	0.15(0.06)	8.50×10^{-3}	0.0019
(472_dom, 920_add)	$-0.27(0.09)$	1.65×10^{-3}	0.0058
(18_dom, 955_add)	$-0.20(0.08)$	3.77×10^{-3}	0.0033
(971_add, 1461_add)	0.27(0.08)	4.75×10^{-4}	0.0075
(620_dom, 1011_add)	$-0.67(0.09)$	3.31×10^{-12}	0.0336
(1035_add, 1224_add)	0.30(0.07)	3.27×10^{-5}	0.0081
(1093_add, 1407_dom)	0.44(0.08)	2.13×10^{-7}	0.0148
(1167_dom, 1168_add)	$-0.47(0.16)$	1.37×10^{-3}	0.0051
(119_dom, 1375_add)	0.61(0.09)	7.83×10^{-11}	0.0289
(1397_add, 1505_add)	0.41(0.09)	1.95×10^{-6}	0.0119
(247_dom, 1505_dom)	$-0.23(0.08)$	1.22×10^{-3}	0.0032
(647_dom, 796_dom)	0.26(0.08)	4.29×10^{-4}	0.0044
Parameter(s)		$a=1, b=1$	
μ		-0.0661	
σ_0^2		0.5317	
$\hat{h}^2$		0.9379	

[a]add: additive effect; dom: dominance effect. If i equals j, then it is a main effect, otherwise, it is an interaction between locus i and locus j. Total number of effects is 90, only 28 effects with a p-value ≤ 0.01 are listed in this table.
[b]The estimated marker effect is denoted by $\hat{\beta}$ and the standard deviation is denoted by $s_{\hat{\beta}}$.
[c]P-value is obtained via t-test.
[d]Phenotypic variation explained.

Effect types and pleiotropic genes

Among the five types of effects (main additive, main dominance effects, additive × additive, additive ×dominance, and dominance × dominance interactions) considered in the EBlasso full models for four traits, no main dominance effects was detected, but several dominance × dominance interactions (one for rice yield, three for the number of panicles per plant, and two for grain weight) were identified. Many additive ×dominance interaction effects were identified, including one for rice yield, 32 for the number of panicles per plant, four for the number of grains per panicle, and 15 for grain weight. Phenotypic variance explained by a single

effect is relatively small for all traits (Tables 1, 3, 5 and 7). For example, the largest effect has $\hat{h}^2 = 7.82\%$ (908_dominance ×994_additive) for the number of panicles per plant, 8.53% (595_dominance×1004_additive) for the number of grains per panicle, 15.48% (729_additive) for grain weight, and 6.08% (1057_additive× 1144_dominance) for yield per plant. Each main effect detected by the main effect model also explained a small percentage of the total phenotypic variance.

Many molecularly characterized genes related to yield are known to play pleiotropic roles in regulating grain productivity [31]. Without surprise, a number of such genes coincide with or close to the QTLs that were identified by our EBlasso for multiple traits, although they did not necessarily have pleiotropic effects. For example, gene *Ghd7*, *OsNRAMP5* and *DEP2* are close to several QTLs common for the four phenotypes, *qSW5/GW5*, *OsEF3* and *LOG* are near the QTLs for three phenotypes except the number of grains per panicle, and *Gn1a*, *OsJAG*, *GS3*, *OsJMT1*, *OsSPL14*, *GW8/OsSPL16*, *SGL1* are associated with QTLs for three phenotypes except yield per plant. Besides *Ghd7*, *OsNRAMP5* and *DEP2*, gene *FZP*, *OsSDR*, and *OsFAD8* was near QTLs for both yield per plant and the number of grains per panicle; 14 genes were close to QTLs for both the number of panicles per plant and the number of grains per panicle; and 62 other genes were associated with QTLs for both the number of grains per panicle and grain weight. While the pleiotropic effect of some genes have been reported [32], our QTL mapping results identified a number of genes associated with multiple phenotypes, implying their possible pleiotropic role worthy of further experimental investigation. Moreover, it is also possible that the QTLs we detected may be closely linked to unknown genes, which, if identified, will yield more insight into the molecular basis of phenotypes [1].

Discussion

Due to its small genome and close relatedness with other grass crops, rice has served as a model plant for investigating genetic factors underlying crop productivity [33,34]. To date, more than 600 rice genes have been experimentally cloned with related traits including yield, biotic and abiotic stresses, grain quality, plant architecture, fertility, etc. [2]. However, there is still a knowledge gap regarding the molecular basis of yield-related biological processes [1], suggesting the importance of systematic tools that can enable to understand functional role of genes [2,35]. In this study, we employed a multiple QTL model that included all additive and dominance main effects of 1,619 markers, and all their pair-wise interactions with a total of more than 5 million possible effects, and then applied our EBlasso algorithm to identify QTLs for four agronomic related traits of rice, including yield, the number of panicles per plant, the number of grains per panicle and grain weight. Our QTL mapping revealed a number of QTLs for four traits, most of which are involved in digenic interactions. Moreover, most of these QTLs have at least one experimentally cloned gene within 20 cM distance.

The same set of markers in the recombinant inbred line (RIL) population where the "immortalized F_2" (IMF$_2$) was derived from were used for QTL mapping, via a composite interval mapping method with a scan window size of five markers [12]. Upon development of the IMF$_2$ population, this dataset was obtained and the ANOVA method was applied to each pair of markers to identify both main and digenic interaction effects from 5,242,322 possible effects [24]. The composite interval mapping identified zero, three (Bin40, Bin446 and Bin1006), seven (Bins 49, 171, 439, 729, 928, 1008, and 1266), and one (Bin1007) QTLs for the

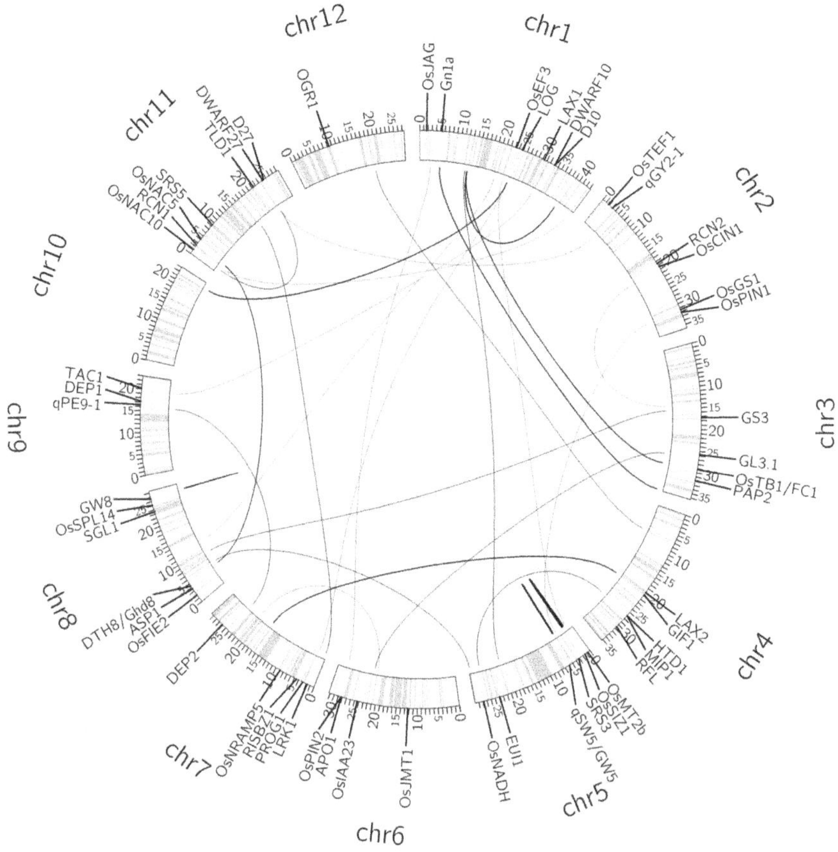

Figure 3. Interaction network of 52 QTLs for grain weight. The circle shows the bin map and columns indicate position of the makers (ticks in million base pairs). The thickness of a link is proportional to the strength of the interaction effect. A short straight line indicates a main effect. Molecularly characterized genes related to yield are also labeled in the appropriate positions of the genome.

number of panicles per plant, the number of grains per panicle, grain weight and yield per plant, respectively. The ANOVA method detected thousands (1432, 2696, 3524 and 2251) of digenic interactions between two bins with a p-value ≤ 0.001; and after those digenic interactions involving adjacent bins were merged, 115, 189, 238, and 204 effects were reported, respectively [24]. In contrast, our EBlasso method identified a reasonable number of effects and QTLs for each trait, and 35%–80% of identified effects for four traits involve at least one QTL that locates within 2 cM distance to at least one gene related to crop yield, which corroborates the reliability of the identified effects.

The list of genes associated with the identified QTLs provides insight into rice yield with respect to yield component traits. First, the number of panicles depends on plant's ability of producing tillers, which is under genetic, developmental and environmental influence. While previous composite interval mapping did not identify any significant effect with the same set of markers in an RIL population [12], we have identified a set of QTLs that have nearby genes known to regulate plant tillering. For example, among genes in Table S2 in File S1, *MOC1/SPA* is the first gene characterized for rice tillering; it initiates axillary buds that grow into lateral braches [36]. *OsTB1/FC1* has been identified as an important gene that negatively regulates lateral branching in rice [37]. *OsSPL14* is a highly expressed gene in the shoot apex and primordial of primary and secondary branches, which promotes panicle branching while reducing tiller number [38]. Through

gene mutations, *D3, D10, D14, D17/HTD1,* and *D27* were found to affect tiller initiation and/or outgrowth [37]. Secondly, the number of grains per panicle is another important trait determining crop yield. While composite interval mapping identified three QTLs (Bin40, Bin446 and Bin1006) close to genes *Gn1a, GS3, OsNRAMP5* and *Ghd7,* our EBlasso also identified these genes in addition to other 13 genes. Among them, *FZP* is known to control spikelet meristem identity [39], *Ghd7* is a pleiotropic gene affecting grain number, plant height and heading date [40], *GW8/OsSPL16, PGL,* and *DEP2* all are known to be essential in regulating cell proliferation or elongation [41–43]. Thirdly, composite interval mapping detected seven QTLs (Bins 49, 171, 439, 729, 928, 1008, and 1266) for grain weight, with nearby genes *Gn1a, LAX1, GS3, GS5, qSW5/GW5, OsJMT1, OsIAA23, Ghd7, OsNRAMP5, TAC1, LGD1* and *SG1.* In addition to these genes, our EBlasso identified many other genes with known effects in controlling grain weight. For example, *GIF1* is a gene encoding a cell-wall invertase required for carbon partitioning during early grain filling, and overexpression of *GIF1* leads to larger and heavier grain weight [44]. Genes *SRS3* and *SRS5* have been found to regulate seed cell elongation [45,46]. Over-expression of *LRK1* gene results in enhanced cellular proliferation and increased grain weight [47]. Finally, yield per plant is the most complex trait and a small number of effects were identified compared with its component traits. While composite interval mapping identified only one QTL (Bin1007) with nearby gene *Ghd7* and *OsNRAMP5,*

Table 6. Estimated QTL effects from the main effect model for grain weight.

locus[a]	$\hat{\beta}(s_{\hat{\beta}})$[b]	p-value[c]	$\hat{h}_j^2$[d]
37$_{add}$	0.40(0.08)	8.22×10^{-7}	0.0262
50$_{add}$	0.21(0.08)	3.14×10^{-3}	0.0072
151$_{add}$	−0.18(0.07)	2.77×10^{-3}	0.0052
173$_{add}$	−0.49(0.07)	6.46×10^{-11}	0.0418
199$_{add}$	0.23(0.06)	1.51×10^{-4}	0.0077
332$_{add}$	0.30(0.06)	1.26×10^{-6}	0.0150
440$_{add}$	−0.98(0.06)	$<10^{-15}$	0.1670
498$_{add}$	−0.35(0.06)	5.18×10^{-8}	0.0188
710$_{add}$	0.17(0.06)	2.13×10^{-3}	0.0047
729$_{add}$	0.81(0.07)	$<10^{-15}$	0.0968
894$_{add}$	−0.18(0.06)	8.21×10^{-4}	0.0053
936$_{add}$	−0.44(0.07)	3.11×10^{-10}	0.0291
1008$_{add}$	−0.34(0.06)	6.63×10^{-8}	0.0177
1110$_{add}$	0.18(0.05)	4.85×10^{-4}	0.0055
1176$_{add}$	−0.26(0.06)	8.77×10^{-6}	0.0108
1251$_{add}$	0.37(0.07)	4.64×10^{-8}	0.0171
1374$_{add}$	0.33(0.06)	2.55×10^{-7}	0.0167
1442$_{add}$	−0.22(0.06)	6.37×10^{-5}	0.0079
1565$_{add}$	0.20(0.06)	2.33×10^{-4}	0.0063
38$_{dom}$	0.29(0.07)	3.18×10^{-5}	0.0066
228$_{dom}$	−0.20(0.07)	1.60×10^{-3}	0.0032
312$_{dom}$	0.17(0.06)	3.20×10^{-3}	0.0024
441$_{dom}$	−0.23(0.07)	7.95×10^{-4}	0.0043
547$_{dom}$	0.15(0.06)	6.68×10^{-3}	0.0019
843$_{dom}$	0.16(0.06)	5.41×10^{-3}	0.0021
1506$_{dom}$	−0.26(0.07)	1.89×10^{-4}	0.0054
Parameter(s)		$a=1, b=1$	
μ		0.1000	
σ_0^2		0.5224	
$\hat{h}^2$		0.8424	

[a]add: additive effect; dom: dominance effect. Total number of effects is 38, only 30 effects with a p-value ≤0.01 are listed in this table.
[b]The estimated marker effect is denoted by $\hat{\beta}$ and the standard deviation is denoted by $s_{\hat{\beta}}$.
[c]P-value is obtained via t-test.
[d]Phenotypic variation explained.

our EBlasso identified this QTL and six other QTLs, four of which have cloned gene within 15 cM distance (Table S11 in File S1).

In conclusion, taking advantage of the powerful EBlasso model for simultaneously accounting for more than 5 million possible effects, we identified a number of QTLs for four traits of the elite rice hybrid Shanyou 63, a vast majority of which are involved in digenic interactions. This set of QTLs not only shed light on the genetic basis of the yield of the rice hybrid, but also provide candidate loci for identification of new genes that may be involved in crop yield.

Materials and Methods

Plant materials and QTLs

The genotype and phenotype data used in this study were obtained from previous studies [12,24]. The mapping plants were created by first crossing between *indica* rice Zhensha 97 and Minghui 63 [7] to produce the elite rice hybrid Shanyou 63 that was the most widely cultivated in China in 1980s –1990s [24]. Then a population of 240 F$_9$ RILs was derived from single-seed descent of Shanyou 63. Next, an "immortalized F$_2$" (IMF$_2$) population consisted of 278 crosses was created by intercrossing RILs for QTL mapping study [7,23]. The crossed population was field tested on the experimental farm of Huazhong Agricultural University in Wuhan, China, in 1999, for traits including yield per plant, the number of panicles per plant, the number of grains per panicle and grain weight.

The RILs were genomic sequenced with an Illumina Genome Analyzer II using the bar-coded multiplexed sequencing approach as described in [25], and 270,820 high quality SNPs were identified. Bin maps were constructed by lumping consecutive SNPs with the same genotype into blocks, masking blocks with less than 250 kb to avoid false double recombinations, and merging recombination bins less than 5 kb, resulting in a map consisting of 1,619 bins without missing data [12]. Genotypes of the IMF$_2$ crosses were deduced according to genotypes of their RIL parents [24]. The three genotypes in each bin were coded as A and B for each parental homozygote genotype and H for the heterozygote. Using the recombinant bins as QTLs, a 1,625.5 cM genetic linkage map was constructed with about 1.0 cM (230 kb) in length per bin (Figure 1).

Bayesian Lasso linear regression model for multiple QTLs

We employed a Bayesian Lasso (BLasso) multiple linear regression model to infer genotypes and quantitative trait associations. The regression model includes main additive and dominance effects of 1,619 SNP bins and all their pair-wise interactions. Let y_i be the phenotypic value of a quantitative trait of the ith individual in a mapping population. In this study we observed y_i, $i=1, \cdots, n$, of $n=278$ individuals and collected them into a vector $\boldsymbol{y}=[y_1, y_2, \cdots, y_n]^T$. In these n individuals, let

Table 7. Estimated QTL effects from the full model for yield per plant.

Loci(i, j)[a]	$\hat{\beta}(s_{\hat{\beta}})$[b]	p-value[c]	$\hat{h}_j^2$[d]
(1014$_{add}$, 1014$_{add}$)	−1.94(0.37)	1.95×10^{-7}	0.0544
(113$_{add}$, 1547$_{add}$)	−2.81(0.53)	1.43×10^{-7}	0.0552
(1057$_{add}$, 1144$_{dom}$)	−2.89(0.53)	5.57×10^{-8}	0.0608
(743$_{dom}$, 1043$_{dom}$)	3.21(0.52)	8.38×10^{-10}	0.0598
Parameter(s)		$a=1, b=1$	
μ		−0.7521	
σ_0^2		22.9734	
$\hat{h}^2$		0.3401	

[a]add: additive effect; dom: dominance effect. If i equals j, then it is a main effect, otherwise, it is an interaction between locus i and locus j. Total number of effects is 4, all with a p-value ≤0.01.
[b]The estimated marker effect is denoted by $\hat{\beta}$ and the standard deviation is denoted by $s_{\hat{\beta}}$.
[c]P-value is obtained via t-test.
[d]Phenotypic variation explained.

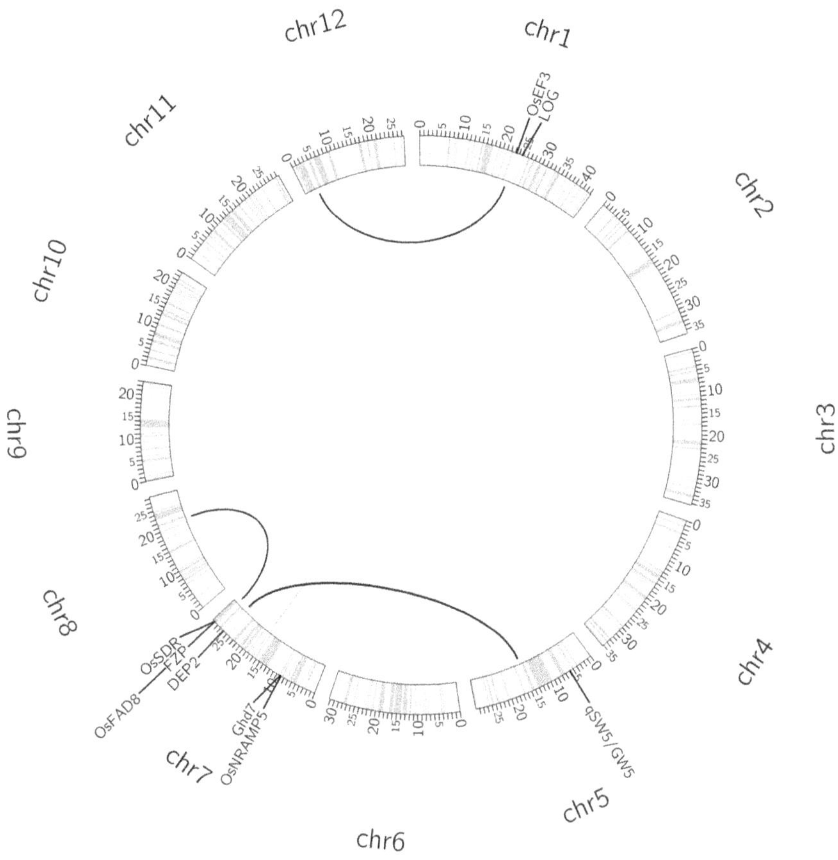

Figure 4. Interaction network of seven QTLs for yield per plant. The circle shows the bin map and columns indicate position of the makers (ticks in million base pairs). The thickness of a link is proportional to the strength of the interaction effect. A short straight line indicates a main effect. Molecularly characterized genes related to yield are also labeled in the appropriate positions of the genome.

$m = 1,619$ denote the number of genetic markers genotyped whose main effects include additive and dominance effects. Let the additive and dominance genotypes of marker j of individual i be x_{Aij} and x_{Dij}, respectively, where x_{Aij} takes on values +1, 0 and −1, and x_{Dij} takes on values 0, +1 and 0, corresponding to genotypes A, H and B, respectively. Let us define $\mathbf{x}_{Ai} = [x_{Ai1}, x_{Ai2}, \cdots, x_{Aim}]^T$ and $\mathbf{x}_{Gi} = [x_{Di1}, x_{Di2}, \cdots, x_{Dim}]^T$. The interactions between any two effects are modeled as element-wise product of the corresponding main effects. Let $\mathbf{x}_{AAi}$, $\mathbf{x}_{ADi}$, $\mathbf{x}_{DAi}$, and $\mathbf{x}_{DDi}$ be $m(m-1)/2 \times 1$ vectors containing $x_{Aij} \cdot x_{Aij'}$, $x_{Aij} \cdot x_{Dij'}$, $x_{Dij} \cdot x_{Aij'}$, and $x_{Dij} \cdot x_{Dij'}$, respectively, where $j = 1, \cdots, m-1$ and $j' > j$. Then we have the following linear regression model for $\boldsymbol{y}$:

$$\boldsymbol{y} = \mu + \mathbf{X}_A \boldsymbol{\beta}_A + \mathbf{X}_D \boldsymbol{\beta}_D + \mathbf{X}_{AA} \boldsymbol{\beta}_{AA} + \mathbf{X}_{AD} \boldsymbol{\beta}_{AD} + \mathbf{X}_{DA} \boldsymbol{\beta}_{DA} + \mathbf{X}_{DD} \boldsymbol{\beta}_{DD} + \varepsilon, \tag{2}$$

where μ is the population mean, vectors $\boldsymbol{\beta}_A$ and $\boldsymbol{\beta}_D$ represent the main additive and dominance effects of all markers, respectively, and vectors $\boldsymbol{\beta}_{AA}, \boldsymbol{\beta}_{AD}, \boldsymbol{\beta}_{DA}$ and $\boldsymbol{\beta}_{DD}$ capture the additive × additive, additive × dominance, dominance × additive, and dominance × dominance interactions, respectively. Matrices $\mathbf{X}_A = [\mathbf{x}_{A1}, \mathbf{x}_{A2}, \cdots, \mathbf{x}_{An}]^T$, $\mathbf{X}_D = [\mathbf{x}_{D1}, \mathbf{x}_{D2}, \cdots, \mathbf{x}_{Dn}]^T$, $\mathbf{X}_{AA} = [\mathbf{x}_{AA1}, \mathbf{x}_{AA2}, \cdots, \mathbf{x}_{AAn}]^T$, $\mathbf{X}_{AD} = [\mathbf{x}_{AD1}, \mathbf{x}_{AD2}, \cdots, \mathbf{x}_{ADn}]^T$, $\mathbf{X}_{DA} = [\mathbf{x}_{DA1}, \mathbf{x}_{DA2}, \cdots, \mathbf{x}_{DAn}]^T$, and $\mathbf{X}_{DD} = [\mathbf{x}_{DD1}, \mathbf{x}_{DD2}, \cdots, \mathbf{x}_{DDn}]^T$ are the corresponding design matrices of different effects, and 1ε is the residual error that follows a normal distribution with zero mean and variance $\sigma_0^2 \mathbf{I}$.

Given m markers, the size of matrix $\mathbf{X}_A$ or $\mathbf{X}_D$ is $n \times m$, and the size of $\mathbf{X}_{AA}$, $\mathbf{X}_{AD}$, $\mathbf{X}_{DA}$, or $\mathbf{X}_{DD}$ is $n \times q$, where $q = m(m-1)/2 = 1,309,771$. Defining $\boldsymbol{\beta} = [\boldsymbol{\beta}_A^T, \boldsymbol{\beta}_D^T, \boldsymbol{\beta}_{AA}^T, \boldsymbol{\beta}_{AD}^T, \boldsymbol{\beta}_{DA}^T, \boldsymbol{\beta}_{DD}^T]^T$, and $\mathbf{X} = [\mathbf{X}_A, \mathbf{X}_D, \mathbf{X}_{AA}, \mathbf{X}_{AD}, \mathbf{X}_{DA}, \mathbf{X}_{DD}]$, we can write (2) in a more compact form:

$$\boldsymbol{y} = \mu + \mathbf{X}\boldsymbol{\beta} + \varepsilon. \tag{3}$$

The size of matrix $\mathbf{X}$ is $n \times k$, where $k = 2m + 4q = 5,242,322$, and we apparently have $k \gg n$. However, we would expect that most elements of $\boldsymbol{\beta}$ are zeros and thus we have a sparse linear model. The Blasso model employs a three-level hierarchical prior distribution to model the sparsity. At the first level, let $\beta_j, j = 1, 2, \cdots, k$, follows an independent normal distribution with mean zero and unknown variance $\sigma_j^2 : \beta_j \tilde{N}(0, \sigma_j^2)$. At the second level, let $\sigma_j^2, j = 1, 2, \cdots, k$, follows an independent exponential distribution with a common parameter λ: $p(\sigma_j^2) = \lambda \exp(-\lambda \sigma_j^2)$. At the third level, we assign a conjugate Gamma prior $Gamma(a, b)$ with a shape parameter a and an inverse scale parameter b to the parameter λ. Finally, we assign non-informative uniform priors to μ and σ_0^2. The three-level hierarchical model has two hyperparameters (a, b) for adjusting the degree of shrinkage, and cross validation (CV) can be applied to choose appropriate values of these parameters.

The QTL model (2) or equivalently (3) includes all main effects and digenic interactions. We refer to this model as the full model throughout the paper. We also performed QTL mapping with the model $y = \mu + \mathbf{X}_A \boldsymbol{\beta}_A + \mathbf{X}_D \boldsymbol{\beta}_D + \boldsymbol{\varepsilon}$, which is referred to as the main effect model, since it includes only the main effects.

Model inference and cross validation

The Blasso model can be inferred efficiently with the empirical Blasso (EBlasso) algorithm [21]. The EBLasso algorithm employs a coordinate ascent method to find $\hat{\sigma}_j^2$, the estimate of $\sigma_j^2, j = 0 \ldots, k$, that maximizes the likelihood function of $\sigma_j^2, j = 0, \ldots, k$. In the iterative process, many σ_j^2 or equivalent β_j are shrunk to zero. The coordinate ascent method along with other algorithmic techniques makes the EBlasso algorithm very efficient. Our previous studies demonstrated that EBlasso outperformed several other multiple QTL mapping methods including the empirical Bayes method [26], the Bayesian hierarchical generalized linear models (BhGLM) [27], HyperLasso [28], and Lasso [29]. Detailed description of the EBLasso algorithm can be found in [21,22] and an efficient C program with the R interface [30] implementing the EBlasso algorithm is available.

The optimal values of two hyperparameters (a, b) of the EBLasso algorithm were obtained with five-fold CV in three steps to minimize the prediction error (PE) calculated from

$$PE = \frac{1}{n} \sum_{i=1}^{n} (y_i - \hat{y}_i)^2,$$

where $\hat{y}_i, i = 1, \cdots, n$, is the estimated phenotypic value. In the first step, $a = b = 0.001, 0.01, 0.1, 1$ were examined and a pair (a_1, b_1) corresponding to the smallest PE was obtained. In the second step, b was fixed at b_1 and a was chosen from the set $[-0.9, -0.8, -0.7, -0.6, -0.5, -0.4, -0.3, -0.2, -0.1, -0.01, 0.01, 0.05, 0.1, 0.5, 1]$, which yielded a value a_2 corresponding to the smallest PE. In the third step, $a = a_2$ was fixed and b varied from 0.01 to 10 with a step size of one for $b > 1$ and a step size of one on the logarithmic scale for $b < 1$. Note that when fixing one of the two parameters, the degree of shrinkage is a monotonic function of the other parameter [21,22]. Therefore, in the second and third steps, the selection did not go through the full path but stopped if the current PE was one standard error larger than the minimum PE in previous steps.

Statistical significance test

One advantage of the EBLasso algorithm relative to Lasso [29] is that it not only outputs a $k' \times 1$ ($k' \ll k$) vector $\hat{\boldsymbol{\beta}}$ as an estimate of nonzero elements of $\boldsymbol{\beta}$, but also gives an estimate of the covariance of $\hat{\boldsymbol{\beta}}, \hat{\boldsymbol{\Sigma}}$. Letting $\hat{\Sigma}_{jj}$ be the jth diagonal element of $\hat{\boldsymbol{\Sigma}}$, we can use the t-statistics $\hat{\beta}_j / \hat{\Sigma}_{jj}^{1/2}$ to test if $\hat{\beta}_j \neq 0$ at a certain significance level.

Supporting Information

File S1 Tables S1–S12. Table S1. Cross-validation for determining hyperparameters (a, b) used in QTL mapping for

Table 8. Estimated QTL effects from the main effect model for yield per plant.

locus[a]	$\hat{\beta}(s_{\hat{\beta}})^{b}$	p-value[c]	$\hat{h}_j^{2d}$
181 $_{add}$	−1.87(0.42)	6.24×10^{-6}	0.0518
1014 $_{add}$	−2.27(0.43)	9.58×10^{-8}	0.0743
1057 $_{add}$	−2.41(0.43)	2.58×10^{-8}	0.0848
1100 $_{add}$	1.17(0.39)	1.35×10^{-3}	0.0205
Parameter(s)		$a = -0.5, b = 0.1$	
μ		−0.1500	
σ_0^2		26.2580	
$\hat{h}^2$		0.2379	

[a]add: additive effect; dom: dominance effect. Total number of effects is 4, all with a p-value ≤ 0.01.
[b]The estimated marker effect is denoted by $\hat{\beta}$ and the standard deviation is denoted by $s_{\hat{\beta}}$.
[c]p-value is obtained via t-test.
[d]Phenotypic variation explained.

the number of panicles per plant. **Table S2.** Experimentally investigated genes near QTLs for the number of panicles per plant identified with the full model. **Table S3.** Experimentally investigated genes near QTLs for the number of panicles per plant identified with the main effect model. **Table S4.** Cross-validation for determining hyperparameters (a, b) used in QTL mapping for the number of grains per panicle. **Table S5.** Experimentally investigated genes near QTLs for the number of grains per panicle identified with the full model. **Table S6.** Experimentally investigated genes near QTLs for the number of grains per panicle identified with the main effect model. **Table S7.** Cross-validation for determining hyperparameters (a, b) used in QTL mapping for grain weight. **Table S8.** Experimentally investigated genes near QTLs for grain weight identified with the full model. **Table S9.** Experimentally investigated genes near QTLs for grain weight identified with the main effect model. **Table S10.** Cross-validation for determining hyperparameters (a, b) used in QTL mapping for yield per plant. **Table S11.** Experimentally investigated genes near QTLs for yield per plant identified with the full model. **Table S12.** Experimentally investigated genes near QTLs for yield per plant identified with the main effect model.

Author Contributions

Conceived and designed the experiments: XC SX. Performed the experiments: AH. Analyzed the data: AH XC SX. Wrote the paper: AH XC SX.

References

1. Xing Y, Zhang Q (2010) Genetic and molecular bases of rice yield. Annu Rev Plant Biol 61: 421–442.
2. Jiang Y, Cai Z, Xie W, Long T, Yu H, et al. (2012) Rice functional genomics research: Progress and implications for crop genetic improvement. Biotechnol Adv 30: 1059–1070.
3. Tester M, Langridge P (2010) Breeding technologies to increase crop production in a changing world. Science 327: 818–822.
4. Ikeda M, Miura K, Aya K, Kitano H, Matsuoka M (2013) Genes offering the potential for designing yield-related traits in rice. Curr Opin Plant Biol 16: 213–220.
5. Lander ES, Botstein D (1989) Mapping mendelian factors underlying quantitative traits using RFLP linkage maps. Genetics 121: 185–199.
6. Zietkiewicz E, Rafalski A, Labuda D (1994) Genome fingerprinting by simple sequence repeat (SSR)-anchored polymerase chain reaction amplification. Genomics 20: 176–183.
7. Hua JP, Xing YZ, Xu CG, Sun XL, Yu SB, et al. (2002) Genetic dissection of an elite rice hybrid revealed that heterozygotes are not always advantageous for performance. Genetics 162: 1885–1895.
8. Li J, Yu S, Xu C, Tan Y, Gao Y, et al. (2000) Analyzing quantitative trait loci for yield using a vegetatively replicated F2 population from a cross between the parents of an elite rice hybrid. Theor Appl Genet 101: 248–254.

9. Xing Y, Tan Y, Hua J, Sun X, Xu C, et al. (2002) Characterization of the main effects, epistatic effects and their environmental interactions of QTLs on the genetic basis of yield traits in rice. Theor Appl Genet 105: 248–257.

10. Tan Y-F, Xing Y-Z, Li J-X, Yu S-B, Xu C-G, et al. (2000) Genetic bases of appearance quality of rice grains in Shanyou 63, an elite rice hybrid. Theor Appl Genet 101: 823–829.

11. Lian X, Xing Y, Yan H, Xu C, Li X, et al. (2005) QTLs for low nitrogen tolerance at seedling stage identified using a recombinant inbred line population derived from an elite rice hybrid. Theor Appl Genet 112: 85–96.

12. Yu H, Xie W, Wang J, Xing Y, Xu C, et al. (2011) Gains in QTL detection using an ultra-high density SNP map based on population sequencing relative to traditional RFLP/SSR markers. PLoS ONE 6: e17595.

13. Zeng ZB (1994) Precision mapping of quantitative trait loci. Genetics 136: 1457–1468.

14. Song X-J, Ashikari M (2008) Toward an optimum return from crop plants. Rice 1: 135–143.

15. Bernardo R (2008) Molecular markers and selection for complex traits in plants: learning from the last 20 years. Crop Sci 48: 1649–1664.

16. Xu S (2003) Estimating polygenic effects using markers of the entire genome. Genetics 163: 789–801.

17. Meuwissen THE, Hayes BJ, Goddard ME (2001) Prediction of total genetic value using genome-wide dense marker maps. Genetics 157: 1819–1829.

18. Zhou X, Carbonetto P, Stephens M (2013) Polygenic modeling with Bayesian sparse linear mixed models. PLoS Genet 9: e1003264.

19. Iwata H, Ebana K, Uga Y, Hayashi T, Jannink J-L (2010) Genome-wide association study of grain shape variation among Oryza sativa L. germplasms based on elliptic Fourier analysis. Mol Breed 25: 203–215.

20. de los Campos G, Pérez P, Vazquez AI, Crossa J (2013) Genome-enabled prediction using the BLR (Bayesian Linear Regression) R-package. Genome-Wide Association Studies and Genomic Prediction. New York: Springer. pp. 299–320.

21. Cai X, Huang A, Xu S (2011) Fast empirical Bayesian LASSO for multiple quantitative trait locus mapping. BMC Bioinformatics 12: 211.

22. Huang A, Xu S, Cai X (2013) Empirical Bayesian LASSO-logistic regression for multiple binary trait loci mapping. BMC Genet 14: 5.

23. Hua J, Xing Y, Wu W, Xu C, Sun X, et al. (2003) Single-locus heterotic effects and dominance by dominance interactions can adequately explain the genetic basis of heterosis in an elite rice hybrid. Proc Natl Acad Sci USA 100: 2574–2579.

24. Zhou G, Chen Y, Yao W, Zhang C, Xie W, et al. (2012) Genetic composition of yield heterosis in an elite rice hybrid. Proc Natl Acad Sci USA 109: 15847–15852.

25. Xie W, Feng Q, Yu H, Huang X, Zhao Q, et al. (2010) Parent-independent genotyping for constructing an ultrahigh-density linkage map based on population sequencing. Proc Natl Acad Sci USA 107: 10578–10583.

26. Xu S (2007) An empirical Bayes method for estimating epistatic effects of quantitative trait loci. Biometrics 63: 513–521.

27. Yi N, Banerjee S (2009) Hierachical generalized linear models for multiple quantitative trait locus mapping. Genetics 181: 1101–1133.

28. Hoggart CJ, Whittaker JC, De Iorio M, Balding DJ (2008) Simultaneous analysis of all SNPs in genome-wide and re-sequencing association studies. PLoS Genet 4: e1000130.

29. Tibshirani R (1996) Regression shrinkage and selection via the lasso. J Roy Stat Soc B Met 58: 267–288.

30. R Development Core Team (2012) R: A language and environment for statistical computing. Vienna, Austria: R Foundation for Statistical Computing.

31. Miura K, Ashikari M, Matsuoka M (2011) The role of QTLs in the breeding of high-yielding rice. Trends Plant Sci 16: 319–326.

32. Yan W-H, Wang P, Chen H-X, Zhou H-J, Li Q-P, et al. (2011) A major QTL, Ghd8, plays pleiotropic roles in regulating grain productivity, plant height, and heading date in rice. Mol Plant 4: 319–330.

33. Yu J, Hu S, Wang J, Wong GK-S, Li S, et al. (2002) A draft sequence of the rice genome (Oryza sativa L. ssp. indica). Science 296: 79–92.

34. Goff SA, Ricke D, Lan T-H, Presting G, Wang R, et al. (2002) A draft sequence of the rice genome (Oryza sativa L. ssp. japonica). Science 296: 92–100.

35. Zhang Q, Li J, Xue Y, Han B, Deng XW (2008) Rice 2020: A call for an international coordinated effort in rice functional genomics. Mol Plant 1: 715–719.

36. Li X, Qian Q, Fu Z, Wang Y, Xiong G, et al. (2003) Control of tillering in rice. Nature 422: 618–621.

37. Minakuchi K, Kameoka H, Yasuno N, Umehara M, Luo L, et al. (2010) FINE CULM1 (FC1) works downstream of strigolactones to inhibit the outgrowth of axillary buds in rice. Plant Cell Physiol 51: 1127–1135.

38. Jiao Y, Wang Y, Xue D, Wang J, Yan M, et al. (2010) Regulation of OsSPL14 by OsmiR156 defines ideal plant architecture in rice. Nat Genet 42: 541–544.

39. Chuck G, Muszynski M, Kellogg E, Hake S, Schmidt RJ (2002) The control of spikelet meristem identity by the branched silkless1 gene in maize. Science 298: 1238–1241.

40. Xue W, Xing Y, Weng X, Zhao Y, Tang W, et al. (2008) Natural variation in Ghd7 is an important regulator of heading date and yield potential in rice. Nat Genet 40: 761–767.

41. Li F, Liu W, Tang J, Chen J, Tong H, et al. (2010) Rice DENSE AND ERECT PANICLE 2 is essential for determining panicle outgrowth and elongation. Cell Res 20: 838–849.

42. Wang S, Wu K, Yuan Q, Liu X, Liu Z, et al. (2012) Control of grain size, shape and quality by OsSPL16 in rice. Nat Genet 44: 950–954.

43. Heang D, Sassa H (2012) Antagonistic actions of HLH/bHLH proteins are involved in grain length and weight in rice. PLoS ONE 7: e31325.

44. Wang E, Xu X, Zhang L, Zhang H, Lin L, et al. (2010) Duplication and independent selection of cell-wall invertase genes GIF1 and OsCIN1 during rice evolution and domestication. BMC Evol Biol 10: 108.

45. Kitagawa K, Kurinami S, Oki K, Abe Y, Ando T, et al. (2010) A novel kinesin 13 protein regulating rice seed length. Plant Cell Physiol 51: 1315–1329.

46. Segami S, Kono I, Ando T, Yano M, Kitano H, et al. (2012) Small and round seed 5 gene encodes alpha-tubulin regulating seed cell elongation in rice. Rice 5: 1–10.

47. Zha X, Luo X, Qian X, He G, Yang M, et al. (2009) Over-expression of the rice LRK1 gene improves quantitative yield components. Plant Biotechnol J 7: 611–620.

14

Australian Wild Rice Reveals Pre-Domestication Origin of Polymorphism Deserts in Rice Genome

Gopala Krishnan S.[1,2], Daniel L. E. Waters[1], Robert J. Henry[3]*

1 Southern Cross Plant Science, Southern Cross University, Lismore, New South Wales, Australia, **2** Division of Genetics, Indian Agricultural Research Institute, New Delhi, India, **3** Queensland Alliance for Agriculture and Food Innovation, The University of Queensland, Brisbane, Queensland, Australia

Abstract

Background: Rice is a major source of human food with a predominantly Asian production base. Domestication involved selection of traits that are desirable for agriculture and to human consumers. Wild relatives of crop plants are a source of useful variation which is of immense value for crop improvement. Australian wild rices have been isolated from the impacts of domestication in Asia and represents a source of novel diversity for global rice improvement. *Oryza rufipogon* is a perennial wild progenitor of cultivated rice. *Oryza meridionalis* is a related annual species in Australia.

Results: We have examined the sequence of the genomes of AA genome wild rices from Australia that are close relatives of cultivated rice through whole genome re-sequencing. Assembly of the resequencing data to the *O. sativa* ssp. *japonica* cv. Nipponbare shows that Australian wild rices possess 2.5 times more single nucleotide polymorphisms than in the Asian wild rice and cultivated *O. sativa* ssp. *indica*. Analysis of the genome of domesticated rice reveals regions of low diversity that show very little variation (polymorphism deserts). Both the perennial and annual wild rice from Australia show a high degree of conservation of sequence with that found in cultivated rice in the same 4.58Mbp region on chromosome 5, which suggests that some of the 'polymorphism deserts' in this and other parts of the rice genome may have originated prior to domestication due to natural selection.

Conclusions: Analysis of genes in the 'polymorphism deserts' indicates that this selection may have been due to biotic or abiotic stress in the environment of early rice relatives. Despite having closely related sequences in these genome regions, the Australian wild populations represent an invaluable source of diversity supporting rice food security.

Editor: David Caramelli, University of Florence, Italy

Funding: The research was funded by the Australian Research Council, the Department of Science and Technology, Government of India under the BOYSCAST Fellowship and the Indian Council for Agricultural Research, New Delhi. The funders had no role in study design, data collection and analysis, decision to publish, or preparation of the manuscript.

Competing Interests: The authors have declared that no competing interests exist.

* E-mail: robert.henry@uq.edu.au

Introduction

Food security depends on sustainable crop production; especially for the major cereals such as rice, which contributes to more than half of the human food across the globe. Enhancing rice production to meet growing food demand requires continuous genetic improvement, especially in response to biotic and abiotic stress that may be intensified by climate change. The wild gene pool of Asian cultivated rice (*Oryza sativa*) is found in Asia and Australia [1]. Much of the Asian wild rice populations have been displaced by cultivated rice since domestication started in China around 7000 years ago [2]. Gene flow from large domesticated Asian rice populations has also impacted the small remaining populations of Asian wild rice.

As in other domesticated plants, the gene pool of cultivated rice has reduced genetic diversity relative to that found in the wild due to the bottleneck imposed by domestication, harbouring genomic regions which are significantly depressed in diversity [3],[4]. Australian wild rice populations have been isolated from domesticated rice and are a reservoir of environmentally adapted genetic diversity which could be exploited for cultivated rice improvement [5].

In the present study, we have undertaken whole shotgun genome sequencing of wild Australian and Asian rices and aligned the sequences with the cultivated rice, *O. sativa* ssp. *japonica* cv. Nipponbare reference genome. Based on the analysis of the variations in DNA polymorphisms across the entire rice genome, we were able to identify 'polymorphism deserts' in the genomic regions possessing genes for adaptive traits. Our analysis shows that this reduction in polymorphism is not restricted to cultivated *Oryza* species but also found in the wild species of *Oryza* from Australia also, suggesting the role of predomestication bottleneck induced due to natural selection.

Materials and Methods

Germplasm, sampling and sequencing

The *Oryza sativa* ssp. *indica* germplasm is an elite parental line used in hybrid breeding programme. The Asian *O. rufipogon* strain was from the Australian Plant DNA Bank Number (AC11-1008369). The sample was collected by Ryuji Ishikawa in a

Table 1. Single nucleotide polymorphisms in cultivated and wild *Oryza* as compared with *O. sativa* ssp. *japonica* cv. Nipponbare.

Oryza species	Whole genome		Chromosome 5		Chromosome 5 low diversity region (8.972 -13.557 Mb)	
	Total	SNPs/kb	Total	SNPs/kb	Total	SNPs/kb
O. sativa ssp.*indica*	978,630	2.56	68,443	2.28*	244	0.053*
O. rufipogon (Asian)	917,738	2.40	70,367	2.34*	6,050	1.302*
O. rufipogon (Australian)	2,564,013	6.71	219,794	6.71*	18,179	3.96*
O. meridionalis	2,418,084	6.33	206,884	6.89*	17,139	3.73*

* Means significantly different (t–test, p<0.01).

collaboration approved by Nguyen Thi Lang, Head of Genetics and Plant Breeding Division, Cuulong Delta Rice Research Institute, Can Tho, Vietnam, from a site located at N9 59.376 E105 39.883. This species is not endangered. The field studies did not involve endangered or protected species. Australian *O. rufipogon* was sourced from the Australian Tropical Crops and Forages Collection, Biloela (AusTRCF 309313; Australian Plant DNA Bank Number - AC01-1002323; collected from site located at N 18.206, E 142.865, about 0.9 K west Gilbert River Bridge in Gulf Development Road). *O. meridionalis* was sourced from the Australian Tropical Crops and Forages Collection, Biloela (AusTRCF 300118_B; originally collected Northern Territory, Australia). Construction of library and sequencing of these germplasm accessions was performed on Illumina Genome Analyser (GAIIx) with detailed procedure as described previously [6]. Paired end reads generated from all the genotypes were deposited in the NCBI sequence read archive (SRA) and can be found under the accession number SRP039365 (*Oryza sativa* ssp. *indica* - SRX480815; Asian *O. rufipogon* - SRX480820; Australian *O. rufipogon* - SRX480822 and *O. meridionalis* - SRX480817).

Mapping reads to the reference and SNP calling

Paired-end sequence reads were trimmed of low-quality data with a quality score limit of 0.01 and adaptor sequence in CLC Genomics Workbench 4.0 (http://www.clcbio.com) and reads of less than 30 base pairs (bp) in length were discarded. Trimmed short-read sequences were first aligned to the published rice organellar genomes (Chloroplast genome: Genbank accession - AY522330.1, mitochondrial genome: Genbank accession - DQ 167400.1) and the unmapped reads were taken up for further assembly against the nuclear genome (IRGSP Pseudomolecules build 4.0, http://rgp.dna.affrc.go.jp/IRGSP/Build4/build4. html). The reads were assembled to the Nipponbare reference with CLC Genomics workbench with the following parameters: mismatch cost - 2, insertion cost - 3, deletion cost - 3, length fraction - 0.5 and similarity - 0.8. Reads that aligned to more than one position of the reference genome were filtered and only unique reads were used for calling the SNPs. For comparison with 93–11 genome, the reads were first aligned to organellar genomes (Chloroplast genome: Genbank accession - AY522329.1, mitochondrial genome: Genbank accession - DQ167399.1) and the unmapped reads were aligned against the nuclear genome

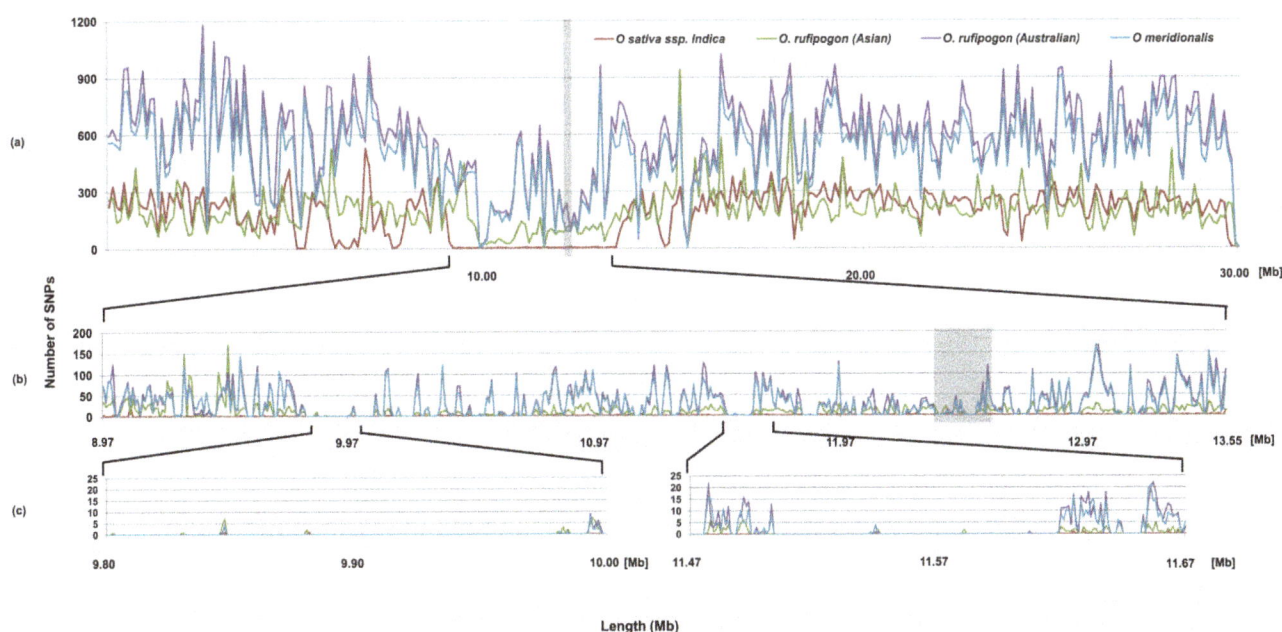

Figure 1. SNP distribution across Chromosome 5. Numbers on vertical axis are SNP/kb. Numbers on vertical horizontal axis are Mb from origin. Vertical grey bar represents the centromere.

(Genbank accession - AAAA02000000) using the above parameters. SNPs in the assembled contigs relative to the reference genome were identified with detailed procedure as described previously [6].

Analysis of variations

To quantify the DNA polymorphisms across the genome in different species, a sliding window of 100 kb intervals was used to analyse each chromosome to determine SNP frequency in each window.

Analysis of 'Polymorphism desert'

Comparison of the mean SNP frequency within the region equivalent to the chromosome 5 'polymorphism desert' of cultivated rice (8.97 and 13.56 Mbp) and the remainder of chromosome 5 was by t-test. The analysis of coverage and SNPs within the 'polymorphism desert' was carried out as described in File S1. Annotated genes within the 'polymorphism desert' equivalent region were retrieved from the IRGSP Pseudomolecules Build 4. RAP-DB ID Converter (http://rapdb.dna.affrc.go.jp/tools/converter) was used for converting the Locus ID of each gene to the corresponding MSU locus identifier. The functions of the genes were retrieved from the Rice Genome Annotation Project website (http://rice.plantbiology.msu.edu).

Analysis of selection sweeps

The SNP variations in 2 Mb region centred around 15 cloned genes in rice namely *Gn1a* [7], *Rd* [8], *qSH1* [9], *sd1* [10] – [12], *GW2* [13], *GS3* [14], *GIF1* [15], *Bh4* [16], *sh4* [17], *qSW5* [18], *wx* [19], *PROG1* [20], *Rc* [21], *GBSSII* [22] and *BAD2* [23] which has undergone selection either during domestication or crop improvement was assessed using a sliding window of 1kb interval and the selective sweeps were determined as reflected by a significant reduction in mean SNPs/kb (Text S1).

Results and Discussion

Whole genome re-sequencing yielded 51859475, 67186809, 62321354 and 46409181 paired reads of raw data in *Oryza sativa* ssp. *indica*, Asian *O. rufipogon*, Australian *O. rufipogon* and *O. meridionalis*, respectively. The reads were 75-bp paired end in case of reads *Oryza sativa* ssp. *indica*, while in case of Asian *O. rufipogon*, Australian *O. rufipogon* and *O. meridionalis*, it was 36-bp paired end reads. After appropriate processing, the short reads were mapped to high quality genomic sequences of *japonica* rice cultivar, Nipponbare using CLC Genome workbench 4.0. A total of

4773330, 6231758, 1013433 and 1644640 reads mapped to the organellar genomes; and 32723087, 32060326, 43606315 and 25523774 reads from *Oryza sativa* ssp. *indica*, Asian *O. rufipogon*, Australian *O. rufipogon* and *O. meridionalis*, respectively were uniquely mapped to the 12 pseudomolecules of the Nipponbare genome. On an average, the sequencing depth of 5.7X, 8.4X, 6.2× and 4.9× across the whole genome, providing genome coverage of 76.2%, 78.2%, 63.2% and 62.6% of the Nipponbare reference genome in case of *Oryza sativa* ssp. *indica*, Asian *O. rufipogon*, Australian *O. rufipogon* and *O. meridionalis*, respectively.

The analysis of genome-wide polymorphisms revealed that the number of SNPs detected in cultivated *O. sativa* ssp. *indica* was only 978,630 as compared to 2,564,013 SNPs in Australian *O. rufipogon* relative to *O. sativa* ssp. *japonica* cv. Nipponbare reference genome (Table S1). There were in the order of 2.5 times more single nucleotide polymorphisms (SNPs) in the Australian wild rice than in the Asian wild rice and cultivated *O. sativa* ssp. *indica* relative to *O. sativa* ssp. *japonica* cv. Nipponbare (Table 1), highlighting the potential value of the Australian wild populations as sources of novel variation for rice improvement. The mean SNPs per kb of genome was 6.7 and 6.3 in Australian *O. rufipogon* and *O. meridionalis*, respectively compared to 2.4 in *O. rufipogon* from Asia and 2.5 in *O. sativa* ssp. *indica*. The mean SNPs per kb of genes observed was 8.06 and 7.84 in Australian wild rice, *O. rufipogon* and *O. meridionalis*, respectively compared to 2.13 in *O. rufipogon* from Asia and 2.12 in cultivated rice (Table S2). The mean number of nonsynonymous SNPs per kb of gene was also up to 3 fold higher in the Australian A genome wild rice (Table S3). *O. rufipogon*, collected from the Mekong delta of Vietnam had fewer SNPs relative to Nipponbare than cultivated *O. sativa* ssp. *indica*, consistent with the hypothesis that *O. sativa* ssp. *japonica* was domesticated directly from *O. rufipogon* and perhaps reflecting the history of pollen flow from cultivated to wild populations [24]. The Mekong delta is the principal rice growing region of Vietnam and it is likely there has been gene flow between the wild *O. rufipogon* and *O. sativa* ssp. *japonica* cultivated in this and many other Asian regions.

Genome wide comparisons between *O. sativa* ssp. *japonica* and *O. sativa* ssp. *indica* cultivars have revealed a low diversity region also referred to as 'polymorphism desert' between 8.97 and 13.56 Mbp on chromosome 5 with less than 10 SNP per 100 kb while the mean SNP rate of Chromosome 5 is comparable to the mean SNP rate across other chromosomes [6]. A similar SNP distribution pattern has been observed in *indica-japonica* [25]–[27], *indica-indica* [6], and *japonica-japonica* [28] comparisons. Analysis of the equivalent region in Asian *O. rufipogon* and the Australian AA

Table 2. Extent of loss in variation in the domestication and plant improvement genes as reflected by SNP distribution (SNPs per kb) in the 2 Mb genomic region surrounding them.

Locus	Length of low polymorphism region (kb)	*O. sativa* ssp. *indica*	*O. rufipogon* (Asian)	*O. rufipogon* (Australian)	*O. meridionalis*
GS3	357	0.07 (25)	2.48 (884)	8.15 (2,910)	7.42 (2,650)
Bh4	374	0.26 (98)	3.01 (1,126)	8.86 (3,313)	7.32 (2,737)
sh4	371	0.48 (178)	1.87 (694)	7.29 (2,711)	8.73 (3,247)
qSW5	151	0.05 (8)	3.46 (522)	10.36 (1,564)	9.61 (1,451)
wx	369	0.07 (27)	2.62 (967)	8.7 (3,209)	8.40 (3,098)
PROG1	267	0.09 (25)	2.42 (647)	6.12 (1,634)	6.17 (1,647)
Rc	488	0.15 (74)	2.78 (1,359)	5.99 (2,922)	4.24 (2,069)

*Numbers in parenthesis is the total number of SNPs detected in the respective regions in comparison with Nipponbare genome.

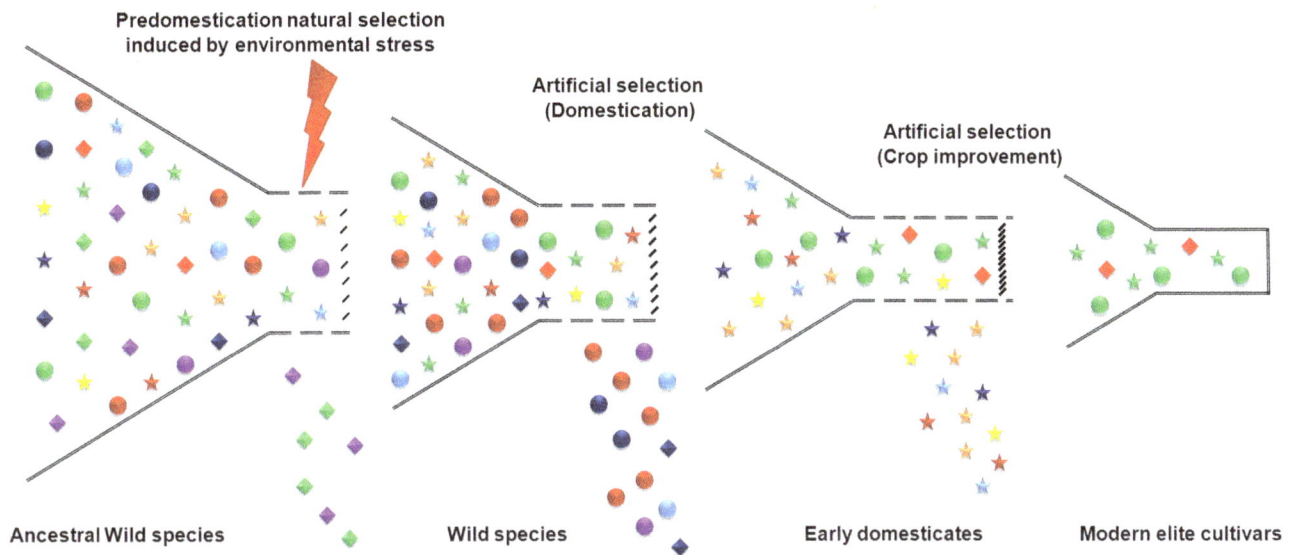

Figure 2. Predomestication bottleneck prior to domestication and crop improvement resulting in 'Polymorphism deserts' in cultivated rice. Different classes of genes are represented by different shapes; dice shape (♦) indicates genes for adaptive traits, circles (●) indicates domestication genes, star shape (*) indicates genes for crop improvement. The allelic forms of genes are represented in different colors. A predomestication bottleneck possibly induced by environmental stress resulted in loss of polymorphisms in the adaptive genes in case of wild rice. Additional selection pressure during domestication and crop improvement resulted in further depauperating the polymorphisms resulting in the 'polymorphism deserts' as in the case of chromosome 5 of rice. Artificial selection during rice domestication in genes such as *sh4*, *PROG1* resulted in reduced diversity in the adjoining genomic regions due to selection sweeps associated with the genes. Further selection during crop improvement in genes such as *GS3*, *Bh4*, *qSW5*, *wx* and *Rc* also reduced the polymorphisms in the regions associated with these genes. While the selections during domestication and crop improvement helped in retaining favourable alleles at these loci, an additional pre-domestication bottleneck has resulted in loss of variation in the genes providing adaptive traits in the 'polymorphism desert' of Chromosome 5.

genome wild rice found divergence from cultivated rice to be reduced by more than 40% in these species relative to the chromosome as a whole. This was observed not only based upon comparisons with *O. sativa* ssp. *japonica* cv. Nipponbare (Table 1 and Figure 1) but also with the *O. sativa* ssp. *indica* (cv. 93-11) reference genome sequences (Figure S1).

The chromosome 5 'polymorphism desert' is in the vicinity of the centromere and it has been observed that recombination in centromeric regions is depressed relative to other chromosomal regions which may influence SNP and InDel frequency [25], [29]. In *O. sativa* ssp. *indica*, analysis of SNP distribution within two Mb centred on the centromere of chromosome 5 showed that the mean SNP rate in this 'polymorphism desert' was only 0.07 per kb which is very low compared to the chromosomal mean of 2.28 SNPs per kb. Comparatively, chromosomes 4 and 8 had an average SNP rate of 1.79 and 1.33, per kb respectively around the centromere, as compared to the mean SNP rate of 2.21 and 2.48 per kb across these chromosomes. This analysis suggests that the presence of the centromere alone is not likely to explain the 'polymorphism desert' on chromosome 5. Additionally, the aligned data was subjected to an analysis which showed that the coverage across chromosome 5 was more than 4 reads ruling out the fact that the 'polymorphism desert' in chromosome 5 is not due to lower coverage (Figure S2 and S3). Further, the analysis of SNPs distribution including SNPs from repetitive sequences in chromosome 5 revealed that the reduction in SNPs is not due to filtering of SNPs from repetitive region (Figure S4).

Rice has been cultivated for seven to eight thousand years [2]. During the process of domestication, a range of favourable alleles have been captured [30]. Each time this has occurred, the rice genome has gone through a bottle neck, the remnants of which are most apparent as selective sweeps surrounding genes which code

for traits that support rice cultivation [31], [32]. A number of these selective sweeps have been investigated and they have been found to be in the range from 0.3 to 1.0Mb [33]. Within the germplasm studied here, selective sweeps ranging from 151 to 488 kb were apparent around seven of 15 candidate domestication/improvement genes (Table 2, Figure S5) while such signatures of domestication were not detected for another eight domestication related genes examined (Table S4, Figures S6 and S7). In contrast, the chromosome 5 'polymorphism desert' in the cultivated rice is significantly larger in extent in comparison to regions of depressed diversity around known domestication genes. The scale of the 'polymorphism desert', 4.58 Mb, suggests this region of the genome may harbour a cluster of several genes important to plant function and perhaps cultivation with overlapping selective sweeps.

Given the low level of polymorphism in the 'polymorphism desert' within *Oryza sativa* sub-species, approaches which rely on analysis of genetic difference would have difficulty in detecting and identifying genes of significance within this region. Reference to the paralogues in the Australian wild AA genome species which have not been in contact with cultivated rice provides clues as to which genes may have been under selection. Of 143 genes annotated within this region, 93 genes had sufficient coverage (atleast 4 reads) across the genomes sequenced to allow comparison. The Australian wild rices had a significantly higher number of SNPs and non-synonymous SNPs (nsSNPs) in the genes compared to cultivated rice (Tables S2 and S3). However, it was observed that the Australian wild rices also had a low number of SNPs in 61 genes. Gene annotations in the rice genome suggest 16 of these genes are involved in signalling, inflorescence and seed development, Fe and P interaction, disease resistance and seed germination (Table S5). The role of the genes within this region in various functions such as aerobic germination, cytokinin response in roots, Fe and P interaction suggests that these genes might have been subjected to natural selection in the wild rice progenitors prior to domestication; and human selection has magnified the effect in cultivated rice.

Conclusions

The present study shows that the Asian cultivated rice has lost variability as a result of selection during domestication and crop improvement, and the diversity within Australian wild rice is of immense value for rice improvement and adaptation to environmental changes especially in the face of climate change. The reduction in variation in certain genomic regions of wild rice populations indicate bottlenecks induced by natural selection prior to domestication has also contributed to reduction in diversity in the rice genome (Figure 2). Biotic or abiotic stress in the environment of wild rice in tandem with reduced recombination [34] associated with the physical distribution of mutations [35] may explain loss of diversity in specific chromosome areas encoding genes contributing to adaptation to these environmental factors. This adaptation may have been important in the evolution of essential features of modern rice such as adaptation to an aquatic environment. Whole genome re-sequencing has enabled the identification of novel polymorphisms preserved in Australian A genome wild rices which would be useful in diversifying the 'polymorphism deserts' of cultivated rice [36].

Acknowledgments

The authors acknowledge technical assistance by Mark Edwards and Stirling Bowen from Southern Cross Plant Genomics, Southern Cross University for Illumina sequencing and Abdul Baten for help with the submission of the sequencing data to SRA.

Author Contributions

Conceived and designed the experiments: GKS DLEW RJH. Performed the experiments: GKS DLEW. Analyzed the data: GKS DLEW RJH. Contributed reagents/materials/analysis tools: GKS DLEW RJH. Wrote the paper: GKS DLEW RJH.

References

1. Waters DLE, Nock CJ, Ishikawa R, Rice N, Henry RJ (2012) Chloroplast genome sequence confirms distinctness of Australian and Asian wild rice. Ecol Evol 2: 211–217. doi:10.1002/ece3.66.
2. Huang X, Kurata N, Wei X, Wang ZX, Wang A, et al. (2012) A map of rice genome variation reveals the origin of cultivated rice. Nature 490: 497–497. doi:10.1038/nature11532.
3. Song ZP, Xu X, Wang B, Chen JK, Lu BR (2003) Genetic diversity in the northernmost Oryza rufipogon populations estimated by SSR markers. Theor Appl Genet 107: 1492–1499. doi:10.1007/s00122-003-1380-3.
4. Tanksley SD, McCouch SR (1997) Seed banks and molecular maps: unlocking genetic potential from the wild. Science 277: 1063–1066. doi:10.1126/science.277.5329.1063.
5. Henry RJ, Rice N, Waters DLE, Kasem S, Ishikawa R, et al. (2010) Australian Oryza: utility and conservation. Rice 3: 235–241. doi:10.1007/s12284-009-9034-y.
6. Subbaiyan GK, Waters DLE, Katiyar SK, Sadananda AR, Satyadev V, et al. (2012) Genome-wide DNA polymorphisms in elite indica rice inbreds discovered by whole-genome sequencing. Plant Biotechnol J 10: 623–634. doi:10.1111/j.1467-7652.2011.00676.x.
7. Ashikari M, Sakakibara H, Lin S, Yamamoto T, Takashi T, et al. (2005) Cytokinin oxidase regulates rice grain production. Science 309: 741–745. doi:10.1126/science.1113373.
8. Furukawa T, Maekawa M, Oki T, Suda I, Lida S, et al. (2007) The Rc and Rd genes are involved in proanthocyanidin synthesis in rice pericarp. Plant J 49: 91–102. doi: 10.1111/j.1365-313X.2006.02958.x.
9. Konishi S, Izawa T, Lin SY, Ebana K, Fukuta Y, et al. (2006) An SNP caused loss of seed shattering during rice domestication. Science 312: 1392–1396. doi:10.1126/science.1126410.
10. Sasaki A, Ashikari M, Ueguchi-Tanaka M, Itoh H, Nishimura A, et al. (2002) Green revolution: a mutant gibberellin-synthesis gene in rice. Nature 416: 701–702. doi:10.1038/416701a.
11. Monna L, Kitazawa N, Yoshino R, Suzuki J, Masuda H, et al. (2002) Positional cloning of rice semidwarfing gene. sd-1: rice 'green revolution gene' encodes a mutant enzyme involved in gibberellin synthesis. DNA Res 9: 11–17. doi: 10.1093/dnares/9.1.11.
12. Spielmeyer W, Ellis MH, Chandler PM (2002) Semidwarf (sd-1), green revolution rice, contains a defective gibberellin 20-oxidase gene. Proc. Natl. Acad. Sci. USA 99: 9043–9048. doi: 10.1073/pnas.132266399.
13. Shomura A, Izawa T, Ebana K, Ebitani T, Kanegae H, et al. (2008) Deletion in a gene associated with grain size increased yields during rice domestication. Nat Genet 40: 1023–1028. doi:10.1038/ng.169.
14. Fan C, Xing Y, Mao H, Lu T, Han B, et al. (2006) GS3, a major QTL for grain length and weight and minor QTL for grain width and thickness in rice, encodes a putative transmembrane protein. Theor Appl Genet 112: 1164–1171. Doi: 10.1007/s00122-006-0218-1.
15. Wang E, Wang J, Zhu X, Hao W, Wang L, et al. (2008) Control of rice grain-filling and yield by a gene with a potential signature of domestication. Nat Genet 40: 1370–1374. doi: 10.1038/ng.220.
16. Zhu BF, Si L, Wang Z, Zhou Y, Zhu J, et al. (2011) Genetic control of a transition from black to straw-white seed hull in rice domestication. Plant Physiol 155: 1301–1311. doi:10.1104/pp.110.168500.
17. Li C, Zhou A, Sang T (2006) Rice domestication by reducing shattering. Science 311: 1936–1939. doi: 10.1126/science.1123604.
18. Song XJ, Huang W, Shi M, Zhu MZ, et al. (2007) A QTL for rice grain width and weight encodes a previously unknown RING-type E3 ubiquitin ligase. Nat Genet 39: 623–630. doi:10.1038/ng2014.
19. Wang ZY, Zheng FQ, Shen GZ, Gao JP, Snustad DP, et al. (1995) The amylose content in rice endosperm is related to the post-transcriptional regulation of the waxy gene. Plant J.7, 613–622. doi: 10.1046/j.1365-313X.1995.7040613.x.
20. Tan L, Li X, Liu F, Sun X, Li C, et al. (2008) Control of a key transition from prostrate to erect growth in rice domestication. Nat Genet 40: 1360–1364. doi: 10.1038/ng.
21. Sweeney MT, Thomson MJ, Pfeil BE, McCouch S. (2006) Caught red-handed: Rc encodes a basic helix-loop-helix protein conditioning red pericarp in rice. Plant Cell 18: 283–294. doi: http://dx.doi.org/10.1105/tpc.105.
22. Hirose T, Terao T (2004) A comprehensive expression analysis of the starch synthase gene family in rice (Oryza sativa L.). Planta 220(1): 9–16. doi: 10.1007/s00425-004-1314-6.
23. Bradbury LMT, Fitzgerald TL, Henry RJ, Jin Q, Waters DL (2005) The gene for fragrance in rice. Plant Biotechnol J 3: 363–370. doi: 10.1111/j.1467-7652.2005.00131.x.
24. Ishii T, Hiraoka T, Kanzaki T, Akimoto M, Shishido R, et al. (2011) Evaluation of genetic variation among wild populations and local varieties of rice. Rice 4: 170–177. doi:10.1007/s12284-011-9067-x.
25. Feltus FA, Wan J, Schulze SR, Estill JC, Jiang N, et al. (2004) An SNP resource for rice genetics and breeding based on subspecies indica and japonica genome alignments. Genome Res 14: 1812–1819. doi:10.1101/gr.2479404.
26. Wang C, Huang H (2004) Development of Genome-Wide DNA Polymorphism Database for Map-Based Cloning of Rice Genes. Plant Physiol 135: 1198–1205. doi:10.1104/pp.103.038463.
27. He Z, Zhai W, Wen H, Tang T, Wang Y, et al. (2011) Two evolutionary histories in the genome of rice: the roles of domestication genes. PLoS Genet 7(6): e1002100. doi:10.1371/journal.pgen.1002100.
28. Yamamoto T, Nagasaki H, Yonemaru J, Ebana K, Nakajima, et al. (2010) Fine definition of the pedigree haplotypes of closely related rice cultivars by means of genome-wide discovery of single-nucleotide polymorphisms. BMC Genomics 11: 267. doi:10.1186/1471-2164-11-267.
29. McMuellen MD, Kresovich S, Villeda HS, Bradbury P, Li H, et al. (2009) Genetic properties of the maize nested association mapping population. Science 325: 737–740. doi:10.1126/science.1174320.
30. Kharabian-Masouleh A, Waters DLE, Reinke RF, Ward R, Henry RJ (2012) SNP in starch biosynthesis genes associated with nutritional and functional properties of rice. Sci Rep 2: 557. doi:10.1038/srep00557.
31. Olsen KM, Caicedo AL, Polato N, McClung A, McCouch S, et al. (2006) Selection under domestication: evidence for a sweep in the rice waxy genomic region. Genetics 173: 975–983. doi:10.1534/genetics.106.056473.
32. Tang H, Sezen U, Paterson AH (2010) Domestication and plant genomes. Curr Opin Plant Biol 13: 160–166. doi:10.1016/j.pbi.2009.10.008.
33. Sweeney MT, Thomson MJ, Cho YG, Park YJ, Williamson SH, et al. (2007) Global dissemination of a single mutation conferring white pericarp in rice. PLoS Genet 3(8): e133. doi:10.1371/journal.pgen.0030133.

34. Lu J, Tang T, Tang H, Huang J, Shi S, et al. (2006) The accumulation of deleterious mutations in rice genomes: a hypothesis on the cost of domestication. Trends Genet 22: 126–131. doi:10.1016/j.tig.2006.01.004.

35. Flowers JM, Molina J, Rubinstein S, Huang P, Schaal BA, et al. (2012) Natural selection in gene-dense regions shapes the genomic pattern of polymorphism in wild and domesticated rice. Mol Biol Evol 29: 675–687. doi:10.1093/molbev/msr225.

36. Kovach MJ, McCouch SR (2008) Leveraging natural diversity: back through the bottleneck. Curr Opin Plant Biol 11: 193–200. doi:10.1016/j.pbi.2007.12.006.

15

Population Genetic Structure of the Cotton Bollworm *Helicoverpa armigera* (Hübner) (Lepidoptera: Noctuidae) in India as Inferred from EPIC-PCR DNA Markers

Gajanan Tryambak Behere[1,4], **Wee Tek Tay**[2]*, **Derek Alan Russell**[3], **Keshav Raj Kranthi**[5], **Philip Batterham**[1]

1 Department of Genetics, Bio21 Molecular Science and Biotechnology Institute, The University of Melbourne, Parkville, Melbourne, Victoria, Australia, **2** CSIRO Ecosystem Sciences, Canberra, Australian Capital Territory, Australia, **3** Department of Agriculture and Food Systems, The University of Melbourne, Parkville, Melbourne, Victoria, Australia, **4** Division of Entomology, Indian Council of Agricultural Research, Research Complex for North Eastern Hill Region, Shilong, Meghalaya, India, **5** Central Institute for Cotton Research, Nagpur, Maharashtra, India

Abstract

Helicoverpa armigera is an important pest of cotton and other agricultural crops in the Old World. Its wide host range, high mobility and fecundity, and the ability to adapt and develop resistance against all common groups of insecticides used for its management have exacerbated its pest status. An understanding of the population genetic structure in *H. armigera* under Indian agricultural conditions will help ascertain gene flow patterns across different agricultural zones. This study inferred the population genetic structure of Indian *H. armigera* using five Exon-Primed Intron-Crossing (EPIC)-PCR markers. Nested alternative EPIC markers detected moderate null allele frequencies (4.3% to 9.4%) in loci used to infer population genetic structure but the apparently genome-wide heterozygote deficit suggests in-breeding or a Wahlund effect rather than a null allele effect. Population genetic analysis of the 26 populations suggested significant genetic differentiation within India but especially in cotton-feeding populations in the 2006–07 cropping season. In contrast, overall pair-wise F_{ST} estimates from populations feeding on food crops indicated no significant population substructure irrespective of cropping seasons. A Baysian cluster analysis was used to assign the genetic make-up of individuals to likely membership of population clusters. Some evidence was found for four major clusters with individuals in two populations from cotton in one year (from two populations in northern India) showing especially high homogeneity. Taken as a whole, this study found evidence of population substructure at host crop, temporal and spatial levels in Indian *H. armigera*, without, however, a clear biological rationale for these structures being evident.

Editor: Daniel Doucet, Natural Resources Canada, Canada

Funding: GTB was supported by grants from the Melbourne International Research Scholarship (MIRS), the Melbourne International Fee Remission Scholarship (MIFRS), and the Albert Shimmins Post Graduate Writing-up Award (Faculty of Science, University of Melbourne). The Australian Research Council (ARC) provided funding to the Special Research Centre for Environmental Stress and Adaptation Research (CESAR). The funders had no role in study design, data collection and analysis, decision to publish, or preparation of the manuscript.

Competing Interests: The authors have declared that no competing interests exist.

* E-mail: weetek.tay@csiro.au

Introduction

The polyphagous nature of the Old World cotton bollworm *Helicoverpa armigera* (Hübner) on a wide range of wild and crop hosts across different ecological zones, its highly variable life-history traits (e.g., number of generations, crop hosts, presence of summer/winter diapause) and seasonal abundance present a unique challenge for ecological and evolutionary studies. The number of generations possible per year is directly influenced by temperature, rainfall and presence of suitable hosts [1]. In India, *H. armigera* is an important pest of cotton, legumes, cereals and vegetables, and presents a unique challenge to those studying its population genetic structure.

The farming landscape in India is predominantly characterised by small farms and mixed cropping systems. The cropping patterns in India normally ensures the presence of five to six different host crops in different proportions for *H. armigera* at any given time of the growing season [2], thereby creating a heterogeneous matrix of hosts which provide ideal platforms for *H. armigera* to move between hosts and geographic areas throughout the year. Furthermore, the presence of three major cropping scenarios in India (in the North, Centre and South) are influenced by the pattern of the monsoons [3] (i.e., southwest monsoons: June to September and northeast monsoons: October to December), and by the sub-tropical nature of the south that allows continuous cropping versus the more continental and temperate climate of the north. India's cropping scenarios therefore provide a range of hosts crops for *H. armigera* all year round in any given region, although cotton represents the main host crop on which this pest species completes three out of possible seven to eight generations annually in 11 states [3], [4], [5], see [6] for a map of cotton states. In the north, facultative pupal diapause is reported in the winter months following the cotton season [7], [8], [9], with synchronous emergence of large numbers of moths frequently triggered by the first heavy rainfall (after the arrival of the monsoon) after prolonged dry periods [10]. The first post-diapause generations in the north are on crops and weeds other than cotton. In the mid-hill regions of Himachal Pradesh in northern India, chickpea is the first crop to be exploited by over-wintered *H. armigera* populations, between March and May [11]. Windborne long-distance migration of *H. armigera* in central India is likely to occur at the end of the cropping season (December–January), while rains prolong the growing season in northern and southern India, with the resulting adult migration in these regions typically occurring around March–April [12]. The temporal pattern of host availability and importance in the agricultural landscape therefore varies in a complex mosaic across India.

Over the past three decades, there has been speculation that Indian *H. armigera* could be categorized into races based on host-feeding preferences and limited inter-mating (e.g. [13], [14]). Such genetic diversity in connection with host plants has been previously shown in *H. armigera* in Australia [15] where there is, for example an identifiable lucerne-preferring 'race'. Variable metabolic mechanisms mediating pyrethroid resistance have been reported with a shift from mixed-function oxidase-mediated pyrethroid resistance to an esterase-mediated mechanism during mid October in central Indian *H. armigera* populations [16], attributable to both the influx of moths from other populations [17], [18] and the emergence from diapause of moth populations with genetic make-ups different from that of the non-diapausing population [19], [20]. Differential responses to pheromones in different populations and variations in parasitoid responses have been reported [2], [21], and can possibly be interpreted as reflecting an influx of populations between different agricultural systems from different ecological zones, although this view has not yet been tested using population genetics data. Recently, genetically modified (GM) cotton varieties which expressed *Bt*-toxins Cry1Ac and Cry2Ab have made important contributions in reducing application frequencies and dosage of insecticides for the control of *H. armigera*. The intense selection with *Bt* proteins may contribute to population substructure, while evolutionary constraints to host crop preferences may further contribute to area-wide gene flow patterns [22]. All these factors may result in genetic patterning in the species across the Indian agricultural landscape. Understanding the movements of *H. armigera* adults between GM and non-GM crops, or between sprayed and unsprayed crops will be crucial to the management of *Bt* and insecticide resistance in this pest.

The only India-wide major polyphagous crop pest thoroughly examined for genetic diversity is the whitefly *Bemisia tabaci* which comprises a polyphagous species complex with ecological niche separation with respect to host plant (and some geographic) preference [23]. It has at least 6 biotypes in India and probably many more. In particular the older Asia I groupings had a preference for eggplant and Asia II for tobacco and cassava. The more recently introduced B-biotype does particularly well on Tomato which only 1 of the 14 Asia 1 'races' does. This is a particularly complex example but it does show the potential for such separations in other widespread polyphagous species.

Studies of *H. armigera* population genetics based on different DNA markers such as random amplified polymorphic DNA [24], isozymes [25], mtDNA [26] and microsatellites (e.g., [27]; [28]) have been reported. These studies found little genetic variation between widely separated populations, supporting the idea that extensive long distance migration was occurring in *H. armigera*. In Australia, studies have revealed small genetic distances between widely separated populations based on isozymes [29], mitochondrial DNA polymorphisms [30], and sodium channel gene alleles [31]. In contrast, studies of Scott et al. [32], [33], [34], [35] based on microsatellites suggested substantial population substructure in Australian populations of *H. armigera*. Endersby et al. [28] applied

markers developed by both Scott et al. [36] and Ji et al. [37] to study Australian *H. armigera* populations collected from the southern and western regions of Australia and found no significant patterns of population substructure. The conflicting findings of Scott et al. [32], [33], [34], [35], Endersby et al. [28] and Weeks et al. [38] were due, at least in part, to factors associated with allele drop-outs (ADO), null alleles caused by mutations at primer annealing sites [28], and microsatellite loci being associated with non-LTR RTE retrotransposable elements (TE's) in Scott et al.'s analyses [39].

Given the wide distribution and migratory ability of *H. armigera*, effective and reliable molecular genetic markers must demonstrate efficiency in PCR amplification in individuals from within and between populations within a country, and between populations from different countries. Although less likely to be affected by TE-induced PCR failures as seen in various lepidopteran microsatellite markers (including three for *H. armigera*, [39]), Exon-Primed Intron-Crossing (EPIC)-PCR markers [40], [41] nevertheless are susceptible to null alleles if exon regions are variable at primer annealing sites, although this is yet to be demonstrated in population genetics studies. This study applies EPIC-PCR markers designed specifically for *H. armigera* [42] to generate data for testing the hypothesis that geographical and host plant components are significant factors underlying genetic variation in Indian *H. armigera*. In the absence of detailed knowledge of gene flow and for the purpose of this study we regard as 'populations', samples of *H. armigera* taken from different crops, areas and/or at different times.

Materials and Methods

Sampling and DNA extraction

A total of 786 *H. armigera* individuals were collected from India in the three cropping seasons 2004–5, 2005–6 and 2006–7, as larvae, or moths (Fig. 1 and Table 1). Collections were made from 14 populations on cotton (*Gossypium hirsutum*, Malvaceae), 5 populations of pigeonpea (*Cajanus cajan*, Fabaceae), 4 populations of chickpea (*Cicer arietum*, Fabaceae) and one of eggplant (*Solanum melongena*, Solanaceae). Larvae were collected by direct sampling from different host plants, either directly into ethanol until needed for gDNA extraction, or kept on artificial diet until the pupal stage. Some of these were taken as samples only after they had emerged into adult moths. Male moths from Nagpur_1 and Nagpur_2 were collected by pheromone traps (Table 1). All pupae and adult moth samples were also preserved in absolute ethanol at −20°C until required for DNA extraction. Only a small portion (5 mm of the posterior portion of larvae and pupae, or half the abdomen of adults) of each sample was used for genomic DNA (gDNA) extraction as previously reported [26] or using the method of Zraket et al. [43] with slight modifications. Absence of cross-contamination during the gDNA extraction process was confirmed by the inclusion of a blank extraction among each gDNA extraction batch. The PCR-RFLP (Restriction Fragment Length Polymorphism) *Helicoverpa* species diagnostic test of Behere et al. [44] was used to confirm that all larvae sampled for this study were *H. armigera*.

RpS2 EPIC marker allele characterisation

EPIC-PCR markers RpL3, RpL12, RpL29, RpS6 and RpS2 [42] were utilised to infer population genetic structures in Indian *H. armigera* populations. Molecular characterisation of RpS2 EPIC-PCR marker allele polymorphisms has not been previously reported and is here investigated using the methods described in Tay et al. [42]. Ten RpS2 EPIC maker alleles from Australian and

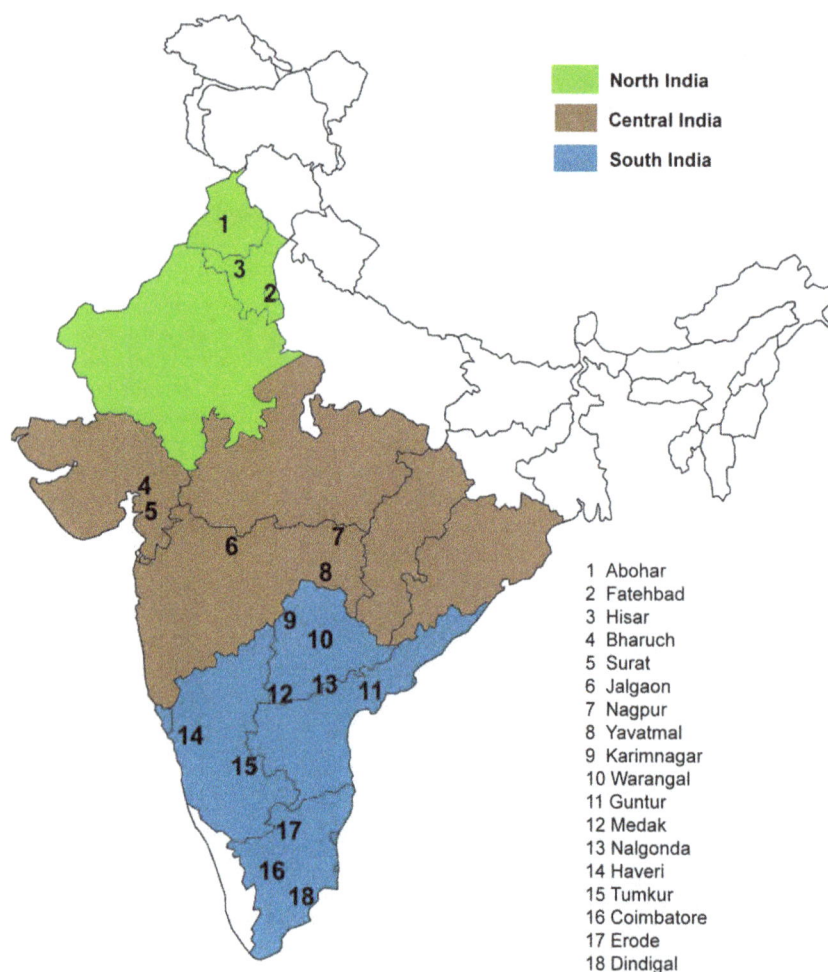

Figure 1. Sampling locations in India.

Chinese *H. armigera* individuals (samples previously used for allele characterisation by Tay et al. [42]) were randomly chosen, cloned and sequenced to ascertain the presence of allele homoplasy and nucleotide insertions/deletions (Indels).

Fluorescence labelling of polymorphic EPIC markers and screening

The forward primer of each EPIC-PCR marker was labelled with a fluorescent dye (FAM, HEX, or TET) to allow detection during electrophoresis. All amplifications were performed in a 15 μL reaction volume consisting of 7.5 μL of 5× GoTaq® Green Master Mix (Promega # M7122); 1.8 mM MgCl₂; 0.5 μM of each labelled forward and reverse primer and 50–75 ng of template DNA. The PCR amplification profile consisted of an initial template denaturation step of 5 min at 95°C (1 cycle); followed by 35 cycles that consisted of template denaturation (95°C, 1 min) – primer annealing at specific temperature for 1 minute [36] – template extension (72°C, 1 min); and a final 10 min template extension at 72°C (1 cycle). PCR amplicons (5 μL) of all five loci were loaded on 1% ethidium bromide-stained 1× tris-borate-EDTA (TBE) agarose gels, run at 90 V for 90 minutes and visualised over a UV-illuminator, prior to individually multiplexed in 96-well plates by pooling 1 μL of PCR product for each of the loci labelled with three different florescent dyes. For genotyping, DNA fragment sizes were determined by a MEGABASE 1000

automated sequencer (Amersham Biosciences) at the Genetic Analysis Facility (GAF, James Cook University, Queensland, Australia). A size standard (400-R) was co-loaded with every sample to allow accurate sizing of DNA fragments. The final volume was adjusted to 10 μL with dH₂O; post-PCR desalting was undertaken according to the protocol provided by GAF. Genotypes were scored manually with the help of marker panel set options implemented in the program GENETIC PROFILER 2.2 (Amersham Biosciences). All genotypes were scored unambiguously and where necessary allele peaks were corrected manually. Individuals which amplified for one locus but failed to amplify in PCR for other loci, were repeat amplified for up to a further two times. If a sample failed to amplify across all loci in at least one of the three rounds of PCR, it was considered as a DNA-extraction failure and discarded from subsequent analyses.

Analysis of null alleles in EPIC-PCR markers

To estimate EPIC-PCR markers null allele frequencies we designed nested (alternative) EPIC-PCR forward and reverse primers for the EPIC-PCR primer pairs used (Table 2). For alt_RpS2 and alt_RpL29 EPIC-PCR primers, we tested 42 randomly selected individuals which were identified as homozygotes using the original RpS2 and RpL29 EPIC markers, as well as six heterozygotes as positive controls. For alt_RpS6, the total number of individuals re-tested was 46 (18 homozygotes, 28

Table 1. Collection details of *H. armigera* populations screened with five EPIC PCR markers.

Regions	Location	n	Latitude	Longitude	Date	Host
Northern India	Abohar_1	15	30°07'N	74°12'E	Jan. 05	Chickpea
	Abohar_2	39	30°07'N	74°12'E	Sept. 06	Cotton
	Hissar	31	29°07'N	75°41'E	Sept. 06	Cotton
	Fatehbad	34	29°30'N	75°27'E	Sept. 06	Cotton
Central India	Bharuch	28	21°42'N	73°00'E	Sept. 05	Cotton†
	Surat	18	21°11'N	72°48'E	Sept. 05	Cotton†
	Jalgaon_1	58	21°01'N	75°33'E	Dec. 05	Cotton†
	Jalgaon_2	44	21°01'N	75°33'E	Jan. 06	Chickpea
	Nagpur_1	18	21°09'N	79°05'E	Jan. 05	Pheromone
	Nagpur_2	40	21°09'N	79°05'E	Sept. 04	Pheromone
	Yavtamal_1	10	20°23'N	78°08'E	Jul. 05	Egg Plant
	Yavtamal_2	23	20°23'N	78°08'E	Oct. 05	Cotton†
	Yavtamal_3	39	20°23'N	78°08'E	Oct. 05	Pigeonpea
	Yavtamal_4	40	20°23'N	78°08'E	Nov. 05	Chickpea
	Yavtamal_3A	31	20°23'N	78°08'E	Nov. 06	Pigeonpea
Southern India	Karimnagar	11	18°26'N	79°08'E	Oct. 05	Cotton
	Warangal	11	18°00'N	79°35'E	Oct. 05	Cotton
	Guntur_1	26	16°17'N	80°25'E	Oct. 05	Cotton
	Guntur_2	42	16°17'N	80°25'E	Dec. 06	Cotton
	Medak	39	18°03'N	78°16'E	Dec. 06	Chickpea
	Nalgonda	30	17°03'N	79°16'E	Dec. 06	Pigeonpea
	Haveri	30	14°47'N	75°24'E	Nov. 06	Pigeonpea
	Tumkur	39	13°20'N	77°04'E	Dec. 06	Cotton
	Coimbatore	22	11°00'N	76°58'E	Jan.05	Pigeonpea
	Erode	34	11°20'N	77°43'E	Jan. 05	Cotton
	Dindigal	34	10°21'N	77°58'E	Dec. 06	Cotton

Numbers after locations represent different collection periods. Known *Bt* cotton hosts are indicated by '†'. All insects were collected as mid to late instar larvae except for the Nagpur populations where moths were captured at pheromone traps.

heterozygotes), and for alt_RpL3 a total of 46 individuals (40 homozygotes, 6 heterozygotes) were re-tested. Null alleles were considered detected if during re-genotyping by nested EPIC-PCR markers individuals previously scored as homozygous were found to be heterozygous and vice versa.

Data analysis

Basic statistics for the EPIC-PCR data (average number of alleles per locus, allelic richness averaged over loci, and Weir and Cockerham's measures of F_{IS} [45]) were calculated using FSTAT version 2.9.3 [46]. F_{IS}, an inbreeding coefficient, measures the reduction in heterozygosity of an individual due to non-random mating within its sub-population. Observed (Ho) and expected (He) heterozygosity were estimated and departures from Hardy-Weinberg equilibrium (HWE) were tested using the probability test as implemented in GENEPOP version 3.2 [47]. The tests for genotypic linkage disequilibrium among pairs of loci were performed in GENEPOP using Fisher's tests [47], with unbiased P values derived by a Markov chain method (10,000 de-memorisations, 1,000 batches and 10,000 iterations/batch). The significance values for multiple significance tests were set using the sequential Bonferroni procedure [48] within the population genetics software FSTAT. To investigate population differentiation, pair-wise F_{ST} estimates [45] (with 95% confidence limits) and

significances (determined with 6,500 permutations) were calculated using FSTAT.

The geographic partitioning regime used by Kranthi et al. [49], [50] was followed. Genetic diversity was partitioned into three model structures according to geographic regions (northern, central and southern India; Fig. 1), host crops (cotton, pigeonpea chickpea and egg plant), and cropping seasons (season 2004–05, 2005–06 and 2006–07). Within each model structure, the genetic variation was further partitioned into three levels: (1) among geographic region/host/cropping season; (2) among populations within geographic regions/hosts/cropping seasons, and (3) within populations. A hierarchical analysis of molecular variance (AMOVA) was carried out using pair-wise F_{ST} as the genetic distance measure using the population genetics software ARLEQUIN 3.1 [51], [52]. In pair-wise F_{ST} estimates and STRUCTURE analysis, the two pheromone-trapped populations (Nagpur_1 and Nagpur_2) were excluded because of the unknown host crops. Erode, the only cotton population in season 1 (2004–05), was also excluded from these analyses.

The program STRUCTURE v2.3.2 [47] that implements a Bayesian clustering method, was used to identify admixed individuals and for assignment to likely membership of population genetic clusters ('K') through the assumption of known source populations and HWE at all loci [53]. To estimate the most likely

Table 2. Primer sequences of four nested EPIC PCR markers used for screening *H. armigera* populations for estimating null allele frequencies.

Locus	Primer sequences (5′ to 3′)	Fluorescent dye	Ta (°C)	Exon size difference from Tay *et al.* 2008 [42]	Expected allele size difference
Alt_RpS2	**F** AGAGGTTACTGGGGTAACAAG	TET	50	−5 bp	
	R GACACAATACCAGTACCACGAG			−3 bp	8 bp
Alt_RpL29	**F** CAAAGTCAAAGAATCACACAAAT	TET	50	−6 bp	
	R GGGTGGATTCGTGCCTTTG			−3 bp	9 pb
Alt_RpS6	**F** CAGGGAGTCCTCACYAAC	TET	50	−20 bp	
	R CTTTGACATCARCARACGA			n.a.†	20 bp
Alt_RpL3	**F** GTGTYACMAAGGGYAAAGGAT	FAM	50	−6 bp	
	R GTGTGCCAACGGGAGGTCAC			−10 bp	16 bp

†Alt_RpS6 EPIC-PCR marker is 5 nucleotides shorter than that reported [36] however it utilised the first 19 bp of the 24 bp original RpS6 EPIC-PCR oligo.

K we evaluated all possible K's (i.e., $K = 1$ representing no genetic structure, to $K = 23$ representing each population being genetically distinct) using simulation of 20 iterations, with each iteration consisted of 50,000 'burnin' followed by 500,000 Markov Chain Monte Carlo (MCMC) replications, with default settings for both the Ancestry Model (Admixture Model) and the Frequency Model (allele frequencies correlated among populations; assumed different F_{ST} values for subpopulations). The ΔK method of Evanno et al. [54] was used to ascertain the most likely K value, although the log probabilities of data (Ln P(D)) for K were also evaluated. The proportions of an individual's genome belonging to particular K population clusters are given a 'Q' score which enables STRUCTURE to assign individuals (or portions of an individual's genome) to a particular cluster [53].

Results

Null alleles in EPIC-PCR markers

Of the five sets of nested EPIC-PCR markers, we failed to design an alternative RpL12 EPIC primer due to the short exon sequence available, and the null allele frequency for this marker was therefore not estimated. Based on the alternative EPIC-PCR markers alt_RpS2, alt_RpL29, alt_RpL3 and alt_RpS6, null allele frequencies for the original EPIC-PCR markers were estimated at 9.4%, 6.5%, 6.3%, and 4.3% respectively which were considered as being at a moderate level [55], and were within the null allele frequency range (i.e., 2.2%–10.3%) of microsatellite DNA markers used by Endersby et al. [28] for inference of Australian *H. armigera* population genetics structure. As we were unable to estimate the null allele frequency of RpL12, analyses of population substructure patterns of our Indian samples were performed by both including or excluding RpL12 (Table 3). F_{IS} estimates from excluding the RpL12 locus remained unchanged for three populations (Abohar_1, Hisar and Karimnagar), reduced in seven populations (Nagpur_2, Yavatmal_2, Yavatmal_3, Warangal, Guntur_1, Nalgonda and Coimbatore), and increased in the remaining 16 populations. Taken as a whole, the inclusion of the RpL12 locus lead to lower and/or no change in F_{IS} estimates in 19 of the 26 populations studied, and did not drastically lower F_{IS} estimates in cotton populations, suggesting that this marker is unlikely to have harboured an excessively high frequency of null alleles.

EPIC marker variability

A total of 155 alleles were scored from the five loci (RpL3, RpL12, RpL29, RpS2 and RpS6) in 26 populations (n = 786) of *H. armigera*. The most polymorphic marker was RpS6 (55 alleles), followed by RpL29 (49 alleles), RpL3 (19 alleles), RpS2 (14 alleles) and RpL12 (15 alleles). The mean number of alleles and the mean observed (*Ho*) and expected (*He*) heterozygosities for each population are shown in Table 3. The average observed heterozygosity value for the five loci was 0.34 (range: 0.18–0.42). Estimates of observed heterozygosity were lower than expected in all populations, and levels of allele richness did not differ significantly between populations. Molecular characterisation of 10 randomly selected alleles (GenBank EU707432–EU707441) from the RpS2 EPIC marker showed no allele homoplasy, with allele length polymorphisms due to Indels within the intron.

Hardy-Weinberg equilibrium

Departures from Hardy-Weinberg equilibrium over all loci were found in all populations of *H. armigera*. This significant deviation from Hardy-Weinberg equilibrium was due to an excess of homozygotes at all loci, and is further reflected by the F_{IS} values (Table 3). Genotypic linkage disequilibrium tests found no significant associations between pairs of loci for any populations or over all populations after Bonferroni corrections for multiple comparisons, indicating independent assortment for these Rp EPIC markers.

Analysis of Molecular Variance (AMOVA) and *F* statistics

AMOVA analysis (Table 4) indicated that as a whole, Indian populations showed similar levels of genetic variability regardless of geographic regions (−0.06%, $F_{CT} = -0.006$, $P = 0.36$), host plants (−0.68%, $F_{CT} = -0.007$, $P = 0.98$) or cropping seasons (−0.23%, $F_{CT} = -0.002$, $P = 0.65$). However significant genetic structures were detected among populations within geographic regions (4.51%, $F_{SC} = 0.045$, $P < 0.001$), within host plants (i.e., within host plant species across sampling sites) (5.07%, $F_{SC} = 0.050$, $P < 0.001$) and within cropping seasons (4.63%, $F_{SC} = 0.046$, $P < 0.001$). Large variations ($F_{ST} > 95\%$) were found within populations in each of three model structures (geographic region, host and cropping season; Table 4). To better understand underlying factors that contributed to within population genetic variations, data from two cropping seasons (2005–06 and 2006–07) were partitioned individually according to geographic region or plant host. The overall trend remained similar (i.e. there

Table 3. Population statistics based on five EPIC PCR markers tested on Indian *H. armigera* populations.

Regions	Locations	n	a	r	Ho	He	F_{IS}	Ho*	He*	F_{IS} *
Northern India	Abohar_1	15	5.8	4.62	0.36	0.51	0.305	0.45	0.64	**0.305**
	Abohar_2	39	5.2	4.07	0.33	0.59	0.455	0.35	0.69	0.498
	Hisar	31	4.4	3.69	0.24	0.50	0.514	0.31	0.62	**0.514**[c]
	Fatehbad	34	6.2	4.00	0.18	0.49	0.628	0.21	0.59	0.644
Central India	Bharuch	28	4	3.14	0.36	0.37	0.034	0.44	0.45	0.035
	Surat	18	4.6	3.79	0.35	0.51	0.31	0.39	0.58	0.343
	Jalgaon_1	58	14	5.91	0.35	0.63	0.445	0.40	0.74	0.463
	Jalgaon_2	44	11	5.53	0.36	0.62	0.422	0.40	0.72	0.454
	Nagpur_1	18	7.2	5.32	0.39	0.59	0.333	0.48	0.72	0.340
	Nagpur_2	40	11	5.33	0.37	0.61	0.396	0.43	0.70	<u>0.386</u>
	Yavatmal_1	10	4.4	4.27	0.27	0.52	0.501	0.31	0.63	0.521
	Yavatmal_2	23	8.6	5.45	0.30	0.60	0.513	0.37	0.73	<u>0.499</u>[c]
	Yavatmal_3	39	13	6.61	0.37	0.65	0.445	0.42	0.75	<u>0.440</u>
	Yavatmal_4	40	11	5.79	0.35	0.64	0.465	0.39	0.74	0.477
	Yavatmal_3A	31	8.6	5.29	0.30	0.56	0.467	0.35	0.67	0.472
Southern India	Karimnagar	11	4.8	4.52	0.36	0.53	0.319	0.45	0.66	**0.319**[c]
	Warangal	11	3.6	3.52	0.36	0.54	0.354	0.45	0.68	<u>0.353</u>[c]
	Guntur_1	26	8.6	5.33	0.28	0.59	0.521	0.35	0.67	<u>0.492</u>[c]
	Guntur_2	42	13	5.99	0.30	0.61	0.517	0.33	0.70	0.537
	Medak	39	10	5.51	0.31	0.60	0.486	0.35	0.72	0.513
	Nalgonda	30	9.4	5.45	0.40	0.59	0.324	0.48	0.70	<u>0.321</u>
	Haveri	30	9.2	5.41	0.38	0.57	0.342	0.45	0.69	0.350
	Tumkur	39	9.2	5.41	0.42	0.62	0.323	0.48	0.71	0.324
	Coimbatore	22	4.8	3.65	0.37	0.50	0.273	0.46	0.61	<u>0.246</u>
	Erode	34	8.8	5.29	0.38	0.55	0.305	0.42	0.63	0.339
	Dindigal	34	9.2	5.29	0.37	0.61	0.396	0.44	0.74	0.408

The number of individuals screened for all five loci is indicated (n); mean values for the number of alleles (a), allelic richness (r), observed heterozygosity (Ho), expected heterozygosity (He), and inbreeding coefficient (F_{IS}) were also estimated for all populations. Analyses excluding the locus RpL12 are indicated by '*'. Cotton populations where F_{IS}* estimates either remained unchanged (values in bold) or decreased (values underlined) through the exclusion of the RpL12 locus are indicated by 'c'.

remained significant variation between populations within geographic regions and within hosts (data not shown)). Therefore, values for these significant genetic variations were examined in detail by estimates of pair-wise F_{ST} values. The overall pair-wise F_{ST} values ranged from −0.001 to 0.431, with higher F_{ST} values in general being associated with cotton populations (Table 5). Cotton populations collected in the 2006–07 cropping season further differed significantly between northern and southern India (Table 5). Within cotton, the highest F_{ST} values were seen in the Bharuch population when compared with all other populations. Pair-wise F_{ST} values for populations from food crops (pigeonpea, chickpea and eggplant) contrasted with those observed in cotton populations and were generally non-significant (except Coimbatore) (range: −0.008 to 0.100; Table 6).

Structure Analyses

Based on the averaged Ln P(D) of 20 simulations for each $K = 1$ to $K = 23$, the best K was identified as $K = 8$ (−9912.76±536.43 s.d.). However, using the Evanno et al. [54] method identified the best ΔK value (614.37) at $K = 2$ followed by the second highest ΔK value of 4.06 at $K = 4$ (Fig. 2, Fig. 3). Although the ΔK value was largest for $K = 2$, this large change probably indicated a shift from the unlikely scenario of our samples showing no structure (i.e.,

$K = 1$) towards more plausible scenario of presence of population substructure (i.e., $K > 1$). The large and positive ΔK value for $K = 4$ (Fig. 2) was therefore selected as the cluster number most likely to assist in visualising significant population substructure. The generalised patterns of population structure for $K = 2$ to $K = 10$ are presented (Fig. 4) to help visualise the selection of $K = 4$. A detailed STRUCTURE bar graph at $K = 4$ across all 23 populations was presented (Fig. 3). No obvious biologically relevant population structure patterns could be inferred from the STRUCTURE analysis, and setting of $K > 4$ clusters progressively introduced greater population heterogeneity and further reduced the power of interpretation, even for the Ln P(D) optimum $K = 8$. Between $K = 2$ and $K = 9$, the Bharuch (pop. 4) and Surat (pop. 5) populations in central India appeared highly homogeneous, while all other populations showed higher levels of admixture (Fig. 4). Detailed examination of the STRUCTURE analysis at $K = 4$ (Fig. 3) showed that substantial substructure within and between populations existed, with populations which were geographically close to each other sorting into very different genetic clusters. Across the three sampling years (Fig. 3), reduced genetic diversity was seen only in some cotton populations (e.g., Year 2: Baruch (pop. 4), Surat (pop. 5), Karimnagar (pop. 9); Year 3: Abohar (pop. 1), Fatehebad (pop. 2)). Substantial substructure between populations

Table 4. Comparisons of genetic variation by AMOVA based on data generated by EPIC-PCR markers from *Helicoverpa armigera* according to three population structure hierarchical models.

Model	Hierarchical levels	Degrees of freedom	Sum of square	Variance components	Fixation indices	Percentage of variation	P- value
Model A Geographic regions	Among regions	2	10.849	−0.0009	−0.0062 F_{CT}	−0.06	0.36
	Among populations within regions	23	122.464	0.0660	0.0451 F_{SC}	4.51	<0.001
	Within populations	1546	2162.19	1.3986	0.0445 F_{ST}	95.55	<0.001
	Total	1571	2295.50	1.4637			
Model B Hosts	Among Hosts (cotton, pigeonpea, chickpea)	2	5.17	−0.0099	−0.007 F_{CT}	−0.68	0.98
	Among populations within host crop	20	118.48	0.0739	0.0504 F_{SC}	5.07	<0.001
	Within populations	1413	1967.81	1.3927	0.0440 F_{ST}	95.60	<0.001
	Total	1435	2091.81	1.4567			
Model C Cropping seasons	Among seasons	2	8.41	−0.0034	−0.0023 F_{CT}	−0.23	0.65
	Among populations within seasons	23	124.90	0.0677	0.0462 F_{SC}	4.63	<0.001
	Within populations	1546	2162.19	1.3986	0.0439 F_{ST}	95.61	<0.001
	Total	1571	2295.50	1.4628			

Model A: compares variation between geographic regions (northern India, central India, and southern India), among populations within regions and within populations. Model B: compares variation between hosts (cotton, pigeonpea and chickpea), among populations within hosts and within populations. Model C: compares genetic variations between cropping seasons (season 2004–05, season 2005–06, season 2006–07), among populations within cropping seasons and within populations.

Table 5. Pair-wise F_{ST} values (below diagonal) and associated P-values (above diagonal) for $H.\ armigera$ populations collected from cotton during cropping seasons 2 (2005–2006) and 3 (2006–2007).

Region			CI	CI	CI	CI	SI	SI	SI	NI	NI	NI	SI	SI	SI
	Cropping Season		2	2	2	2	2	2	2	3	3	3	3	3	3
		Sampling locations	Bharuch	Surat	Yavatmal_2	Jalgaon_1	Karimnagar	Warangal	Guntur_1	Abohar_2	Hissar	Fatehbad	Guntur_2	Tumkur	Dindigal
CI	2	Bharuch	-	0.000	0.000	0.000	0.000	0.000	0.000	0.000	0.000	0.000	0.000	0.000	0.000
CI	2	Surat	**0.192**	-	0.001	0.000	0.000	0.001	0.002	0.000	0.000	0.000	0.005	0.000	0.001
CI	2	Yavatmal_2	**0.275**	0.061	-	0.188	0.007	0.002	0.741	0.000	0.000	0.004	0.030	0.000	0.031
CI	2	Jalgaon_1	**0.142**	**0.033**	0.007	-	0.000	0.074	0.233	0.000	0.000	0.000	0.638	0.000	0.396
SI	2	Karimnagar	**0.431**	**0.202**	0.052	**0.092**	-	0.010	0.003	0.000	0.000	0.000	0.001	0.000	0.002
SI	2	Warangal	**0.244**	0.097	0.054	0.020	0.074	-	0.003	0.000	0.000	0.000	0.007	0.001	0.001
SI	2	Guntur_1	**0.228**	0.033	-0.007	0.003	0.112	0.060	-	0.000	0.000	0.000	0.174	0.000	0.012
NI	3	Abohar_2	**0.240**	**0.084**	0.014	0.033	0.065	0.065	0.037	-	0.074	0.060	0.000	0.000	0.000
NI	3	Hissar	**0.200**	**0.041**	**0.048**	**0.025**	**0.140**	**0.075**	**0.035**	**0.074**	-	0.017	0.000	0.000	0.000
NI	3	Fatehbad	**0.247**	**0.055**	0.044	**0.038**	**0.103**	**0.055**	**0.045**	**0.060**	**0.017**	-	0.000	0.000	0.000
SI	3	Guntur_2	**0.116**	0.028	0.031	-0.002	0.128	0.043	0.016	**0.044**	**0.036**	**0.052**	-	0.000	0.450
SI	3	Tumkur	**0.141**	**0.044**	**0.042**	**0.016**	**0.087**	0.030	**0.040**	**0.048**	**0.040**	**0.034**	**0.013**	-	0.012
SI	3	Dindigal	**0.164**	0.042	0.008	-0.001	0.073	0.033	0.015	**0.019**	**0.038**	**0.039**	0.003	**0.012**	-

P-values obtained after 1,560 permutations. Values in bold are significant at $P < 0.00064$ after sequential Bonferroni correction. Geographic regions are northern India (NI), central India (CI), and southern India (SI).

Table 6. Pair-wise F_{ST} values (below diagonal) and associated *P*-values (above diagonal, values obtained from 720 permutations) for *Helicoverpa armigera* populations collected from food crops (eggplant (ep), chickpea (cp) and pigeonpea (pp)).

Crops and regions			ep (CI)	cp (NI)	pp (SI)	cp (CI)	pp (CI)	cp (CI)	pp (CI)	cp (SI)	pp (SI)	pp (SI)
	Cropping season		1	1	1	2	2	2	3	3	3	3
	Sampling locations		Yavatmal_1	Abohar_1	Coimbatore	Jaglaon_2	Yavatmal_3	Yavatmal_4	Yavatmal_3A	Medak	Nalgonda	Haveri
ep (CI)	1	Yavatmal_1	-	0.037	0.000	0.001	0.262	0.006	0.026	0.029	0.007	0.062
cp (NI)	1	Abohar_1	0.041	-	0.003	0.064	0.243	0.008	0.392	0.014	0.410	0.649
pp (SI)	1	Coimbatore	0.148	0.069	-	0.001	0.001	0.011	0.003	0.004	0.001	0.001
cp (CI)	2	Jaglaon_2	0.038	0.006	**0.066**	-	0.007	0.003	0.075	0.003	0.069	0.114
pp (CI)	2	Yavatmal-3	0.018	0.022	**0.086**	0.006	-	0.013	0.040	0.001	0.001	0.197
cp (CI)	2	Yavatmal_4	0.048	0.016	0.029	0.004	0.009	-	0.086	0.006	0.011	0.011
pp (CI)	3	Yavatmal_3A	0.032	-0.008	0.058	-0.002	0.009	0.003	-	0.022	0.474	0.339
cp (SI)	3	Medak	0.066	0.037	0.042	0.021	**0.023**	0.006	0.019	-	0.003	0.001
pp (SI)	3	Nalgonda	0.041	-0.006	**0.074**	0.001	**0.018**	0.009	-0.005	0.019	-	0.006
pp (SI)	3	Haveri	0.031	0.000	**0.101**	0.007	0.010	0.018	0.012	**0.047**	0.006	-

Values in bold are significant at $P<0.0013$ after sequential Bonferroni correction; sampling regions are (NI) northern India, (CI) central India, and (SI) southern India.

can be seen, for example, in Year 2 cotton populations between Baruch (pop. 4)/Surat (pop. 5) and Guntur (pop 11)/Karimnagar (pop. 9)/Yvatmal (pop. 8). Substructure of populations is further seen between populations which are in close geographic proximity (e.g. in cotton Year 3: Abohar (pop. 1), Fatehbad (pop. 2) and Hissar (pop. 3); food crops Year 3: Medak (pop. 12) and Nalgonda (pop. 13)). This contrasted with the genetic patterns of food crop populations (Table 6), which did not show such strong substructure.

Discussion

H. armigera population genetics inferred from EPIC markers

Population genetic analysis of Indian *H. armigera* samples using EPIC markers clearly indicated no obvious population substructure with geographic region, year or crop, and indicated significant genetic differentiation between northern and central/southern Indian cotton-feeding populations in the cropping season of 2006–07. Within the populations HW non-equilibrium was detected, with a likely contributing factor being the presence of null alleles within our DNA marker system. Null allele frequency estimates for four of these EPIC-PCR markers were moderate (4.3% to 9.4%). The conclusion on population substructure patterns remained overall unaltered regardless of the inclusion or exclusion of the untested RpL12 locus from analysis. In the population genetic study involving predominantly Australian *H. armigera* populations, Endersby et al. [28] excluded 3 pairs of SSR markers with the greatest null allele frequencies (i.e., 19.2%, 31.6% and 47.4%), and demonstrated that DNA markers with low to moderate levels of null allele frequencies were sufficiently powerful to enable meaningful interpretation of gene flow patterns, a conclusion consistent with population genetics simulation studies [55], [56]. Chapuis and Estoup [55] also concluded that the most accurate F_{ST} estimates were obtained when null alleles in marker systems were not excluded (i.e. by scoring only visible alleles within study populations), although within populations with substantial gene flow (i.e., non-cotton populations in this study), F_{ST} estimates were less biased than in populations with limited gene flow (i.e., cotton populations).

Our study further demonstrated that null allele events in *H. armigera* population genetic markers are relatively common, despite attempts to minimise their occurrence by designing markers at conserved gene coding regions. The effort in designing EPIC-PCR markers has other advantages such as enabling nested PCR markers to be developed for retesting of populations at the same loci, as well as the overall lower null allele frequencies as compared with null allele frequencies from random SSR markers which may be associated with TE's [39], [57]. Null alleles or inbreeding may also contribute to heterozygote deficiency. A locus-specific heterozygote deficit is an indication of null alleles, rather than inbreeding or other population processes which are generally reflected across all loci. Significant heterozygote deficiencies were detected at all five loci in most *H. armigera* populations tested, further indicating only low to moderate levels of null alleles affecting our EPIC markers.

A Wahlund effect (i.e., fine-scale heterogeneity versus large-scale homogeneity; e.g., see [58]) as indicated in our high F_{IS}, may further affect interpretations of population structure of *H. armigera* in India. The sample size of n = 786 allowed us to detect significant departures from HWE for all five EPIC markers. Although the STRUCTURE analysis assumed that all potential source populations were sampled when assigning genetic clusters to individuals, our study was aimed at understanding broader *H. armigera* population

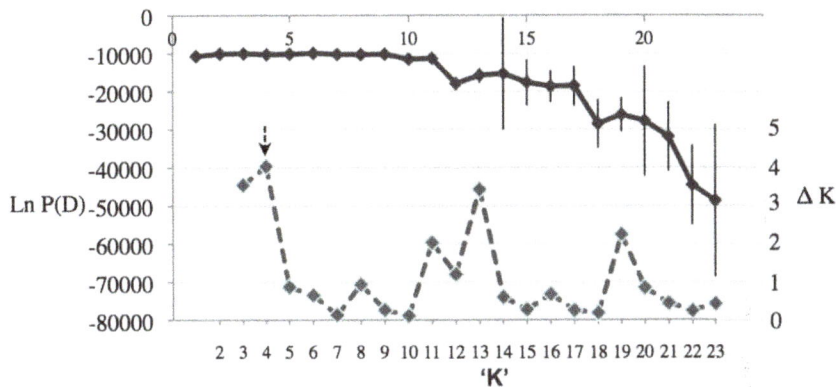

Figure 2. STRUCTURE **analysis of 23** *Helicoverpa armigera* **populations collected from India from cotton crop (years 2 and 3) and food crops (years 1, 2, 3).** Locations and populations are as in Fig. 1 and Table 1: northern India (1) Abohar, (2) Fatehbad, (3) Hissar; central India (4) Bharuch, (5) Surat, (6) Jalgaon, (8) Yavatmal; southern India (9) Karimnagar, (10) Warangal, (11) Guntur, (12) Medak, (13) Nalgonda, (14) Haveri, (15) Tumkur, (16) Coimbatore, and (18) Dindigal. Average log likelihood of data Ln P(D) (primary axis, solid line) and ΔK (secondary axis, dashed line) are shown, with best K = 4 indicated an arrow.

patterns in Indian's highly heterogeneous agricultural landscape rather than inferring individuals' specific origins. The lack of interpretable and biologically meaningful STRUCTURE results as well as the general patterns of low population substructure in our study could potentially be due to our markers being in non-HWE (i.e., STRUCTURE analysis assumes HWE in loci, [53]), although various authors (e.g., De Barro [59], Brown et al. [60]) reported no apparent significant effects to STRUCTURE results in loci that did not demonstrate HWE. Our finding that the Bharuch and Surat populations are generally highly homogeneous may therefore either be an indication of a lack of gene flow and/or due to selection.

Departures from HWE due to homozygous excess may represent true biological phenomena in *H. armigera* such as those due to strong inbreeding caused by frequent bottlenecks, the Wahlund effect (e.g., Nielsen et al. [61]), or be due to extrinsic factors such as insecticide selection pressure, *Bt* proteins and/or plant secondary chemicals, or other environmental selectors. In addition to possible effects of utilising microsatellite DNA families/

TE-associated loci [57], [39], the generally small-scale heterogeneous Indian cropping landscape, intense selection pressure from heavy insecticide applications, exposure to *Bt* toxins from GM cotton, and/or exposure to host plant secondary compounds could cause *H. armigera* populations to deviate from HWE (although we should note that the later two factors may be similar in Australia).

The different cropping systems to which *H. armigera* is exposed to may also be important underlying factors that contributed to population substructure differences. Host crops with short flowering periods (e.g., food crops such as eggplant, chickpea and pigeonpea) generally support no more than one or two *H. armigera* generation; while hosts such as cotton with prolonged flowering periods are capable of supporting ≥ three consecutive generations [5], [3]. Populations that feed on cotton are under tremendous selection pressure from insecticide applications [49], and from varying levels of the allelochemical gossypol associated with different life stages and specific cotton varieties [62]. Although host crop species are generally the same across India, the temporal pattern of availability and of the size of the *H.*

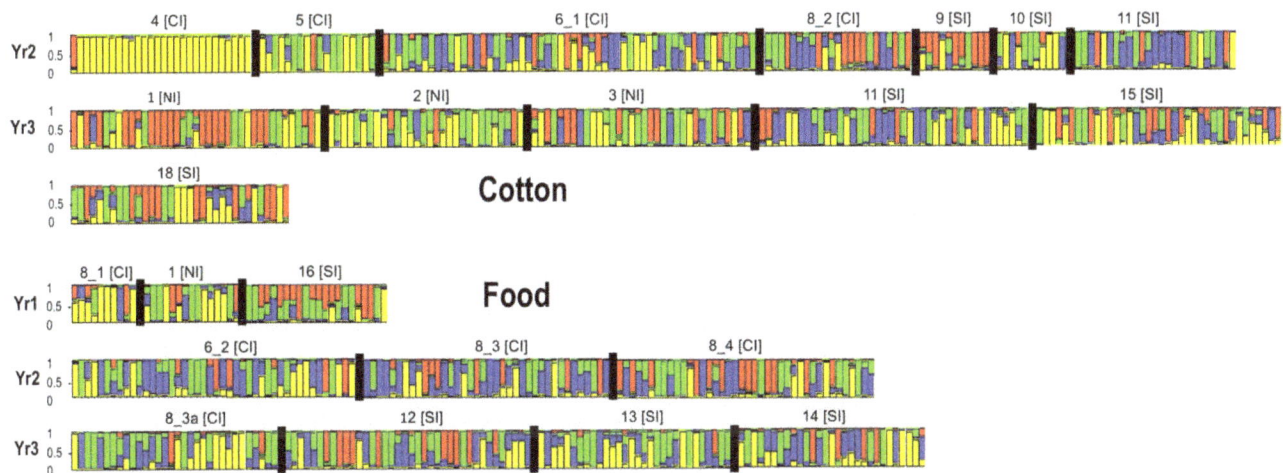

Figure 3. *Helicoverpa armigera* **population structure as inferred using the Bayesian clustering algorithm implemented in the program** STRUCTURE **2.3.** Bar graph of *K* = 4 for all individuals from 23 populations collected across three sampling years from either cotton or food crop plant hosts are shown. Locations and populations are as in Fig. 1 and Table 1, Indian agricultural regions are northern Indian (NI), central India (CI) and southern Indian (SI).

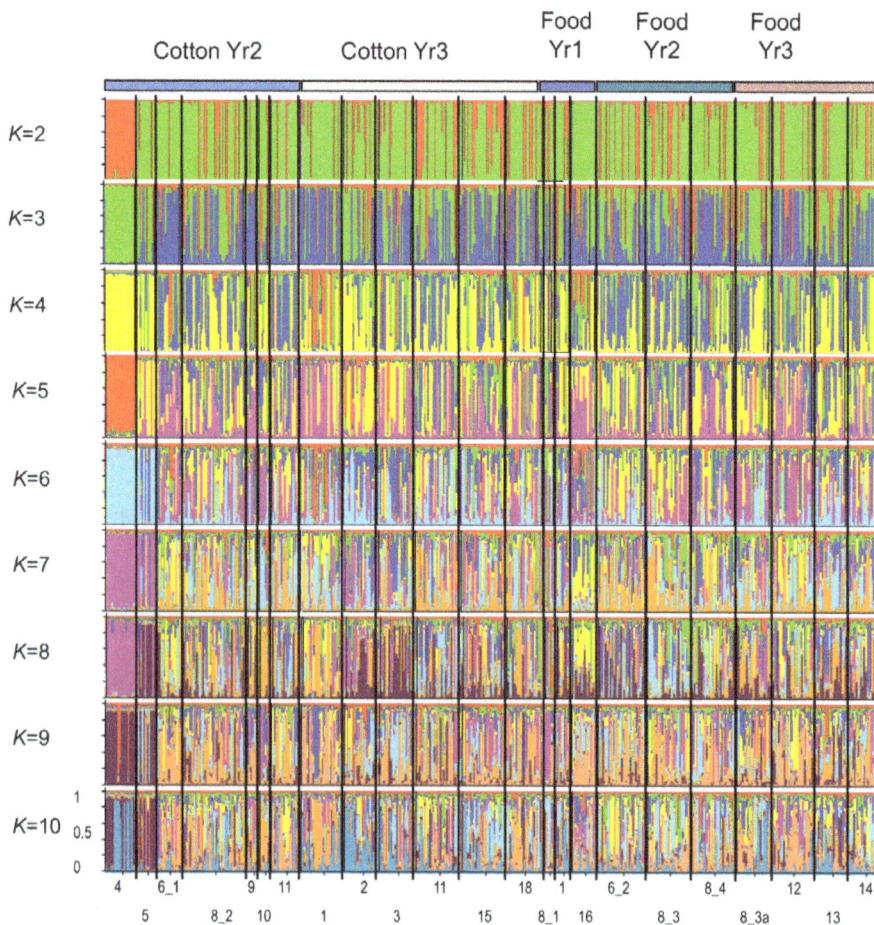

Figure 4. Population structure of 23 *H. armigera* populations inferred using the STRUCTURE** program after 50,000 'burnin' followed by 500,000 MCMC replications.** Each individual is represented by a single line, with black lines separating between populations. The estimated membership fractions for K = 2 to K = 10 clusters for all individuals are shown.

armigera populations varies greatly, with relatively small populations on only one or a few crops at some times of year in some regions. In northern India, *H. armigera* is known to occur initially on food crops (chickpea, sunflower, some vegetable crops) during February to July prior to feeding on cotton from August (see [63]). A large proportion of the population in the north may enter diapause, to emerge after the partial break between cropping seasons, and re-mix with the smaller, non-diapause population which has been subsisting for 1–2 generations on other, less intensively managed, hosts. The switch from food crops (i.e., with low insecticide exposure) each typically supporting a single *H. armigera* generation, to cotton hosts with increased insecticide and *Bt* toxin exposure capable of supporting multiple generations, can lead to intense selection on sedentary cotton populations. These cotton populations are accompanied by an increase in population densities, with peak infestations typically recorded during September to November in the north (see [63]). These peaks are accompanied by a significant increase in population density in non-GM cotton.

In contrast, central and southern Indian *H. armigera* populations initially feed on cotton (August to October in central India, September to December in southern India) prior to switching over to food crops. Central and southern Indian *H. armigera* populations are therefore likely to experience less consistent insecticide/*Bt* protein selection pressure approaching the end of each cropping

season as host crops change from cotton to food crops. This may promote gene exchange between populations as they move between crops. This scenario is the reverse of that in northern India, where populations sampled near the end of the cropping season might be expected to show more differentiation, as selection by insecticides and *Bt* toxins in cotton on such populations potentially operates over several generations with reduced migration, creating a mosaic of genetically different populations across the various cotton types and management systems. This may be part of the explanation for the significant pairwise F_{ST} values from cotton populations (especially in northern India) (Table 5) suggesting raised levels of substructures in cotton populations.

H. armigera population structure on cotton

Population genetic analysis of Indian *H. armigera* populations showed a high degree of differentiation between collections within cropping seasons. Based on F_{ST} values the observed genetic variation was most marked with populations collected from cotton in cropping season 3 (2006–07). The genetic differentiation revealed by pair-wise estimates of F_{ST} values suggested that there existed seasonal and geographical variation within *H. armigera* populations collected from cotton (Table 5). Further, populations from northern India were significantly different from southern

Indian populations in cropping season 3, although the underlying factors responsible for this remain unclear.

Overall estimates of pairwise F_{ST} values for cotton populations in cropping season 2 between central and southern India were non-significant (Table 5). The exception being the Bharuch population which showed significant F_{ST} values against all other populations. The distance between Bharuch and Surat is only 60 km and there are no geographical barriers, differences in cropping practices, or climatic conditions between these areas. Nevertheless these populations were significantly different from each other ($F_{ST} = 0.192$). Possible reasons contributing to such strong genetic differentiation may include different H. armigera generations being collected (see Table 1 for collection dates), and/or sampled populations feeding on different cotton types (e.g., Bt or non-Bt cotton; cotton varieties with different levels of gossypol contents). It is also possible that some of the diversity/structure observed may be associated with the mosaic of Bt versus non-Bt fields, and this would warrant further study. Pairwise F_{ST} values between seasons 2 and 3 in the cotton crop broadly indicated population substructure differences between northern and central/southern Indian, and may reflect underlying differences in cropping patterns (i.e., 'cotton first' or 'cotton last' in the cropping season).

Fluctuations in host availability may influence H. armigera populations and could result in genetic differentiation among local populations, but this assumption is only valid when there is substantial genetic isolation between populations. In India, H. armigera population substructure has been further suggested based on feeding preferences [14], insecticide resistance [16], differential response to pheromones [21] and to parasitoids [64]. The abundance, movement and distribution of H. armigera were found to be associated with rainfall and humidity in Australia (e.g., [65]) and suggested for India [66]. H. armigera is a facultative migrant [1], responding largely to local environmental conditions and host availability (i.e., moths remain sedentary where food resources such as flowering plant hosts are available). Cropping and landscape patterns, as well as insecticide application practices and resistance pest management strategies in Bt cotton differ greatly between Australia and India, which therefore limits meaningful comparison between findings from this study and Australian H. armigera population genetic structure. For example, the agricultural landscape in India is typically of low acreage, highly diverse and fragmented in the pattern of crop hosts growing at any given time. As such, it generally enables the presence of more than five alternate hosts of H. armigera at any given time of the cropping season [2], [67], while there would frequently be only a single major host over the corresponding period in Australia across comparatively large cropping areas within each production region.

H. armigera population structure on food crops

Pigeonpea is cultivated all over India, where it is commonly grown alongside cotton or cultivated as an inter-crop within the cotton agro-ecosystem [68]. Furthermore, flowering periods of cotton and pigeonpea overlap which may facilitate population movements between cotton and pigeonpea, and may explain the high pairwise F_{ST} values associated with pigeonpea (Table 6). Pulse crops (i.e., pigeonpea and chickpea) are preferred hosts of H. armigera compared to cotton and are planted on larger areas than cotton (Directorate of Economics and Statistics, Department of Agriculture and Cooperation, Ministry of Agriculture, Govern-

ment of India [69], [70], [71]). Pesticide applications on food crops are, however, less intense than on cotton and H. armigera populations are therefore expected to experience less selection for specific genotypes, which may result in a lower level of apparent genetic sub-structure than is seen in cotton. Overall estimates of pairwise F_{ST} values among food crops were by and large non-significant in all three cropping seasons (Table 6), although within cropping seasons the number of populations sampled was relatively low.

If the ideas presented above are correct, patterns of population structure analysed in one season should be reflected in the analysis of subsequent seasons. Consistent patterns observed between seasons, hosts and regions would thus support there being host- and/or region-associated micro-population structuring in Indian H. armigera. Such patterns have not been clearly seen in this study. Unsurprisingly, the situation is dynamic. For example population 8_3 was collected from Yvatmal on pigeonpea in years 2 and 3 (pop. 8_3a) and shows some shift of genetic profile (Fig. 3). A similar situation exists with cotton populations from Guntur (pop. 11) in years 2 and 3. Although the overall number of Indian populations sampled in this study is comparable to that in other lepidopteran population genetic studies (e.g., H. armigera, [28]; P. xylostella, [72]; C. pomonella, [73]), the complexity of Indian cropping systems nevertheless means that additional populations from northern, central and southern India over the season and on various hosts will be needed to enable more detailed interpretation. H. armigera populations analysed in this study do not cover all desired sampling locations and hosts for all seasons, and the patterns seen may also be influenced by factors such as sampling errors. In order to explain this intra-seasonal or host crop-associated genetic differentiation, further analysis of samples collected several times from the same host crops and sites over multiple cropping seasons will be needed.

Understanding the observed genetic structuring in Indian H. armigera populations from cotton may be further advanced with research on the detoxification capabilities and ecological aspects of this highly polyphagous pest insect species. Variables such as insecticide usage, Bt/non-Bt cotton, different hybrids of cotton, different hosts, climatic conditions, and diapause should be considered separately to better ascertain their importance to the population genetic structure of cotton-feeding H. armigera in India.

Acknowledgments

C. Robin and N. Endersby (Dept. Genetics, University of Melbourne, Australia), Gordon Luikart (Montana Conservation Genetics Laboratory, The University of Montana, USA), Paul Sunnucks (School of Biological Sciences, Monash University, Melbourne, Australia), and Karl Gordon (CSIRO Ecosystem Sciences, Canberra, Australia) provided helpful discussions during the course of the project and/or critically read earlier versions of this manuscript. Field populations of Helicoverpa armigera were collected by Sandhya Kranthi (Central Institute for Cotton Research, Nagpur, India), Rahul Wadaskar (International Crops Research Institute for the Semi-Arid Tropics, Patancheru, India) and Vilas Tajane (Jain Irrigation, Jalgaon, India).

Author Contributions

Contributed overall knowledge of the Indian agricultrual landscape: KRK. Conceived and designed the experiments: DAR KRK PB. Performed the experiments: GTB WTT. Analyzed the data: GTB WTT. Contributed reagents/materials/analysis tools: PB KRK DAR WTT. Wrote the paper: GTB WTT DAR PB.

References

1. Fitt GP (1989) The ecology of *Heliothis* species in relation to agroecosystems. Ann Rev Entomol 34: 17–52.

2. Manjunath TM, Bhatnagar VS, Pawar CS, Sithanantham S (1989) Economic importance of *Heliothis* spp. in India and an assessment of their natural enemies and host plants; 11–15 November, 1989; New Delhi, India. pp. 197–228.

3. Singh J, Bains SS (1986) Role of plants in population build up of *Heliothis armigera* (Hübner) in Punjab. Indian J Ecol 13: 113–119.

4. Bhatnagar VS (1980) A report on research on the *Heliothis* complex at ICRISAT (India) 1974–79; 24–26 April, 1980; Udaipur, India.

5. Jayaraj S (1982) Biological and Ecological studies of *Heliothis*. In: Reed W, editor, 15–20 November 1981; ICRISAT Center, Patancheru, A.P., India. pp. 17–28.

6. Kranthi KR, Jadav DR, Wanjari RR, Shaki-Ali S, Russell DA (2001) Carbamate and organophosphate resistance in cotton pests in India, 1995–1999. Bull Entomol Res 91: 37–46.

7. Sachan JN (1987) Status of gram pod borer *H. armigera* (Hb.) in India and its management. In: Mathour YK, editor. Recent Advances in Entomology. Parade, Kanpur: Gopal Parkashan. pp. 91–110.

8. Singh J, Sandhu SS, Singla ML (1990) Ecology of *Heliothis armigera* (Hub.) on chickpea in Punjab. J Insect Sci 3: 47–52.

9. Singh J, Sidhu AS (1992) Present status of *Helicoverpa armigera* in cotton and strategies for its management in Punjab, Haryana and Rajasthan. In: Sachan JN, editor. Directorate of Pulse Research, Kanpur. pp. 92–98.

10. Durairaj C, Subbaratnam GV, Singh TVK, Shannower TG (2005) *Helicoverpa* in India: Spatial and temporal dynamics and management options. In: Sharma H, editor. *Heliothis/Helicoverpa* management – Emerging trends and strategies for future research Oxford&IBH Publishing Co, New Delhi. pp. 91–117.

11. Verma AK, Sankhyan S (1993) Pheromone monitoring of *Heliothis armigera* (Hubner) and relationship of moth activity with larval infestation on important cash crops in mid-hills of Himachal Pradesh. Pest Manag Eco Zool 1: 43–49.

12. Pedgley DE, Tucker MR, Pawar CS (1987) Windborne migration of *Heliothis armigera* (Hübner) (Lepidoptera: Noctuidae) in India. Int J Trop Insect Sci 8: 599–604.

13. Bhattacherjee NS (1972) *Heliothis armigera* (Hübner) a polytypic species. Ent Newsl 2: 3–4.

14. Reed W, Pawar CS (1982) *Heliothis*: A global problem. In: Reed W, editor. 15–20 November 1981; ICRISAT Center, Patancheru, A.P., India. pp. 9–14.

15. Jallow MFA, Cunningham JP, Zalucki MP (2004) Intra-specific variation for host plant use in *Helicoverpa armigera* (Hübner) (Lepidoptera: Noctuidae): implications for management. Crop Prot 23: 955–964.

16. Kranthi KR, Armes NJ, Rao NGV, Sheo R, Sundarmurthy VT (1997) Seasonal dynamics of metabolic mechanisms mediating pyrethroid resistance in *Helicoverpa armigera* in central India. Pestic Sci 50: 91–98.

17. Armes NJ, Jadhav DR, DeSouza KR (1996) A survey of insecticide resistance in *Helicoverpa armigera* in the Indian subcontinent. Bull Entomol Res 86: 499–514.

18. Madden AD, Holt J, Armes NJ (1995) The role of uncultivated hosts in the spread of pyrethroid resistance in *Helicoverpa armigera* populations in Andhra Pradesh, India: a simulation approach. Ecol Model 82: 61–74.

19. Armes NJ, Jadhav DR, Bond GS, King ABS (1992) Insecticide resistance in *Helicoverpa armigera* in South India. Pestic Sci 43: 355–364.

20. De Souza K, Holt J, Colvin J (1995) Diapause, migration and pyrethroid-resistance dynamics in the cotton bollworm, *Helicoverpa armigera* (Lepidoptera: Noctuidae). Ecol Entomol 20: 333–342.

21. Tamhankar AJ, Rajendran TP, Hariprasad Rao N, Lavekar RC, Jeyakumar P, et al. (2003) Variability in response of *Helicoverpa armigera* males from different locations in India to varying blends of female sex pheromone suggests male sex pheromone response polymorphism. Curr Sci 84: 448–450.

22. Zalucki MP, Cunningham JP, Downes S, Ward P, Lange C, et al. (2012) No evidence for change in oviposition behaviour of *Helicoverpa armigera* (Hübner) (Lepidoptera: Noctuidae) after widespread adoption of transgenic insecticidal cotton. Bull Entomol Res 102: 468–476.

23. Chowda-Reddy RV, Kirankumar M, Seal SE, Muniyappa V, Valand GB, et al. (2012) *Bemisia tabaci* phylogenetic groups in India and the relative transmission efficacy of Tomato leaf curl Bangalore virus by an indigenous and an exotic population. J Integrative Agric 11(2): 235–248.

24. Zhou X, Faktor O, Applebaum SW, Coll M (2000) Population structure of the pestiferous moth *Helicoverpa armigera* in the eastern Mediterranean using RAPD analysis. Heredity 85: 251–256.

25. Nibouche S, Bües R, Toubon J-F, Poitout S (1998) Allozyme polymorphism in the cotton bollworm *Helicoverpa armigera* (Lepidoptera: Noctuidae): comparison of African and European populations. Heredity 80: 438–445.

26. Behere GT, Tay WT, Russell DA, Heckel DG, Appleton BR, et al. (2007) Mitochondrial DNA analysis of field populations of *Helicoverpa armigera* (Lepidoptera: Noctuidae) and of its relationship to *H. zea*. BMC evolutionary biology 7: 117.

27. Vassal JM, Brevault T, Achaleke J, Menozzi P (2008) Genetic structure of the polyphagous pest *Helicoverpa armigera* (Lepidoptera: Noctuidae) across the Sub-Saharan cotton belt. Commun Agric Appl Biol Sci 73: 433–437.

28. Endersby NM, Hoffmann AA, McKechnie SW, Weeks AR (2007) Is there genetic structure in populations of *Helicoverpa armigera* from Australia? Entomol Exp Appl 122: 253–263.

29. Daly JC, Gregg P (1985) Genetic variation in *Heliothis* in Australia: species identification and gene flow in the two pest species *H. armigera* (Hübner) and *H. punctigera* Wallengren (Lepidoptera: Noctuidae). Bull Entomol Res 75: 169–184.

30. McKechnie SW, Spackman ME, Naughton NE, Kovacs IV, Ghosn M, et al. (1993) Assessing budworm population structure in Australia using the A-T rich region of mitochondrial DNA; New Orleans. Proceedings of Beltwide Cotton Conference. pp. 838–840.

31. Stokes NH, McKechnie SW, Forrester NW (1997) Multiple allelic variation in sodium channel gene from populations of Australian *Helicoverpa armigera* (Hübner) (Lepidoptera: Noctuidae) detected via temperature gel electrophoresis. Aust J Entomol 36: 191–196.

32. Scott KD, Lawrence N, Lange CL, Scott LJ, Wilkinson KS, et al. (2005) Assessing moth migration and population structuring in *Helicoverpa armigera* (Lepidoptera: Noctuidae) at the regional scale: example from the Darling Downs, Australia. J Econ Entomol 98: 2210–2219.

33. Scott KD, Wilkinson KS, Lawrence N, Lange CL, Scott LJ, et al. (2005) Gene-flow between populations of cotton bollworm *Helicoverpa armigera* (Lepidoptera: Noctuidae) is highly variable between years. Bull Entomol Res 95: 381–392.

34. Scott KD, Wilkinson KS, Merritt MA, Scott LJ, Lange CL, et al. (2003) Genetic shifts in *Helicoverpa armigera* Hübner (Lepidoptera: Noctuidae) over a year in the Dawson/Callide Valleys. Aust J Agr Res 54: 739–744.

35. Scott LJ, Lawrence N, Lange CL, Graham GC, Hardwick S, et al. (2006) Population dynamics and gene flow of *Helicoverpa armigera* (Lepidoptera: Noctuidae) on cotton and grain crops in the Murrumbidgee Valley, Australia. J Econ Entomol 99: 155–163.

36. Scott KD, Lange CL, Scott LJ, Gahan LJ (2004) Isolation and characterization of microsatellite loci from *Helicoverpa armigera* Hübner (Lepidoptera: Noctuidae). Mol Ecol Notes 4: 204–205.

37. Ji Y-J, Zhang D-X, Hewitt GM, Kang L, Li D-M (2003) Polymorphic microsatellite loci for the cotton bollworm *Helicoverpa armigera* (Lepidoptera: Noctuidae) and some remarks on their isolation. Mol Ecol Notes 3: 102–104.

38. Weeks A, Endersby NM, Lange CL, Low A, Zalucki MP, et al. (2010) Genetic variation among *Helicoverpa armigera* populations as assessed by microsatellites: a cautionary tale about accurate allele scoring. Bull Entomol Res 100: 445–450.

39. Tay WT, Behere GT, Batterham P, Heckel DG (2010) Generation of microsatellite repeat families by RTE retrotransposons in lepidopteran genomes. BMC Evol Biol 10: 144.

40. Lessa EP (1992) Rapid survey of DNA sequence variation in natural populations. Mol Biol Evol 9: 323–330.

41. Palumbi SR, Baker CS (1994) Contrasting population structure from nuclear intron sequences and mtDNA of Humpback whales. Mol Biol Evol 11: 426–435.

42. Tay WT, Behere GT, Heckel DG, Lee SF, Batterham P (2008) Exon-Primed Intron-Crossing (EPIC) PCR markers of *Helicoverpa armigera* (Lepidoptera: Noctuidae). Bull Entomol Res 98: 509–518.

43. Zraket CA, Barth JL, Heckel DG, Abbott AG (1989) Genetic linkage mapping with restriction fragment length polymorphisms in the tobacco budworm, *Heliothis virescens*. In: Hagedorn HH, Hildebrand JG, Kidwell MG, Law JH, editors, 22–27 October 1989; Tucson, Arizona. Plenum Press. pp. 13–20.

44. Behere GT, Tay WT, Russell DA, Batterham P (2008) Molecular markers to discriminate among four pest species of *Helicoverpa* (Lepidoptera: Noctuidae). Bull Entomol Res 98: 599–603.

45. Weir BS, Cockerham CC (1984) Estimating *F*-statistics for the analysis of population structure. Evol 38: 1358–1370.

46. Goudet J (1995) FSTAT (version 1.2): A computer program to calculate F-statistics. J Hered 86: 485–486.

47. Raymond M, Rousset F (1995) GENEPOP (version 1.2): Population genetics software for exact tests and ecumenicism. J Hered 86: 248–249.

48. Rice WR (1989) Analyzing tables of statistical tests. Evol 43: 223–225.

49. Kranthi KR, Jadhav DR, Kranthi S, Wanjari RR, Ali SS, et al. (2002) Insecticide resistance in five major insect pests of cotton in India. Crop Protec 21: 449–460.

50. Kranthi S, Dhawad CS, Naidu S, Bharose A, Chaudhary A, et al. (2009) Susceptibility of the cotton bollworm, *Helicoverpa armigera* (Hübner) (Lepidoptera: Noctuidae) to the *Bacillus thuringiensis* toxinCry2Ab before and after the introduction of Bollgard-II. Crop Protec 28: 371–375.

51. Excoffier L, Smouse PE, Quattro JM (1992) Analysis of molecular variance inferred from metric distances among DNA haplotypes: application to human mitochondrial DNA restriction data. Genet 131: 479–491.

52. Schneider S, Roessli D, Excoffier L (2000) Arlequin: A Software for Population Genetics Data Analysis, Version 2.000. 2.000 ed. Geneva: Genetics and Biometry Laboratory, Dept. of Anthropology, University of Geneva, Switzerland.

53. Pritchard JK, Stephens M, Donnelly P (2000) Inference of population structure using multilocus genotype data. Genet 155: 945–959.

54. Evanno G, Regnaut S, Goudet J (2005) Detecting the number of clusters of individuals using the software STRUCTURE: a simulation study. Mol Ecol 14: 2611–2620.

55. Chapuis M-P, Estoup A (2007) Microsatellite null alleles and estimation of population differentiation. Mol Biol Evol 24: 621–631.

56. Carlsson J (2008) Effects of microsatellite null alleles on assignment testing. J Hered 99: 616–623.

57. Gordon K, Tay WT, Collinge D, Williams A, Batterham P (2009) Genetics and molecular biology of the major crop pest genus *Helicoverpa*. In: Goldsmith MR, Marec F, editors. Molecular Biology and Genetics of the Lepidoptera: CRC Press. pp. 219–238.

58. Johnson MS, Black R (1984) The Wahlund effect and the geographical scale of variation in the intertidal limpet *Siphonaria* sp. Marine Biol 79: 295–302.

59. De Barro PJ (2005) Genetic structure of the whitefly *Bemisia tabaci* in the Asia-Pacific region revealed using microsatellite markers. Mol Ecol 14: 3695–3718.

60. Brown JE, McBride CS, Johnson P, Ritchie S, Paupy C, et al. (2011) Worldwide patterns of genetic differentiation imply multiple 'domestications' of *Aedes aegypti*, a major vector of human diseases. Proc Biol Sci 278: 2446–2454.

61. Nielsen EE, Hansen MM, Ruzzante DE, Meldrup D, Gronkjaer P (2003) Evidence of a hybrid-zone in Atlantic cod (*Gadus morhua*) in the Baltic and the Danish Belt Sea revealed by individual admixture analysis. Mol Ecol 12: 1497–1508.

62. Stipanovic RD, Lopez JD Jr, Dowd MK, Puckhaber LS, Duke SE (2006) Effect of racemic and (+)- and (−)-gossypol on the survival and development of *Helicoverpa zea* larvae. J Chem Ecol 32: 959–968.

63. Kranthi KR, Kranthi NR (2004) Modelling adaptability of cotton bollworm, *Helicoverpa armigera* (Hübner) to *Bt*-cotton in India. Curr Sci 87: 1096–1107.

64. Manjunath TH, Phalak VR, Subramanian S (1970) First records of egg parasites of *Heliothis armigera* (Hubner) (Lepidoptera: Noctuidae) in India. Comm Inst Biol Contr Tech Bull 13: 111–115.

65. Zalucki MP, Furlong MJ (2005) Forecasting *Helicoverpa* populations in Australia: A comparison of regression based models and a bio-climatic based modeling approach. Insect Sci 12: 45–56.

66. Fakrudin B, Prakash SH, Krishnareddy KB, Vijaykumar, Badari Prasad PR, et al. (2004) Genetic variation of cotton bollworm, *Helicoverpa armigera* (Hübner) of South Indian cotton ecosystem using RAPD markers. Curr Sci 87: 1654–1657.

67. Khadi BM, Kulkarni VN, Katageri IS, Vamadevaiah HM, Maralappanavar M (2003) Strengths, weaknesses, opportunities and threats (SWOT) analysis of *Bt* cotton in India. In: Swanepoel A, editor. 9–13 March 2003; Cape Town, South Africa. Agricultural Research Council - Institute for Industrial Crops. pp. 393–406.

68. Ravi KC, Mohan KS, Manjunath TM, Head G, Patil BV, et al. (2005) Relative abundance of *Helicoverpa armigera* (Lepidoptera: Noctuidae) on different host crops in India and the role of these crops as natural refuge for *Bacillus thuringiensis* cotton. Env Entomol 34: 59–69.

69. Directorate of Economics and Statistics, State Wise Land Use Statistics, Department of Agriculture and Cooperation, Ministry of Agriculture, Government of India (2007) Section 9. Area under crops (pulses, food grains & sugarcane) 1996–97 & 2005–06; Section 14. Area under crops (fibers) 1996–97 & 2005–06. Available: http://eands.dacnet.nic.in/state.htm.

70. Firempong S, Zalucki MP (1990) Host plant preferences of populations of *Helicoverpa-Armigera* (Hubner) (Lepidoptera, Noctuidae) from different geographic locations. Aust J Zool 37: 665–673.

71. Jallow MFA, Zalucki MP (1996) Within- and between-population variation in host-plant preference and specificity in Australian *Helicoverpa Armigera* (Hubner) (Lepidoptera: Noctuidae). Aust J Zool 44: 503–519.

72. Endersby NM, McKechnie SW, Ridland PM, Weeks AR (2006) Microsatellites reveal a lack of structure in Australian populations of the diamondback moth, *Plutella xylostella* (L.). Mol Ecol 15: 107–118.

73. Franck P, Reyes M, Olivares J, Sauphanor B (2007) Genetic architecture in codling moth populations: comparison between microsatellite and insecticide resistance markers. Mol Ecol 16: 3554–3564.

High-Throughput Sequencing and Mutagenesis to Accelerate the Domestication of *Microlaena stipoides* as a New Food Crop

Frances M. Shapter[1]*, **Michael Cross**[1], **Gary Ablett**[1], **Sylvia Malory**[1], **Ian H. Chivers**[1,2], **Graham J. King**[1], **Robert J. Henry**[3]

1 Southern Cross Plant Science, Southern Cross University, Lismore, New South Wales, Australia, **2** Native Seeds Pty Ltd, Sandringham, Victoria, Australia, **3** Queensland Alliance for Agriculture and Food Innovation, University of Queensland, Brisbane, Queensland, Australia

Abstract

Global food demand, climatic variability and reduced land availability are driving the need for domestication of new crop species. The accelerated domestication of a rice-like Australian dryland polyploid grass, *Microlaena stipoides* (Poaceae), was targeted using chemical mutagenesis in conjunction with high throughput sequencing of genes for key domestication traits. While *M. stipoides* has previously been identified as having potential as a new grain crop for human consumption, only a limited understanding of its genetic diversity and breeding system was available to aid the domestication process. Next generation sequencing of deeply-pooled target amplicons estimated allelic diversity of a selected base population at 14.3 SNP/Mb and identified novel, putatively mutation-induced polymorphisms at about 2.4 mutations/Mb. A 97% lethal dose (LD_{97}) of ethyl methanesulfonate treatment was applied without inducing sterility in this polyploid species. Forward and reverse genetic screens identified beneficial alleles for the domestication trait, seed-shattering. Unique phenotypes observed in the M2 population suggest the potential for rapid accumulation of beneficial traits without recourse to a traditional cross-breeding strategy. This approach may be applicable to other wild species, unlocking their potential as new food, fibre and fuel crops.

Editor: Gen Hua Yue, Temasek Life Sciences Laboratory, Singapore

Funding: Funding for this research was provided by the Australian Research Council Linkage program (Project number LP0776409,www.arc.gov.au). The funders had no role in study design, data collection and analysis, decision to publish, or preparation of the manuscript.

Competing Interests: The authors have the following interests. Dr Ian Chivers is employed by Native Seeds Pty Ltd, the company that supplied the seventh generation, predominantly inbred, breeding line of M. stipoides, cv AR1 used in this study.

* E-mail: frances.shapter@scu.edu.au

Introduction

Cereal production is the major source of carbohydrate in human diets, with most provided by eight genera of the Poaceae [1]. Increasing world population, climate variability and reduced agricultural land, water and associated inputs drive a need to develop new food crops. New cereal species, able to be cultivated in more marginal environments should be considered. Wild grass species intrinsically adapted to marginal environments and climatic variability, provide an excellent target for domestication.

Australia is a unique source of under-utilised germplasm, due to its short agricultural history, geographic isolation and relative lack of arable land, plant domestication. The Australian Poaceae have evolved independently of other world environments. Moreover, they are adapted to a wide range of resource-limited environments, providing a novel genetic resource for plant pre-breeding. To date, no commercial cereal crops have been developed from this gene pool, although there has been a history of traditional use by indigenous Australians [2].

Approximately 35 million years ago *Microlaena stipoides* (weeping rice grass) shared a common ancestor with cultivated rice, *Oryza sativa* (Figure 1; [3–8]). *M. stipoides* was one of the first Australian grasses identified as having potential to be domesticated as a new cereal crop [9]. Its large grain size, plant architecture, suite of adaptations to marginal and variable environments, and high level of intra-species diversity have been widely recognised [10,11]. Additionally, *M.stipoides* has the same base chromosome number as rice (n = 12) and its tetraploid genome size (880 Mbp) is approximately double that of diploid *O. sativa* (394 Mbp) [3,12].

Increased availability of genomic data has contributed to a better understanding of the genetic basis of the so-called "domestication syndrome" in commercially cultivated Poaceae [13–16]. Indeed, for many domestication traits such as seed shedding upon ripening (shattering), grain color, awn length, dwarfing, grain size, grain number and panicle shape, quantitative trait loci and gene sequences have been identified [17,18]. Although domestication traits may often be controlled by a network of genes, the loss of function of a single component may result in phenotypic modification. For example, seed shattering in rice may be eliminated due to a loss of function of either *qSH1* [19] or *sh4* [20]/SHA1 [21]. Loss of function may result from single base polymorphisms, either naturally occurring or induced [22]. Therefore in most cases the domesticated phenotype results from the cumulative loss of function of multiple genes.

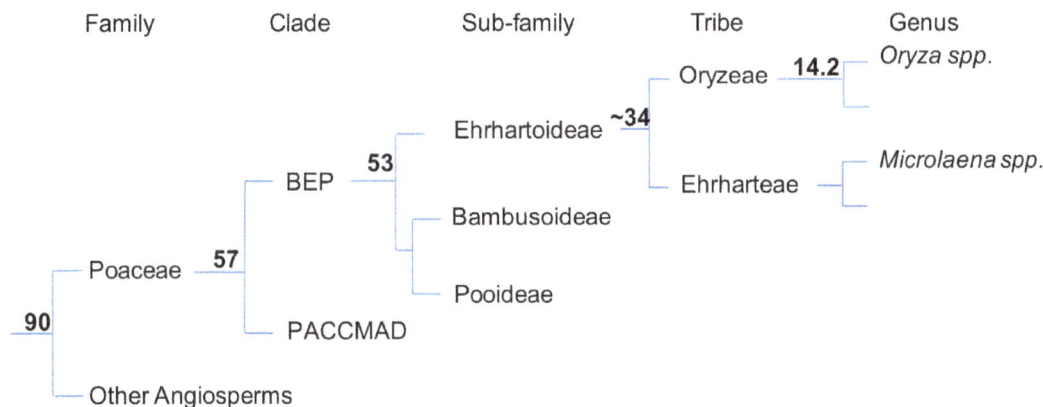

Figure 1. Composite figure outlining the evolutionary relationship between _Microlaena stipoides_ and cultivated rice (_Oryza sativa_). Nodal numbers reflect the estimated million years since each bifurcation [4–6]. Although _O. sativa_ and _M. stipoides_ shared their last common ancestor approximately 34 million years ago they retain the same base chromosome number, n = 12, similar individual genome sizes [3], endosperm morphology and characteristics [8] and genetic homology [7].

By 2004, mutant-derived _O. sativa_ lines had an estimated global value of over US$20 billion [23]. Ethyl methanesulfonate (EMS), a water soluble mutagen alkylates the DNA nucleobase, guanine. This results in randomly distributed point mutations throughout the genome, which in the majority of cases are GC→AT transitions [24]. Targeting Induced Local Lesions IN Genomes (TILLING) has enabled reverse genetic screening of EMS-mutagenised populations [24–27], with several recent modifications to the protocol [28,29]. More recently, Next Generation Sequencing (NGS) using 'short read' platforms such as the Illumina GAII provide a cost effective and informative option for reverse genetic screening of large mutant populations [30,31].

We aimed to accelerate the process of crop domestication by identifying variation in specific traits and their underlying component genes. We harnessed the sensitivity of NGS to characterise both natural and EMS-induced variation within bulked amplicon pools, identifying candidate alleles for improved breeding of the semi-domesticated species, _M. stipoides_.

Methods

Plant material and EMS treatment protocol

A seventh generation, predominantly inbred, breeding line of _M. stipoides_, cv AR1, was supplied by Native Seeds Pty Ltd (nativeseeds.com.au, last accessed 27/7/11) as our base material. Due to poor germination rates in _M. stipoides_ when dehusked, seeds where treated with husks intact and all florets included in the trial were manually checked to ensure the husk contained a filled seed. Imbibition was conducted at a concentration of approximately 10 g of seed per 40 ml of solution (adapted from [32]). Seeds were pre-soaked in water for six hours, then imbibed in either de-ionised water, 40, 60, 80, 100, 115, 130, 145, 160, 175 or 200 mM aqueous solutions of ethyl methanesulfonate (EMS) for 18 hours on a Bio-line orbital shaker at 160 cycles/minute at 22 degrees centigrade in 200 ml Schott bottles. The treatment solution was decanted and 40 mL of de-ionised water was added as a wash solution. This was repeated every 15 mins for four hours. Seed was then immediately planted into germination trays containing Searle's premium potting mix (http://www.searle.com.au/PottingMixes.html, last accessed 01/10/13) and grown under glass house conditions with average minimum and maximum daily temperatures of 15°C and 24°C respectively.

Optimisation of EMS treatment for _M. stipoides_

Based on effective doses used in other Poaceae, an initial dose response curve with a control and four treatments of 40, 60, 80 and 100 mM EMS (200 seeds/treatment) was generated. A second dose response experiment was then conducted to ascertain the efficacy endpoints for _M. stipoides_ with a control and nine treatments of 60, 80, 100, 115,130,145,160,175 and 200 mM EMS (150 seeds/treatment). Germination was monitored and recorded every 2–3 days for a month for the mutated, $M1_d$, and control seedlings, $S1_d$, (Figure 2; generational nomenclature as per [25], where $_d$ denotes dosage trial, M – EMS treated and S-selfed control plants). In order to maximise mutation density in the polyploid background, an LD_{95} was targeted with final percentage germination calculated at 6 weeks post treatment resulting in an LD_{97} being achieved. Seedlings were then monitored on an ongoing basis for any notable phenotypic variations. Seedlings were transplanted into individual 6 cm×6 cm×15 cm forestry tubes of Searle's premium potting mix and grown to maturity on a semi shaded roof top with water applied as required to keep the potting mix moist. $M2_d$ and $S2_d$ seed was collected individually from all non-sterile mutants. Due to the perennial habit of _M. stipoides_ both $M1_d$ and $M2_d$ individuals from the dosage trial which displayed a promising or novel phenotype could be retained and were transplanted to a field site for final evaluation.

Development of the EMS mutant breeding population

For the 145 mM EMS screening population, both the control (200) and treatment (10,000) seeds were screened for grain fill prior to the water or 145 mM EMS treatment. _M. stipoides_ has a predominantly cleistogamous (selfing) reproductive system, though it can also exhibit opportunistic chasmogamous (outcrossing) breeding cycles [33]. The latter was rarely observed and always recorded during the experiment. The M1 population and its control plants (S1) were evaluated under glass house conditions until maturity when M2 and S2 seed was harvested with novel phenotypes recorded at harvest. M1 and S1 plants were then trimmed and transplanted into the field site for phenotypic observation. Three M2 and S2 control seeds per M1 or S1 plant were planted and subsequent germination percentages recorded. Leaf tissue was collected from all individuals in the 145 mM M2 and S2 population followed by transplantation into the field site. At transplantation to the field site all mature plants were trimmed

A

┌─────────────────────────────────┐   ┌──────────────────┐
│   **1°Dosage Trial**   │   │ **M1_d and S1_d** │
│ M0_d=200 seeds, S0_d=200 seeds   │────────▶│ Harvested and   │
│   0,40,60,80,100 mM EMS   │   │ fertility checked│
└─────────────────────────────────┘   └──────────────────┘
                                                                      ┌──────────────────┐
                                                                      │   **M2_d**   │
┌─────────────────────────────────┐   ┌──────────────────┐   │   Selected   │
│   **2°Dosage Trial**   │   │ **M1_d and S1_d** │   │   phenotypes   │
│ M0_d=150seeds, S0_d=150seeds   │────────▶│ Harvested and   │   │ transplanted to │
│ 0, 60, 80,100, 115, 130, 145, 160, 175, 200 mM EMS │ │ fertility checked│ │   field site   │
└─────────────────────────────────┘   └──────────────────┘   └──────────────────┘

B

┌──┐
│ **Established ~LD_{95} as 145mM EMS for 18hrs** │
│ Treatment **M0**=10,000seeds, **S0**=200seeds   │
└──┘
                        ↓
┌──┐
│   **M1 and S1 populations**   │
│ Grown to maturity, phenotyped and harvested in glasshouse, │
│ and transplanted to field. All seed catalogued and stored. │
└──┘
                        ↓
┌──┐
│   **M2 and S2 population**   │
│ Three M2 and S2 seeds planted per surviving M1 and selected S1 plants │
└──┘
                        ↓
┌────────────────────────────────┐   ┌──────────────────────────────┐
│   **M2 and S2 population**   │   │   **DNA extraction**   │
│ M2=754 seedlings, S2=109 seedlings at transplant │──▶│ All sampled individually as seedlings │
│   │   │ prior to transplant to field plot │
└────────────────────────────────┘   └──────────────────────────────┘
                        ↓   ↓
┌────────────────────────────────┐   ┌──────────────────────────────┐
│   **M2 and S2 population**   │   │ **Next Generation Screening** │
│ Phenotypic selection of improved │   │ DNA samples pooled and PCR amplified │
│ lines made from total field plot after │◀─│ for SNP discovery post massively parallel │
│ plants reached maturity.   │   │ Sequencing on Illumina GAII │
│ Selected lines harvested   │   │   │
└────────────────────────────────┘   └──────────────────────────────┘
                        ↓   ↓
┌────────────────────────────────┐   ┌──────────────────────────────┐
│   **M3 and S3 Seed**   │   │   **SNP confirmation**   │
│ Catalogued and stored   │──────▶│ DNA used for Sequenom MassARRAY │
│ for future breeding work   │   │ of a subset of SNP identified by Illumina │
└────────────────────────────────┘   └──────────────────────────────┘

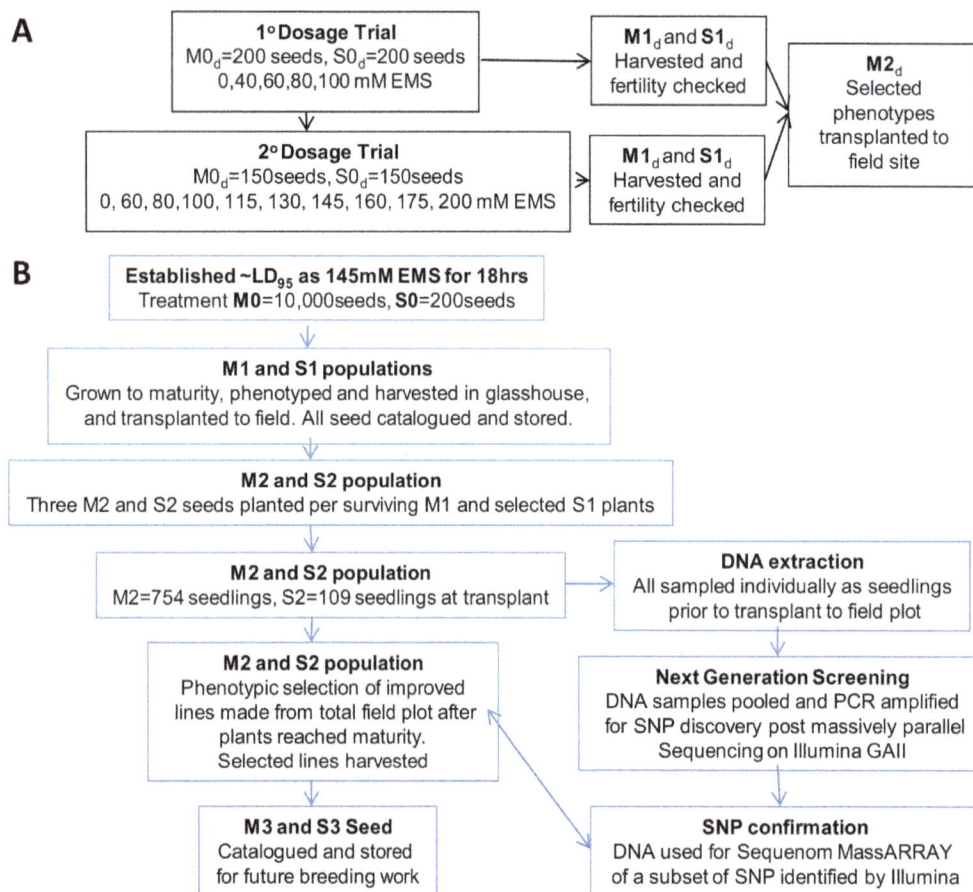

Figure 2. An overview of experimental flow identifying initial optimisation of appropriate EMS treatment (A) and development of the control and mutant populations for reverse genetic screening for SNP discovery and validation. Note, mutant generational nomenclature is as per McCallum et al 2000, where _d_ denotes dosage trial, M indicates EMS treated seed and S indicates the selfed control line.

to approximately 5 cm above the culm. This decreased the stress on the plant and reduced the phenotypic variability resulting from the glasshouse environment.

Illumina sequencing of pooled amplicons

The M2 and S2 populations were used as the basis for genotypic screening. Leaves were collected from 754 juvenile M2 seedlings and 109 S2 individuals. DNA was extracted from fresh leaf tissue using a modified MagAttract 96 DNA Plant Protocol (Qiagen, Frankfurt, Germany) with one additional reverse osmosis purified water wash prior to being quantified using UV spectrophotometry at a wavelength of 260 nm and 280 nm (MWG Sirius Plate reader, MWG Biotech, Ebersberg, Germany). The DNA was normalised using Gibco Nuclease Free water to a concentration of 2 ng/μl using the MWG Theonyx (MWG Biotech, Ebersberg, Germany). Prior to amplification DNA was pooled from five individuals and 10 ng of pooled template was used per PCR. Stringent quality controls were applied during sample preparation. DNA was quantified, normalised and pooled in equimolar proportions at each step in an attempt to maintain relative allele frequencies in the subsequent GAII sequence data.

Four candidate domestication related homologues were targeted for PCR amplification (Table S1 in File S1). Homologues of granule bound starch synthase 1 (GBSS1), encoded by the _Waxy_ gene [7], the _Isa_ gene [34] and two gene homologues controlling seed shattering in rice, _sh4/SHA1_ and _qSH1_ [35] identified in

M.stipoides were targeted. PCR products were quantified by gel electrophoresis using Scion image (http://softwaretopic.informer.com/scion-image-free-software/, last accessed 01/10/13). Amplification products were combined in equimolar amounts to form homologue-specific pools of 109 and 754 M2 (mutant) individuals, in addition to a pool of 109 S2 (control) individuals. The homologue-specific pools were then quantified by pico-green and combined in equimolar amounts to form megapools representing two mutant and one control population. These three megapools were run as individual lanes on the Illumina GAIIx platform (Illumina, San Diego, CA, USA) using a paired-end strategy with a fragment size of 400 bp and a read length of 75 bp.

Sequence data were trimmed using CLC Genomics Workbench version 4.0.3 (www.clcbio.com, last accessed 02/010/13). Reads with a quality score of less than 0.001 were discarded and paired-end reads were trimmed to a minimum of 30 bp. Reference assembly against _M. stipoides_ sequence (Genbank accessions; EF600044, HQ008270, HQ008271, HQ008272) was undertaken with a mismatch cost of 2, insertion and deletion costs of 3, length fraction of 0.8 and similarity of 0.8, minimum distance for paired end reads of 180 bp with a maximum of 340 bp, and non-specific matches ignored. SNP detection parameters of; window length 21, maximum number of gaps or mismatches 2, SNP minimum quality score 30 and quality score for the surrounding bases 30, minimum coverage required 1×, with a minimum variant frequency of 0.000001%, was designed to capture all high read quality polymorphisms.

SNP discovery

Analysis of the CLC SNP discovery output was conducted using Microsoft Excel 2007 following parameters in line with the currently reported error limitations of the Illumina GAII platform for pooled rare SNP discovery [30]. A minimum coverage requirement was set at $400\times$ (approximately $10\times$ the effective pool size for the 109 pools). Based on alignment of these gene homologues and their splice junction sites to rice, putative exon/intron boundaries were assigned to the *M. stipoides* reference sequence [7] and this was used to assign putative functionality of the SNPs.

We carried out an assessment of site-specific variability by calculating the information-content at each nucleotide position [36,37] to test an error threshhold of 0.5% for these data. The work of Tsai et al 2011 indicated that SNP calls with a frequency $>0.5\%$ are unlikely to be false positives and that in all cases the predicted frequency from the Illumina GAII data will be higher than a SNPs actual or theoretical frequency, due to the addition of erroneous 'noise' inherent at all reference positions in Illumina GAII data.

Sequenom MassARRAY SNP confirmation

A subset of 24 SNPs of interest was incorporated into a Sequenom SNP assay. PCR and single base extension primers for each SNP investigated were designed using Assay Design software, version 4.0 (Sequenom Inc., San Diego, CA). The genotyping was performed according to the iPLEX Gold SNP protocol on the Sequenom MassARRAY Compact platform and analysed using Typer 4.0.

Results

Determination of optimal dose of EMS

We established an optimal EMS treatment for *M. stipoides* using a two stage experiment based on final germination frequency (Figure S1). Figure 2 provides an overview of the experimental methodology. The closest approximate to a LD_{95} dose determined was a 145 mM EMS treatment, which resulted in 3% germination (LD_{97}). To ensure that sterility had not been induced [38] seed from plants within both the $M1_d$ and $S1_d$ (where M# (mutant) and S# (selfed control) refer to the generation and $_d$ denotes the dosage trial) populations were harvested. Germination tests confirmed the viability of the M2 seed at this dose.

Phenotypic analysis of the mutant population

Survival at 42 days post treatment was 9% in the mutant population (10,000 seeds), compared with 82.5% in the control population (200 seeds). Novel phenotypic variations amongst the surviving mutant seedlings were observed throughout development (Figure 3). At harvest two M1 individuals underwent anthesis, with the remaining population exhibiting the typical cleistogamous breeding cycle. Seed was harvested from multiple tillers per plant. Overall M2 seedling survival was 82% compared with 91% in the S2 control line. Chlorophyll aberrations were observed in the M2 seedlings at low frequency (1.2%), but were absent in the S2 control population.

Field based evaluation of M2 mutants identified 50 plants (Table 1) with component traits contributing to a more 'domesticated' phenotype than the original base material. These component traits included higher grain yield, plant dry matter, erect seed head architecture, reduced- or non-shattering seed

heads and larger grain size. Twenty four M2 plants possessed our primary domestication target, the non-shattering phenotype (Figure 4). M3 seed and phenotypic data were collected and are currently being evaluated in growth trials, as a grain crop for human consumption.

Next generation sequencing (NGS) for SNP discovery

The four target genes, (*isa*, *qSH1*, *sh4* and *waxy*) were selected for their impact on domestication in other cereals and sequence availability in *M. stipoides*. Preliminary Sanger sequencing (data not shown) of wild type individuals identified within-individual polymorphisms assumed to result from multi-locus variation due to either tetraploidy, and/or heterozygosity. In addition, SNP variation was found between individuals in the base population suggesting a degree of out-crossing.

A PCR pooling strategy was used for NGS analysis, creating cost effective, single lane experiments characterising SNP type and frequency for each gene. Three amplicon pools (109 control plants, 109 mutant plants and a screening pool of 754 mutant plants) were sequenced on the Illumina GAII. After stringent trimming the three pools retained the following read numbers and average read lengths (ARL); Control pool (~58 million reads, ARL – 67 bp), 109 mutants pool (~66 million reads, ARL – 63 bp) and 754 mutants pool (~48 million reads, ARL – 55 bp). Reads which assembled to the reference genes were submitted to the NCBI database (Bioproject ID: SRP030218, Biosample ID: SRS486800; Control_109: SRR1001453, Mutant_109: SRR1001454, Mutant_754: SRR1001455 (http://www.ncbi.nlm.nih.gov/biosample/2361099, last accessed 30/10/13)). Subsequent to reference assembly and application of a minimum coverage threshhold of $400\times$, an average coverage was calculated for each pool; 109 control plants - $27118\times$, 109 mutants -$32289\times$ and 754 mutants -$10171\times$. Coverage was both gene and pool dependent, as previously reported for the Illumina GA platform [39]. Site specific assessment of the SNPs [36], indicated the use of a 0.5% minimum SNP frequency threshold for sequencing error (noise). This also clearly identified SNPs which were shared between pools or unique to a single pool (Figure S2).

NGS identified SNPs in each of the four genes examined. The shattering genes, *qSH1* (69 SNPs) and *sh4* (111 SNPs) had a greater number of SNPs identified than the *waxy* gene (49 SNPs) or the *isa* gene (5 SNPs). Each of the control/mutant pools had a distinct distribution of SNPs either unique to a single pool, or shared between multiple pools (Figure 5). In total, 234 SNPs were identified across the four target genes (Table S2 in File S1), with 229 designated as natural variation, corresponding to a wildtype SNP density in the base population of 14.3 SNP/Mb. Five putatively EMS-induced SNPs were identified as unique to the M109 pool, at the theoretical allele frequency predicted for an EMS-induced mutation (0.5%–1.5%), calculated on the assumption of 1–3 individuals in the pool having a unique homozygous $G/C{\rightarrow}A/T$ transition SNP per genome. This would correspond to an induced mutation density of 2.4 mutations/Mb, in addition to the polymorphism found in the wildtype population.

Of the 234 SNPs identified, 46 were predicted to cause nonsynonymous amino acid changes, of which three are putative stop codons. A further 43 synonymous amino acid changes were also predicted, with the remaining 145 SNPs occurring in introns. This is based on the premise that both homeologues are potentially functional and carry the polymorphism, and that no other indels or polymorphisms have disrupted the reading frame upstream of a target SNP. Sequenom MassARRAY of selected target SNP confirmed 11 of the SNP loci. Notably, the assay confirmed a wildtype C/A SNP, predicted to cause a premature stop codon in the

Figure 3. Examples of mutant plant phenotypes with differing doses of EMS and developmental stages. A. 145 mM EMS treated chlorophyll aberration, only observed in M2 145 mM population, and not observed across all sibling M2 seedlings **B**. 130 mM treated mutant seedling **C**. 115 mM EMS treated mature plants showing root variation within pot trial **D**. 145 mM treated dwarf, no seed produced **E**. control **F–I**. 145 mM mature plants showing mutant phenotypes not seen in control populations; variations to plant architecture, leaf width, length and color, plant vigor, panicle shape, seed production and synchrony of maturity, and inter-nodal span length **I**. Individuals with this plant architecture did not produce seed. Other novel phenotypes observed in the mutant population included rhizome production, crooked nodes, non-surviving dwarfs, and sectoring as variegated leaves.

Figure 4. Panicle shattering habit and awn length variations observed in *Microlaena stipoides*. **A**. wild-type shattering habit with individual grains dehiscing as they reach maturity, and lodging seed heads **B**. Typical wild-type seed head showing empty panicle (↓) by the time the lower seeds have reached maturity **C**. Non-shattering panicle with all seeds retained at maturity (→) **D**. Short versus long awned grains. Short awned varieties are highly desirable as they minimise difficulties associated with handling, processing and mechanisation of the production system.

Table 1. Total *Microlaena stipoides* population numbers and treatment distribution in final field site compared with top 50 mutant phenotypes selected for their amenability to domestication and the component phenotypes observed in these groups.

Treatment	Number transplanted to field	Number of mutants selected with improved phenotypic traits	Phenotypic characteristics selected for, because of their role in accelerating the domestication process
S1 control	148		
S2 control	174		
M1 145 mM	462	11	HTN, EH, NS, HH, DS, BL, CP, LM, TT, LL, EP, LM, CSH, LG
M2 40 mM	90	1	EH, HH
M2 60 mM	135	7	HTN, EH, NS, HH, LG, HY
M2 80 mM	105	1	CP, EH, LG, NS,
M2 100 mM	100	1	NS, EH, HH, HTN, BU
M2 115 mM	43	2	VHH, EH, HTN, ST, CSH
M2 130 mM	29	3	HTN, EH, NS, HH, FO, HY
M2 145 mM	872	23	HTN, EH, NS, HH, DS, BL, CP, LM, TT, LL, EP, LM, CSH, LG, CL, PA
M2 175 mM	4	2	CP, HTN, NS, HH
Total	2162	50	

Legend: BL-broad leaves, BU-blue green leaf color, CP-compact plant, CSH-compact seed head, DS-delayed shattering, EH-erect habit, FO-forage application, HH-high herbage (for grazing applications), HTN-high tiller number, LG-large grains, LL-low leaf material (for grain applications), LM- late maturing, NS-Non-shattering, PA-purple awn, TT-thick tillers, V-very. **Note:** All 145 mM treated plants were progressed to the final evaluation while the S1/S2 controls and M1/M2 dosage trial populations underwent selection for potentially beneficial phenotypes, particularly reduced shattering, prior to transplant.

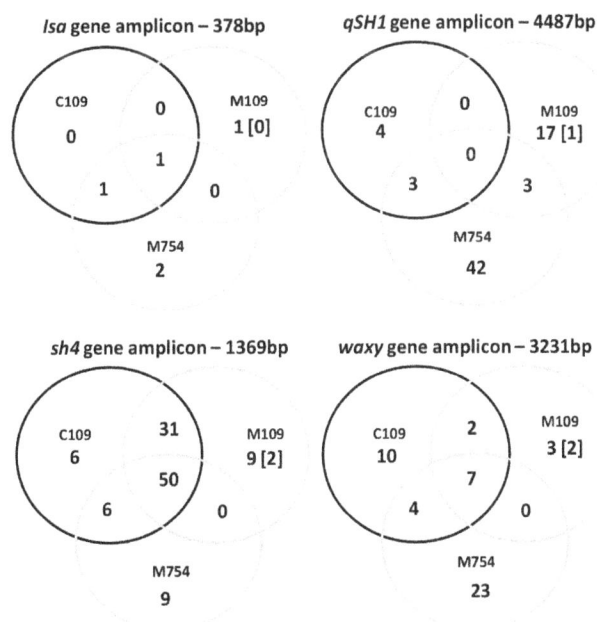

Figure 5. Venn diagram identifying the number of SNP identified in each of the three pools: C109 pool containing 109 wild-type individuals, M109 pool containing 109 145 mM EMS treated mutants and M754 pool containing 754 145 mM EMS treated individuals. The number of putatively EMS induced SNP (rare G/A or C/T polymorphism only found in mutant pools) is in square brackets. Full descriptions of SNPs are available in Table S2 in File S1.

sh4 shattering gene. This was confirmed for two individuals which had been noted as having a non-shattering habit.

Discussion

Determining an effective EMS treatment to induce functional point mutations depends on the species of plant, tissue type, ploidy, and the level of mutation load sustainable without inducing lethality or sterility [26,40]. When the target species' genome contains functional redundancy (due to polyploidy, or ancient genome duplication events), a higher EMS LD can be tolerated [41]. This is supported by our study, which used a LD_{97} without inducing significant levels of sterility in the M1 population. We determined the effective dose using low numbers of seed, followed by a larger-scale generation of mutants and selection. The use of EMS as a mutagen has proven a cost and time effective method for creating new combinations of desirable phenotypic component traits in *M. stipoides*.

Where seed is treated, each genetically effective cell (GEC) will be independently mutagenised [25]. In *M. stipoides*, the GEC number is unknown. Similarly, the pattern of differentiation of each genetically effective M1 sector cannot be tracked. Although M2 populations are often constructed using only a single M2 seed from each M1 plant [27,41], we sampled multiple seeds from each M1 plant, hence our M2 population may capture heritable traits from more than one unique reproductive sector.

The range of phenotypic variation observed allowed the selection of 50 enhanced mutant plants to be used as a pre-breeding population. The selected plants exhibited a unique composite of improvements to plant architecture, reduction of shattering at grain maturity and an increased grain size and/or yield not observed in individuals within the wild-type AR1 base material.

The *M. stipoides* AR1 base material is a semi-domesticated facultative cleisotogamous polyploid with the capacity to outcross, though this was rarely observed in this study. Cloning of another undomesticated, yet cultivated polyploid Poaceae, *Echinochloa ssp.*, identified multiple homeologues of *sh4*, with sequence polymorphism confirmed between the individual's genomic copies [42]. Similarly, in the current study we expected significant levels of wild-type polymorphism both between individuals, and between genomes within each individual.

The use of a pooled amplicon-based NGS approach was effective for detecting wild-type polymorphisms and putatively, EMS-induced mutations. However, there are limitations to using the Illumina platform for this purpose. These include the need to account for and robustly identify low frequency alleles, the non-uniformity of coverage and end bias [39], and the determination of adequate pooling and coverage to distinguish true SNPs from sequencing error [30,43,44]. Similarly, minimisation of the potential effects of PCR error during amplicon and library preparation [45] needs to be addressed. Accurate quantification of amplicons is also crucial to ensure all individuals comprising a pool are represented in equimolar amounts [43,46,47].

Site-specific analysis determined that SNPs identified from the Illumina data with an allele frequency greater than 0.5% were above the error threshold for this analysis. Subsequent Sequenom analysis confirmed the presence of SNP with an allele frequency as low as 0.7%. Similarly, Tsai *et al.* (2011) reliably identified heterozygous mutations in pools of 96 diploid individuals (allele frequency of 0.52%) and confirmed that utilizing a 0.5% minimum allele frequency greatly reduced the risk of false positives. Hence the lowest frequency SNPs identifiable from our 109 pools would be an individual with a homozygous SNP in a single genome or wild-type heterozygous loci on both (allele frequency of 0.46%), which is then expected to be slightly over-represented in the Illumina data [30]. As DNA from up to three M2 siblings was included in the pools, these SNPs may in fact occur at a frequency of up to 1.38%. Within our pool of 754 mutants, the homozygous SNP frequency for an individual (0.07%–0.20%) lies well below the imposed error threshold. Although we were able to identify some wild-type SNPs unique to this pool, there are likely to be more unidentified true SNPs (false negatives), which were disregarded due to their low frequency in the population.

As expected, the frequency and distribution of the 234 SNPs was gene dependent (Figure 5). This variation may be partially due to the different amplicon sizes, proportion of intron sequence screened and the sequence composition. With an abundance of wild-type variation (14.3 SNP/Mb), it is to be expected that the sub-sample represented in each pool would not capture all the variability in the base population. Since a relatively small number of control samples were screened, it is not surprising that we identified many SNPs unique to the mutant populations which are not the result of mutagenesis. The use of an LD$_{97}$ may have created a genetic bottleneck amongst the mutant population. It is therefore to be expected that some non-EMS induced SNPs were unique to the control populations.

The wild-type polymorphism in the AR1 line (14.3 SNP/Mb) indicates that there is considerable diversity within the base material which has potential to be captured, though the greater proportion of these polymorphisms appear to be functionally neutral. If the estimated EMS induced mutation density generated in the *M. stipoides* population is accurate (2.4 mutations/Mb), it is lower than mutation densities previously reported for hexaploid (42 mutations/Mb) and tetraploid (25 mutations/Mb) wheat [41] and in mesopolyploid *Brassica* species (17 mutations/Mb) [48]. However our data is in accord with the tenet that polyploids are

capable of withstanding high dose chemical mutation (LD$_{97}$) without inducing significant levels of M2 lethality or sterility.

Many traits associated with the domestication syndrome are often the result of a loss of function of a recessive gene, such as seed shattering, which in the wild is advantageous and maintained by heterozygosity and natural selection [18]. Such genes are highly desirable targets when domesticating a new species. Both natural polymorphisms and induced mutations can cause such a loss of function of these genes, resulting in a 'domesticated phenotype'. NGS facilitates the screening of large populations for polymorphisms which may induce loss of function. This approach can contribute to identification of candidate alleles for selection and pre-breeding programs.

The non-shattering phenotype was observed in the control lines, but was more prevalent in the mutant population. A non-shattering phenotype in rice can result from loss of function of either the *qSH1* gene, controlling the formation of the grain abscission layer, or the *sh4* gene, a putative transcription factor [19–21]. We identified a wild-type SNP in exon one of *qSH1*, putatively causing a premature stop codon. The *sh4* amplicon screened was 52% intronic DNA, and the majority of wild-type SNPs occurred in this non-coding region. However two SNPs (a C/A and a G/T) identified as causing putative premature stop codons were identified at the 5′ end of exon two. These SNPs may be responsible for the low numbers of non-shattering plants observed in the control lines. Subsequent screening using Sequenom MassARRAY analysis for the C/A SNP confirmed its presence in an individual recorded to have a non-shattering phenotype at harvest.

Loss of function of the *waxy* gene, encoding Granule Bound Starch Synthase I, causes high amylose (waxy) starch to form in grain endosperm in cultivated hexaploid wheat, where a gene dosage effect has been identified. This gene has now been shown to be the major determinant of endosperm starch composition in rice [49]. This was the only gene we examined in *M. stipoides* which had species-specific UTR based primers. As starch composition of the endosperm is important for seed germination, SNPs affecting this gene's function may be under strong selective pressure. Of the 49 SNPs identified in this gene the majority occurred in non-coding regions. Only two potentially non-synonymous SNPs were found, one early in the transit peptide and the other at the end of exon 13.

The *Isa* gene, first characterised in barley, encodes bi-functional amylase/subtilisin inhibitor which acts as part of a seed's defense mechanism against fungal and bacterial pathogens [50]. This locus is reported as a small single copy gene with no introns in rice, barley and wild barley (*Hordeum*) species. Sequence diversity within this gene has been positively correlated with increasing environmental variability [34,51]. Only two of the five SNPs identified in the *Isa* gene of the wild populations were putatively functional, and neither would necessarily cause a loss of function [34]. Two non-synonymous SNPs identified in wild populations sampled close to the provenance of the AR1 base population were also identified in our AR1 control pool and/or the M2 754 pool. In both cases the minor allele from the wild population was the consensus sequence in the Illumina data.

In species where the breeding system is well understood, reverse genetic information provides the opportunity for a renaissance in mutation breeding, precisely because it can pin-point and isolate an independent series of component traits and their alleles. This information may then be used to guide breeding programs over a relatively short number of generations. Where pooled samples are analyzed using short read NGS technologies, it is critical that rare natural or mutant alleles are distinguishable from sequencing error

[39,43,44,46]. It is encouraging that we were able to identify unique SNPs in the pool of 754 M2 individuals. With the increasing throughput, read-length, specificity and sensitivity of NGS platforms, the associated reduction in error thresholds will contribute to more efficient and accurate screening of large mutant populations, attainable at greater pooling depths.

We have successfully accelerated the process of domestication in *M. stipoides* and demonstrated value in both forward and reverse genetic screening of the population. The reverse genetic screen has added valuable knowledge about the extent of the diversity within this base population, while screening only a very limited proportion of such a large genome (~0.001%). Continued technological developments in DNA sequencing will allow greater efficacy of deep pooled screening, and genome-wide screening to include a comprehensive set of domestication genes. Although mutation breeding has been widely used over the past 80+ years to introduce novel alleles into many major crops, it has only had limited use in accelerated domestication. Here we have been able to select mutants with a set of component phenotypes representing multiple beneficial traits without the use of a cross breeding strategy. These phenotypes were identified during the forward screening of the mutant population and included beneficial combinations of plant size, improved architecture, compact panicle structure, increased seed size, non-shattering habit, rhizome production, increased tillering, shorter awn length, increased dry matter yield and greater seed production.

With the rapidly advancing field of molecular genomics and a growing understanding of the genetic events behind domestication, the utilization of molecular techniques in conjunction with mutation breeding should make it possible to accelerate the domestication of other wild plants as new environmentally sustainable crop species.

References

1. Henry RJ (2010) Plant Resources for Food, Fuel and Conservation. London UK: Earthscan.
2. Tindale NB (1977) Adaptive significance of the Panara or grass seed culture of Australia. In: Wright RVS, editor. Stone Tools as Cultural Markers. New Jersey, , USA: Humanities Press.
3. Murray BG, De Lange PJ, Ferguson AR (2005) Nuclear DNA variation, chromosome numbers and polyploidy in the endemic and indigenous grass flora of New Zealand. Ann Bot 96 1293–1305.
4. Kellogg EA (2009) The Evolutionary History of Ehrhartoideae, Oryzeae, and Oryza. Rice 2: 1–14.
5. Bouchenak-Khelladi Y, Verboom GA, Hodkinson TR, Salamin N, Francois O, et al. (2009) The origins and diversification of C-4 grasses and savanna-adapted ungulates. Glob Change Biol 15: 2397–2417.
6. Bouchenak-Khelladi Y, Verboom GA, Savolainen V, Hodkinson TR (2010) Biogeography of the grasses (Poaceae): a phylogenetic approach to reveal evolutionary history in geographical space and geological time. Bot J Linn Soc 162: 543–557.
7. Shapter FM, Eggler P, Lee LS, Henry RJ (2009) Variation in Granule Bound Starch Synthase 8 (GBSS8) loci amongst Australian wild cereal relatives (Poaceae). J Cereal Sci 49: 4–11.
8. Shapter FM, Lee LS, Henry RJ (2008) Endosperm and starch granule morphology in wild cereal relatives. Plant Genet Res 6: 85–97.
9. Turner F (1895) Australian Grasses. Sydney: Charles Potter, Government Printer.
10. Whalley RBD, Brown RW (1993) A method for the collection and transport of native grasses from the field to the glasshouse. 26: 376–377.
11. Davies CL, Waugh DL, Lefroy EC (2005) Perennial grain crops for high water use; the case for *Microlaena stipoides*. Canberra: Rural Industries Research and Development Corporation. RIRDC publication number 05/024 RIRDC publication number 05/024. 1–50 p.
12. Project IRGS (2005) The map-based sequence of the rice genome. Nature 436: 793–800.
13. Izawa T, Konishi S, Shomura A, Yano M (2009) DNA changes tell us about rice domestication. Curr Opin Plant Biol 12: 185–192
14. Paterson AH, Lin Y-R, Li Z, Schertz KF, Doebley JF, et al. (1995) Convergent Domestication of Cereal Crops by Independent Mutations at Corresponding Genetic Loci. Science 269: 1714–1718.
15. Ross-Ibarra J, Morrell PL, Gaut S (2007) Plant domestication, a unique opportunity to identify the genetic basis of adaptation. Proc Natl Acad Sci USA 104: 8641–8648.
16. Doebley JF, Gaut BS, Smith BD (2006) The molecular genetics of crop domestication. Cell 127: 1309–1321.
17. Sweeney MT, McCouch S (2007) The complex history of the domestication of rice. Ann Bot 100: 951–957.
18. Vaughn DA, Balazs E, Heslop-Harrison JS (2007) From crop domestication to super-domestication. Ann Bot 100: 893–901.
19. Konishi S, Izawa T, Lin SY, Ebana K, Fukuta Y, et al. (2006) A SNP caused loss of seed shattering during rice domestication. Science 312: 1392–1396.
20. Li CB, Zhou AL, Sang T (2006) Rice domestication by reducing shattering. Science 311: 1936–1939.
21. Lin Z, Griffith ME, Li x, Zhu Z, Tan L, et al. (2007) Origin of seed shattering in rice (*Oryza sativa*). Planta 226: 11–20.
22. Li CB, Zhou AL, Sang T (2006) Genetic analysis of rice domestication syndrome with the wild annual species, *Oryza nivara*. New Phytol 170: 185–193.
23. Ahloowalia BS, Maluszynski M, Nichterlein K (2004) Global impact of mutation-derived varieties. EUPHYTICA 135: 187–204.
24. Greene EA, Codomo CA, Taylor NE, Henikoff JG, Till BJ, et al. (2003) Spectrum of chemically induced mutations froma large-scale reverse-genetic screen in aribidopsis. Genetics 164: 731–740.
25. McCallum CM, Comai L, Greene EA, Henikoff S (2000) Targeting Induced Local Lesions IN Genomes (TILLING) for plant functional genomics. Plant Physiol 123: 439–442.
26. Till BJ, Reynolds S, Greene EA, Codomo CA, Enns LC, et al. (2003) Large-scale discovery of induced point mutations with high-throughput TILLING. Genome Res 13: 524–530.
27. Henikoff JG, Comai L (2003) Single-nucleotide mutations for plant functional genomics. Annu Rev Plant Biol 54: 375–401.
28. Cordeiro G, Eliott FG, Henry RJ (2006) An optimized ecotilling protocol for polyploids or pooled samples using a capillary electrophoresis system. Anal Biochem 355: 145–147.
29. Dong C, Dalton-Morgan J, Vincent K, Sharp P (2009) A modified TILLING method for wheat breeding. Plant Genome 2: 39–47.
30. Tsai H, Howell T, Nitcher R, Missirian V, Watson B, et al. (2011) Discovery of Rare Mutations in Populations: TILLING by Sequencing. Plant Physiol 156: 1257–1268.

Supporting Information

Figure S1 EMS dosage effect on frequency of seed germination. Note: a single fertile seedling was produced by the 175 mM EMS treatment, though there was complete lethality at both 160 mM and 200 mM EMS treatments.

Figure S2 Sequence variability as measured by the method of Shenkin *et al.* (1991), for the *Microlaena stipoides waxy* gene. The information-theoretical complexity measure *S* is plotted for each nucleotide position. **A.** Control pool; **B.** pool of 109 mutant individuals; **C.** pool of 754 mutant individuals.

File S1. Table S1: Primers and PCR conditions for gene homologues amplified for next generation SNP discovery. **Table S2**: Complete details of SNP loci identified in the four target genes in *Microlaena stipoides*, within the Illumina sequence data above the error threshold (an allele frequency >0.5%).

Acknowledgments

Thanks to Stirling Bowen, Dr Peter Bundock, Dr Nicole Rice, Dr Tim Sexton, Dr Cathy Nock, Dr Mark Edwards, Dr Martin Elphinstone, Dr Abdul Baten and Dr Dan Waters for their technical input.

Author Contributions

Conceived and designed the experiments: FMS IHC RJH. Performed the experiments: FMS MC GA SM. Analyzed the data: FMS GJK. Contributed reagents/materials/analysis tools: IHC. Wrote the paper: FMS MC GJK RJH.

31. Abe A, Kosugi S, Yoshida K, Natsume S, Takagi H, et al. (2012) Genome sequencing reveals agronomically important loci in rice using MutMap. Nat Biotechnol 30: 174–178.

32. Caldwell DG, McCallum CM, Shaw P, Muehlbauer GJ, Marshall DF, et al. (2004) A structured mutant population for forward and reverse genetics in Barley (*Hordeum vulgare* L.). Plant J 40: 143–150.

33. Huxtable CHA (1990) Ecological and embryological studies of Microlaena stipoides (Labill) RBr [Dissertation]. Armidale, NSW, , Australia: University of New England.

34. Fitzgerald TL, Shapter FM, McDonald S, Waters DLE, Chivers IH, et al. (2011) Genome diversity in wild grasses under environmental stress. Proc Natl Acad Sci USA 108: 21140–21145.

35. Malory S, Shapter FM, Elphinstone MS, Chivers IH, Henry RJ (2011) Characterising homologues of crop domestication genes in poorly described wild relatives by high-throughput sequencing of whole genomes. Plant Biotechnol J 9: 1131–1140.

36. King GJ, Lynn JR (1995) Constraints on mutability in a multiallelic gene family. J Mol Evol 41: 732–740.

37. Shenkin PS, Erman B, Mastrandrea LD (1991) Information-theoretical entropy as a measure of sequence variability. Proteins Struct Funct Genet 11: 297–313.

38. Mesken M, Van der Veen JH (1968) The problem of induced sterility: A comparison between EMS and X-rays in *Arabidopsis thaliana*. Euphytica 17: 363–370.

39. Harismendy O, Ng P, Strausberg R, Wang X, Stockwell T, et al. (2009) Evaluation of next generation sequencing platforms for population targeted sequencing studies. Genome Biol 10: R32.

40. van Harten AM (1998) Mutation Breeding: theory and practical applications. Cambridge, UK.: Cambridge University Press.

41. Slade A, Fuerstenberg SI, Loeffler D, Steine MN, Facciotti D (2005) A reverse genetic, nontransgenic approach to wheat crop improvement by TILLING. Nat Biotechnol 23: 75–81.

42. Aoki D, Yamaguchi H (2009) Oryza *sh4* gene homologue represents homoeologous genomic copies in polyploid *Echinochloa*. Weed Biol Manag 9: 225–233.

43. Druley TE, Vallania FLM, Wegner DJ, Varley KE, Knowles OL, et al. (2009) Quantification of rare allelic variants from pooled genomic DNA. Nat Methods 6: 263–265.

44. Out AA, van Minderhout IJHM, Goeman J, Ariyurek Y, Ossowski S, et al. (2009) Deep sequencing to reveal new variants in pooled DNA samples. Hum Mutat 30: 1703–1712.

45. Pienaar EM, Theron M, Nelson M, Viljoen HJ (2006) A quantitative model of error accumulation during PCR amplification. Comput Biol Chem 30: 102–111.

46. Kim SY, Li YR, Guo YR, Li RQ, Holmkvist J, et al. (2010) Design of Association Studies with Pooled or Un-pooled Next-Generation Sequencing Data. Genet Epidemiol 34: 479–491.

47. Sexton T, Shapter FM (2013) Amplicon sequencing for marker discovery. In: Henry RJ, editor. Molecular Markers in Plants, First Edition John Wiley & Sons, Inc. pp. 35–56.

48. Stephenson P, Baker D, Girin T, Perez A, Amoah S, et al. (2010) A rich TILLING resource for studying gene function in *Brassica rapa*. BMC Plant Biol 10: Http;//www.biomedcentral.com/1471-2229/1410/1462.

49. Kharabian-Masouleh A, Waters DLE, Reinke RF, Ward R, Henry RJ (2012) SNP in starch biosynthesis genes associated with nutritional and functional properties of rice. Scientific Rep 2: 557.

50. Mundy J, Svendsen IB, Hejgaard J (1983) Barley α-amylase/subtilisin inhibitor; Isolation and characterization. 48: 81–90.

51. Cronin JK, Bundock PC, Henry RJ, Nevo E (2007) Adaptive Climatic Molecular Evolution in Wild Barley at the ISA Defense Locus. Proc Natl Acad Sci USA 104: 2773–2778.

Diversifying Selection on Flavanone 3-Hydroxylase and Isoflavone Synthase Genes in Cultivated Soybean and Its Wild Progenitors

Hao Cheng❾, Jiao Wang❾, Shanshan Chu, Hong-Lang Yan, Deyue Yu*

National Center for Soybean Improvement, National Key Laboratory of Crop Genetics and Germplasm Enhancement, Nanjing Agricultural University, Nanjing, China

Abstract

Soybean isoflavone synthase (IFS) and flavanone 3-hydroxylase (F3H) are two key enzymes catalyzing the biosynthesis of isoflavonoids and flavonoids, both of which play diverse roles in stress responses. However, little is known about the evolutionary pattern of these genes in cultivated soybean and its wild progenitors. Herein, we investigated the nucleotide polymorphisms in Isoflavone synthase (*IFS1*, *IFS2*) and Flavanone 3-hydroxylase (*F3H2*) genes from 33 soybean accessions, including 17 cultivars (*Glycine max*) and 16 their wild progenitors (*Glycine soja*). Our data showed that the target genes shared the levels of nucleotide polymorphism with three reference genes involved in plant-microbe interactions, but possessed a much higher nucleotide polymorphism than other reference genes. Moreover, no significant genetic differentiation was found between cultivated soybean and its wild relatives in three target genes, despite of considering bottleneck and founder effect during domestication. These results indicate that *IFS* and *F3H* genes could have experienced gene introgressions or diversifying selection events during domestication process. Especially, *F3H2* gene appears to evolve under positive selection and enjoy a faster evolutionary rate than *IFS1* and *IFS2* genes.

Editor: Dorian Q. Fuller, University College London, United Kingdom

Funding: This work was supported in part by National Basic Research Program of China (973 Program) (2010CB125906, 2009CB118400), National Natural Science Foundation of China (31000718, 31171573, 31201230), Jiangsu Provincial Support Program(BE2012328, BK2012768) and Ph.D. Programs Foundation of Ministry of Education of China (No. 20090097120016). The funders had no role in study design, data collection and analysis, decision to publish, or preparation of the manuscript.

Competing Interests: The authors have declared that no competing interests exist.

* E-mail: dyyu@njau.edu.cn

❾ These authors contributed equally to this work.

Introduction

Cultivated soybean [*Glycine max* (L.) Merr.] was domesticated from its annual wild relative [*Glycine soja* Sieb. and Zucc.] in East Asia more than 3,000 years ago [1,2]. Recent research showed that cultivated soybean has lost many rare sequence variants in wild soybean and has undergone numerous allele frequency changes throughout its cultivated history, indicating a severe genetic bottleneck during domestication [3]. Therefore, wild soybean could serve as a vast genetic reservoir and invaluable germplasm resource for both broadening genetic diversity and improving important agronomic traits in soybean breeding.

Soybean isoflavone synthase (IFS) and flavanone 3-hydroxylase (F3H) genes encode two key enzymes involved in the phenylpropanoid pathway (see Figure S1). Phenylpropanoid products, isoflavonoids and flavonoids, play diverse roles in the responses of plants to different biotic and abiotic stresses, particularly in plant-environment interactions [4]. In all cases the unique aryl migration reaction to create isoflavones is mediated by IFS, a legume specific enzyme. IFS enzyme belongs to the CYP93C subfamily of cytochrome P450 monooxygenases and its encoding gene has been identified [5–7]. There were two copies of IFS genes in soybean genome, *IFS1* and *IFS2*, both of which function in isoflavones synthesis. The flavanones in this reaction, including naringenin and liquiritigenin, are also substrates for various other flavonoid biosynthesis processes. The most common enzyme competing with IFS for these common substrates is flavanone 3-hydroxylase (F3H), a 2-oxoglutartate-dependent dioxygenase that initiates syntheses of the majority of flavonoid compounds, including flavonols, anthocyanins, and proanthocyanidins [8]. There were also two tandem duplicates *F3H1* and *F3H2* in soybean, all of which might contribute to isoflavone accumulation in soybean seeds [9,10]. In addition, soybean isoflavones are associated with many health benefits of soy consumption. Because of their biological importance, the genes of isoflavone synthase i.e., *IFS1*, *IFS2* and flavanone 3-hydroxylase *F3H2* gene, were chosen as target genes in this study.

The related genes of IFS and F3H enzyme have been decoded and their DNA sequences have been investigated. As for IFS genes, *IFS1* and *IFS2* genes from 18 Korean soybeans were isolated and sequenced [11]. Variations at the amino acid level among the isoflavone synthases (IFS1, IFS2) were uncovered, yet the detailed nucleotide diversity and selection information were not presented. Turning to *F3H* gene, Aguadé et al (2001) surveyed the pattern of nucleotide variation at *F3H* gene by sequencing a sample of 20 worldwide *Arabidopsis thaliana* ecotypes with one *Arabidopsis lyrata* spp. *petraea* stock as outgroup. They found that *F3H* gene in *Arabidopsis* was compatible with a neutral model with no recombination [12]. However, so far there is no parallel study of nucleotide polymorphisms in soybean *F3H* gene as to the knowledge of the authors.

As functionally important and environment response related genes, *IFS1*, *IFS2* and *F3H2* could show the same or a different evolutionary pattern from other functional genes. Analyzing patterns of DNA polymorphisms and interspecific divergence of a given gene can offer important messages for elucidating the selective forces acting on this gene due to its functional requirement. Comparative investigation of polymorphism between wild and cultivate soybean could provide the evidence for the genetic variation origin and maintenance. Therefore, in our study, we investigated the level and pattern of diversity along the sequences of *IFS* and *F3H* genes in both cultivated and wild soybean accessions. Our results revealed that the three target genes expressed consistent evolutionary pattern with environment response related genes, with much higher level of nucleotide polymorphism than the other reference genes. Particularly, *F3H2* gene was under positive selection and evolved at a faster rate than *IFS* genes. In summary, *F3H* gene and *IFS* genes respond to diverse environments and could be applied for the association analyses of SNPs corresponding to various stress.

Results

Patterns of Polymorphism in Target and Reference Genes

Several genetic estimators, including π, H_d, θ, θ_{sil} and K_{sil} etc., were employed to investigated the patterns of polymorphism in both target and reference genes. Overall, our results consistently suggest a distinctive evolution pattern in *IFS1*, *IFS2* and *F3H2* genes during soybean domestication compared with the reference genes (Table 1).

Data in table 1 showed that the average nucleotide diversities (π) of the target genes (0.44% in cultivated soybean, 0.44% in wild soybean) were similar to those (0.32% in cultivated soybean, 0.37% in wild soybean) of R1 genes in both cultivated and wild soybeans, while more than 5 fold higher than those in R2 genes (0.07% in wild soybean, 0.08% in wild soybean). The target genes were known to be involved in response to various environmental stresses [4]. And R1 genes play roles in plant-microbe interaction. Therefore, we analyzed these genes together as environmental response genes. Obviously, environmental response genes have higher nucleotide diversities than those in R2 genes in both cultivated and wild soybeans (*t*-test, $P<0.05$). Similarly, this result was also observed when using haplotype diversity (H_d) as a estimator, regardless of using the cultivated or wild samples (Table 1). These results indicate that more abundant polymorphisms were maintained in the environmental response genes than other genes, which might be due to diversify selection acting on these genes.

In addition, we compared the values of θ (nucleotide diversity per sequence calculated based on the number of segregating sites) between cultivar and wild soybeans. The values of θ in cultivated soybeans were significantly lower than those of wild soybeans for R2 genes, which were only 54.5% of the levels in their wild relatives (*t*-test, $P<0.01$). It indicates the existence of bottleneck and founder effect in soybean genomes during the domestication process. In contrast, the levels of polymorphism in cultivars were 92.1% (θ) of those found in wild soybean for the target and R1 genes, reflecting the introgression or diversifying selection effect on these environmental response genes to respond to changes of the environment. All these results also indicate that *IFS* and *F3H* genes have a distinctive evolutionary pattern during soybean domestication.

To further investigate this interesting contrast, intra-specific silent diversity (θ_{sil}) was employed in our analysis. Under neutral evolution model, it is expected that intra-specific silent diversity (θ_{sil}) and inter-specific silent divergence (K_{sil}) are correlated with each other. In both cultivated and wild soybeans, we found significant positive correlations between θ_{sil} and K_{sil} in R2 genes ($P<0.01$). However, the correlation was undetectable in environmental response genes, also suggesting different evolutionary pattern involved in these environmental response genes.

High Nucleotide Polymorphim in F3H2 Gene

Among the three target genes, *F3H2* has the highest nucleotide diversity (0.91–0.95%), which is four to five fold higher than that in *IFS1* and *IFS2* genes (0.16–0.23%). Meanwhile, the value of nucleotide divergence (D_{xy}) between wild and cultivated soybean was also much higher in *F3H2* (1.00%) than that of *IFS* genes (0.17–0.23%). Given the same divergence time between cultivated and wild soybeans, *F3H2* genes must enjoy a much higher evolutionary rate. The phylogenetic tree of *F3H2* alleles was constructed using *F3H1* as an outgroup (Figure 1b). Using a cutoff of nucleotide-divergence value 0.008 and significant genetic differentiation between each group (Table S5), we revealed five major groups of *F3H* alleles (Figure 1b). According to the results, the nucleotide diversity within each group ranged from 0.048% to 0.264%, much lower than nucleotide divergence between each two groups (0.813% to 3.502%). Spontaneously, it is reasonable to assume that these grouping might be the result of demographic factors. However, the facts that the alleles from same population scattered in different clades of the phylogenetic tree and the alleles in the same clade contain individuals from different populations reject the hypothesis. On the other hand, the phylogenetic tree showed that group 4 was apparently an intermediate group locating between *F3H1* and *F3H2* clades.

Subsequently, we examine the distribution pattern of polymorphism sites along *IFS1* (Figure S2a), *IFS2* (Figure S2b) and *F3H2* (Figure 2). According to our results, distinct polymorphism distribution patterns displayed among these three genes. In *IFS1* and *IFS2* genes, most of the polymorphism sites were singletons scattered across the coding sequences of genes, although four non-linked parsimony informative sites detected in *IFS1* gene. In contrast, *F3H2* gene (Figure 2) presents a different polymorphism pattern from two *IFS* genes. We found several apparent groups of linked polymorphism sites across the whole gene region. Most of these groups cover majorly cultivated or wild soybean varieties. Especially, an intermediate group between *F3H1* and *F3H2* was detected, sharing nucleotide sequences with both copies. These phenomena provided further evidence that genetic introgression or sequence exchange exist not only between wild and cultivated soybean but also among paralogs. In addition, there was no clear relationship between patterns of polymorphism sites and areas where those soybean samples come from, suggesting that gene introgression or recombination frequently happened at this locus.

Furthermore, our previous study has shown that single nucleotide polymorphisms were associated with SMV strain resistance for *F3H2* locus [13]. Therefore, we combined the phenotypes for each soybean accessions to SMV strains SC-3 and SC-7 with the phylogenetic tree of *F3H2*. Both resistance and susceptible phenotypes were discovered in each group, suggesting diverse resistance responses were preserved at this locus. Moreover, the resistance rates for each group were defined as the number of resistance accessions to the total number of accessions. The resistance rate was lowest in group 1 (36.4%) comparing to those values in group 2, 3 and 4, ranging from 71.4–100%. These suggest accessions in group 1 were more susceptible to SMVs.

Table 1. Nucleotide variation at *IFS1*, *IFS2*, *F3H* genes and reference genes, only CDS were considered.

Gene	π (%)		Hd		θ (%)		θ_{sil} (%)		K_{sil}(%)	Dxy(%)
	C	W	C	W	C	W	C	W	C	
IFS1	0.21	0.23	0.97	0.97	0.32	0.5	0.81	1.07	0.59	0.23
IFS2	0.16	0.17	0.79	0.97	0.4	0.39	0.9	0.49	0.32	0.17
F3H2	0.95	0.91	0.84	0.92	1	0.77	0.67	0.9	0.92	1
Average	0.44	0.44	0.87	0.95	0.57	0.55	0.79	0.82	0.61	0.47
R1										
EU450800	0.27	0.41	1.00	0.99	0.40	0.66	0.55	1.07	0.48	0.35
L20310	0.43	0.51	0.87	0.92	0.41	0.50	0.22	0.43	0.20	0.46
D13505	0.24	0.18	0.98	0.93	0.35	0.31	1.13	0.73	0.64	0.22
Average	0.32	0.37	0.95	0.95	0.39	0.49	0.64	0.74	0.44	0.34
R2										
K00821	0	0.05	0	0.33	0	0.10	0	0.14	0.06	*0.03*
AF083880	0	0.01	0	0.12	0	0.03	0	0	0	0.01
AF124148	0.16	0.10	0.91	0.66	0.12	0.12	0.23	0.22	0.26	0.14
U13987	0	0.08	0	0.52	0	0.11	0	0.23	0.08	0.04
E00532	0.05	0.05	0.25	0.22	0.07	0.06	0	0	0	0.17
L10292	0.05	0.07	0.34	0.42	0.04	0.08	0	0.16	0.07	0.06
AB030491	0	0.03	0	0.32	0	0.06	0	0	0	0.02
AF079058	0.04	0	0.42	0	0.03	0	0.11	0	0.11	0.02
D31700	0.04	0.03	0.13	0.23	0.08	0.08	0.37	0.35	0.14	0.03
J01298	0.06	0.14	0.59	0.82	0.11	0.13	0.12	0.44	0.27	0.11
AF089850	0.12	0.07	0.39	0.17	0.08	0.15	0.29	0.29	0.35	0.10
M94012	0.11	0.12	0.52	0.41	0.05	0.15	0	0	0	0.15
AB004062	0.04	0.21	0.18	0.66	0.06	0.12	0	0	0	0.18
M11317	0.28	0.21	0.65	0.56	0.19	0.38	0.32	0.64	0.36	0.25
Average	0.07	0.08	0.31	0.39	0.06	0.11	0.10	0.18	0.12	0.09

R1 represents plant-environment interaction related reference genes, which include EU450800, the disease resistance gene *Rps1-k-1*; L20310, nodulin (nod-20) gene and D13505, early nodulin related gene; R2 signifies other fourteen reference genes.
C, Cultivated soybean; W, Wild soybean.
π, Nucleotide diversity with Jukes and Cantor correction.
Hd, Haplotype diversity.
θ, Watterson's estimator of θ per basepair calculated on the total number of polymorphic sites.
θ_{sil}, Watterson's estimator of θ per basepair calculated on the silent sites.
K_{sil}, Average divergence per silent site (with the Jukes and Cantor correction) between cultivars and wild soybean.
Dxy, Nucleotide divergence with Jukes and Cantor correction between cultivars and wild soybean.

Genetic Structure and Differentiation between Cultivated and Wild Soybean

To clarify the phylogenetic relationship of the three genes between cultivated and wild soybean, NJ trees were constructed based on nucleotide variations (Figure 1a, 1b). In the phylogenetic trees of *IFS1* and *IFS2* (Figure 1a), and *F3H2* (Figure 1b) genes, cultivated and wild soybean were not clearly separated to form different clades. Instead, there were always some mixed clades including homolog genes from both cultivars and their wild relatives (Figure 1a, 1b), suggesting that there is no apparent differentiation in these three genes between cultivated and wild soybeans. For the 17 reference genes, the mixtures between the cultivated and wild soybeans can also be detected from their phylogenetic trees (see Figure S3). Meanwhile, the nucleotide divergence (*Dxy*) of three target genes and the reference genes between cultivated and wild soybean are similar to the intra-specific nucleotide diversity (π) in wild soybean (Table 1). Besides, there were no fixed polymorphisms

but a much larger number of shared polymorphisms detected between cultivated and wild soybeans (Table 2).

Three different genetic statistics (Chi-square statistic, F_{st} and S_{nn}) were employed to further examine genetic differentiation between wild and cultivated soybeans (Table 2). In our analysis, genetic differentiation was considered significant only when all of the three statistic test gave significant results ($P<0.05$). According to this criterion, none of these three genes showed significant genetic differentiation between cultivated and wild soybean. No significant genetic differentiation was detected for most of the reference genes expect for four genes, which showed significant genetic differentiation between cultivated and wild soybean. Collectively, both phylogenetic analysis and genetic differentiation tests showed no significant divergence between cultivated and wild soybeans for both the target and the reference genes, which could result from the short domestication time of soybean.

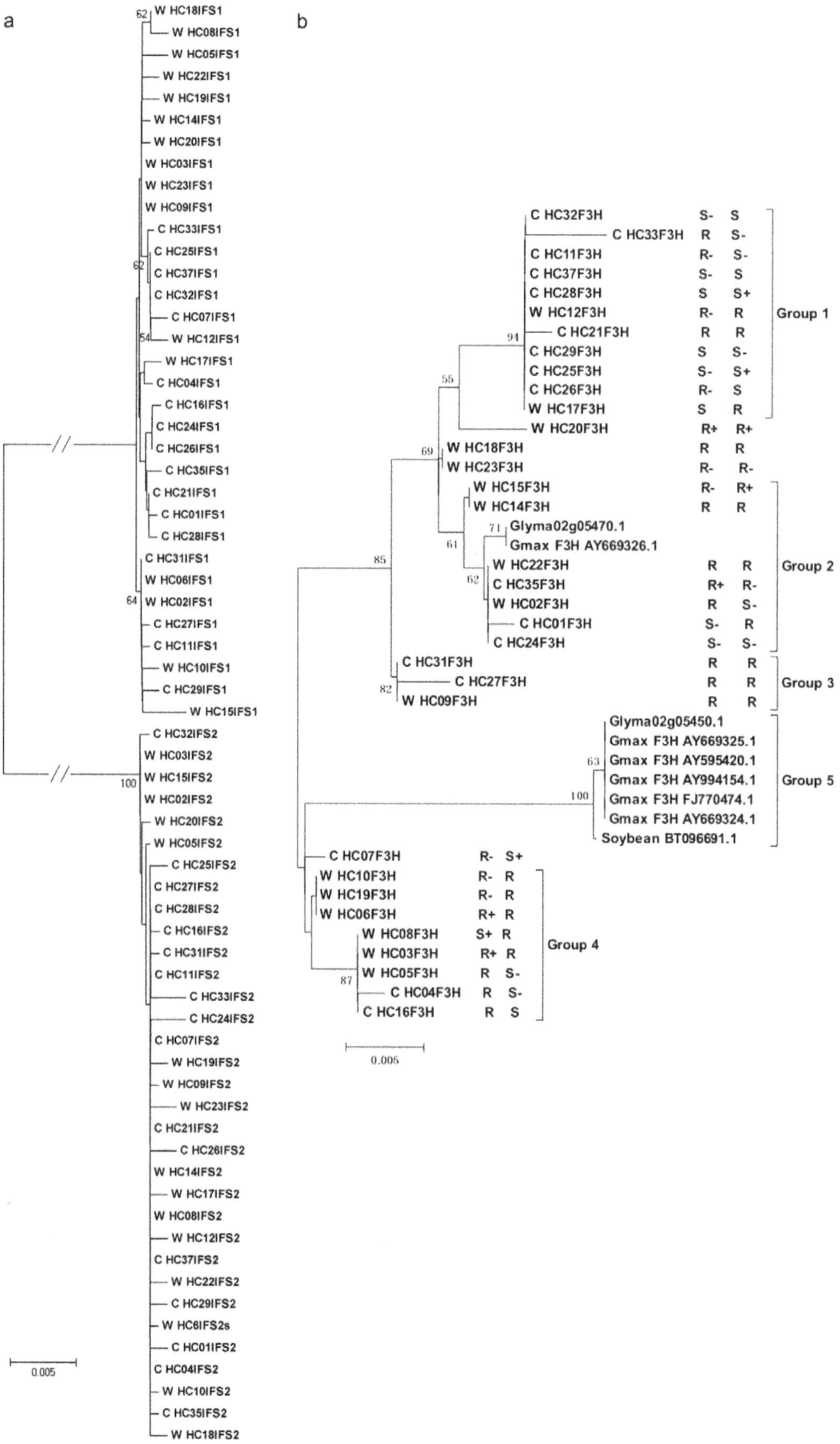
a
W HC18IFS1
W HC08IFS1
W HC05IFS1
W HC22IFS1
W HC19IFS1
W HC14IFS1
W HC20IFS1
W HC03IFS1
W HC23IFS1
W HC09IFS1
C HC33IFS1
C HC25IFS1
C HC37IFS1
C HC32IFS1
C HC07IFS1
W HC12IFS1
W HC17IFS1
C HC04IFS1
C HC16IFS1
C HC24IFS1
C HC26IFS1
C HC35IFS1
C HC21IFS1
C HC01IFS1
C HC28IFS1
C HC31IFS1
W HC06IFS1
W HC02IFS1
C HC27IFS1
C HC11IFS1
W HC10IFS1
C HC29IFS1
W HC15IFS1
C HC32IFS2
W HC03IFS2
W HC15IFS2
W HC02IFS2
W HC20IFS2
W HC05IFS2
C HC25IFS2
C HC27IFS2
C HC28IFS2
C HC16IFS2
C HC31IFS2
C HC11IFS2
C HC33IFS2
C HC24IFS2
C HC07IFS2
W HC19IFS2
W HC09IFS2
W HC23IFS2
C HC21IFS2
C HC26IFS2
W HC14IFS2
W HC17IFS2
W HC08IFS2
W HC12IFS2
C HC37IFS2
W HC22IFS2
C HC29IFS2
W HC6IFS2s
C HC01IFS2
C HC04IFS2
W HC10IFS2
C HC35IFS2
W HC18IFS2
62
62
54
64
100
0.005
b
C HC32F3H S- S
C HC33F3H R S-
C HC11F3H R- S-
C HC37F3H S- S
C HC28F3H S S+
W HC12F3H R- R
C HC21F3H R R
C HC29F3H S S-
C HC25F3H S- S+
C HC26F3H R- S
W HC17F3H S R
W HC20F3H R+ R+
W HC18F3H R R
W HC23F3H R- R-
W HC15F3H R- R+
W HC14F3H R R
Glyma02g05470.1
Gmax F3H AY669326.1
W HC22F3H R R
C HC35F3H R+ R-
W HC02F3H R S-
C HC01F3H S- R
C HC24F3H S- S-
C HC31F3H R R
C HC27F3H R R
W HC09F3H R R
Glyma02g05450.1
Gmax F3H AY669325.1
Gmax F3H AY595420.1
Gmax F3H AY994154.1
Gmax F3H FJ770474.1
Gmax F3H AY669324.1
Soybean BT096691.1
C HC07F3H R- S+
W HC10F3H R- R
W HC19F3H R- R
W HC06F3H R+ R
W HC08F3H S+ R
W HC03F3H R+ R
W HC05F3H R S-
C HC04F3H R S-
C HC16F3H R S
Group 1
Group 2
Group 3
Group 5
Group 4
94
55
69
71
61
62
85
82
63
100
87
0.005

Figure 1. Phylogenetic trees of *IFS1* and *IFS2* (a), *F3H2* (b). The trees were estimated by neighbor-joining (NJ) method based on multiple CDS sequence alignments. CDS of *F3H1* (Glyma02g05450.1) alleles were used as outgroup for soybean *F3H2* in this study. The synthetic disease index (SDI) [32] was used in evaluation of soybean resistance to soybean mosaic virus. If the SDI was under 0, 20 and 35, the accession was classified as R+, R and R-. Meanwhile, we defined the accession as S+, S and S- respectively, if the SDI was above 70, 51 and 36. The resistance responses of 33 soybean accessions to SMV strains SC-3 and SC-7 were listed on the former and latter columns respectively. Bootstrap values >50% are indicated on the branches.

Pseudogenes in the Two IFS Genes

The full *IFS1* alignment spanned 2,748bp including 64 sites with alignment gaps (indel polymorphisms). In the first exon, 13 replacement changes were identified. Among these replacement changes, there is a 'T' to 'C' mutation which causes the start codon, 'ATG', of *IFS1* gene of accession C_HC35 change to 'ACG'. This may lead to no translation of *IFS1* gene in accession C_HC35 (Figure S4a).

The full *IFS2* alignment spanned 2,956bp including 24 sites with alignment gaps. In the first exon, four indels of 1 or 2bp were identified. These four indels were all singletons, two found in accession C_HC16, one in C_HC28 and one in C_HC31 (Figure S4b). These indels all caused frameshift mutations, leading to probable function loss of related genes.

One start codon mutation of *IFS1* in accession C_HC35 and a total of four frameshift mutations of *IFS2* gene in accession C_HC16, C_HC28 and C_HC31 may result in no function of the *IFS1* or *IFS2* in the corresponding accession plants. However, though IFS acts as the key metabolic entry point for the formation of all isoflavonoids [4], the silence of these genes showed no significant impact on isoflavone content of the corresponding accessions [14], indicating that these two IFS genes may be able to compensate for each other. Also, this was supported by previous study which showed both IFS1 and IFS2 enzyme could convert naringenin and liquiritigenin to genistein and daidzein respectively [6]. Due to such complementary effect of *IFS1* and *IFS2* genes, the harm caused by deleterious mutations can be overcome effectively.

Different Selection Contexts of IFS Genes and F3H Gene

Tajima's *D* was used to determine allele frequency changes by comparing the two *IFS* genes and the *F3H* gene with reference genes in cultivated and wild soybean. Interestingly, the results (Table 3) showed that *D* values at the three genes (−1.33 on average) in cultivars were slightly lower than that (−1.22 on average) in wild soybean, suggesting an excess of low-frequency nucleotide polymorphisms in both cultivated and wild soybean. However, *D* values at the fourteen R2 genes (0.03 on average) in cultivars were higher than that (−0.65 on average) in wild soybean, suggesting an excess of low-frequency nucleotide polymorphisms in wild soybean while more intermediate-frequency polymorphisms in cultivars, which was consistent with the theoretical expectation that some low-frequency variants have been preferentially lost in cultivars because of the recent bottleneck during domestication.

The ratio of nonsynonymous (*Ka*) to synonymous (*Ks*) nucleotide substitutions, is widely used to evaluate selection effect. Under neutral evolution, there should be *Ka/Ks* = 1. In contrast, *Ka/Ks* < 1 indicates a negative or purifying selection and *Ka/Ks* > 1 on the

Groups	SeqNames	15	23	28	44	48	71	90	93	96	122	124	169	173	191	194	213	216	217	232	234	253	272	333	387	391	393	410	414	454	466	492	524	573
1	C_HC29F3H	A	A	C	C	T	C	A	G	A	T	T	C	A	G	T	G	G	C	G	G	C	A	C	G	A	C	A	C	T	G	T	C	T
1	C_HC28F3H	.	.	.	.	.	.	.	.	.	.	.	.	.	.	.	.	.	.	.	.	.	.	.	.	.	.	.	.	.	.	.	.	.
1	C_HC32F3H	.	.	.	.	.	.	.	.	.	.	.	.	.	.	.	.	.	.	.	.	.	.	.	.	.	.	.	.	.	.	.	.	.
1	C_HC11F3H	.	.	.	.	.	.	.	.	.	.	.	.	.	.	.	.	.	.	.	.	.	.	.	.	.	.	.	.	.	.	.	.	.
1	C_HC26F3H	.	.	.	.	.	.	.	.	.	.	.	.	.	.	.	.	.	.	.	.	.	.	.	.	.	.	.	.	.	.	.	.	.
1	C_HC25F3H	.	.	.	.	.	.	.	.	.	.	.	.	.	.	.	.	.	.	.	.	.	.	.	.	.	.	.	.	.	.	.	.	.
1	C_HC37F3H	.	.	.	.	.	.	.	.	.	.	.	.	.	.	.	.	.	.	.	.	.	.	.	.	.	.	.	.	.	.	.	.	.
1	W_HC17F3H	.	.	.	.	.	.	.	.	.	.	.	.	.	.	.	.	.	.	.	.	.	.	.	.	.	.	.	.	.	.	.	.	.
1	W_HC12F3H	.	.	.	.	.	.	.	.	.	.	.	.	.	.	.	.	.	.	.	.	.	.	.	.	.	.	.	.	.	.	.	.	.
1	C_HC21F3H	.	.	G	.	.	.	.	.	.	.	.	.	.	.	.	.	.	.	.	.	.	.	.	.	.	.	.	.	.	.	.	.	.
1	C_HC33F3H	.	.	.	.	.	T	.	.	.	.	.	.	.	.	.	.	.	.	.	.	G	.	.	.	.	.	.	.	C	.	.	.	.
2	Glyma02g05470.1	G	.	.	.	.	T	.	.	.	.	.	.	.	.	C	.	.	.	T	.	T	.	.	.	.	.	.	.	A	.	.	.	.
2	Gmax_F3H_AY669326.1	G	.	.	.	.	T	.	.	.	.	.	.	.	.	C	.	.	.	T	.	T	.	.	.	.	.	.	.	A	.	.	.	.
2	C_HC01F3H	.	.	.	.	.	T	.	.	.	.	.	.	.	.	C	.	.	.	T	.	T	.	.	.	G	.	.	.	A	.	.	.	.
2	C_HC24F3H	.	.	.	.	.	T	.	.	.	.	.	.	.	.	C	.	.	.	T	.	T	.	.	.	.	.	.	.	A	.	.	.	.
2	W_HC02F3H	.	.	.	.	.	T	.	.	.	.	.	.	.	.	C	.	.	.	T	.	T	.	.	.	.	.	.	.	A	.	.	.	.
2	C_HC35F3H	.	.	.	.	.	T	.	.	.	.	.	.	.	.	C	.	.	.	T	.	.	.	.	.	.	.	.	.	A	.	.	.	.
2	W_HC22F3H	.	.	.	.	.	T	.	.	.	.	.	.	.	.	C	.	.	.	T	.	T	.	.	.	.	.	.	.	A	.	.	.	.
2	W_HC14F3H	.	.	.	.	.	T	.	.	.	.	.	.	.	.	C	.	.	.	T	.	.	.	.	.	.	.	.	.	A	.	.	.	.
2	W_HC15F3H	.	.	.	.	.	T	.	.	.	.	.	.	.	.	C	.	.	.	T	.	.	.	.	.	.	.	.	.	A	.	.	.	.
3	W_HC09F3H	.	.	.	.	.	T	.	.	.	.	.	.	.	.	C	.	A	.	T	.	.	.	.	.	.	.	.	.	A	.	.	.	.
3	C_HC31F3H	.	.	.	.	.	T	.	.	.	.	.	.	.	.	C	.	A	.	T	.	.	.	.	.	.	.	T	.	A	.	.	.	.
3	C_HC27F3H	.	.	.	T	.	T	.	.	.	.	.	.	.	.	C	.	A	.	T	.	.	.	.	.	.	.	.	.	A	.	.	.	.
4	W_HC10F3H	.	.	.	.	.	.	.	.	.	.	.	G	C	C	.	C	A	G	T	.	.	.	.	.	.	.	.	.	A	.	.	.	.
4	W_HC19F3H	.	.	.	.	.	.	.	.	.	.	.	G	C	C	.	C	A	G	T	.	.	.	.	.	.	.	.	.	A	.	.	.	.
4	W_HC06F3H	.	.	.	.	.	.	.	.	.	.	.	G	C	C	.	C	A	G	T	.	.	.	.	.	.	.	.	.	A	.	.	.	.
4	C_HC07F3H	.	.	.	.	.	.	.	.	.	.	.	G	C	C	.	C	A	G	T	.	.	.	.	.	.	.	.	.	A	.	.	.	.
4	W_HC08F3H	.	.	.	.	.	T	.	.	.	.	.	G	C	C	.	C	A	G	T	.	.	.	.	.	T	T	.	.	A	.	.	.	.
4	W_HC05F3H	.	.	.	.	.	T	.	.	.	.	.	G	C	C	.	C	A	G	T	.	.	.	.	.	T	T	.	.	A	.	.	.	.
4	C_HC04F3H	.	.	.	.	.	T	.	.	.	C	.	G	C	C	.	C	A	G	T	.	.	.	.	.	T	T	.	.	A	.	.	.	.
4	W_HC03F3H	.	.	.	.	.	.	.	.	.	.	.	G	C	C	.	C	A	G	T	.	.	.	.	.	T	T	.	.	A	.	.	.	.
4	C_HC16F3H	.	.	.	.	.	T	.	.	.	.	.	G	C	C	.	C	A	G	T	.	.	.	.	.	T	T	.	.	A	.	.	.	.
5	Glyma02g05450.1	.	.	.	.	C	T	G	A	G	C	.	G	C	C	.	C	A	G	T	C	T	.	.	.	T	.	T	.	A	.	C	G	C
5	Gmax_F3H_FJ770474.1	.	.	.	.	C	T	G	A	G	C	.	G	C	C	.	C	A	G	T	C	T	.	.	.	T	.	.	.	A	.	C	G	C
5	Gmax_F3H_AY669324.1	.	.	.	.	C	T	G	A	G	C	.	G	C	C	.	C	A	G	T	C	T	.	.	.	T	.	.	.	A	.	C	G	C
5	Gmax_F3H_AY595420.1	.	.	.	.	C	T	G	A	G	C	.	G	C	C	.	C	A	G	T	C	T	.	.	.	T	.	.	.	A	.	C	G	C
5	Soybean_BT096691.1	.	.	.	.	C	T	G	A	G	C	.	G	C	C	.	.	G	G	T	C	T	.	.	.	T	.	.	.	A	.	C	G	C
5	Gmax_F3H_AY669325.1	.	.	.	.	C	T	G	A	G	C	.	G	C	C	.	C	A	G	T	C	T	.	.	.	T	.	.	.	A	.	C	G	C
5	Gmax_F3H_AY994154.1	.	.	.	.	C	T	G	A	G	C	.	G	C	C	.	C	A	G	T	C	T	.	.	.	T	.	.	.	A	.	C	G	C

Figure 2. Nucleotide polymorphism sites of *F3H2* genes. The locations of the polymorphism sites were shown in the above line. The polymorphic sites were highlighted by red color.

Table 2. Genetic differentiation between cultivated and wild soybean at *IFS1*, *IFS2*, *F3H* genes and reference genes, only CDS were considered.

Gene	Polymorphisms between C and W				R_M			Genetic differentiation between W and C				
	Fixed	In C	In W	Shared	C	W	All	S_{nn}	F_{st}	Chi-square statistic		
										χ^2	df	P-value
IFS1	0	14	23	3	1	3	3	0.78**	0.05*	30.33	25	0.21
IFS2	0	22	20	0	0	2	2	0.57**	0.01	26.59	21	0.18
F3H2	0	8	3	12	2	5	5	0.60	0.07	17.76	13	0.17
Average	0	14.33	15.33	5.00	1.00	3.33	3.33					
R1												
EU450800	0	17	52	31	8	13	20	0.45	0.01	29.32	29	0.45
L20310	0	0	2	8	2	4	4	0.40	−0.02	11.07	12	0.52
D13505	0	12	10	11	6	2	9	0.56	0.01	26.26	24	0.34
Average	0	9.67	21.33	16.67	5.33	6.33	11.00					
R2												
K00821	0	0	3	0	0	0	0	0.50	0.02	2.92	3	0.40
AF083880	0	0	1	0	0	0	0	0.49	−0.01	0.91	2	0.34
AF124148	0	1	1	6	4	3	4	0.48	0.07	14.60	14	0.41
U13987	0	0	10	0	0	1	1	0.53	0.09*	5.23	5	0.39
E00532	0	0	0	1	0	0	0	0.77**	0.70**	17.95	1	0**
L10292	0	0	1	1	0	1	1	0.47	−0.03	2.12	3	0.55
AB030491	0	0	2	0	0	0	0	0.52	0.03	2.92	2	0.23
AF079058	0	1	0	0	0	0	0	0.57*	0.23*	5.18	1	0.02*
D31700	0	0	0	2	0	0	1	0.47	−0.07	2.92	3	0.40
J01298	0	0	3	2	0	2	2	0.58	0.07	9.00	5	0.11
AF089850	0	0	1	1	0	0	0	0.52	0.05	5.25	2	0.07
M94012	0	0	2	1	0	0	0	0.58**	0.21**	13.16	3	0.00**
AB004062	0	0	1	1	0	0	0	0.69**	0.28**	26.52	2	0***
M11317	0	0	3	3	0	0	0	0.49	0.01	6.00	6	0.42
Average	0	0.14	2.00	1.29	0.29	0.50	0.64					

R1 represents plant-environment interaction related reference genes, which include EU450800, the disease resistance gene *Rps1-k-1*; L20310, nodulin (nod-20) gene and D13505, early nodulin related gene; R2 signifies other fourteen reference genes.
C, Cultivars; W, Wild soybean.
Fixed, the number of fixed differences between cultivated and wild soybeans; In W, Mutations that are polymorphic in wild soybeans, but monomorphic in cultivars; In C, Mutations that are polymorphic in cultivars, but monomorphic in wild soybean; Shared, the total number of shared mutations.
R_M, the minimum number of recombination events [25], both coding and noncoding sequences were considered.
*P<0.05,
**P<0.01,
***P<0.001.

other hand is a strong evidence showing positive or diversifying selection.

We calculated *Ka/Ks* ratio for the full length coding sequences of *IFS1*, *IFS2*, *F3H2* and the reference genes in cultivated and wild samples (Table 3). We excluded those loci with synonymous substitution rate <0.1%, because *Ks* (as denominator) was too small to get a reliable estimate. According to our results, no obvious positive selection was detected in the two *IFS* genes. However, for the *F3H2* gene in the cultivated accessions, *Ka/Ks* = 1.19, which suggested positive selection or relaxation of selection pressure during and after domestication process. Interestingly, similar selection pattern was found in the three R1 genes. At one of these three R1 genes (L20310), there was significantly positive selection detected in both cultivated (*Ka/Ks* = 5.51) and wild (*Ka/Ks* = 1.82) soybeans. While in the fourteen R2 genes, there was no positive selection detected in both cultivated and wild soybeans.

Discussion

Unique Evolution Scenario of IFS Genes and F3H2 Gene During Soybean Domestication

Our results reveal two remarkable phenomena: consistent polymorphism pattern between the target genes and three plant-environment interaction related reference genes (R1), contrastive evolutionary characteristics between environmental response related genes and other functional genes (R2). Previous studies have indicated that these three targeted genes play important roles in soybean's resistance to diverse environmental stresses [4]. As well as, three reference genes are involved in pathogens or rhizobiums interaction. Therefore, similar functional genes might share unique evolutionary patterns, e.g. shared evolutionary parameters were detected in these genes, including high nucleotide diversity, frequent sequence exchange and prone to positive

Table 3. Test for evolutionary forces shaping *IFS1, IFS2, F3H2* gene, only CDS were considered.

Gene	Tajima's D		K_a/K_s	
	C	W	C	W
IFS1	−1.38	−2.20**	0.16	0.27
IFS2	−2.39***	−2.22**	0.25	0.56
F3H2	−0.21	0.76	1.19	0.81
Average	−1.33	−1.22		
R1				
EU450800	−1.39	−1.61	0.68	0.57
L20310	0.14	0.03	5.51	1.82
D13505	−1.28	−1.65	0.10	0.17
Average	−0.84	−1.07		
R2				
K00821	/	−1.38	0/0	0.33
AF083880	/	−1.16	0/0	1.53e−4/0
AF124148	0.96	−0.49	0.33	0.43
U13987	/	−1.19	0/0	0.34
E00532	−0.40	−0.49	6.93e−4/0	6.18e−4/0
L10292	0.24	−0.26	6.17e−4/0	0.46
AB030491	/	−1.07	0/0	0/0
AF079058	0.74	/	0.00	0/0
D31700	−1.49	−1.50	0.00	0.00
J01298	0.22	0.24	6.40e−4/5.07e−4	0.15
AF089850	0.85	−1.06	0.00	0.12
M94012	1.65	−0.40	1.39e−3/0	1.65e−3/0
AB004062	−0.45	1.61	5.26e−4/0	2.69e−3/0
M11317	1.02	−1.27	0.58	0.60
Average	0.33	−0.65		

R1 represents plant-environment interaction related reference genes, which include EU450800, the disease resistance gene *Rps1-k-1*; L20310, nodulin (nod-20) gene and D13505, early nodulin related gene; R2 signifies other fourteen reference genes.
C, Cultivars; W, Wild soybeans; All, Cultivars and Wild soybeans.
*0.01<P<0.05;
**0.001<P<0.01;
***P<0.001.

selection. These environmental response related genes face a variety of biotic or abiotic factors, making them co-evolve with pathogens or other stress [15]. On the contrary, other reference genes appear critical to soybean's growth and some of them are connected with important agricultural characters of domesticated soybean. Hence, they are more conservative in nucleotide sequences.

In addition, inconsistent evolutionary signatures between environmental response related genes and R2 were detected in cultivated and wild soybeans. For example, much higher nucleotide diversity (θ), more frequent recombination and more negative Tajima's D values were detected in wild soybeans than those of cultivated soybeans for R2 genes. In the environmental response related genes, however, shared level of those values between cultivated and wild soybeans were detected. Considering the environmental stress faced by the environmental response related genes, either biotic or abiotic, may have not changed much

for either wild or cultivated soybean, the high level of nucleotide diversity therefore may be important to such functional requirements for both, especially considering many polymorphism sites which shared between cultivated and wild soybean (Table 2). The nucleotide diversity (θ) at these environmental response related genes showed no significant difference (*t*-test, P = 0.30) between the cultivated and wild ones. This disparity is inconsistent with the commonly believed diversity loss after domestication. There must be some forces that can contribute to the quick recovery of diversity in these three target genes, for example intentional or accidental gene introgression during soybean domestication. In contrast, for R2 genes, the nucleotide diversity (θ) in cultivated soybeans were significantly (*t*-test, P<0.01) lower than those in the wild ones. These fourteen reference genes appear critical to soybean's growth and some of them are connected with important agricultural characters of domesticated soybean. Therefore, both bottleneck effect and probable artificial selection may play important role in the reduction of their nucleotide diversity in cultivated soybean relative to wild soybean [3].

The allele frequency changes determined by Tajima's D showed similar conclusion. The Tajima's D values detected in the environmental response related genes were similarly minus in both two populations (Table 3), indicating that the mutations in these genes were mostly low-frequency changes which had vague relationship with population structure. As for the reference genes, however, the Tajima's D values were much higher in the cultivated than the wild soybean, suggesting much higher mutation frequency in the reference genes of the cultivated soybean, which could be the results of the removal of the rare alleles in the process of artificial selection.

Based on above observations, the evolution scenario of these three target genes can be inferred as follows: for *IFS1, IFS2* and especially *F3H2* genes, frequent gene introgression and recombination may have occurred in the domestication history of soybean, introducing considerable amount of diversities. And these relative large diversities are kept either simply by hitchhiking effects or due to the unique functional requirements of these three genes.

F3H2 Gene Evolves Faster than the Two IFS Genes

According to our results, we found that *F3H2* gene enjoys a faster evolution rate than those two *IFS* genes based on the observation that the nucleotide diversity level of *F3H2* gene is higher than that of *IFS1* and *IFS2*, both in wild and cultivated soybean populations. Understanding the heterogeneity of evolutionary rate would shed light on the origin of large number of polymorphisms.

The recombination events found in *F3H2* gene is more frequent than those of the two *IFS* genes, either in wild and cultivated soybean populations, suggesting frequent sequence exchange between homologs was an important mechanism for generating diversity. The frequency of recombination between two homologs was commonly negatively correlated with distance and positively correlated with sequence similarity [16]. The higher nucleotide similarity and closer distance between *F3H1* and *F3H2* resulted in more frequent recombination than *IFS1* and *IFS2*, which in turn cause the higher level of diversity in *F3H2* than *IFS* genes. And this point is further confirmed in the linked polymorphism sites and phylogenetic analysis, in which we observed a group of *F3H2* genes are mixed with *F3H1* gene.

Furthermore, positive selection was observed in *F3H2* gene (Table 3), especially in the cultivated population, but not detected in *IFS1* and *IFS2* (Table 3), which shows a considerable diversifying selection force driving the evolution of *F3H2* gene. Theoretically, certain types of mutations are preferentially preserved for specific characters in adaption. Therefore, such diversifying selection of *F3H2* gene may reflect its functional importance in soybean's adaption to the environment. As introduced above, *IFS* genes serve as the entry point of isoflavonoids biosynthesis pathway in legume plants and some other non-legume crops. According to previous study, these two genes are very conservative even between distant species. For instance, both *IFS* genes encode proteins in sugarbeet show a surprisingly >95% similarity to soybean IFS1 protein [6]. In contrast, *F3H* gene, which plays a major role in biosynthesis of flavonoids, seems to enjoy a different context of evolution. Previous studies indicated that flavonoids have key roles in diverse functions, a major part of which relates to providing defense to various microbes and environmental stress, such as UV light etc. Our previous association analysis showed that some SNPs in *F3H* were associated with SMV resistance [13]. The SMV response association analysis in our study revealed that both resistance and susceptible reactions composite in most clearly separate genetic groups, implying long-time coexistence of these phenotypes. One explanation for the result was that this defense function is critical for plants to live under various biotic and abiotic stresses in nature, but may also have considerable fitness costs as resistance genes (R genes) in plants. Therefore, balancing selection may operate on plant *F3H* genes and maintain different phenotypes in a frequency-dependent manner for a long time.

Materials and Methods

Plant Materials

The seeds of 33 Chinese soybean accessions consisted of 16 wild and 17 cultivated accessions were provided by Germplasm Storage of Chinese National Center for Soybean Improvement (Nanjing Agricultural University, Nanjing, China). These accessions were distributed in 19–49°N and 106–131°E. This sampling strategy was designed to not only cover all the six soybean ecological habitats in China [17] but also take into account the broad spectrum of different levels of seed isoflavone and flavonoids concentration in Chinese soybeans. Plants were grown at Jiangpu Experimental Station of Nanjing Agricultural University (Nanjing, China). The plant materials were listed in Online Resource (Table S1).

Selection of Reference Gene Loci

Based on the facts that *IFS* and *F3H* genes are involved in stress response, we chose three genes interacting with microbes, for example EU450800, the disease resistance gene *Rps1-k-1*; L20310, nodulin (nod-20) gene and D13505, early nodulin related gene, grouping as R1. Meanwhile, fourteen other genes were randomly selected to represent a range of functions [3], classified as R2 (Table S2). Sequences of reference genes were obtained from 31 re-sequencing wild and cultivated soybean genomes [18]. To test whether there is a genetic differentiation between 33 accessions in our study and the 31 re-sequencing ones, four reference genes were sequenced in 33 accessions used in this study and compared with the sequences from the re-sequencing database. The four reference genes include MAT9 gene (M94012), A5A4B3 glycinin gene (AB004062), urate-degrading peroxidase gene (AF089850) and low molecular weight heat shock protein gene (M11317). Three types of statistical tests, (Chi-square statistic, F_{st} and S_{nn}, as described in "Analysis of genetic structure") were applied to detect genetic differentiation between accessions used in our study and the ones from re-sequencing project. The results demonstrated that there was no significant genetic difference between the two populations for the four loci tested (Table S3).

Cloning and Sequencing

Total genomic DNA was extracted from bulk leaf tissue of 8–10 *G. soja* or *G. max* plants as described by Keim et al. [19]. For the two *IFS* genes, we obtained partial promoter region, 5′-UTR, complete CDS, all introns and partial 3′-UTR. For the *F3H2* gene, most CDS and all introns were obtained. For the four reference genes, we used the primers published by Hyten et al. [3]. The primers were listed in Table S4. The process of PCR, cloning and sequencing were described in our previous study [13,14]. Sequences of *IFS* and *F3H* genes were deposited in GenBank under accession number from EU391427 to EU391525. Sequences of the four reference genes have also been deposited in GenBank accession number JQ660375-JQ660504.

Sequence Analysis

The sequencing results were assembled using BioLign software (http://en.bio-soft.net/dna/BioLign.html) and aligned by ClustalX software version 1.83 by manual check [20]. The nucleotide alignments were analyzed using DnaSP version 5.0 [21]. Indels were excluded from all estimates. Haplotype diversity (H_d) was calculated using Equation 8.4 [22], except that n was used instead of $2n$. Nucleotide diversity was estimated by π with Jukes and Cantor correction [23] and by θ from the number of polymorphic segregating (S) sites [24]. The divergences between species were estimated by D_{xy} and divergence at silent sites (K_{sil}) with the Jukes and Cantor correction [22]. The minimum numbers of recombination events (R_M) were estimated [25]. Phylogenetic trees were constructed based on the bootstrap neighbor-joining (NJ) method with a Kimura 2-parameter model by MEGA version 4.0 [26]. Coding sequences (CDSs) of *F3H1* (Glyma02g05450.1) alleles were download and employed as outgroup for soybean *F3H2* phylogenetic study (GenBank accession numbers: AY669324, AY669325, AY595420, AY994154, BT096691 and FJ770474). The stability of internal nodes was assessed by bootstrap analysis with 1,000 replicates.

DnaSP was also used to perform a statistical test of neutrality, Tajima's D test [27]. In order to evaluate the power of selection, we estimated the ratio of nonsynonymous (*Ka*) to synonymous (*Ks*) nucleotide substitutions (*Ka/Ks*) according to Nei and Gojobori's equations [28]. Protein sequences were initially aligned by ClustalX [20], and the resulting amino acid sequence alignments were then used to guide the alignments of nucleotide CDSs using MEGA version 4.0.

Analysis of Genetic Structure

Two distinct classes of test, haplotype-based statistical test (Chi-square statistic) and sequence-based statistical test (F_{st} and S_{nn}), were applied to detect genetic differentiation between wild and cultivated soybean accessions. The F_{st} statistic [29] was calculated by ARLEQUIN version 3.11 software with permuting the data 1,000 times [30]. The nearest-neighbors statistic (S_{nn}) measures how often the "nearest-neighbors" (in sequence space) of sequences and appears to be the most powerful statistic or nearly as powerful as the best statistic method under all conditions examined [31]. The statistical significance of pairwise S_{nn} values was determined by permuting the data 1000 times in DnaSP v 5.0. The genetic differentiation were also estimated by using χ^2 test [22], which can be directly adapted to use with nucleotide variation by treating each distinct haplotypes as an allele.

Supporting Information

Figure S1 Biosynthesis of isoflavonoids and flavonoids in soybean. F3H, flavonone-3-hydroxylase; FLS, Flavonol synthase; FNS, Flavone synthase; IFS, isoflavanone synthase. Hollowed arrows represent multiple or uncertain steps.

Figure S2   Nucleotide polymorphism sites of *IFS1* (a) and *IFS2* (b) genes. The locations of the polymorphism sites were shown in the above line. The polymorphic sites were highlighted by red color.

Figure S3   Phylogenetic trees of the seventeen reference genes constructed by neighbor-joining (NJ) method.

Figure S4   (a) Patial (from start codon to 121bp) alignment of *IFS1* coding sequence, start codon mutant of accession C_HC35 was shown (red line). (b) Patial aliagnment of *IFS2* coding sequence, four indels of 1 or 2bp, two found in accession C_HC16, one in C_HC28 and one in C_HC31, were shown (red arrow).

Table S1   Thirty-three soybean accessions included in this study.

Table S2   GenBank accession numbers, functional description and Gene Ontology for reference genes.

Table S3   Genetic differentiation between soybean accessions in our study and the re-sequencing population.

Table S4   Primer information for *IFS1*, *IFS2* and *F3H2* gene and four reference genes.

Table S5   Nucleotide diversity (%) within and between groups at *F3H2*.

Acknowledgments

We thank Dr. Si-Hai Yang of Nanjing University for his critical comments and helpful discussions. We also thank Jia-Xing Yue of Nanjing University for his help on data analysis.

Author Contributions

Conceived and designed the experiments: DY HC. Performed the experiments: HC JW SC H-LY. Analyzed the data: HC JW. Contributed reagents/materials/analysis tools: HC JW SC H-LY. Wrote the paper: DY HC JW.

References

1. Lee G, Crawford G, Liu L, Sasaki Y, Chen X (2011) Archaeological soybean (*Glycine max*) in East Asia: does size matter? PloS one 6: e26720.
2. Xu D, Abe J, Gai J, Shimamoto Y (2002) Diversity of chloroplast DNA SSRs in wild and cultivated soybeans: evidence for multiple origins of cultivated soybean. Theor Appl Genet 105: 645–653.
3. Hyten DL, Song Q, Zhu Y, Choi IY, Nelson RL, et al. (2006) Impacts of genetic bottlenecks on soybean genome diversity. Proc Natl Acad Sci USA 103: 16666–16671.
4. Yu O, McGonigle B (2005) Metabolic engineering of isoflavone biosynthesis. Adv Agron 86: 147–190.
5. Akashi T, Aoki T, Ayabe S (1999) Cloning and functional expression of a cytochrome P450 cDNA encoding 2-hydroxyisoflavanone synthase involved in biosynthesis of the isoflavonoid skeleton in licorice. Plant Physiol 121: 821–828.
6. Jung W, Yu O, Lau SM, O'Keefe DP, Odell J, et al. (2000) Identification and expression of isoflavone synthase, the key enzyme for biosynthesis of isoflavones in legumes. Nat Biotechnol 18: 208–212.
7. Steele CL, Gijzen M, Qutob D, Dixon RA (1999) Molecular Characterization of the Enzyme Catalyzing the Aryl Migration Reaction of Isoflavonoid Biosynthesis in Soybean. Arch Biochem Biophys 367: 146–150.
8. Springob K, Nakajima J, Yamazaki M, Saito K (2003) Recent advances in the biosynthesis and accumulation of anthocyanins. Nat Prod Rep 20: 288–303.
9. Zabala G, Vodkin LO (2005) The wp mutation of Glycine max carries a gene-fragment-rich transposon of the CACTA superfamily. Plant Cell 17: 2619–2632.
10. Gutierrez-Gonzalez JJ, Wu X, Gillman JD, Lee JD, Zhong R, et al. (2010) Intricate environment-modulated genetic networks control isoflavone accumulation in soybean seeds. BMC Plant Biol 10: 105.
11. Kim HK, Jang YH, Baek IS, Lee JH, Park MJ, et al. (2005) Polymorphism and expression of isoflavone synthase genes from soybean cultivars. Mol Cells 19: 67–73.
12. Aguadé M (2001) Nucleotide sequence variation at two genes of the phenylpropanoid pathway, the *FAH1* and *F3H* genes, in *Arabidopsis thaliana*. Mol Biol Evol 18: 1–9.
13. Cheng H, Yang H, Zhang D, Gai J, Yu D (2010) Polymorphisms of soybean isoflavone synthase and flavanone 3-hydroxylase genes are associated with soybean mosaic virus resistance. Molecular Breeding 25: 13–24.
14. Cheng H, Yu O, Yu D (2008) Polymorphisms of *IFS1* and *IFS2* gene are associated with isoflavone concentrations in soybean seeds. Plant Sci 175: 505–512.
15. Bergelson J, Kreitman M, Stahl EA, Tian D (2001) Evolutionary dynamics of plant R-genes. Science 292: 2281–2285.
16. Shen P, Huang HV (1986) Homologous recombination in Escherichia coli: dependence on substrate length and homology. Genetics 112: 441–457.
17. Wang Y, Gai J (2002) Study on the ecological regions of soybean in China II. Ecological environment and representative varieties. Ying Yong Sheng Tai Xue Bao 13: 71–75.
18. Lam HM, Xu X, Liu X, Chen W, Yang G, et al. (2010) Resequencing of 31 wild and cultivated soybean genomes identifies patterns of genetic diversity and selection. Nat Genet 42: 1053–1059.
19. Keim P, Olson TC, Shoemaker RC (1988) A rapid protocol for isolating soybean DNA. Soybean Genet Newsl 15: 150–152.
20. Thompson JD, Gibson TJ, Plewniak F, Jeanmougin F, Higgins DG (1997) The CLUSTAL_X windows interface: flexible strategies for multiple sequence alignment aided by quality analysis tools. Nucleic Acids Res 25: 4876–4882.
21. Rozas J, Sanchez-DelBarrio JC, Messeguer X, Rozas R (2003) DnaSP, DNA polymorphism analyses by the coalescent and other methods. Bioinformatics 19: 2496–2497.
22. Nei M (1987) Molecular Evolutionary Genetics. New York: Columbia Univ. Press.
23. Lynch M, Crease TJ (1990) The analysis of population survey data on DNA sequence variation. Mol Biol Evol 7: 377–394.
24. Watterson GA (1975) On the number of segregating sites in genetical models without recombination. Theor Popul Biol 7: 256–276.
25. Hudson RR, Kaplan NL (1985) Statistical properties of the number of recombination events in the history of a sample of DNA sequences. Genetics 111: 147–164.
26. Tamura K, Dudley J, Nei M, Kumar S (2007) MEGA4: Molecular Evolutionary Genetics Analysis (MEGA) Software Version 4.0. Mol Biol Evol 24: 1596.
27. Tajima F (1989) The effect of change in population size on DNA polymorphism. Genetics 123: 597–601.
28. Nei M, Gojobori T (1986) Simple methods for estimating the numbers of synonymous and nonsynonymous nucleotide substitutions. Mol Biol Evol 3: 418–426.
29. Weir B (1996) Genetic data analysis; Xu J, editor. Beijing: Publishing House of Agricultural Science.
30. Excoffier L, Laval G, Schneider S (2005) Arlequin (version 3.0): an integrated software package for population genetics data analysis. Evolutionary bioinformatics online 1: 47.
31. Hudson RR (2000) A new statistic for detecting genetic differentiation. Genetics 155: 2011–2014.
32. Zhi HJ, Gai JY, He XH (2005) Study on methods of classification of quantitative resistance to soybean mosaic virus in soybean. Soybean science 24: 5–11.

Influence of Ethnolinguistic Diversity on the Sorghum Genetic Patterns in Subsistence Farming Systems in Eastern Kenya

Vanesse Labeyrie[1]*, Monique Deu[1], Adeline Barnaud[3], Caroline Calatayud[1], Marylène Buiron[1], Peterson Wambugu[2], Stéphanie Manel[4,5], Jean-Christophe Glaszmann[1], Christian Leclerc[1]

1 UMR AGAP, CIRAD, Montpellier, France, 2 National Genebank of Kenya, KARI, Nairobi, Kenya, 3 UMR DIADE, IRD, Montpellier, France, 4 UMR LPED, Université Aix-Marseille/IRD, Marseille, France, 5 UMR AMAP, CIRAD, Montpellier, France

Abstract

Understanding the effects of actions undertaken by human societies on crop evolution processes is a major challenge for the conservation of genetic resources. This study investigated the mechanisms whereby social boundaries associated with patterns of ethnolinguistic diversity have influenced the on-farm distribution of sorghum diversity. Social boundaries limit the diffusion of planting material, practices and knowledge, thus shaping crop diversity *in situ*. To assess the effect of social boundaries, this study was conducted in the contact zone between the Chuka, Mbeere and Tharaka ethnolinguistic groups in eastern Kenya. Sorghum varieties were inventoried and samples collected in 130 households. In all, 297 individual plants derived from seeds collected under sixteen variety names were characterized using a set of 18 SSR molecular markers and 15 morphological descriptors. The genetic structure was investigated using both a Bayesian assignment method and distance-based clustering. Principal Coordinates Analysis was used to describe the structure of the morphological diversity of the panicles. The distribution of the varieties and the main genetic clusters across ethnolinguistic groups was described using a non-parametric MANOVA and pairwise Fisher tests. The spatial distribution of landrace names and the overall genetic spatial patterns were significantly correlated with ethnolinguistic partition. However, the genetic structure inferred from molecular makers did not discriminate the short-cycle landraces despite their morphological distinctness. The cases of two improved varieties highlighted possible fates of improved materials. The most recent one was often given the name of local landraces. The second one, that was introduced a dozen years ago, displays traces of admixture with local landraces with differential intensity among ethnic groups. The patterns of congruence or discordance between the nomenclature of farmers' varieties and the structure of both genetic and morphological diversity highlight the effects of the social organization of communities on the diffusion of seed, practices, and variety nomenclature.

Editor: John P. Hart, New York State Museum, United States of America

Funding: This study was supported by Agropolis Fondation in the framework of the ARCAD project No.0900-001. V. Labeyrie was supported by Agropolis Fondation through a Ph.D. grant and received financial support from CIRAD for this study. The funders had no role in study design, data collection and analysis, decision to publish, or preparation of the manuscript.

Competing Interests: The authors have declared that no competing interests exist.

* E-mail: vanesse.labeyrie@gmail.com

Introduction

Identifying factors involved in crop evolution is of great importance for genetic resource conservation and crop improvement. Crop genetic diversity patterns result from selection, migration and genetic drift processes which are strongly influenced by human action. Recent studies combining linguistic, archeological and genetic data have unraveled the past domestication and diversification processes of crops such as banana [1] and sweet-potatoes [2], on a large time-space scale, by linking global diversity patterns to human migrations. However, the evolution of crops is still ongoing in smallholder farming systems under the pressure of agro-ecological conditions and farmers' management practices [3]. The study of these processes at the community scale is complementary to large time-space approaches and contributes to the general understanding of the *in situ* genesis of crop genetic patterns.

Social boundaries contribute to the evolution of crop populations both directly, by determining seed flows, and indirectly, by inducing the divergence of seed selection practices [4]. Previous studies notably showed that the ethnic organization of farming communities plays an important role in differentiating the domesticated populations of allogamous crops [5], vegetatively-propagated crops [6] and animals [7].

Sorghum (*Sorghum bicolor* L. Moench) is an annual cereal extensively cultivated in smallholder farming systems because of its ability to grow under harsh climatic conditions. De Wet and Huckabay [8] and Harlan et al. [9] suggested that the spatial distribution of sorghum botanical races in Africa was related to that of the ethnic groups, but this hypothesis was not further tested. In a study undertaken in Niger, Deu et al. [10] suggested that human ethnic diversity has probably a greater impact on sorghum diversity than recent environmental constraints. However, the authors were not able to assess this hypothesis as the spatial localization of the different ethnic groups in Niger corresponded to different agro-ecological regions. Thus, deciphering how the social organization of farmers affects the structure of sorghum diversity remains a challenge.

This article addresses the role of social boundaries in sorghum evolution and diversification processes. It set out to identify the mechanisms whereby social boundaries, associated with ethnolinguistic diversity patterns, shape sorghum genetic diversity on-farm. To study only the main effect of social boundaries, this study focused on an ethnolinguistic contact zone where both geographical distance between ethnic groups and agro-ecological variability were limited. If social boundaries do not limit seed-mediated gene flows and the diffusion of selection practices, then no relation should be observed between ethnic diversity patterns and both the genetic and morphological structure of sorghum diversity. Otherwise, it would reflect the impact of social boundaries on the evolutionary mechanisms that shape sorghum diversity *in situ*.

Farmers' varieties are relevant units for studying on-farm crop diversity as they are consciously defined and named by farmers for management, selection, seed exchanges and knowledge transmission purposes [11]. Farmer's nomenclature and taxonomy of crop varieties is a marker of knowledge diffusion and exchanges across communities [12], while the distribution of the genetic and morphological diversity of crop populations reflects gene flows and selection forces [5]. This study thus used molecular markers to estimate genetic diversity and compared the spatial distribution of varieties with genetic spatial patterns according to ethnic groups. These patterns were then discussed regarding the congruence between farmer's varieties and the structure of their genetic and morphological diversity. Combining these three approaches enabled us to investigate the influence of social boundaries on the evolutionary mechanisms that shape sorghum diversity *in situ*.

Clarifying the effect of social boundaries on crop evolutionary mechanisms has important applications for crop genetic resource collection, characterization and conservation. This study hence contributes to increasing the overall understanding of on-farm crop diversification processes. By highlighting the overall role of societies in shaping crop diversity, it stresses the relevance of multidisciplinary approaches for crop genetic diversity studies.

Materials and Methods

Ethics statement

This was a collaborative study between CIRAD and KARI-National Genebank of Kenya. KARI has the national mandate for the collection and conservation of all plant genetic resources and documentation of all accompanying information. Under this framework and mandate, the study was mounted and all laid down institutional and administrative procedures were carefully followed prior to undertaking the study. Based on the aforementioned mandate given to KARI, no specific permission was required to undertake the study. Though KARI does not have a body designated as ethical review board, it has equivalent committees and administrative organs which review proposed research activities before granting approval. Research clearance was therefore sought from these organs at all levels including the institutional legal office. Local government administrative as well as agricultural extension officers were informed of the study and kept updated of the activities.

During the survey, the mandate given to KARI as well as the importance of the study, both nationally and globally, was explained to the farmers and concurrence was sought before undertaking the study activities. According to KARI's procedures governing genetic resources collection and documentation, prior informed consent was obtained verbally and not recorded, all with the understanding that the process would only involve collection of genetic resources and no sensitive traditional knowledge. Where such consent was not granted, the germplasm collectors stopped any more activities in that particular household. In each household, we interacted with the female household head. Upon granting consent, they were interviewed mainly on their ethnicity and the sorghum varieties they grew. The survey was conducted by the authors among them V.L, A.B, P.W, and C.L with questions being translated by a local field assistant. We confirm that sorghum, the studied crop, is neither endangered nor protected.

Study site: Agro-ecological conditions and ethnic organization

This study was conducted on the eastern slope of Mount Kenya (0°24'27.88"S, 37°46'35.59"E), in an ethnolinguistic contact zone between Chuka, Tharaka and Mbeere groups (Figure 1). The three ethnolinguistic groups (hereafter ethnic) live within the same agro-ecological zone, as defined by Jaetzold et al. [13]. The study site was 15 km-square, and the elevation ranged from 810 to 946 m above sea level, so rainfall and temperature variability was limited. The mean temperature on the area ranges between 21.7°C and 23.9°C. The mean rainfall is about 700–800 mm per year, distributed across two rainy seasons with the Long Rains occurring from March to May and the Short Rains from October to December [14]. Soil characteristics are homogeneous in the area occupied by the three ethnic groups, corresponding to well drained Ferralsols, with a loamy-sand texture and moderate fertility [13].

The three ethnic groups, Chuka, Tharaka and Mbeere, migrated to the study area by the end of the 19th century, either because of a population increase or because of recurrent drought [15]. Social boundaries exist between Chuka, Tharaka and Mbeere groups as revealed by their distinct ethnic identity, and their current cultural and linguistic differences [16,17]. The Mbeere are closely related to the Embu group [18], while the Chuka and Tharaka are related to the Meru group. The Mbeere and Chuka had conflictual mutual relationships in the past [19], while the Chuka and Tharaka maintain strong social ties and consider they are kin [20]. Intermarriage is usual between the Chuka and Tharaka, while it is very uncommon between the Mbeere and Chuka or Tharaka (unpublished data). Men usually settle near their father's compound once they get married. The residence is thus patrilocal [16]. The three ethnic groups present a non-random spatial distribution. The Mbeere households are located in the southern part of the study area, the Tharaka mostly on the north-eastern side, and the Chuka on the north-western side (Figure 1). Consistently with the social relationships between groups depicted above, a clear spatial boundary was found between the Mbeere and both the Chuka and Tharaka, while the Chuka and Tharaka appeared to be spatially more mixed.

The three ethnic groups manage low-input cropping systems that harbor high specific and infra-specific crop diversity. Cropping systems are based on cereals and legumes that are usually intercropped. Sorghum (*Sorghum bicolor*), cowpeas (*Vigna unguiculata*), maize (*Zea mays*), mungo bean (*Vigna radiata*) and pearl millet (*Pennisetum glaucum*) are the main crop species grown in the area. Sowing is done either by hand-dibbling or by drilling, while plowing is done with animals. The different sorghum varieties are either grown in separate plots or mixed together within farmers' fields. Improved varieties, mainly disseminated by the extension services of the Kenyan Ministry of Agriculture, have also been adopted by the farmers. They are cultivated together in the same

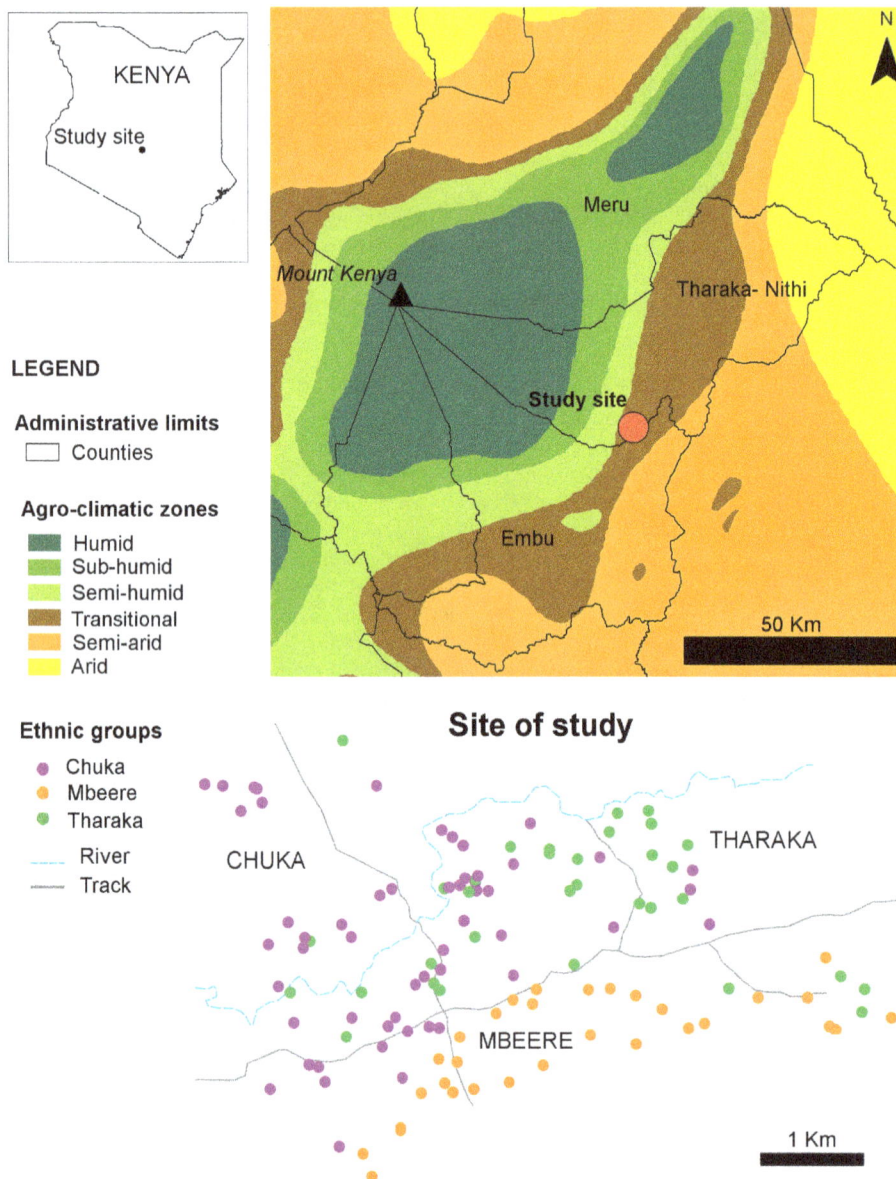

Figure 1. Study site location. Map of the eastern side of Mount Kenya and location of the farms where sorghum samples were collected (colors correspond to the ethnic identity of the male house-head).

field with the local varieties (or landraces). Farmers distinguish between short-cycle varieties that can be grown either from October to January or from March to June, and long-cycle varieties that are subjected to the ratooning practice [21] (Figure S1). These long-cycle varieties are sown in October, the vegetative part being cut before the grains are mature to stimulate regrowth from basal buds, and panicles are finally harvested in July.

Data collection

Sorghum inventory and germplasm collection. The field work consisted of two stages. A preliminary survey was carried out to estimate the frequency of varieties in the three ethnic groups. The strategy for on-farm germplasm collection was then based on that estimation of diversity, as it aimed at representing the diversity and frequencies of each variety in each ethnic group.

The preliminary inventory survey was conducted in both January (Short Rains cropping season) and June 2011 (Long Rains cropping season), just before harvesting and prior to germplasm collection. The inventory of sorghum varieties was based on the local names as reported by women farmers who were in charge of sorghum selection in each of the 124 households surveyed. Indeed, grain crop farming comprising seed sowing, harvesting, selection and trading is ensured by women ([16], personal observation). The ethnolinguistic identity of male house-heads or single women was also recorded, women becoming members of the family of their husband when they get married in this patrilinear society.

Sorghum panicles were then collected from a total of 130 households selected randomly. In 22 of these households, panicles were collected in both January and July, while the rest of the households were visited only in January (34 households) or in July (74 households). Half of these 130 households were visited during the preliminary survey described above. They represented about half of the total number of households in the area, hence insuring a good representativeness. 60 households belonged to the Chuka ethnic group, 35 to the Mbeere and 35 to the Tharaka. In order to be representative of the sorghum population of each ethnic group, all the varieties grown in each household were collected, except a highly dominant variety of improved origin (*Kaguru*). As this variety was much more abundant than the others, we limited the number of samples collected. We thus sampled *Kaguru* variety randomly in a maximum of 19 households per ethnic group. One or two individual panicles of each variety were collected in each household cultivating it. The mean number of varieties collected per household was 1.5 (min: 1, max: 6). It was similar across ethnic groups, as well as the mean number of panicles of each variety sampled per household (Table S1). The fraction of households where each variety was collected for the study of genetic diversity was correlated to the fraction of households where each variety was previously inventoried (Linear regression R^2: 0.77, Figure S2). In all, 290 samples were collected on-farm after harvest, each consisting of a single panicle. About 47% of the individual plants were sampled from the Chuka ethnic group, 30% from the Tharaka and 23% from the Mbeere. Information concerning the names, the origin (local or improved) and the cycle length of each sampled panicle was recorded from women house-heads, and we recorded the geographic coordinates of each household using a global positioning system (GPS).

DNA extraction and SSR genotyping. Seeds from the 290 panicles collected on-farm were sown in an experimental field *in-situ*, and the leaves of one sibling randomly chosen for each mother plant were collected and stored on silicagel. Leaves from seven individuals grown from certified seeds of the improved varieties *Serendo* and *Gadam* were also collected as controls. In total 297 individual plants were thus used for the genetic diversity study.

Twenty-two pairs of primers were selected for their high polymorphism in central Kenya (unpublished data) and West Africa [10]; twenty of them were part of a set of reference microsatellite markers proposed by Billot and colleagues ([22], http://sat.cirad.fr/sat/sorghum_SSR_kit/). Loci were distributed over the 10 chromosomes. DNA was extracted from dried leaves and the polymerase chain reaction amplifications were done following the procedure described previously [10,22]. The fluorescent dye–labeled PCR products from differentially labeled primers and with non-overlapping size were pooled and subjected together to capillary electrophoresis using a 24-capillary 3500xL System (Applied Biosystems). GeneMapper v 4.1 (Applied Biosystems) was used for genotype scoring. GeneScan 600 LIZ Size Standard v2.0 was added to each well, and three control samples were used to facilitate allele scoring [22]. Genotyping was done at the Montpellier Languedoc-Roussillon Genopole platform located on the CIRAD campus in Montpellier (France).

Four markers presenting either a high number of missing data, or low polymorphism (at a 99% threshold) were discarded from the analysis, so eighteen markers were kept, covering 9 chromosomes out of 10. The percentage of missing data for the 18 markers kept was 1%.Table S2 provides a list of these 18 markers and their description.

Panicle morphological characterization. Fifteen qualitative morphological traits were measured on the panicles of the 297 individuals that were genotyped (Table S3). Eight morphological descriptors were selected from the IPGRI descriptors [23] and were completed by seven additional descriptors for seeds and glumes characteristics that showed variability on the sorghum collected in our study area. Descriptors covered the characteristics of the whole panicle (panicle shape), seeds (color, presence of sub-coat, pericarp thickness, shape, endosperm texture and shattering)

and glumes (color, adherence, covering, opening, texture, hairiness, awning and transversal wrinkle). Only qualitative traits were kept for these analyses because they are stable characteristics on which farmers base their nomenclature and classification [24]. Multiple characterizations of randomly sampled individuals enabled to check for morphological trait scoring consistency.

Data analysis

Comparing sorghum assemblages between ethnic groups. We characterized each household by its sorghum assemblage, which is the panel of co-occurring sorghum varieties that are cultivated by the household. The differentiation of sorghum assemblages across ethnic groups was tested using a non-parametric Multivariate Analysis of Variance (perMANOVA, [25]). The PerMANOVA was implemented under the *adonis* function in the R package *vegan* [26]. The presence/absence matrix for sorghum varieties in each household was transformed into a distance matrix using the Bray-Curtis index [27]. The *adonis* function partitions the distance matrix according to grouping factors (ethnic groups) and compares the sum of squared distances within groups (which is the sum of squared distances from individual replicates to their group centroid) and between groups (which is the sum of squared distances from group centroids to the overall centroid). A pseudo F-ratio is then computed and compared to its distribution under the null hypothesis simulated using 4000 random permutations of the raw data. Pairwise Fisher exact tests implemented in the R package *fmsb* [28] were then used to compare the occurrence frequencies of the most frequent varieties across the Chuka, Mbeere and Tharaka ethnic groups. The calculation of p-values was corrected for multiple comparisons using the False Discovery Rate (FDR) procedure [29] implemented in the *p.adjust* function.

Genetic diversity and genetic structure of sorghum populations. Genetic diversity within sampling populations. The genetic diversity of sorghum populations sampled in each ethnic group was assessed using several indexes. The observed number of alleles and the observed heterozygosity were calculated using GENETIX 4.05.2 software [30]. The allelic richness corrected for sample size [31], the unbiased gene diversity (expected heterozygosity) corrected for small sample size [32], and the F_{IS} [33] of multi-locus genotypes were estimated using the procedures implemented in FSTAT 2.9.3.2 software [34]. These indexes were compared among ethnic groups using paired Pairwise Wilcoxon tests with False Discovery Rate (FDR) correction implemented in R (package *stats*, *pairwise.wilcox.test* function).

Genetic structure assuming sampling populations. Pairwise F_{ST} [33] were computed among the sorghum populations collected in the three ethnic groups. The significance of the differences was assessed using a permutation test (3000 permutations) and corrected using a Bonferroni procedure [35]. A multilocus G-test of differentiation, known to be accurate for measuring the genetic differentiation between populations with unbalanced sizes [36], was used to test the genetic differentiation between the populations sampled in each ethnic group (10000 permutations). Calculations were carried out using FSTAT 2.9.3.2. Pairwise G-tests implemented in GENEPOP 4.2 [37] were used to estimate the genotypic differentiation among pairs of populations and p-values were corrected for multiple tests using FDR correction (*p.adjust* function in the R package *stats*).

Analysis at individual level. Two complementary approaches, Bayesian clustering and Neighbor-Joining tree, were used to assess the genetic structure without defining a-priori populations. First, the genetic structure of sorghum populations was characterized using the Bayesian clustering algorithm implemented in STRUCTURE 2.3.3 software [38] and run on the Bioportal server (http://www.bioportal.uio.no). The admixture model with correlated allele frequencies was used, assuming that the genome of each individual resulted from the mixture of K ancestral populations. The estimated proportions of each individual's genotype originating from each of the K ancestral populations (q) was calculated for K ranging from 2 to 10 ancestral populations (or clusters), with twenty runs for each K value. The burn-in period was set at 500 000 and 1 000 000 iterations were performed. The criterion suggested by Evanno et al. [39], based on the rate of change in the log probability of data between successive K values, was used to determine the most likely number of clusters (K). Second, a Neighbor-Joining tree [40] was built from a simple matching genetic dissimilarity index [41] using Darwin V5 software [42]. The results of both the Bayesian clustering and Neighbor-Joining methods were then compared to check for the consistency of the clusters. This led to what we refer to as an MMb (molecular-marker-based) classification scheme.

For further analysis, individuals whose estimated proportion of genome originating from one population (q, hereafter admixture coefficient) was below a 0.8 threshold were considered as resulting from admixture between the populations. Individuals whose q value was equal to or above 0.8 for a population were assigned to that population (hereafter cluster). To explain the MMb genetic structure, the assignment of individuals to clusters thus defined was crossed with information concerning their origin and cycle length as reported by farmers during the collection of samples *in situ*. The occurrence frequencies of each MMb genetic cluster were then compared across ethnic groups using Pearson's Chi-squared test, and pairwise Fisher exact tests with False Discovery Rate (FDR) correction for multiple comparisons.

To test a potential isolation-by-distance effect in cultivated sorghum, we applied Mantel test between pairwise genetic distances and geographical distances. The matrix of geographical distances among individuals was computed. The kinship coefficient of Loiselle et al. [43] was computed using SPAGeDI software [44] for each pair of cultivated sorghum individuals, producing a matrix of individual pairwise genetic distance. A stratified Mantel test implemented in the R package Vegan was used to test the significance of the correlation between the logarithm of the pairwise geographical distances [45] and sorghum individuals' pairwise genetic relatedness. 4000 permutations of the locations of samples were done within the genetic clusters previously identified using STRUCTURE software (stratified test), as recommended by Meirmans [46] for populations presenting a strong genetic structure.

Morphological structure of sorghum populations

To describe the structure of individual panicle morphological diversity, a dissimilarity matrix was computed on the basis of the 15 morphological traits coded through a total of 43 modalities using the simple matching index. The morphological similarity between individuals was then assessed using a Principal Coordinates Analysis (PCoA) using the R package *ade4*.

Results

Differences in variety assemblages across ethnic groups

On the basis of their local names, seventeen different varieties were inventoried among the 124 households visited during both the January and June surveys. 14 different varieties were respectively inventoried in the Chuka and Tharaka groups, and 10 in the Mbeere group, out of which 9 were shared by the three

ethnic groups. The mean number of varieties inventoried in both cropping seasons per household was similar across ethnic groups (2.77, SE: 0.17 for the Chuka, 2.65, SE: 0.17 for the Mbeere, 3.02, SE: 0.21 for the Tharaka). The most frequent variety was *Kaguru*, (76% of the households), followed by *Gadam* (48% of the households), both of which are improved varieties. *Ngirigacha*, *Mugeta*, *Mbura imwe*, *Muruge mbura ciiri*, and *Muruge mbura imwe* were the most frequent local varieties (landraces) (Figure 2).

The non-parametric perMANOVA showed that sorghum variety assemblages differed significantly between ethnic groups (Table S4), even though the ethnic partition explained a limited part of variability (pseudo-$F_{2,121}$ = 4.971, p-value = 0.0002, R^2 = 0.076). Pairwise Fisher exact tests confirmed that the frequency of three out of the five most frequent landraces differed significantly between ethnic groups, while the frequency of improved varieties (*Gadam*, *Kaguru* and *Serendo*) did not differ significantly between ethnic groups. *Muruge mbura imwe* and *Mugeta* were significantly less frequent in the Mbeere group than in the Chuka and Tharaka groups while *Ngirigacha* was significantly more frequent in the Mbeere group.

Genetic and morphological structure of cultivated sorghum

The most likely number of populations (K) identified by STRUCTURE was $K = 4$. Indeed, the log-probability of data increased up to $K = 4$, where it reached a plateau. This was congruent with Evanno's ΔK curve which presented a clear peak for $K = 4$. The populations (clusters) inferred by STRUCTURE for $K = 4$ (Figure 3.A) corresponded to distinct groups on the Neighbor-Joining tree (Figure 4.A). Cluster A and C were distinct and showed higher genetic uniformity than cluster B and D. Most of the individuals sampled (88%) showed an admixture coefficient (q) above or equal to $q = 0.8$, and they were thus assigned to the corresponding cluster. The remaining 12% of the individuals were considered to result from admixture between clusters.

The MMb genetic structure was found to be strongly related to the improvement status of the germplasm – improved varieties or local landraces, and by differences in growth-cycle length (Figure 3.B). Individuals assigned to the uniform clusters A and C were mostly improved varieties introduced by the extension services, while individuals assigned to the broader clusters B and D were mainly classified by farmers as local landraces. Moreover, almost all individuals assigned to cluster D were identified by farmers as long-cycle varieties (ratoon) while those individuals assigned to clusters A, B and C were mainly identified as short-cycle varieties.

Despite this global coherence, the characteristics of varieties reported by farmers showed some divergence from the MMb genetic classification. Twenty-two percent (22%) of the individual plants that were identified as long-cycle landraces by farmers during the collection were assigned to cluster B by STRUCTURE (Figure 3.B). A substantial proportion of individuals identified by farmers as short-cycle landraces were assigned to clusters A (13%) or C (10%). Indeed, young farmers may consider as local the varieties that were introduced a long time ago, perhaps before they began farming. Conversely, 14% of the individuals identified by farmers as improved varieties were assigned to cluster B.

The morphological diversity was summarized by the PCoA (Figure 4.B). The two first axes accounted for 29 and 13% of the variation, respectively. Axis 1 isolated a clear group on its positive side (II), corresponding to the major share of individuals assigned to MMb cluster D, while the rest of the individuals were broadly distributed along axes 1 and 2. Individuals assigned to MMb clusters A and C displayed narrow distributions indicating uniform morphological types, which is consistent with their improved

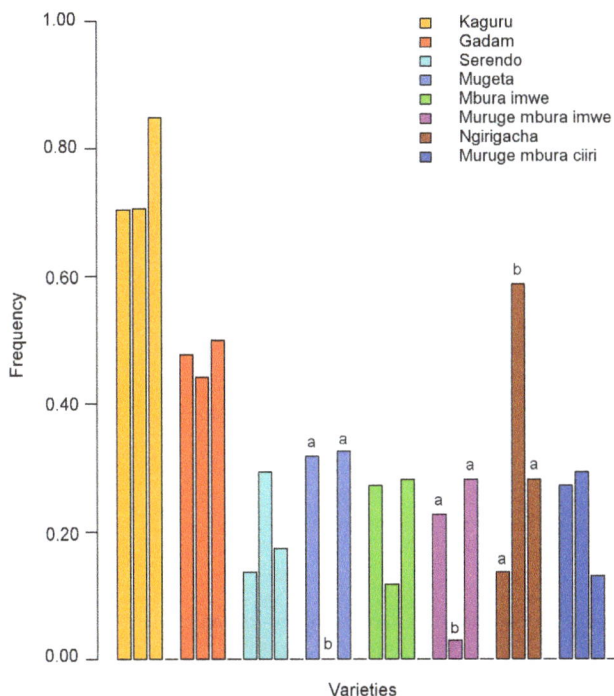

Figure 2. Frequency of the eight major varieties in each ethnic group. The vertical axis displays the percentage of farms where each variety was cultivated. Ethnic groups are present in the following order for each variety: Chuka, Mbeere, Tharaka. The letters (a, b) on top of the bars indicate the statistical significance of differences (Fisher test) at a 5% level after correction for multiple testing (FDR). For a given variety, ethnic groups with the same letter did not present significantly different frequencies.

origin and recent introduction. Individuals assigned to cluster C formed a distinct morphological group (I), discriminated on the third axis of the PCoA (expressing 10.7% of the total variation, data not shown). Individuals assigned to MMb cluster B displayed

a broad distribution, reflecting high variability and continuous distribution across diverse morphotypes. It is noteworthy that some of the individuals assigned to MMb genetic clusters A and B displayed morphological similarity (Figure 4.B.), which may induce possible confusion in naming the recent improved variety and the local landraces (homonymy). Nevertheless, part of the individuals assigned to the MMb cluster B clustered in a separate morphological groups (III).

Genetic differentiation of sorghum populations across ethnic groups

Various indexes were used to characterize the diversity displayed within each ethnic group (Table 1). The unbiased gene diversity estimates (H_e) of Chuka and Tharaka sorghum populations were significantly higher than that of the Mbeere (Wilcoxon test: p-value < 0.01). Similar results were found for the unbiased allelic richness. F_{IS} was very high in the three groups, yet it was significantly lower in the Chuka population as compared to those of both the Tharaka and the Mbeere (Wilcoxon test: p-value < 0.05 for both pairwise comparisons), in relation with the higher heterozygosity found within the Chuka sorghum population (0.033) compared to the other two populations (0.022 for the Mbeere and 0.023 for the Tharaka).

An exact G-test of genetic differentiation of sorghum across ethnic groups was significant (p-value = 0.0205). The differentiation was clearer (G-test p-value = 0.0026) when removing from the analysis the individuals assigned to cluster A, derived from the recent introduction of the *Gadam* improved variety. The Pairwise G-tests showed that genetic differentiation was highly significant between the sorghum populations of the three groups, being highest between the Chuka and both the Tharaka (p-value < 0.0001) and Mbeere (p-value = 0.0002) populations and lowest between the Tharaka and Mbeere populations (p-value = 0.0083). The F_{ST} values between the sorghum populations of the three ethnic groups were low: 0.027 between the Chuka and Mbeere sorghum populations and 0.019 between the Chuka and Tharaka populations, both significant; and non significant between the Mbeere and Tharaka populations (F_{ST} = 0.010).

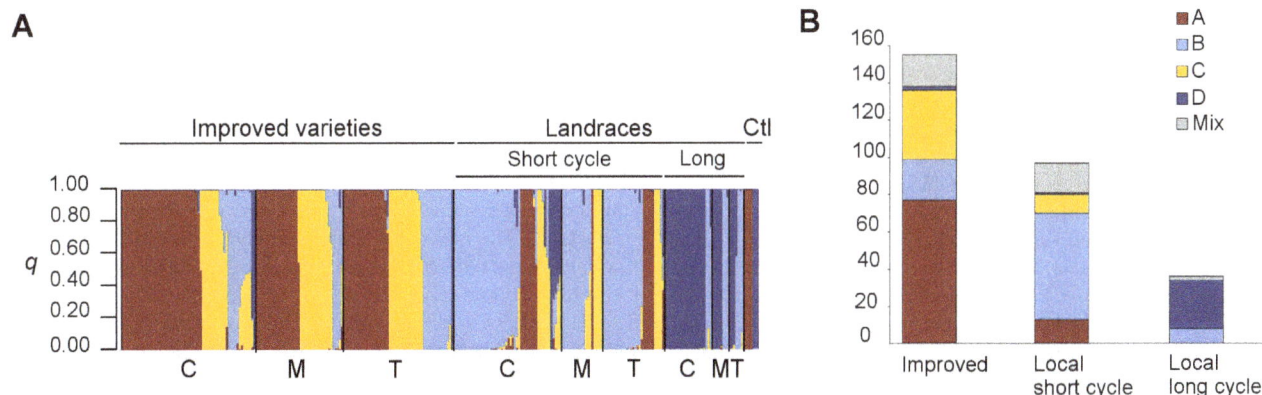

Figure 3. Genetic structure of the sorghum cultivated on the area of study. (A) Cluster assignment of 297 sorghum individuals estimated using STRUCTURE for $K = 4$. The genome of each individual is represented by a vertical line, which is partitioned into K colored segments that represent the admixture coefficient (q), i.e the estimated proportion of membership of its genome in each of the K clusters (Red: cluster A, light blue: cluster B, yellow: cluster C, dark blue: cluster D). Thick black lines separate the individuals identified by farmers as improved varieties, short-cycle landraces or long-cycle landraces, and control individuals (Ctrl), as labeled above the figure. Thin black lines separate individuals sampled in the different ethnic groups (Chuka: C, Mbeere: M, Tharaka: T, as labeled below the figure. The figure shown is based on the highest probability run at $K = 4$. (B) Number of individuals classified according to their origin and cycle length (farmers' information) assigned to each MMb genetic cluster. The vertical axis indicates the number of individuals assigned to each cluster. Individuals were assigned to a cluster when their estimated admixture coefficient (q) for this cluster was equal to or over 0.8. Admixed individuals are represented in gray.

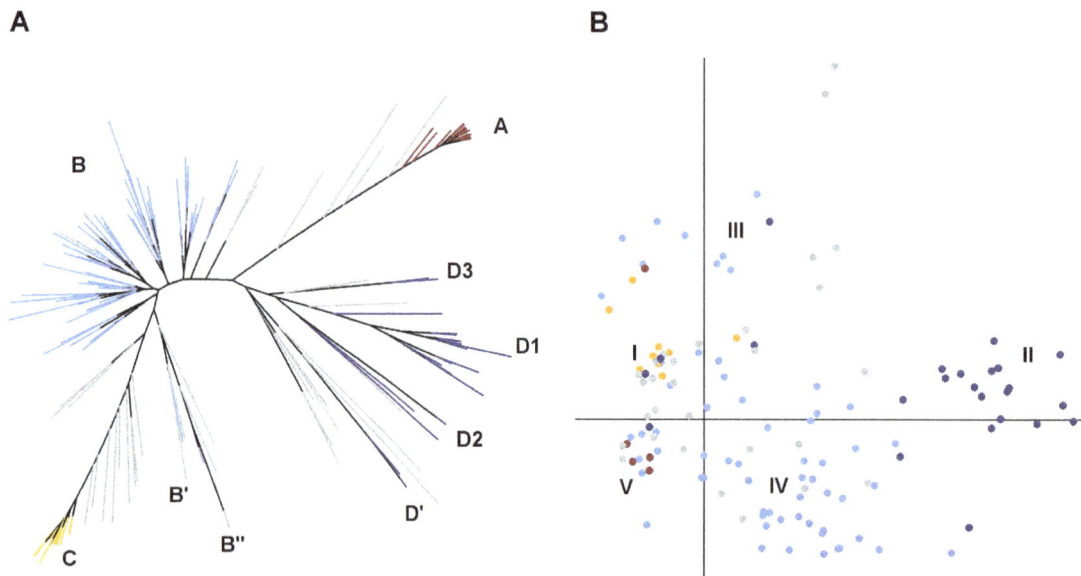

Figure 4. Genetic and morphological structure of the sorghum cultivated on the study area. (A) Neighbor-Joining tree based on 18 SSRs among sorghum plants using the simple matching index. Genetic clusters inferred by STRUCTURE are displayed using different colors (Cluster A: red, B: light blue; C: yellow, D: dark blue). Sub-clusters are identified by letters followed by a number. (B) Plot of the two first axes of the Principal Coordinates Analysis (PCoA) based on 15 panicle morphological traits using the simple matching index. The first axis (x) expresses 29.3% of the total variation and the second axis 13.2%. The main morphological groups are indicated by roman numerals and the MMb genetic assignment of individuals for $K=4$ is displayed with the same colors as in figure 4.A.

No significant relationship was found between the genetic relatedness of individuals and their geographical distance. The partial Mantel test was not significant ($r = -0.38$, p-value $= 0.130$), and the correlogram did not display any significant spatial structure. Nevertheless, the spatial distribution of the four MMb genetic clusters was not uniform (Figure 5.C) and they were not evenly distributed across the three ethnic groups (Table 2). Pearson's Chi-squared test led to rejecting independence between the genetic clusters and the ethnic groups (p-value $= 0.003$).

Correspondence between the genetic structure and farmers' variety names

The MMb cluster A was clearly separated from the others, as illustrated by the Neighbor-Joining tree. It included the four control individuals stemming from certified seeds of the *Gadam* improved variety, which has been disseminated in the area since 2009. Most of the other individuals assigned to cluster A were identified by farmers as *Gadam* (Chuka: 50%, Mbeere: 71%, Tharaka: 48%), confirming the cluster A – *Gadam* correspondence. Yet cluster A also included 46% of varieties collected under other names, mainly *Ngirigacha* (Chuka: 29%, Mbeere: 24%, Tharaka: 28%) and *Mbura-imwe* (Chuka: 11%, Tharaka: 16%). As a result, cluster A was distributed throughout the study area and its spatial distribution appeared more uniform than that of the individuals designated by farmers as *Gadam* (Figure 5.A & C).

The MMb cluster C was also clearly separated from the others, yet with an array of individuals that appeared as intermediates (along the branch of the Neighbor-Joining tree). The major share of the individuals assigned to cluster C was identified by farmers as an improved variety called *Kaguru*, which was introduced in the area about ten years ago (Chuka: 83%, Mbeere: 94%, Tharaka: 100%). *Kaguru* individuals originated uniformly from the study area (Figure 5.A) and in the three ethnic groups (Figure S1), but were less frequent in the Chuka area (Figure 5.C). The proportion of the Chuka sorghum individuals assigned to cluster C was significantly

smaller (9%) as compared to the Tharaka (21%) population (Fisher test: p-value $= 0.023$), and the Mbeere (25%) population (Fisher test: p-value $= 0.009$). Half of the individuals (52%) collected under the name *Kaguru* in the Chuka farms were admixed, while this proportion was significantly lower in the Mbeere farms (15%, Pairwise Fisher test p-value: 0.0300) and in the Tharaka farms (10%, p-value $= 0.0190$). Accordingly, the genetic diversity parameter estimates calculated for the *Kaguru* individuals collected in the Chuka farms were significantly higher than those for the Mbeere and Tharaka farms (Table 3). Altogether, these observations suggest that more admixture occurred between the *Kaguru* population and local landraces within the Chuka cropping systems than within the Tharaka and Mbeere systems.

The MMb cluster D appears clearly separated but rather heterogeneous on the Neighbor-Joining tree. On a morphological basis, these varieties mostly fall in a clearly distinct group. Most individuals assigned to cluster D were identified as long-cycle landraces by farmers (*Muruge mbura ciiri, Mugana, Muthigo, Mucuri, Kathirigwa*) and a few as short-cycle improved varieties (*Serendo* and *Musalama*). The latter individuals identified as improved varieties, both collected on-farm and stemming from certified seeds, formed a distinct genetic sub-group D' on the Neighbor-Joining tree and STRUCTURE confirmed these results for $K=5$. The rest of the individuals assigned to cluster D were distributed across three major sub-clusters (Figure 4.A). Most *Muruge mbura ciiri* individuals clustered together in a separate branch on the Neighbor-Joining tree (D1). *Mugana* and *Kathirigwa* formed another branch (D2), and *Mucuri* a third one (D3). Hence, there was a clear correspondence between the farmers' nomenclature and the genetic structure of individuals assigned to MMb cluster D, as well as with the structure of panicle morphological diversity (Figure S4). Cluster D was mainly observed in the Chuka area, as seen on Figure 5.C and confirmed with pairwise Fisher tests (p-value < 0.05). Interestingly, the few Tharaka households where we collected individuals assigned to cluster D were located in the Chuka area. Moreover,

Table 1. Summary of the genetic polymorphism indexes of sorghum individuals sampled in each ethnic group.

Ethnic group	N_i	N_{hh}	$N_{Al.}$	$R_{Al.}$	H_e	H_o	F_{IS}
Chuka	135	60	6.8	6.1[a]	0.590[a]	0.033	0.943[a]
Mbeere	68	35	4.7	4.6[b]	0.544[b]	0.022	0.959[b]
Tharaka	87	35	6.1	5.9[a]	0.569[a]	0.023	0.961[b]
Total	290	130	7.7	7.7	0.574	0.028	0.952

N_i: number of samples, N_{hh}: number of households, $N_{Al.}$: Mean number of observed alleles over the 18 loci, $R_{Al.}$: unbiased allelic richness corrected for sample size, H_e: unbiased gene diversity, H_o: observed heterozygosity, F_{IS}: fixation index. The letters (a, b) next to the $R_{Al.}$, H_e and F_{IS} values indicate the statistical significance of their differences between ethnic groups (Wilcoxon test) at a 5% level after correction for multiple testing (FDR). For a given index, ethnic groups with the same letter did not present significant differences.

one household located on the eastern side presented several individuals assigned to cluster D, but it was a Chuka household settled in the Tharaka area (Figure 5).

The MMb cluster B is both central and diverse on the basis of molecular markers as well as morphological traits. The individuals assigned to cluster B were mainly identified as local landraces bearing various local names, whose occurrence frequency differed across ethnic groups (Table 4). Most of those collected in the Chuka and Tharaka farms were named *Muruge mbura imwe*, *Mugeta* and *Mbura imwe* while no or very few individuals collected in the Mbeere farms were named as such. Moreover, most of those collected in the Mbeere farms were named *Ngirigacha* (61%), while fewer individuals bore that name in the Chuka (8%) and Tharaka (18%) populations. Cluster B accounted for a uniformly large share among the farmers of the Tharaka (39%), the Chuka (27%) and the Mbeere (27%) ethnic groups. It showed little internal substructure with no clear correspondence to farmers' varieties, and a morphological differentiation between *Muruge mbura imwe* and *Mugeta* (Figure S4). As the only individuals with peculiar features, four individuals assigned to MMb cluster B for $K = 4$ formed a separate branch on the genetic Neighbor-Joining tree (B') and their difference was confirmed by STRUCTURE for $K = 5$. It could be explained by their foreign origin, as farmers reported purchasing these seeds at a lowland market. A fifth individual assigned to cluster B for $K = 4$ formed a long branch (B'') indicative of a marked genetic differentiation. It was identified as *Muthigo wa mwimbi* which means that it was introduced from another ethnic group (*Mwimbi*).

Discussion

Our study showed that in a uniform agro-ecological environment, social boundaries associated with ethnolinguistic diversity patterns have impacted the distribution of sorghum varieties and their genetic spatial patterns. If seeds, knowledge and practices were freely exchanged across the three ethnic groups, we would expect their sorghum varieties to be similar and, because of their geographical proximity and similar environmental conditions, to display no genetic differentiation. Quite the contrary, we showed that ethnic groups maintained different sorghum landraces, whereas improved varieties were uniformly distributed across groups.

Factors structuring the distribution of sorghum genetic diversity

The genetic diversity of sorghum in the area of study, as assessed with molecular markers, is organized in four major groups. These groups reflected the influence of improved variety dissemination and a differentiation in terms of cycle duration and phenology. The improved varieties (groups A and C) and short-cycle landraces (group B) collected on the area of study clustered with the Caudatum accessions from eastern Africa and central Africa of a reference set representing the worldwide sorghum genetic diversity ([47]; Figure S3). The long-cycle landraces (group D) clustered with accessions from various origins (eastern & central Africa, India, Middle-East) and races (Durra, Caudatum, Bicolor and intermediates). Some new alleles, absent from the global reference set, were found in the local pool and notably among the long-cycle landraces, which could hence complement the reference set.

The overall distribution patterns of sorghum diversity on our study site were clearly associated with the farmers' ethnic partition. This genetic differentiation did not appear to result from isolation-by-distance as no significant relationship between the geographic distance and the genetic relatedness of individuals was detected.

Table 2. Number of individuals sampled in each ethnic group and assigned to each MMb genetic cluster.

MMb cluster	Chuka	Mbeere	Tharaka	Total	Chi²	P-value
A	44 (33%) [a]	21 (31%) [a]	25 (29%) [a]	90 (31%)	0.37	0.832
B	36 (27%) [a]	18 (27%) [a]	34 (39%) [a]	88 (30%)	4.49	0.106
C	12 (9%) [a]	17 (25%) [b]	18 (21%) [b]	47 (16%)	10.5	0.005
D	21 (15%) [a]	5 (7%) [ab]	4 (4%) [b]	30 (11%)	7.7	0.021
Mix	22 (16%) [a]	7 (10%) [a]	6 (7%) [a]	35 (12%)	4.7	0.097
Total	135 (100%)	68 (100%)	87 (100%)	290 (100%)		

Individuals with a q value equal to or above the threshold of 0.8 for a cluster were assigned to that cluster. The Chi-Square statistics and p-value compare, for each MMb cluster, the observed and the expected frequencies under the null hypothesis of independence. For each cluster, the letters indicate the statistical significance of the differences in its frequency between ethnic groups (Fisher test) at a 5% level after correction for multiple testing (FDR). For a given cluster, ethnic groups with the same letter did not present significant differences.

Long-cycle landraces formed a genetically distinct cluster which was more frequently encountered in the Chuka sorghum population than in the Tharaka sorghum population. The improved *Kaguru* variety showed more admixture with the local landraces in the Chuka sorghum population than in the Mbeere and Tharaka ones. As a result of the unbalanced frequency of the different genetic clusters across ethnic groups, the genetic differentiation of their sorghum populations was significant. The uneven distribution of named landraces across the Chuka, Tharaka and Mbeere ethnic groups is consistent with the results of Baco et al. [48], who reported that different ethnic groups in Benin cultivated different varieties of yam. A similar relationship between the structure of the genetic diversity of domesticated populations and farmers' social organization was found in taro populations across linguistic groups in Vanuatu [6], and in goat populations across ethnic groups in Vietnam [7]. However, a common caveat to such crop diversity studies conducted on large spatial scales is the difficulty involved in assessing whether the spatial patterns of crop diversity are related to variations in agro-ecological conditions, geographical distances, or to socio-cultural differences between human societies [10].

The field setting adopted in our study enabled us to limit the interference between socio-cultural factors and other environmental factors. Notably, climate and soil variations can influence the distribution of crop diversity. The climatic variation was neglectable on our study site regarding the limited gradient of altitude. In addition, we conducted a survey which did not highlight significant differences of soils' physical properties among the areas inhabited by the three ethnic groups (data not shown). Furthermore, farmers did not report that some varieties were better adapted to particular types of soils. Hence, the interference between socio-cultural factors and other uncontrolled environmental factors remains much unlikely, even though it cannot be totally left out.

The community-scale approach we used in this study revealed that social boundaries have contributed to the differentiation of sorghum populations across spatially-close ethnic groups living in the same agro-ecological environment. Such an approach is thus complementary to country or regional-scale studies. In addition, such an approach makes it possible to investigate the mechanisms behind the relationship by jointly analyzing the distribution of varieties and the structure of genetic and morphological diversity in relation to the social organization of the communities concerned.

The ethnic identity of human groups is maintained by social boundaries that impede their cultural homogenization [49]. Our results suggest that these social boundaries also maintain differences between crop populations across ethnic groups. Indeed, gene flows in crop populations greatly depend on the exchange of seed, which is facilitated by social relationships and limited by social boundaries [4]. In addition, farmers' seed selection practices have a strong impact on crop populations and can differ considerably across communities [5,50]. The comparison of the structure of the genetic and morphological diversity of sorghum populations provides information concerning gene flows and selection forces, while the study of the nomenclature given to farmers' varieties tracks the diffusion of knowledge across farming communities. Thus, by combining the two approaches it is possible to investigate the respective influence of seed exchanges and the diffusion of selection practices across ethnic groups on sorghum genetic diversity patterns.

Limited diffusion of long-cycle landraces across ethnic groups

Long-cycle landraces formed a distinct MMb genetic cluster, whose frequency differed across ethnic groups. It was more frequent among the Chuka than among the Tharaka, and, interestingly, these results confirmed farmers' reports stating that long-cycle landraces were "*Muvia wa Chuka*", the sorghum of Chuka people. Moreover, certain sub-types within this cluster (sub-clusters D2 and D3) were not present in the Mbeere population and corresponding landraces were not inventoried in that ethnic group. The relation between the spatial distribution of the MMb genetic clusters and that of ethnic groups suggests that social boundaries limit the diffusion of planting material. Indeed, in most rural societies, seed exchanges depend on social networks as trust is required for seed transactions [51]. On the one hand, social relationships directly shape the seed exchanges because they facilitate access to seed [52,53,54]. On the other hand, the social network is the major pathway for information exchange [55] and indirectly helps shape seed exchanges, as farmers tend to imitate relatives [56]. The joint action of these two mechanisms can thus explain the uneven distribution of long-cycle landraces across ethnic groups. In addition, the small grains and the bitter taste could explain the low economic value of these landraces, which probably helps limit their diffusion.

Management practices of improved varieties differ across ethnic groups

In contrast to the case of some landraces, improved varieties were uniformly distributed and their frequencies did not differ between ethnic groups. The recently introduced *Gadam* variety was

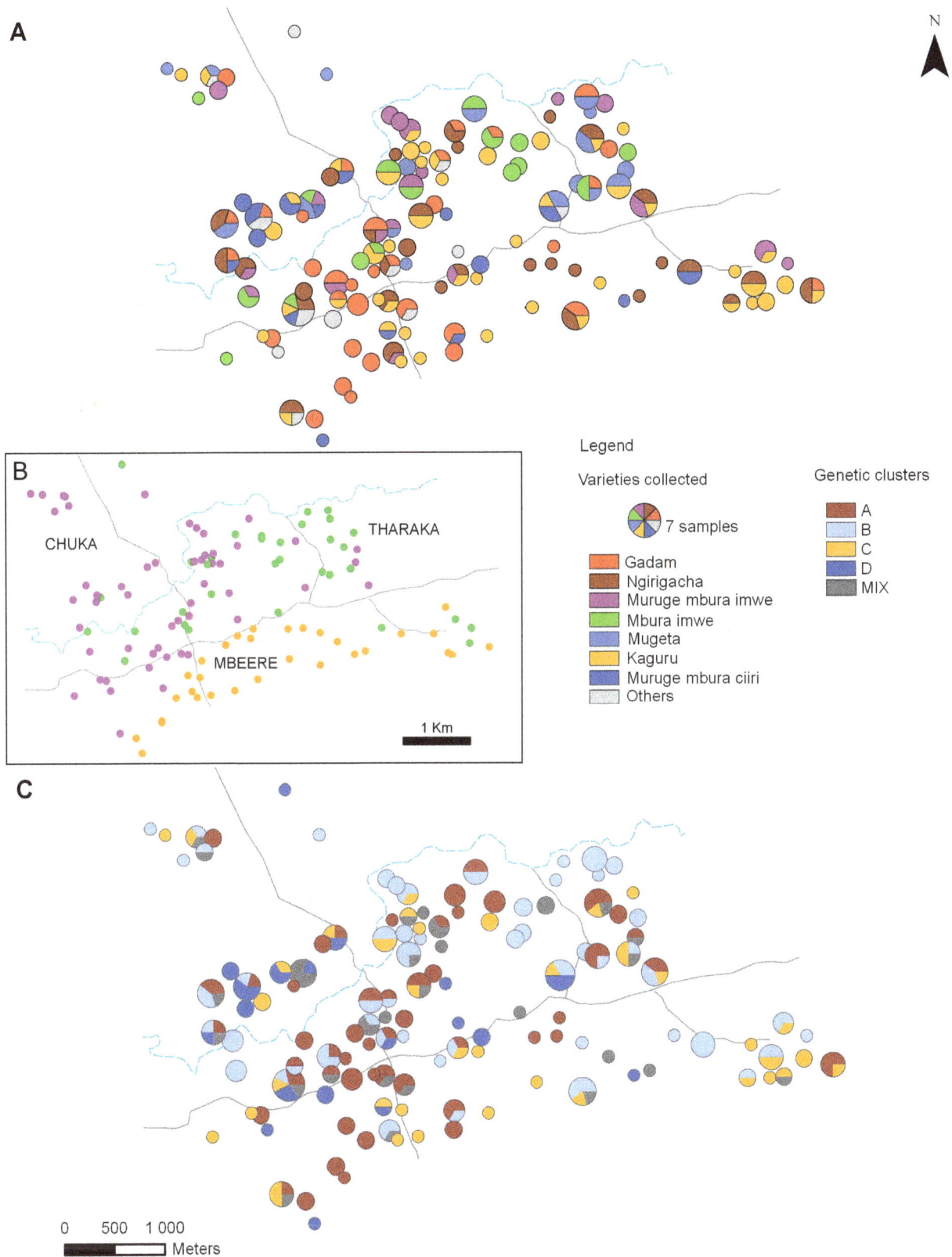

Figure 5. Spatial distribution of the sorghum varieties, ethnic groups, and sorghum genetic clusters. (A) Map of the named varieties collected in each ethnic group. Pie charts represent the number of samples of each variety collected in each household. The size of each circle is proportional to the number of individuals sampled. (B) Location of the ethnic groups (Purple: Chuka, Green: Tharaka, Orange: Mbeere). (C) Map of the number of sorghum individuals in each household assigned to each of the four MMb genetic clusters. Individuals were assigned to a cluster if their estimated genome fraction to that cluster, i.e. admixture coefficient (q), was higher than 0.8.

Table 3. Summary of the genetic polymorphism indexes of the *Kaguru* variety sampled in the three ethnic groups.

Ethnic group	N_{hh}	N_i	$R_{Al.}$	He	Ho	F_{IS}
Chuka	19	22	3.29 [a]	0.339 [a]	0.049	0.857
Mbeere	17	20	2.33 [b]	0.184 [b]	0.003	0.985
Tharaka	13	20	1.83 [b]	0.091 [c]	0.006	0.939

N_i: number of samples, N_{hh}: number of households, R_{Al}: unbiased allelic richness. H_e: unbiased gene diversity, H_o: observed heterozygosity, F_{IS}: fixation index. The letters (a, b, c) next to the R_{Al} and H_e values indicate the statistical significance of their differences between ethnic groups (Wilcoxon test) at a 5% level after correction for multiple testing (FDR). For a given index, ethnic groups with the same letter did not present significant differences.

genetically distinct from the landraces and showed limited introgression from the other genetic clusters. It was genetically uniform and complied with certified control. However, farmers also gave the names of local and already known variety to individuals that have the same genetic profile as *Gadam*, an improved variety. This can be explained by a morphological similarity. Yet it raises the question of the consequences this will have for the on-farm evolution of the improved variety. *Kaguru*, for instance, which was introduced in the area 10–15 years ago, seems to have evolved differently across ethnic groups. High admixture was detected between *Kaguru* and the local landraces in the Chuka population, resulting in a range of genetically diverse materials still called *Kaguru*, while this variety was found to be genetically more uniform in the Mbeere and Tharaka populations. As a result, the genetic diversity of *Kaguru*, as identified by farmers, was greater in the Chuka population than in the Mbeere and Tharaka populations. According to the farmers, the variety was introduced simultaneously in the three ethnic groups but little information is available concerning the origin of the seed lots. The divergence of the *Kaguru* variety across ethnic groups within a few decades could thus be the result of differences in their management practices, be it in planting (spatial arrangement of the varieties) or in seed selection. The higher admixture rate between *Kaguru* and the local landraces among the Chuka could be due to more intense gene flows within fields or to less stringent selection practices. As our observations suggest that the cropping systems used by the three ethnic groups were similar, the hypothesis of different selection practices is more likely. Cases of divergent selection practices between geographically close communities were observed by Pressoir and Berthaud [50], and by Perales et al. [5], who hypothesized that social boundaries impede the homogenization of

selection practices. However, the hypothesis of the introduction of seed lots with different genetic characteristics in the three ethnic groups cannot be excluded.

Divergence in the nomenclature of the landraces between ethnic groups

Comparing the genetic structure of short-cycle landraces, their morphological characteristics and the farmers' nomenclature raises interesting questions concerning the relation between farmers' nomenclature and the diffusion of planting material. Indeed, the frequencies of the majority of named short-cycle landraces differed significantly between ethnic groups even though they were assigned to the same genetic pool and no clear correlation was detected between named landraces and the MMb genetic sub-structure. The molecular markers used did not discriminate the three major short-cycle landraces whose frequency varied markedly across ethnic groups and which display different morphological characteristics. The short-cycle landraces grown by the different ethnic groups thus appeared to belong to the same genetic pool. Yet, the analysis of the morphological characteristics of the panicles suggests that the landraces presented morphological differences that were not detected with neutral genetic markers. *Mugeta* and *Muruge mbura imwe*, mainly grown by the Chuka and Tharaka, corresponded to two distinct morphological groups while *Ngirigacha*, which is mainly grown by the Mbeere, was distributed over the entire PCoA plot (Figure S4.A). These results suggest that ethnic groups use different names for landraces with similar morphotypes: the Chuka and Tharaka appear to identify and name two main short-cycle landraces corresponding to distinct morphotypes while the Mbeere mainly

Table 4. Proportion of individuals of each variety assigned to MMb cluster B regarding their collection ethnic group.

Variety	Chuka	Mbeere	Tharaka
Muruge mbura imwe	14 (39%)	1 (5%)	10 (29%)
Mugeta	10 (28%)	-	8 (24%)
Ngirigacha	3 (8%)	11 (61%)	6 (18%)
Mbura imwe	7 (19%)	-	8 (23%)
Muthigo wa mwimbi	1 (3%)	-	-
Others :			
Gadam	1 (3%)	2 (12%)	2 (6%)
Kaguru	-	1 (5%)	-
Muruge mbura ciiri	-	3 (17%)	-
Total	36 (100%)	18 (100%)	34 (100%)

Percentages in brackets.

use the name *Ngirigacha* for all the morphotypes corresponding to the short-cycle landraces group.

This difference in folk-nomenclature and classification between the Mbeere and both the Chuka and Tharaka groups may result from limited knowledge diffusion. This is consistent with a number of observations concerning the conflictual relationship between the Chuka and Mbeere groups [19]. The impact of social relationships on the diffusion of folk-taxonomy and nomenclature among farming communities was demonstrated by Boster [57], who showed that the cassava nomenclature used by kin-related women was more similar than that used by non-kin in the Aguaruna community in Peru. Nuijten and Almekinders [12] also reported that the naming of rice varieties was more consistent within villages than between villages in Gambia. They pointed out that information concerning varieties, such as names, is not necessarily passed on with the seed lots. Hence, seed exchanges between communities can be more intense than knowledge diffusion, leading to the use of different names for similar morphotypes and genotypes. Further comparison of farmers' nomenclature and taxonomy between ethnic groups is required to confirm this hypothesis.

Previous studies showed that different sorghum varieties may display no genetic differentiation despite being morphologically distinct. Notably, in Cameroon, Barnaud et al. [3] showed that considerable gene flows existed between Guinea sorghum landraces while farmers kept on selecting them for their morphological distinctiveness. Rabbi et al. [58] reported similar results in western Kenya, while varieties collected in eastern Sudan were clearly genetically distinct. He explained these results by the varietal isolation practiced in Sudan, while Kenyan farmers mixed varieties within their fields. Soler et al. [59] found that landraces were distinct genetic units, but in that study each landrace was sampled in a single field belonging to one farmer, which considerably limits the variability. As farmers' taxonomy and nomenclature is based on morphological traits with a simple genetic determinism, morphological differences can be maintained even though gene flows occur within farmers' fields. The 18 SSRs used in our study were selected because they revealed high polymorphism in previous diversity studies, and they proved to be adequate for characterizing the genetic sub-structure of long-cycle landraces. However, their resolution power may not be sufficient to reveal a finer-scale genetic sub-structure in short-cycle landraces. The use of high-density markers may help to evidence finer-scale genetic structure and could hence contribute to decipher the evolutionary mechanisms that molded the landraces.

Effect of community social organization on the diffusion of seeds, knowledge and farmers' practices

According to local elders, the three ethnic groups migrated to the study area about a century ago. Our results suggest that even though they have lived in proximity since then, the way knowledge, practices and seeds are diffused has maintained differences between sorghum populations across ethnic groups. Ethnographic observations of community social organization provide explanations for such limited exchanges across geographically close communities. Indeed, information transmission and diffusion appear to be confined within the residential groups (parents and married sons) first, which is common in patrilineal and patrilocal societies [60], and next within the neighborhood group, which is a major social institution among eastern Kenyan Bantu communities [16,61], (Linsig pers.com). The way knowledge is transmitted and diffused is very conservative and favors cultural differentiation between communities [62,63]. It thus probably plays a major role in maintaining differences in

nomenclature and practices between ethnic groups, and maybe also in limiting seed exchanges.

Conclusion

Our study highlights the importance of local short-scale studies to investigate farm crop evolution processes. To date, emphasis has been placed on the effect of agro-ecological conditions on crop evolution processes, as in the study of the evolution of wild plants. The influence of the cultural diversity and social organization of farming communities has consequently been neglected, although the major role of smallholders in the management of crop diversity has been acknowledged [64]. Crop evolution is still ongoing in smallholder farming systems and such systems occupy a substantial proportion of croplands in developing countries, especially in Africa [65]. Most of these rural communities have retained pre-colonial social institutions that continue to shape the relationships between people. Sixty-eight living language groups were inventoried in Kenya and about 2146 linguistic groups in Africa [66], so the situation of ethnic co-existence described in this paper is not an isolated case. This study confirms the influence of the ethnolinguistic patterns of rural communities on gene flows and on farmers' selection practices that shape crop diversity *in situ*. Crop diversity patterns, thus result not only from an interaction between genetic and environmental factors, $G \times E$, but from a three-way interaction $G \times E \times S$, where "S" stands for effects of the social boundaries [4]. Investigating this relation in other communities, with different social organizations and rules for the transmission of knowledge, would thus help gain a clearer picture of crop evolution dynamics in subsistence farming systems.

A further study is now needed to probe the mechanisms involved. Notably, the link between seed exchange networks and social organization deserves more investigation to confirm whether seed exchanges are confined within ethnic groups. This would explain why the diffusion of long-cycle landraces is more limited than that of short-cycle landraces. Moreover, further comparison of the local sorghum nomenclature and classification systems (folk taxonomy) across ethnic groups would make it possible to test whether their definition of landraces differs, and whether it influences their seed selection practices.

The uneven distribution of the genetic clusters across ethnic groups within a restricted geographic area highlights the need to take the social relationship and exchanges into account in the characterization, collection, and conservation of crop diversity. Accounting for the impact of human practices on crop populations would help capture their diversity more efficiently and, to this end, ethnic contact zones are of major interest for their potentially high genetic diversity. This study paves the way for participatory plant breeding as it shows that farmers' individual choices concerning planting material are not only determined by agro-ecological conditions or economic interest, but also by their cultural background.

Supporting Information

Figure S1   Diagram displaying the rain seasons and the growth-cycle of sorghum on our study site. Inventories' dates are symbolized by the letter I (orange points) and collections' dates by the letter C (red points).

Figure S2   Comparison between the inventory of varieties and their sampling. (A) Percentage of households where each variety was sampled for the genetic diversity study on a total of 130 households. (B) Linear correlation between the proportions of households where each variety was inventoried (vertical axis, 124 households) and where it was collected (horizontal axis, 130 households) in each ethnic group.

Figure S3 Neighbor-Joining tree based on the genetic dissimilarity among the individuals sampled on our study site (in Black) and the accessions of a global reference set (Billot et al. 2013). The genetic dissimilarities were calculated on 16 SSRs using the simple matching index. The sorghum individuals sampled on our study site are displayed in black. The genetic assignment (A, B, C, D - q>0.8) or unassignment (Unassigned - q≤0.8) of our individuals is indicated on the figure. Colors represent the ten genetic groups identified in Billot et al. 2013, and described as following by the authors: "Group 1 [Dark orange] included Caudatum, Caudatum-Bicolor and Durra from Eastern Asia; Group 2 [Light orange] encompassed Durra and Bicolor from the Indian subcontinent, while Group 3 [Light green] exhibited Durra from Eastern Africa. Bicolor and Durra-Bicolor from Eastern Africa were assigned in Group 4 [Light blue]. Group 5 [Dark blue] included Guinea and Guinea margaritiferum from Western Africa and Bicolor from North America. Group 6 [Red] appeared as a well-separated group made predominantly of Guinea accessions from western Africa, accompanied by intermediate race Durra-Caudatum materials from western Africa while Group 7 [Magenta] was made essentially of materials collected from eastern Africa and central Africa generally classified as race Caudatum (visible along FA axis 3). Group 8 [Dark green] was a small and heterogeneous group made of Durra and Caudatum race accessions from central Africa. Group 9 [Pink] was made essentially of Guinea race accessions from the Indian subcontinent and southern/eastern Africa with Guinea-Caudatum (GC) intermediate race accessions from various parts of Africa. Group 10 [Purple] was made almost exclusively of accessions from southern Africa of race Kafir or intermediate race Kafir-Caudatum (KC)." Unassigned individuals in the global reference set are displayed in grey.

Figure S4 Structure of the morphological and genetic diversity within the MMb clusters B (top) and D (bottom). (A) Plot of the two first axes of the Principal Coordinates Analysis (PCoA) done on the sorghum plants assigned to the MMb cluster B and based on 15 panicle morphological traits. The first axis (x) expresses 35.1% of the total variation and the second axis 13.1%. Varieties are displayed using the following color code: Blue: *Mugeta*, purple: *Muruge mbura imwe*, green: *Mbura imwe*, brown: *Ngirigacha*, Red: *Gadam*, yellow: *Kaguru*, salmon: *Muthigo wa mwimbi*. (B) Neighbor-Joining tree based on the genetic dissimilarity among individuals assigned to the MMb cluster B calculated on 18 SSRs using the simple matching index. (C) Plot of the two first axes of the Principal Coordinates Analysis (PCoA) done on the sorghum plants assigned to the MMb cluster D and

based on 15 panicle morphological traits. The first axis (x) expresses 51.5% of the total variation and the second axis 18.0%. Varieties are displayed using the following color code: Yellow: *Serendo*, orange: *Musalama*, light-pink: *Kathirigwa*, Fushia: *Mugana*, Greenish blue: *Mucuri*, dark-blue: *Muruge mbura ciiri*, black: *Muthigo*, blue: *Mugeta*. (D) Neighbor-Joining tree based on the genetic dissimilarity among individuals assigned to the MMb cluster D calculated on 18 SSRs using the simple matching index.

Table S1 Summary of the sampling of planting material. Mean number of varieties collected per household (Mean no. varieties/household) and mean number of samples of each variety collected per household (Mean no. samples/variety/household) in each ethnic group, followed by their standard error (SE).

Table S2 Summary of information and genetic diversity estimates per locus. Minimum and maximum size of alleles (Size), chromosome where the locus is located (Ch), percentage of missing data per locus (Miss), number of sampled alleles (N_{Al}), He: unbiased gene diversity, F_{IS}: Fixation index.

Table S3 Morphological descriptors used for panicle description.

Table S4 Results from the perMANOVA comparing the effect of ethnic groups on sorghum variety assemblages. Df: degrees of freedom, Ssq: sequential sum of squared distance between individuals and their group's centroïd, Mean Ssq = Ssq/ Df, F.Model: pseudo F ratio, R^2: coefficient of determination [Ssq Etnic group/Ssq Total].

Acknowledgments

We wish to thank Joseph I. Kamau and Catherine Wanjira for their contribution to field work, the staff of the National Genebank of Kenya and its director Zachary Muthamia, as well as Philippe Letourmy and Xavier Perrier for their help with statistical analysis. Thanks are also due to the participating farmers for their collaboration in our study. We feel indebted to the headmaster, teachers and students of Kabururu primary school for hosting our experimental trials on their school farm. We are also deeply grateful to the extension and local administrative staff in the study area. The support of the chiefs and assistant chiefs is particularly acknowledged. We thank three anonymous reviewers and the editor for their helpful comments.

Author Contributions

Conceived and designed the experiments: CL VL AB MD. Performed the experiments: CL VL PW AB MB CC. Analyzed the data: VL MD SM AB CL. Contributed reagents/materials/analysis tools: SM AB PW. Wrote the paper: VL CL MD JCG AB SM PW.

References

1. Perrier X, De Langhe E, Donohue M, Lentfer C, Vrydaghs L, et al. (2011) Multidisciplinary perspectives on banana (*Musa* spp.) domestication. Proc Nat Acad Sci USA 108: 11311–11318.

2. Roullier C, Benoit L, McKey DB, Lebot V (2013) Historical collections reveal patterns of diffusion of sweet potato in Oceania obscured by modern plant movements and recombination. Proc Nat Acad Sci USA 110: 2205–2210.

3. Barnaud A, Deu M, Garine E, McKey D, Joly H (2007) Local genetic diversity of sorghum in a village in northern Cameroon: structure and dynamics of landraces. Theor Appl Genet 114: 237–248.

4. Leclerc C, Coppens d'Eeckenbrugge G (2012) Social organization of crop genetic diversity. The G × E × S interaction model. Diversity 4: 1–32.

5. Perales H, Benz B, Brush S (2005) Maize diversity and ethnolinguistic diversity in Chiapas, Mexico. Proc Nat Acad Sci USA 102: 949–954.

6. Sardos J, Noyer JL, Malapa R, Bouchet S, Lebot V (2012) Genetic diversity of taro (Colocasia esculenta (L.) Schott) in Vanuatu (Oceania): an appraisal of the distribution of allelic diversity (DAD) with SSR markers. Genet Resour Crop Evol 59: 805–820.

7. Berthouly C, Do Ngoc D, Thévenon S, Bouchel D, Van T, et al. (2009) How does farmer connectivity influence livestock genetic structure? A case-study in a Vietnamese goat population. Mol Ecol 18: 3980–3991.

8. De Wet JMJ, Huckabay JP (1967) The origin of Sorghum bicolor. II. Distribution and domestication. Evolution 21: 787–802.

9. Harlan JR, De Wet JMJ, Stemler ABL (1976) Origins of African plant domestication. The Hague: De Gruyter Mouton. 498 p.

10. Deu M, Sagnard F, Chantereau J, Calatayud C, Hérault D, et al. (2008) Niger-wide assessment of *in situ* sorghum genetic diversity with microsatellite markers. Theor Appl Genet 116: 903–913.

11. Bellon MR, Brush SB (1994) Keepers of maize in Chiapas, Mexico. Econ Bot 48: 196–209.

12. Nuijten E, Almekinders CJ (2008) Mechanisms explaining variety naming by farmers and name consistency of rice varieties in the Gambia. Econ Bot 62: 148–160.

13. Jaetzold R, Schmidt H, Hornetz B, Shisanya C (2007) Farm management handbook. Vol II, Part C, East Kenya. Subpart C1, Eastern Province. Nairobi Ministry of Agriculture/GTZ.

14. Camberlin P, Boyard-Micheau J, Philippon N, Baron C, Leclerc C, et al. (2012) Climatic gradients along the windward slopes of Mount Kenya and their implication for crop risks. Part 1: climate variability. Int J Climatol. Available: http://onlinelibrary.wiley.com/doi/10.1002/joc.3427/full. Accessed 2 July 2013.

15. Ambler C (1988) The Great Famine, 1897–1901. In: Ambler C, editor. Kenyan Communities in the Age of Imperialism: The central region in the late Nineteenth Century. New Haven, London: Yale University Press. pp.123–149.

16. Middleton J (1953) The central tribes of the north-eastern Bantu; Forde D, editor. London: International African Institute. 105 p.

17. Moehlig WJG, Guarisma G, Platiel S (1980) La dialectométrie: une méthode de classification synchronique en Afrique. In: Guarisma G, Platiel S editors. Dialectologie et comparatisme en Afrique Noire. Paris: SELAF. pp. 27–45.

18. Chesaina C (1997) Chapter One: Historical and cultural background. In: Chesaina C editor. Oral literature of the Embu and Mbeere. Nairobi: East African Educational Publishers. pp. 3–10.

19. Glazier J (1970) Ritual and social conflict: circumcision and oath-taking in Mbeere. Nairobi: Institute of Development Studies, Nairobi University. 17 p.

20. Fadiman JA (1993) When we began there were witchmen: an oral history from Mount Kenya. Los Angeles: University of California Press. 395 p.

21. Plucknett DL, Evenson JP, Sanford WG (1970) Ratoon cropping. Adv Agron 22: 285–330.

22. Billot C, Rivallan R, Sall M, Fonceka D, Deu M, et al. (2012) A reference microsatellite kit to assess for genetic diversity of Sorghum bicolor (Poaceae). Am J Bot 99: e245–e250.

23. International Plant Genetic Resources Institute (IPGRI) (1993) Descriptors for Sorghum [*Sorghum bicolor* (L.) Moench]. Rome: IBPGR/ICRISAT. 38 p.

24. Gibson RW (2009) A review of perceptual distinctiveness in landraces including an analysis of how its roles have been overlooked in plant breeding for low-input farming systems. Econ Bot 63: 242–255.

25. Anderson MJ (2001) A new method for non-parametric multivariate analysis of variance. Austral Ecol 26: 32–46.

26. Oksanen J, Blanchet FG, Kindt R, Legendre P, Minchin PR, et al. (2012) Vegan: Community Ecology Package. 2.0-5 ed.

27. Bray JR, Curtis JT (1957) An ordination of the upland forest communities of southern Wisconsin. Ecol Monogr 27: 325–349.

28. Nakazawa M (2013) fmsb: Functions for medical statistics book with some demographic data. 0.3.8 ed.

29. Benjamini Y, Hochberg Y (1995) Controlling the false discovery rate: a practical and powerful approach to multiple testing. J R Stat Soc Series B Stat Methodol 1: 289–300.

30. Belkhir K, Borsa P, Chikhi L, Raufaste N, Catch F (2004) GENETIX, software under Windows for the genetics of populations. 4.05 ed. Montpellier: University of Montpellier.

31. Petit RJ, El Mousadik A, Pons O (1998) Identifying populations for conservation on the basis of genetic markers. Conserv Biol 12: 844–855.

32. Nei M (1987) Molecular Evolutionary Genetics. New York: Columbia University Press.

33. Weir BS, Cockerham CC (1984) Estimating F-statistics for the analysis of population structure. Evolution 38: 1358–1370.

34. Goudet J (2001) FSTAT, a program to estimate and test gene diversities and fixation indices 2.9.3 ed. Lausanne: Institute of Ecology.

35. Rice WR (1989) Analyzing tables of statistical tests. Evolution 43: 223–225.

36. Goudet J, Raymond M, de Meeüs T, Rousset F (1996) Testing differentiation in diploid populations. Genetics 144: 1933–1940.

37. Raymond M, Rousset F (1995) GENEPOP version 1.2: population genetics software for exact tests and ecumenicism. J Hered 86: 248–249.

38. Pritchard JK, Stephens M, Donnelly P (2000) Inference of population structure using multilocus genotype data. Genetics 155: 945–959.

39. Evanno G, Regnaut S, Goudet J (2005) Detecting the number of clusters of individuals using the software STRUCTURE: a simulation study. Mol Ecol 14: 2611–2620.

40. Saitou N, Nei M (1987) The Neighbor-Joining method: a new method for reconstructing phylogenetic trees. Mol Biol Evol 4: 406–425.

41. Sokal R, Michener CD (1958) A statistical method for evaluating systematic relationships. University of Kansas Science Bulletin 38: 1409–1438.

42. Perrier X, Jacquemoud-Collet JP (2006) DARwin software 5.0.156 ed. Montpellier: CIRAD.

43. Loiselle BA, Sork VL, Nason J, Graham C (1995) Spatial genetic structure of a tropical understory shrub, *Psychotria officinalis* (Rubiaceae). Am J Bot 82: 1420–1425.

44. Hardy OJ, Vekemans X (2002) SPAGeDi: a versatile computer program to analyse spatial genetic structure at the individual or population levels. Mol Ecol notes, 2: 618–620.

45. Rousset F (2000) Genetic differentiation between individuals. J Evol Biol 13: 58–62.

46. Meirmans PG (2012). The trouble with isolation by distance. Mol Ecol 21: 2839–2846.

47. Billot C, Ramu P, Bouchet S, Chantereau J, Deu M, et al. (2013). Massive sorghum collection genotyped with SSR markers to enhance use of global genetic resources. PLOS ONE 8: e59714.

48. Baco MN, Biaou G, Pham JL, Lescure JP (2008) Geographical and social factors of cultivated yam diversity in northern Benin. Cah Agric 17: 172–177.

49. Barth F (1969) Introduction In: Barth F, editor. Ethnic groups and boundaries: The social organization of culture difference. Boston: Little, Brown. pp. 9–38

50. Pressoir G, Berthaud J (2004) Population structure and strong divergent selection shape phenotypic diversification in maize landraces. Heredity 92: 95–101.

51. Badstue LB, Bellon MR, Berthaud J, Ramírez A, Flores D, et al. (2007) The dynamics of farmers' maize seed supply practices in the central valleys of Oaxaca, Mexico. World Dev 35: 1579–1593.

52. McGuire S (2008) Securing access to seed: Social relations and sorghum seed exchange in eastern Ethiopia. Hum Ecol 36 217–229.

53. David S, Sperling L (1999) Improving technology delivery mechanisms: lessons from bean seed systems research in Eastern and Central Africa. Agric Human Values 16: 381–388.

54. Bellon MR (2004) Conceptualizing interventions to support on-farm genetic resource conservation. World Dev 32: 159–172.

55. Van den Broeck K, Dercon S (2011) Information flows and social externalities in a Tanzanian banana growing village. J Dev Stud 47: 231–252.

56. Bandiera O, Rasul I (2006) Social networks and technology adoption in northern Mozambique. Econ J 116: 869–902.

57. Boster JS (1986) Exchange of varieties and information between Aguaruna manioc cultivators. Am Anthropol 88: 428–436.

58. Rabbi IY, Geiger HH, Haussmann BI, Kiambi D, Folkertsma R, et al. (2010) Impact of farmers' practices and seed systems on the genetic structure of common sorghum varieties in Kenya and Sudan. Plant Genet Resour 8: 116–126.

59. Soler C, Saidou AA, Hamadou TVC, Pautasso M, Wencelius J, et al. (2013) Correspondence between genetic structure and farmers' taxonomy - a case study from dry-season sorghum landraces in northern Cameroon. Plant Genet Resour 11: 36–49.

60. Herbich I, Dietler M (2008) The long arm of the mother-in-law: Learning, postmarital resocialization of women, and material culture style. In: Stark MT, Bowser BJ, Horne L, editors. Breaking down boundaries: Anthropological approaches to cultural transmission and material culture. Tucson: University of Arizona Press. pp. 223–244.

61. Labeyrie V, Rono B, Leclerc C (2013) How social organization shapes crop diversity: an ecological anthropology approach among Tharaka farmers of Mount Kenya. Agric Human Values. In press.

62. Reyes-García V, Broesch J, Calvet-Mir L, Fuentes-Peláez N, McDade TW, et al. (2009) Cultural transmission of ethnobotanical knowledge and skills: an empirical analysis from an Amerindian society. Evol Hum Behav 30: 274–285.

63. Cavalli-Sforza LL, Feldman MW (1981) Cultural transmission and evolution: a quantitative approach. Princeton: Princeton University Press. 388 p.

64. Brush SB (2000) Genes in the field: on-farm conservation of crop diversity; Brush SB, editor. Rome: Copublished by International Plant Genetic Resources Institute. 288 p.

65. International Fund for Agricultural Development (IFAD) (2001) Rural Poverty Report 2001: The Challenge of Ending Rural Poverty. Oxford: Oxford University Press for IFAD.

66. Lewis MP, Gary FS, Fennig CD (2013) Ethnologue: Languages of the World, Seventeenth edition. Dallas: SIL International. Available: http://www.ethnologue.com. Accessed 25 August 2013.

Massive Sorghum Collection Genotyped with SSR Markers to Enhance Use of Global Genetic Resources

Claire Billot[1][*][◑], Punna Ramu[2][◑], Sophie Bouchet[1][¤a], Jacques Chantereau[1], Monique Deu[1], Laetitia Gardes[1][¤b], Jean-Louis Noyer[1], Jean-François Rami[1], Ronan Rivallan[1], Yu Li[3], Ping Lu[3], Tianyu Wang[3], Rolf T. Folkertsma[2][¤c], Elizabeth Arnaud[4], Hari D. Upadhyaya[2], Jean-Christophe Glaszmann[1], C. Thomas Hash[2]

1 Cirad, UMR AGAP, Montpellier, France, 2 International Crops Research Institute for the Semi-Arid Tropics (ICRISAT), Hyderabad, Andhra Pradesh, India, 3 Institute of Crop Science, Chinese Academy of Agricultural Sciences (CAAS), Beijing, China, 4 Bioversity International, Montpellier, France

Abstract

Large *ex situ* collections require approaches for sampling manageable amounts of germplasm for in-depth characterization and use. We present here a large diversity survey in sorghum with 3367 accessions and 41 reference nuclear SSR markers. Of 19 alleles on average per locus, the largest numbers of alleles were concentrated in central and eastern Africa. Cultivated sorghum appeared structured according to geographic regions and race within region. A total of 13 groups of variable size were distinguished. The peripheral groups in western Africa, southern Africa and eastern Asia were the most homogeneous and clearly differentiated. Except for Kafir, there was little correspondence between races and marker-based groups. Bicolor, Caudatum, Durra and Guinea types were each dispersed in three groups or more. Races should therefore better be referred to as morphotypes. Wild and weedy accessions were very diverse and scattered among cultivated samples, reinforcing the idea that large gene-flow exists between the different compartments. Our study provides an entry to global sorghum germplasm collections. Our reference marker kit can serve to aggregate additional studies and enhance international collaboration. We propose a core reference set in order to facilitate integrated phenotyping experiments towards refined functional understanding of sorghum diversity.

Editor: Wengui Yan, National Rice Research Center, United States of America

Funding: The authors thank the Generation Challenge Programme (GCP) for their financial support to this project. The research fellowship provided to PR by the Council of Scientific and Industrial Research (CSIR), New Delhi, India, is gratefully acknowledged. Part of this work was carried out by using the resources of the Computational Biology Service Unit from Cornell University partially funded by Microsoft Corporation. The funders had no role in study design, data collection and analysis, decision to publish, or preparation of the manuscript.

Competing Interests: Part of our work was carried out by using the resources of Cornell University's Computational Biology Service Unit. It was partially funded by Microsoft Corporation.

* E-mail: claire.billot@cirad.fr

◑ These authors contributed equally to this work.

¤a Current address: UMR de Génétique Végétale, INRA – Université Paris-Sud – CNRS, Gif-sur-Yvette, France
¤b Current address: Cirad, UMR CMAEE, Montpellier, France
¤c Current address: Plant Health Lead EMEA/India, Monsanto Holland B.V., Bergschenhoek, The Netherlands

Introduction

Crop domestication is characterised by human selection on wild species for traits useful for food production. This continuous process made possible the development of agriculture and of civilizations. While migrating, man moved together with his crops and spread agriculture worldwide. It led to global development as well as occasional harsh competitions. While many industrial crops have a recent domestication history intermingled with that of colonization, food crops present distributions that have little relation with their domestication place.

Recent global planetary constraints create a new threatening situation; plant breeding is currently faced with unprecedented challenges, which call for global cooperation. Plant genetic resources conceal the matter for future improvement and adaptation. They bear thousands of years of genetic adaptation to multiple conditions and usages by Man. In times when 1) food

security is dramatically challenged by population growth, shortage of input supply and climate changes, and 2) genomic tools and methodologies bring about unprecedented capacities of scientific investigation, they are and will remain a stake, matter of competition as well as cooperation.

Sorghum [*Sorghum bicolor* (L.) Moench, $2n = 2x = 20$] is the fifth most important cereal crop in the world. Its use as staple food and fodder confers it the status of a 'failsafe' crop in global agro-ecosystems. It is widely adapted to harsh environmental conditions, and more specifically to arid and semi-arid regions of the world. It is currently a model crop for tropical grasses that employ C_4 photosynthesis because of the availability of its complete genome sequence [1] [2], http://genome.jgi-psf.org/Sorbi1/Sorbi1.info.html).

There are several identified collections of sorghum genetic resources (for example core-collections [3] [4], US converted tropical and breeding lines described in [5], US sweet sorghum collection [6], mutant populations [7], Japanese collection [8], as well as accessions available at ICRISAT). Sorghum's center of diversity lies in the northeastern quadrant of Africa and it is thought sorghum was domesticated there over 5,000 years before present [9]. Based on spikelet and grain morphology, Harlan and de Wet [9] developed a simplified classification of traditional sorghum cultivars into five basic races: Bicolor (B), Caudatum (C), Durra (D), Guinea (G) and Kafir (K), and ten intermediate races (in all pair-wise combinations of basic races).

Biochemical genetic markers provided the first assessment of neutral genetic variation and enabled demarcation of groups by race and origin [10]. Several generations of DNA-based molecular markers were then used and refined the assessment. In the early 1990 s, restriction fragment length polymorphism (RFLP) markers were effectively utilized for sorghum diversity analysis (reviewed in [11], [12]), genetic mapping (e.g. [13], [14], [15], [16], [17]) and comparative genome mapping ([18], [19], [20], [21]). Later other marker systems were tried, including randomly amplified poly-morphic DNAs (RAPDs), simple sequence repeats (SSRs), and amplified fragment length polymorphisms (AFLPs). These markers systems, independently or in combination with others, were efficiently used for sorghum genetic diversity analysis. Diversity array technology (DArT) markers have recently been developed and utilized for genetic diversity assessment and mapping [22], as well as SNPs [23].

SSRs were developed independently by several different research groups (see review by [24], and [25], [26], [27]) and were exploited for genetic diversity analysis. Many of these diversity analyses focused on local collections (e.g. [28], [29], [30], [31]), trait-specific genotypes [e.g. aluminum tolerance [32], sweet stalks ([33], [6]), disease resistance [34]], or a particular race (Guinea [35]).

In front of the large size of the collections available and the diversity of interests expressed in the various studies, we undertook this study in order to provide a better insight into global sorghum genetic diversity and to set a reference, which can attract interest, stimulate cooperation and coordination and enhance interactions and connections among all initiatives. A large collection of sorghum (global composite germplasm collection, GCGC) including over 3300 accessions was thus genotyped with highly polymorphic markers (41 SSRs) providing coverage across all 10 chromosome pairs in the nuclear genome of *Sorghum bicolor*. This was performed in the frame of the Generation Challenge Programme (GCP, www.generationcp.org). It may provide a foundation for more efficient management and utilization of available genetic resources in this crop, as well as a tool for mining alleles of genes controlling important agronomic traits.

Methods

Plant Material

Sorghum material studied was mainly selected among ICRI-SAT's collection ([4]), since ICRISAT has one of the largest crop germplasm collections held in trust by the Consultative Group for International Agricultural Research (CGIAR). ICRISAT's collection includes germplasm of staple food crops of the semi-arid tropics including sorghum, pearl millet, groundnut, pigeonpea, chickpea and several small millets (foxtail millet, finger millet, etc). Chinese material was under-represented in ICRISAT's collection; so it was complemented with material provided by CAAS. It also included a previously defined core collection, mainly from ICRISAT's collection and extensively studied ([3]). A total of 3367 sorghum accessions were thus studied in this paper, representing cross-compatible sorghum germplasm of broad initial

taxonomic status (passport information available in Table S1). This GCP sorghum GCGC included 280 breeding lines and elite cultivars from public sorghum breeding programs, 68 wild and weedy accessions, and over 3000 landrace accessions from collections held by CIRAD or ICRISAT that were selected either from previously defined core collections ([3], [4]), for resistance to various biotic stresses, and/or for variation in other agronomic and quality traits. All three labs, CAAS-China, CIRAD-France and ICRISAT-India, contributed accessions to the study. CIRAD contributed 225 well-characterized genotypes that constitute a mini-core collection representing a very broad range of diversity [11], CAAS contributed 250 accessions comprising sweet sorghums, grain sorghums and glutinous sorghums from China, and the remaining accessions were contributed by ICRISAT. All accessions from this sorghum GCGC collection are publically available, except the 250 provided by CAAS. This collection included representation of all 5 basic races of cultivated sorghum [Bicolor (B), Caudatum (C), Durra (D), Guinea (G) and Kafir (K)] and their ten intermediate collected from different parts of the world (Table 1). All together one third of the accessions were provided by all ten intermediate races (1159 accessions), while the largest numbers of basic races were represented by Durra (651 accessions) and Caudatum (577 accessions).

DNA Extraction

DNA extraction was carried out in the labs contributing the sorghum entries to this study, with a single representative plant providing the DNA for each accession, following a protocol described by [36] for accessions contributed by ICRISAT and as described in [37] for accessions contributed by CIRAD and CAAS. Extracted DNA samples were exchanged between the labs for SSR marker genotyping.

SSR Markers

All 48 markers used were part of a sorghum SSR kit [24] (http://sorghum.cirad.fr/SSR_kit), which provides reasonable coverage across the sorghum nuclear genome. Marker genotyping at CIRAD was performed on the Genotyping Platform of the Montpellier Languedoc-Roussillon Genopole (GPTR. http://www.gptr-lr-genotypage.com/) for markers *gpsb*067, *gpsb*089, *gpsb*123, *mSbCIR*246, *mSbCIR*262, *mSbCIR*300, *mSbCIR*329, *Sb5-206 = Xgap*206, *Sb6-84 = Xgap*084, *SbAGB*02, *Xcup*02, *Xcup*14, *Xcup*53, *Xcup*61, *Xcup*63, *Xtxp*010, *Xtxp*015, *Xtxp*040, *Xtxp*057, and *Xtxp*145. The forward primer was designed with a 5′-end M13 extension (5′-CACGACGTTGTAAAACGAC-3′). IRDye® 700 or IRDye®800-labeled PCR products were diluted 10-fold and 4-fold, respectively, and subjected to electrophoresis in 6.5% poly-acrylamide gels with a Licor IR2 system (Licor, USA).

Markers *Xisep*0107, *Xisep*0310, *mSbCIR*223, *mSbCIR*238, *mSbCIR*240, *mSbCIR*248, *mSbCIR*276, *mSbCIR*283, *mSbCIR*286, *mSbCIR*306, *Sb4-72 = Xgap*072, *Xcup*11, *Xtxp*012, *Xtxp*021, *Xtxp*114, *Xtxp*136, *Xtxp*141, *Xtxp*265, *Xtxp*273, *Xtxp*278, *Xtxp*320, *Xtxp*321 and *Xtxp*339 were genotyped at ICRISAT. Amplified PCR products, according to their multiplexes, along with internal ROX-400 size standard, were separated by capillary electropho-resis using an ABI 3700 sequencer (Applied Biosystems, USA).

Markers *gpsb*069, *gpsb*148, *gpsb*151, *Xcup*62 and *Xtxp*295, were genotyped at CAAS according to the same protocol used at ICRISAT, except that amplification products, along with ROX-400 size standard, were separated by capillary electrophoresis in single-marker runs.

In all three labs, three control panel DNA samples were used as standard checks ([24], http://sorghum.cirad.fr/SSR_kit), in every PCR and electrophoresis run to facilitate accurate allele calling.

Table 1. Distribution of accessions in the sorghum Global Composite Germplasm Collection (GCGC).

Accession status, race or passport origin	Number of accessions (% of total)
Status	
Wild or weedy	68 (2.0%)
Landrace	3013 (89.5%)
Breeding lines or advanced cultivars	280 (8.3%)
Unknown	6 (0.1%)
Race	
Bicolor	195 (5.8%)
Caudatum	577 (17.2%)
Durra	656 (19.4%)
Guinea	365 (10.8%)
Kafir	239 (7.1%)
Intermediate	1159 (34.4%)
Unknown	115 (3.3%)
Passport origin	
Africa	1926 (57.3%)
Central Africa	224 (6.6%)
Eastern Africa	570 (16.9%)
Southern Africa	735 (21.8%)
Western Africa	397 (11.8%)
Asia	1010 (30.1%)
Eastern Asia	441 (13.1%)
Indian subcontinent	449 (13.3%)
Middle East	120 (3.6%)
North America	227 (6.7%)
Latin America	21 (0.6%)
Unknown	138 (4.0%)
Other	45 (1.3%)

Data Analysis

SSR markers used in this study showed high reproducibility in PCR amplification and ABI/Licor runs based on the allele sizes produced by control panel entries that were included in every PCR run. SagaGT software (Licor, USA) was used for allele scoring for the markers genotyped at CIRAD. At ICRISAT and CAAS, fragment analysis of PCR products was carried out using GeneScan and Genotyper 3.7 software packages (Applied Biosystems, USA). PCR amplicon sizes were scored in base pairs (bp) based on migration relative to the internal ROX-400 size standard. At ICRISAT these raw allele calls were further processed through the AlleloBin software program (available at http://www.icrisat.org/bt-software-d-allelobin.htm) to provide adjusted allele calls. AlleloBin uses a standard repeat motif length (following the step-wise mutation model [38]) and a least squares algorithm to call allele sizes to integer values as suggested by Idury and Cardon [39], adjusting for imperfections in the co-migration of size standards and PCR products.

Marker data for 7 SSR markers (*gpsb069, gpsp089, gpsb148, gpsb151, Xcup62, Xtxp295* and *Xtxp33*) were removed from the final analysis due to incomplete data or low quality genotyping. Finally, 3367 accessions were retained for further analysis across 41 markers (Table 1).

Data files were assembled in a database (Sagacity v.10, Rami, in preparation) and allele sizes were checked for congruency and adjusted according to the allelic references provided in the SSR kit [24].

Descriptors of observed genetic diversity, such as allele number per marker, observed heterozygosity (Ho) and gene diversity (expected heterozygosity, He) were calculated using PowerMarker v3.25 software [40]. Allelic richness and private alleles by locus were estimated using ADZE software [41]. Genetic distance between groups, estimated by F_{st} statistics, was calculated with hierfstat R package [42]. Mann-Whitney (MW) tests were used to determine whether estimates were significantly different between groups.

To identify the pair-wise genetic relationships between the accessions of this sorghum global composite germplasm collection, a genetic dissimilarity matrix was calculated using simple matching with DARwin v5 software [43] (available at http://darwin.cirad.fr/darwin/Home.php). An overall representation of the diversity structure was obtained by a factorial analysis using the distance matrix, while individual relations were analyzed with a tree construction based on Neighbor Joining (NJ) method, as implemented in DARwin v5.

In order to test for sample clustering in conjunction with admixture between sub-groups, Bayesian statistics based on Monte

Carlo Markov Chain algorithm were used. Although the Instruct software package [44] was developed to handle specifically species with a high level of inbreeding, as expected for sorghum, it was not used here because it cannot handle such a large number of samples. STRUCTURE software v.2.3.3 [45] was thus preferred. One hundred replicates were performed for each K, the number of clusters considered. Each run used a burn-in period of 100,000 iterations followed by 200,000 iterations. For each K, the 10 runs presenting the highest maximum likelihood value were kept, and sample assignation to groups was performed with CLUMPP software (up to K = 6, greedy algorithm, 1000 repeats, over K = 6, large K greedy algorithm, 1000 permutations) in order to deal with label switching or multimodalities. Estimate of the best cluster number was performed following [46] with a R (http://www.r-project.org/) script modified from [47]. It was compared to information given by each cluster, and identified when no new individual presented a majority of ancestry in a new cluster (threshold 0.7). Genome plot representations were performed using a specifically developed R script (available upon request).

A Reference Set of 383 sorghum accessions including *S. bicolor* subspecies *bicolor* and wild *S. bicolor* subspecies *verticilliflorum* was chosen among the publically available accessions to best represent genetic diversity as well as geographic origins. Maximum Length Subtree function of DARwin v5 software [43] was used to deal with genetic diversity. It is based on successive elimination of samples, each eliminated sample presenting a minimal reduction of overall diversity, measured as branch length of a tree. Since in the GCGC collections, phenotyping data were already available on a subset of diverse accessions ([11], [4]), this subset was first analyzed to reduce redundancy. Widely used breeding lines completed it. A first run of completion of these accessions was performed on *S. bicolor* only, checking that all geographic origins are conserved. The same process was performed for wild accessions, and both datasets were merged to represent the Sorghum Reference Set.

Results

Global Variation

Level of polymorphism. All 41 SSR markers used detected polymorphism in the sorghum GCGC. A total of 783 SSR marker alleles were detected, with an average of 19.2 alleles per marker. Numbers of alleles per marker ranged from three (*Xtxp*136) to 39 (*SbAGB*02), with an average of 3.44% of missing data (Table 2).

A mean gene diversity (expected heterozygosity, *He*) of 0.67 was observed across the sorghum global composite collection, with values ranging from 0.24 (*mSbCIR*246) to 0.94 (*Sb*5-206) for individual markers (Table 2). Even though *SbAGB*02 produced the highest number of alleles (39), it presented an intermediate *He* value of 0.67 because 92% of these alleles can be considered as rare (74% below 1% frequency). With the exception of *mSbCIR*248, which had an unusually high observed heterozygosity (*Ho*) value of 0.23, the *Ho* values ranged from 0.01 (*mSbCIR*246) to 0.06 (*Xtxp*015) with a mean of 0.03. Its outstanding *Ho* value suggests that marker *mSbCIR*248 may have detected more than one polymorphic locus, but this is not confirmed yet by *in-silico* hybridisation to the complete reference sorghum sequence.

Allelic distributions among taxonomic components. Allele number distribution and genetic diversity in sorghum GCGC according to biological status, race, and geographic origin is reported in Table 3. All 41 SSR markers used detected polymorphism in all compartments. The 3013 landrace accessions (87% of total accessions) contributed 94% of SSR marker alleles detected, all breeding lines (including advanced

cultivars, 280 accessions, 8%) and wild and weedy accessions (68 entries, 2%) captured 57% and 65% of the detected alleles, respectively. Allelic richness of standardized sample sizes of 100 haploid genomes showed that breeding lines tended to present less genetic diversity compared to landraces and wild samples, and that wild samples appeared more diverse (MW test, non-significant P values, P = 0.15 for breeding-landrace comparison and P = 0.08 for landrace-wild comparison). This is confirmed for private alleles (MW test, P<0.05 and P<0.01, respectively), with three times more private allele numbers in wild and weedy samples than in landraces (3.25 vs 1.04) and larger average expected heterozygosity values (MW test, P = 0.017).

Except of Kafir, the other four basic races exhibited no significant difference in allele numbers per marker. Kafir presented the smallest numbers of alleles per marker and private alleles (almost 3 alleles per marker less than the four others, MW tests, P<0.001) and a lower genetic diversity (*He* = 0.41 versus *He* of 0.60–0.67 for the other four basic races). The Guinea race encompassed the Guinea margaritiferum (Gma) accessions (at least 12), for which two markers (*mSbCIR*240 and *Xcup*53) were found to be monomorphic, whereas allelic richness of same sample sizes of all races, including other Guineas, ranged 1.58–7.02 and 1.56–3.04, respectively.

Highest numbers of alleles 680 (86.8%) were detected among the accessions of African origin. When correcting for sample sizes at the continent level, North American accessions (all originally introduced from elsewhere, or derived from such introduced materials) tended to be more diverse both in terms of total numbers and private alleles, but the MW tests were not conclusive. In Africa, Eastern Africa exhibited the largest gene diversity, followed by Central Africa while Southern Africa was the poorest (MW test, P = 0.02). In Asia, Middle East origins presented a higher genetic diversity than India and East Asia (MW test, private alleles, P = 0.05).

Allele specificity. Among the 783 alleles detected, 35% (280) were observed only in cultivated sorghum accessions and 5% (40) only in wild/weedy accessions.

Among the 41 SSR markers analyzed, 17 markers produced alleles unique to wild/weedy accessions, three (*mSbCIR*276, *Xisep*0107 and *Xtxp*136) for cultivated accessions, and *Xisep*0310 did not detect alleles unique to either the cultivated or wild/weedy accessions. Among these 17 SSRs, eight markers (*gpsb*067, *gpsb*123, *mSbCIR*223, *mSbCIR*238, *Sb*5-206, *Xcup*02, *Xcup*53 and *Xtxp*265) detected only one allele unique to wild/weedy accessions and a maximum of six such alleles were detected for marker *Xtxp*273. Out of the 68 wild/weedy accessions included in this study, 37 accessions produced these 40 alleles that were not detected in cultivated accessions. Wild accession IS 18931 alone contributed six alleles that were not found among the cultivated accessions and IS 18818 (of the *aethiopicuum* group within *S. bicolor* subspecies *verticilliflorum*) contributed five such alleles. Three alleles that were not detected among the cultivated accessions were detected in the only accession of *S. propinquum* (IS 18933) included in this global composite germplasm collection. Among the 3299 cultivated accessions, 40 of 41 SSR markers detected alleles not found among the 68 wild/weedy accessions. This is probably related to sample sizes differences and to the fact that SSR markers used in this study were chosen for their genome-wide distribution, based on existing maps built from crosses of cultivated accessions only, representing thus a diversity compartment different from wild/weedy entries.

The largest number of alleles unique to cultivated accessions was detected for *mSbCIR*240, for which 24 out of 35 alleles detected in the global collection were detected only in cultivated

Table 2. Marker characteristics and genetic diversity of the sorghum Global Composite Germplasm Collection (GCGC).

SSR marker	Forward primer sequence (5′-3′)	Reverse primer sequence (5′-3′)	Repeat	Chr	Allele Number	Gene diversity (*He*)	Observed heterozygosity (*Ho*)
gpsb067	TAGTCCATACACCTTTCA	TCTCTCACACACATTCTTC	(GT)10	8	15	0.681	0.032
gpsb123	ATAGATGTTGACGAAGCA	GTGGTATGGGACTGGA	(CA)7+(GA)5	8	14	0.720	0.030
mSbCIR223	CGTTCCAATGACTTTTCTTC	GCCAATGTGGTGTGATAAAT	(AC)6	2	10	0.703	0.023
mSbCIR238	AGAAGAAAAGGGGTAAGAGC	CGAGAAACAATTACATGAACC	(AC)26	2	27	0.859	0.027
mSbCIR240	GTTCTTGGCCCTACTGAAT	TCACCTGTAACCCTGTCTTC	(TG)9	8	35	0.746	0.034
mSbCIR246	TTTTGTTGCACTTTTGAGC	GATGATAGCGACCACAAATC	(CA)7.5	7	13	0.237	0.010
mSbCIR248	GTTGGTCAGTGGTGGATAAA	ACTCCCATGTGCTGAATCT	(GT)7.5	5	13	0.659	0.226
mSbCIR262	GCACCAAAATCAGCGTCT	CCATTTACCCGTGGATTAGT	(CATG)3.25	10	30	0.663	0.044
mSbCIR276	CCCCAATCTAACTATTTGGT	GAGGCTGAGATGCTCTGT	(AC)9	3	10	0.559	0.023
mSbCIR283	TCCCTTCTGAGCTTGTAAAT	CAAGTCACTACCAAATGCAC	(CT)8 (GT)8.5	10	24	0.810	0.020
mSbCIR286	GCTTCTATACTCCCCTCCAC	TTTATGGTAGGATGCTCTGC	(AC)9	1	19	0.795	0.026
mSbCIR300	TTGAGAGCGGCGAGGTAA	AAAAGCCCAAGTCTCAGTGCTA	(GT)9	7	11	0.689	0.031
mSbCIR306	ATACTCTCGTACTCGGCTCA	GCCACTCTTTACTTTTCTTCTG	(GT)7	1	5	0.616	0.015
mSbCIR329	GCAGAACATCACTCAAAGAA	TACCTAAGGCAGGGATTG	(AC)8.5	5	13	0.746	0.028
Sb4-72	TGCCACCACTCTGGAAAAGGCTA	CTGAGGACTGCCCCAAATGTAGG	(AG)16	6	24	0.699	0.021
Sb5-206	ATTCATCATCCTCATCCTCGTAGAA	AAAAACCAACCCGACCCACTC	(AC)13/(AG)20	9	34	0.941	0.046
Sb6-84	CGCTCTCGGGATGAATGA	TAACGGACCACTAACAAATGATT	(AG)14	2	32	0.859	0.027
SbAGB02	CTCTGATATGTCGTTGTGCT	ATAGAGAGGATAGCTTATAGCTCA	(AG)35	7	39	0.668	0.033
Xcup02	GACGCAGCTTTGCTCCTATC	GTCCAACCAACCCACGTATC	(GCA)6	9	10	0.656	0.030
Xcup11	TACCGCCATGTCATCATCAG	CGTATCGCAAGCTGTGTTTG	(GCTA)4	3	6	0.501	0.050
Xcup14	TACATCACAGCAGGGACAGG	CTGGAAAGCCGAGCAGTATG	(AG)10	3	22	0.542	0.019
Xcup53	GCAGGAGTATAGGCAGAGGC	CGACATGACAAGCTCAAACG	(TTTA)5	1	11	0.577	0.021
Xcup61	TTAGCATGTCCACCACAACC	AAAGCAACTCGTCTGATCCC	(CAG)7	3	6	0.474	0.024
Xcup63	GTAAAGGGCAAGGCAACAAG	GCCCTACAAAATCTGCAAGC	(GGATGC)4	2	11	0.316	0.033
Xisep0107	GCCGTAACAGAGAAGGATGG	TTTCCGCTACCTCAAAAACC	(TGG)4	3	5	0.556	0.014
Xisep0310	TGCCTTGTGCCTTGTTTATCT	GGATCGATGCCTATCTCGTC	(CCAAT)4	2	10	0.252	0.019
Xtxp10	ATACTATCAAGAGGGGAGC	AGTACTAGCCACACGTCAC	(CT)14	9	15	0.778	0.055
Xtxp12	AGATCTGGCGGCAACG	AGTCACCCATCGATCATC	(CT)22	4	30	0.935	0.039
Xtxp15	CACAAACACTAGTGCCTTATC	CATAGACACCTAGGCCATC	(TC)16	5	23	0.863	0.062
Xtxp21	GAGCTGCCATAGATTTGGTCG	ACCTCGTCCCACCTTTGTTG	(AG)18	4	33	0.625	0.036
Xtxp40	CAGCAACTTGCACTTGTC	GGGAGCAATTTGGCACTAG	(GGA)7	7	21	0.380	0.021
Xtxp57	GGAACTTTTGACGGGTAGTGC	CGATCGTGATGTCCCAATC	(GT)21	6	29	0.823	0.058
Xtxp114	CGTCTTCTACCGCGTCCT	CATAATCCCACTCAACAATCC	(AGG)8	3	11	0.597	0.040
Xtxp136	GCGAATAGCATCTTACAACA	ACTGATCATTGGCAGGAC	(GCA)5	5	3	0.457	0.022
Xtxp141	TGTATGGCCTAGCTTATCT	CAACAAGCCAACCTAAA	(GA)23	10	22	0.887	0.035
Xtxp145	GTTCCTCCTGCCATTACT	CTTCCGCACATCCAC	(AG)22	6	32	0.917	0.055
Xtxp265	GTCTACAGGCGTGCAAATAAAA	TTACCATGCTACCCCTAAAAGTGG	(GAA)19	6	26	0.919	0.058
Xtxp273	GTACCCATTTAAATTGTTTGCAGTAG	CAGAGGAGGAGGAAGAGAAGG	(TTG)20	8	21	0.689	0.030
Xtxp278	GGGTTTCAACTCTAGCCTACCGAACTTCCT	ATGCCTCATCATGGTTCGTTTTGCTT	(TTG)12	7	25	0.474	0.027
Xtxp320	TAAACTAGACCATATACTGCCATGATAA	GTGCAAATAAGGGCTAGAGTGTT	(AAG)20	1	19	0.847	0.046
Xtxp321	TAACCCAAGCCTGAGCATAAGA	CCCATTCACACATGAGACGAG	(GT)4+(AT)6 +(CT)21	8	30	0.934	0.033
				Mean	19.244	0.674	0.037
				Min	3	0.237	0.010
				Max	39	0.941	0.226

Na: Number of alleles, He: unbiased genetic diversity, according to Nei (1987), Ho: observed heterozygosity.
Availability of marker data ranged from 88% (*gpsb*123) to 99% (*Xcup*63). On average 3.44% of data was missing.

Table 3. Genetic diversity in the sorghum Global Composite Germplasm Collection (GCGC) and in the Reference Set, partitioned into biological status, races and geographic origins as indicated in passport data.

	Global Composite Germplasm Collection							Reference Set				
	N	Na	MeanNa	Arich (100)	PrivA (100)	Gene Diversity	Ho	N	Nall	MeanNa	Gene Diversity	Ho
Overall	3367	783	19.10	10.04 (0.95)[1]		0.674	0.037	383	613	14.95	0.712	0.048
Status												
Wild or weedy	68	508	12.39	11.91 (0.95)	3.25 (0.41)	0.743	0.234	23	355	8.66	0.748	0.216
Landrace	3013	736	17.95	10.06 (0.95)	1.04 (0.17)	0.671	0.032	332	576	14.05	0.707	0.035
Breeding lines or advanced cultivars	280	443	10.80	8.53 (0.78)	0.58 (0.09)	0.630	0.042	28	263	6.41	0.621	0.058
Unknown	6	163	3.98			0.536	0.153	0	0	0	0.000	0.000
Race												
Bicolor	195	483	11.78	9.65 (0.92)	0.74 (0.10)	0.669	0.041	36	334	8.15	0.695	0.045
Caudatum	577	539	13.15	9.16 (0.94)	0.59 (0.18)	0.626	0.029	76	378	9.22	0.633	0.040
Durra	656	521	12.71	8.91 (0.88)	0.47 (0.07)	0.600	0.043	44	312	7.61	0.655	0.024
Guinea	365	476	11.61	8.62 (0.80)	0.65 (0.15)	0.628	0.025	64	331	8.07	0.661	0.027
Kafir	239	327	7.98	5.97 (0.59)	0.15 (0.04)	0.410	0.021	23	191	4.66	0.444	0.031
Intermediate	1159	629	15.34	9.78 (0.94)	0.48 (0.06)	0.661	0.029	104	450	10.98	0.703	0.039
Unknown	116	376	9.17			0.610	0.085	18	236	5.76	0.626	0.100
Passport origin												
Africa	1853	680	16.59	9.83 (0.91)	1.27 (0.18)	0.654	0.032	257	558	13.61	0.697	0.040
Central Africa	219	444	10.83	8.68 (0.85)	0.66 (0.15)	0.630	0.037	35	281	6.85	0.645	0.040
Eastern Africa	537	571	13.93	9.86 (0.98)	1.06 (0.15)	0.670	0.036	85	431	10.51	0.688	0.046
Southern Africa	718	508	12.39	7.49 (0.72)	0.59 (0.10)	0.511	0.026	74	372	9.07	0.592	0.038
Western Africa	379	512	12.49	8.82 (0.71)	0.99 (0.14)	0.611	0.038	63	351	8.56	0.674	0.034
Asia	976	594	14.49	8.91 (0.89)	1.12 (0.19)	0.587	0.043	71	372	9.07	0.644	0.048
Eastern Asia	439	438	10.68	7.49 (0.81)	1.05 (0.22)	0.474	0.052	18	181	4.41	0.466	0.027
Indian Subcontinent	417	454	11.07	7.98 (0.77)	1.16 (0.15)	0.576	0.022	35	278	6.78	0.623	0.030
Middle East	120	400	9.76	8.64 (0.82)	1.82 (0.28)	0.602	0.085	18	238	5.8	0.614	0.106
Europe	1	42	1.02			–	–	1	42	1.02	–	–
Mediterranean Basin	29	271	6.61			0.637	0.031	7	161	3.93	0.649	0.037
North America	185	506	12.34	10.10 (0.91)	1.51 (0.18)	0.690	0.042	34	330	8.05	0.710	0.093
South America	21	201	4.90			0.587	0.038	0	0	0	0.000	0.000
Australia	13	166	4.05			0.486	0.038	2	82	2	0.396	0.195
Unknown	3	93	2.27			0.311	0.008	0	0	0	0.000	0.000

[1]calculated over 383 diploid genomes.
Partition into geographic origins is limited to landraces and wild samples for which geographic origin relates to reality.
N: number of accessions; Na: total number of alleles; MeanNa: mean number of alleles per marker, ARich: allelic richness calculated according to Petit et al. (1998) as in [41] for standard sample sizes of 100 genomes; PrivA: Private allele number per marker calculated according to [41] for standard sample sizes of 100 genomes (allelic richness and private allele number per marker were calculated for those classes which presented more than 100 genomes, by continent – Africa, Asia and Northern America – and in each sub-continent – Africa and Asia –), before the pipe are the values observed inside the continent, each continent being analyzed separately, after the pipe are the values observed when all 3 continents are analyzed together at the sub-continent level; Ho: observed heterozygosity.

accessions, but no alleles of this marker were detected only in wild/weedy accessions. The overall frequency of rare marker alleles in the sorghum GCGC was very high. Across the 3367 accessions, 428 rare alleles (54.2%) below 1% frequency and 621 rare alleles (78.7%) below 5% frequency were detected.

Patterns of Multi-locus Diversity

Factorial analysis. Factorial analysis (FA) of the SSR-based dissimilarity matrix of the complete sorghum GCGC (3367 accessions) showed that the first four axes were to be considered (See plot in Figure S1). The first axis enabled the separation of accessions collected in Africa versus more eastern origins (including some of eastern Africa) (6.05% of the global inertia) (Figure 1). The second (4.09%) and third axes (2.92%) refined the situation of Africa by separating southern Africa and western Africa from central and eastern Africa. Finally the fourth axis (2.35%) enabled the separation of origins from the Indian subcontinent, the Middle East, and eastern Asia. The reference to the racial classification (Figure 2)

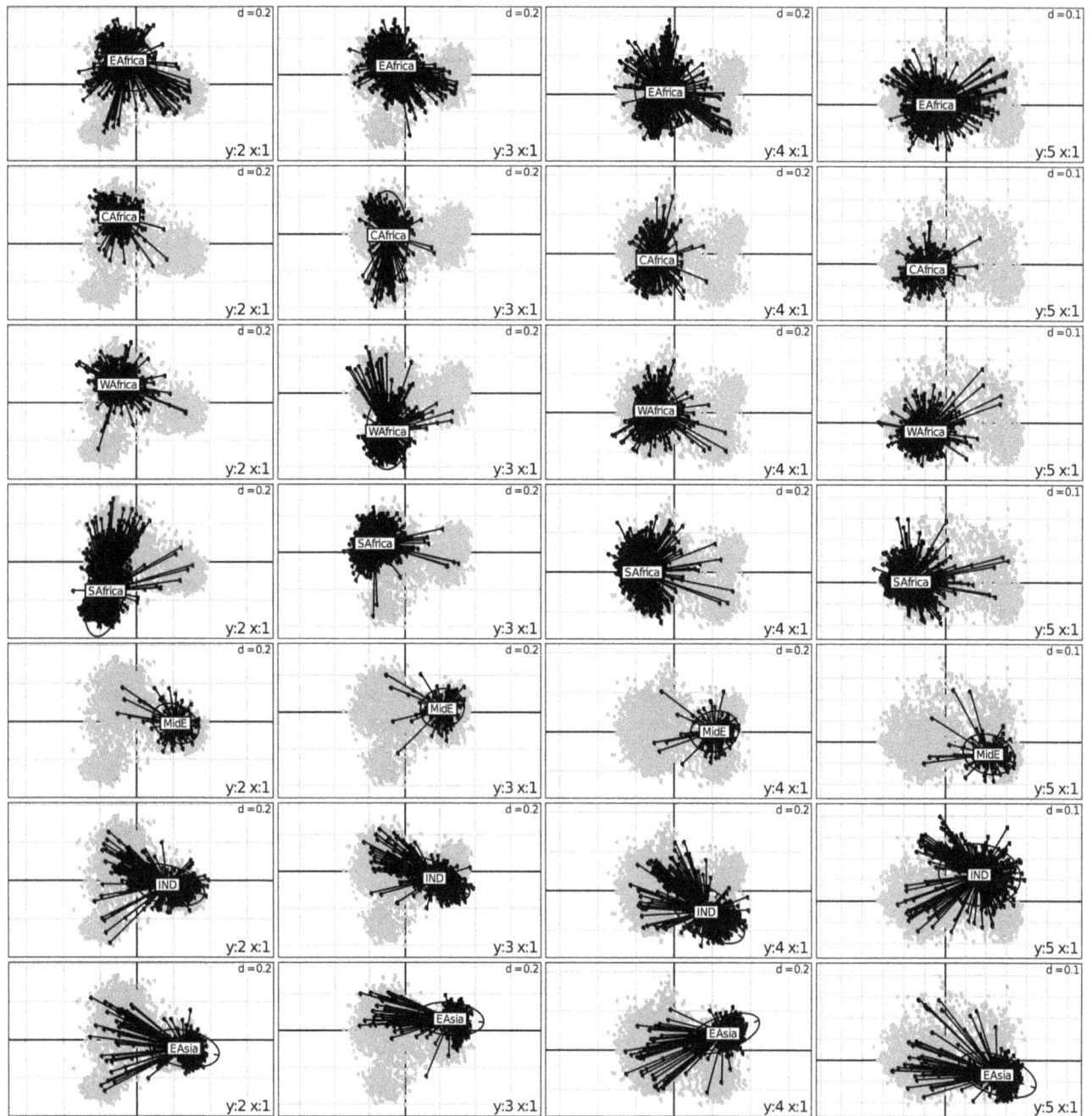

Figure 1. Factorial analysis of the simple matching distance matrix. Representation of the first four axes with accessions characterized by the seven main geographic origins. EAfrica: Eastern Africa, CAfrica: Central Africa, WAfrica: Western Africa, MidE: Middle East countries, IND: Indian subcontinent, EAsia: Eastern Asia, SAfrica: Southern Africa.

yields a much less coherent picture, with most races distributed over the whole planes of the FA, with the sole exception of Kafir, clearly separated on plane (1, 2).

Classification using bayesian assignations (STRUCTURE). Bayesian assignations to sub-groups were performed for 2 to 10 populations (Figure 3), after which no new group was detected with an admixture threshold of 0.7. Two-thirds (2190) of the accessions could be assigned to one of these 10 groups. The unassigned accessions presented genomes scattered among different groups. They included 85% of the wild (58), 54% of the breeding accessions (151), half of the intermediate or

unknown races (552), half of the Caudatum (265) and one fifth of the Durra (142) mainly from Eastern Africa. Analysis of assignation rate showed, however, that accessions could be grouped primarily in three populations followed by a sub-division into seven populations (Figure S2).

The first three subdivisions obtained by Bayesian assignment (Figure 3, K = 2, 3 and 4) reflected the main features revealed by the first three axes of the FA and the fourth subdivision (Figure 3, K = 5) reflected axis 4. The subsequent subdivisions also corroborated patterns that appeared through the FA. Thus, Group 1 included Caudatum, Caudatum-Bicolor and Durra from Eastern

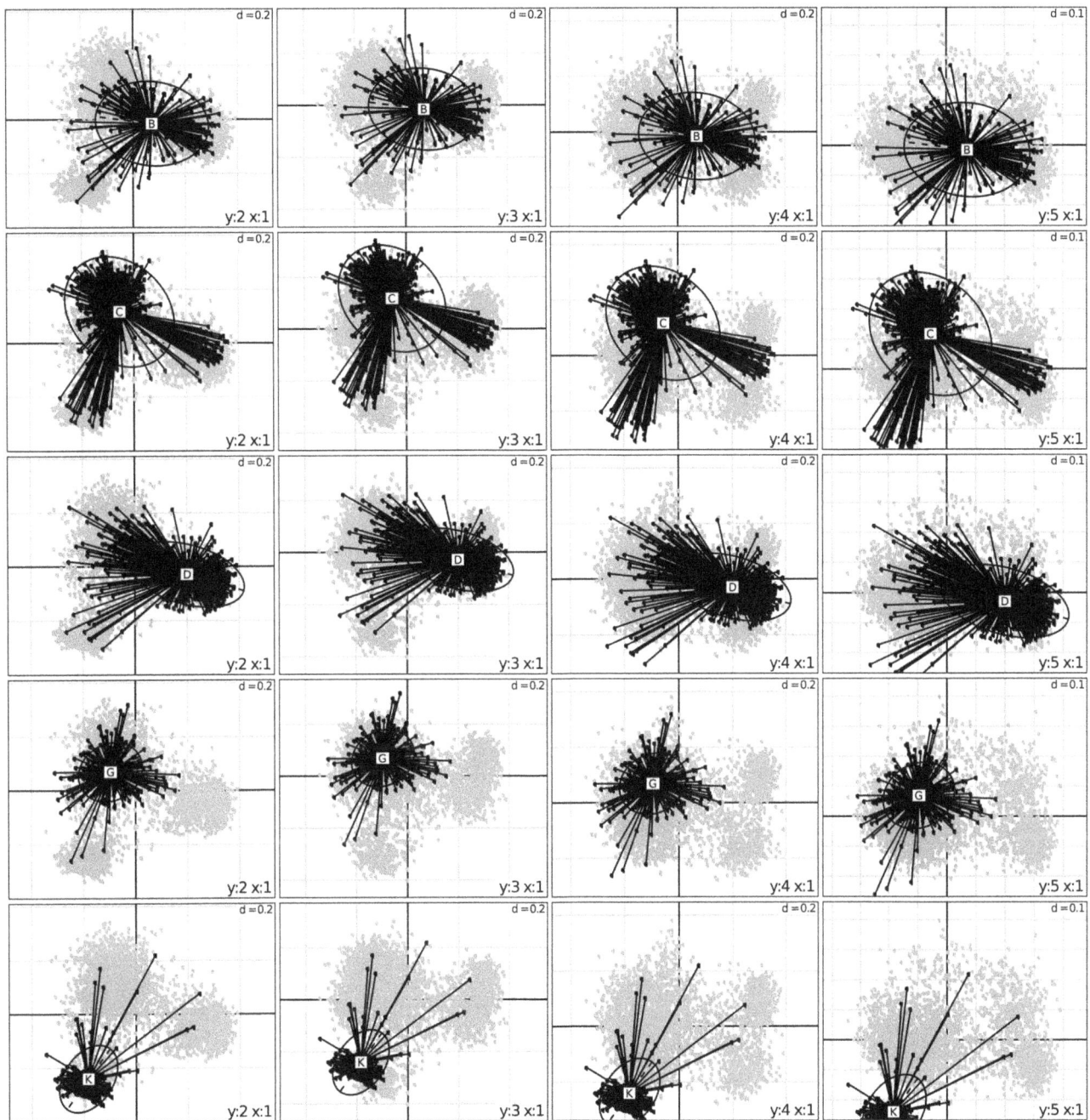

Figure 2. Factorial analysis of the simple matching distance matrix. Representation of the first four axes with accessions characterized by races. B: Bicolor, C: Caudatum, D: Durra, G: Guinea, K: Kafir. Based on allelic richness, there is a trend for Bicolor to be more diverse, followed by Caudatum, Durra, Guinea and finally Kafir being significantly less diverse.

Asia; Group 2 encompassed Durra and Bicolor from the Indian subcontinent, while Group 3 exhibited Durra from Eastern Africa. Bicolor and Durra-Bicolor from Eastern Africa were assigned in Group 4. Group 5 included Guinea and Guinea margaritiferum from Western Africa and Bicolor from North America. Group 6 appeared as a well-separated group made predominantly of Guinea accessions from western Africa, accompanied by intermediate race Durra-Caudatum materials from western Africa while Group 7 was made essentially of materials collected from eastern Africa and central Africa generally classified as race Caudatum (visible along FA axis 3). Group 8 was a small and heterogeneous

group made of Durra and Caudatum race accessions from central Africa. Group 9 was made essentially of Guinea race accessions from the Indian subcontinent and southern/eastern Africa with Guinea-Caudatum (GC) intermediate race accessions from various parts of Africa. Group 10 was made almost exclusively of accessions from southern Africa of race Kafir or intermediate race Kafir-Caudatum (KC).

Neighbor joining analysis. The NJ dendrogram representation on all samples revealed global congruence with the Bayesian assignment with a few apparent discrepancies (Figure 4).

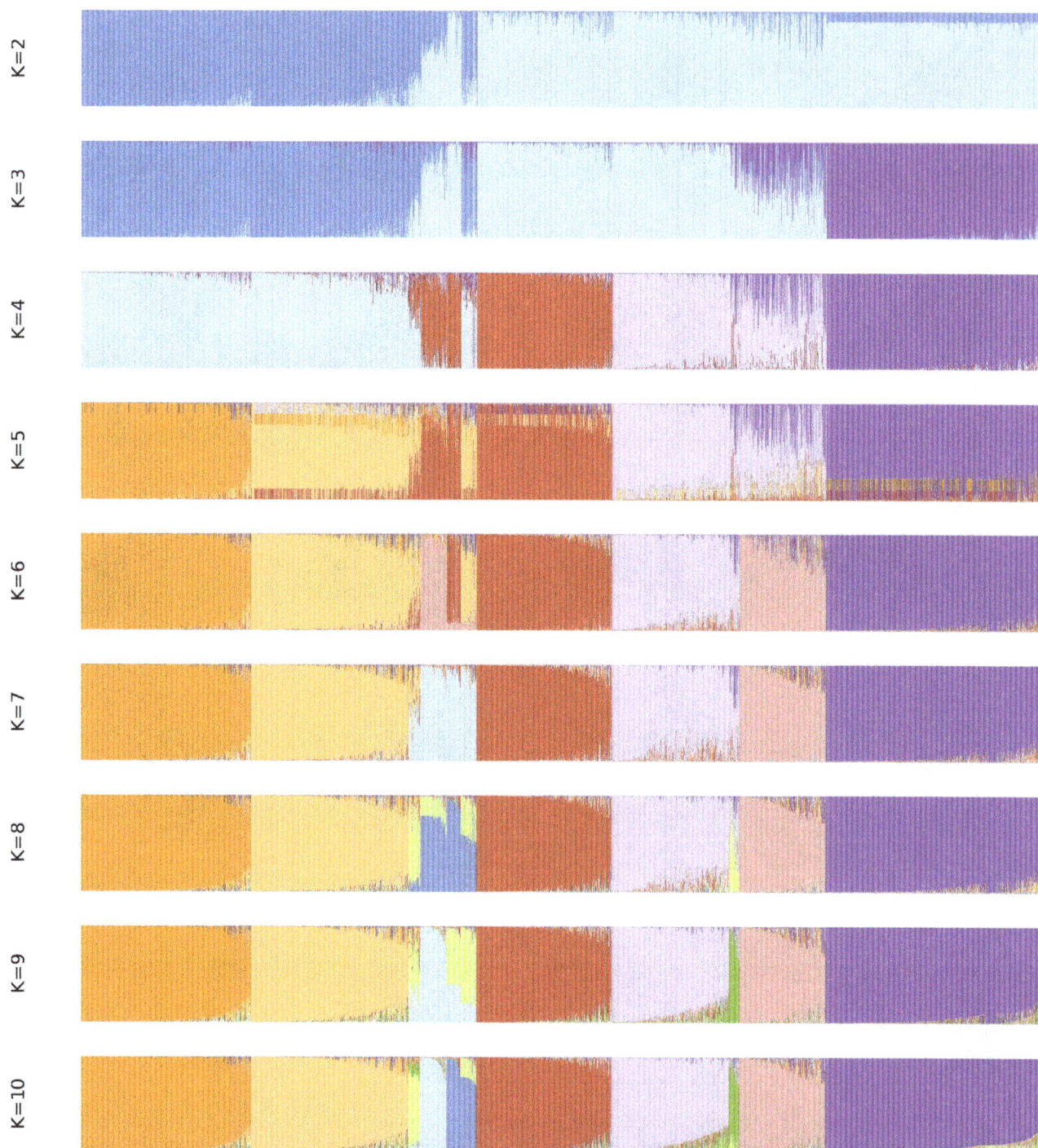

Figure 3. Genome representation of the sorghum GCGC collection, obtained from the assignation by STRUCTURE software at a 0.7 threshold (Pritchard et al. 2000) of each sample in K hypothetical sub-groups. In this study, K varied from 2 (top) to 10 (bottom). Each accession on the X-axis is represented by K colours ordered according to a decreasing genome fraction on the Y-axis. At K = 10, Group 1 in orange: C, CB and D from Eastern Asia, Group 2 in light orange: D and B from the Indian subcontinent, Group 3 in light green: D from Eastern Africa, Group 4 in light blue: B and DB from Eastern Africa, Group 5 in dark blue: G and Gma from Western Africa and B from North America, Group 6 in red: D, DC and G from Western Africa, Group 7 in light purple: C from Central and Eastern Africa, Group 8 in dark green: C and GC from Southern Africa, Group 9 in pink: G from Asia and Southern Africa and C from Eastern Africa, and Group 10 in purple: GC, K and KC from Southern Africa.

The main discrepancies were the splits of Group 5 and Group 9 into distinct dendrogram sectors. Within Group 5, this split corresponded well with a Bicolor vs Guinea differentiation and led to the distinction of 5a and 5b. Group 9 split into three components 9a, 9b and 9c, 9a and 9b being essentially made of Guinea varieties from South Asia and eastern and southern Africa, respectively, and 9c made of a few Caudatum varieties from eastern Africa. The NJ analysis also threw light on an array of

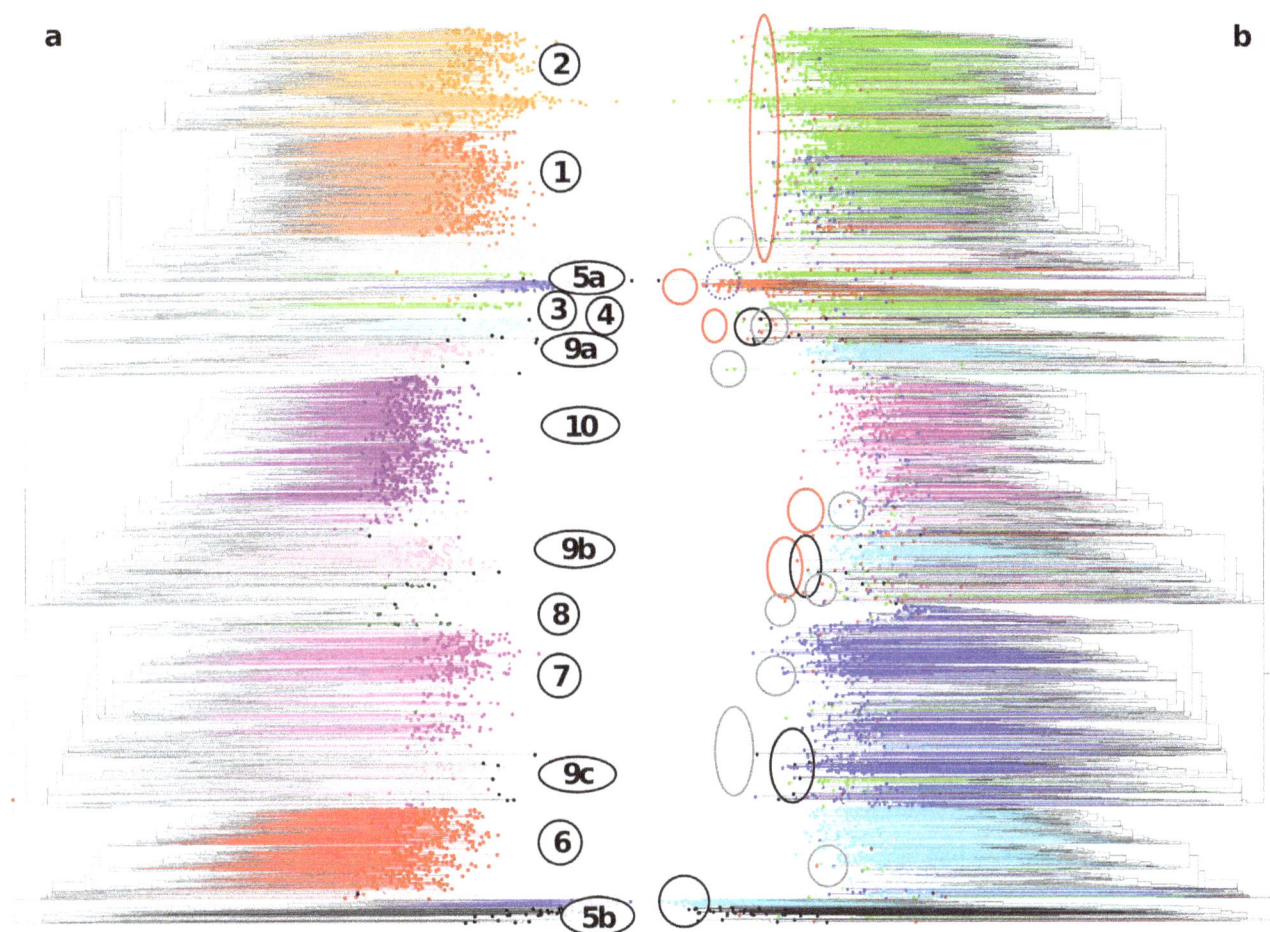

Figure 4. Hierarchical NJ cluster analysis of 3367 sorghum accessions of a global composite germplasm collection based on allelic data from 41 SSR markers (simple matching distance). a- Accessions grouped by Bayesian analysis (Figure 3, K = 10) are represented in color, corresponding to Group 1 in orange, Group 2 in light orange, Group 3 in light green, Group 4 in light blue, Group 5 in dark blue, Group 6 in red, Group 7 in light purple, Group 8 in dark green, Group 9 in pink, and Group 10 in purple. NJ clustering enabled finer resolution of these groups, leading to subdivisions into Group 5a and Group 5b in dark blue, Group 9a, Group 9b and Group 9c in pink. Unassigned accessions are presented in grey. Wild accessions are presented in black. b- Accessions coloured according to their classification in various taxonomic components: Bicolor in red, Caudatum in dark blue, Durra in green, Guinea in light blue, Kafir in purple, unclassified in grey, wild in black. The dendrogram sectors including dispersed components accessions (wild/weedy, Bicolor, and unclassified) are highlighted by circles of the corresponding colours (black, red, grey).

unclassified accessions in the periphery of groups 1 and 2, consisting predominantly of Durra and DC accessions from the Middle East. Group 3 was also challenged by the NJ representation, with most accessions in one dendrogram sector but several of them in another; the size of this group was, however, too small for justifying internal sub-divisions.

Distribution of taxonomic components. The classification derived from the STRUCTURE analysis complemented by the NJ dendrogram enabled analyzing the distribution of the various *a priori* taxonomic components. The NJ dendrogram further helped locating all the unassigned materials in relation to the groups that it supported or revealed.

Wild and weedy sorghum accessions were mainly found in four dendrogram sectors (Figure 4). Almost two-thirds (40) of accessions of *S. bicolor* subspecies *verticilliflorum* (belonging to races *aethiopicuum*, *arundinaceum*, *verticilliflorum*, and *virgatum*) of diverse origins, as well as weedy intermediate *S. bicolor* subspecies *drummondii* clustered around Group 5b. A separate group of *drummondii* and *verticilliflorum* accessions from eastern Africa was observed around Group 9c, associated with cultivated materials from Sudan and Uganda.

Another group of *drummondii* accessions from Tanzania, Kenya and Zimbabwe were clustered around Group 9b materials from southern and eastern Africa. Finally, a group of wild and weedy accessions from eastern Africa clustered around Group 4 in close proximity to intermediate race Durra-Bicolor accessions from that region.

The 195 accessions classified as race Bicolor were scattered across many dendrogram sectors and no distinct Bicolor cluster was observed, other than Group 5a, comprised of accessions specifically collected to represent "broom sorghum". However, the periphery of Groups 1, 2, 4, 5b, 9a and 10 appeared Bicolor-enriched. The four Bicolor accessions close to Group 5b fell among wild/weedy accessions.

Guinea accessions were mainly grouped into four separate dendrogram sections (Figures 3 and 4). Some Guinea accessions, mainly *roxburghii* sub-race materials from the Indian subcontinent and southern Africa, were in Group 9a. A large number of Guinea race accessions from southern Africa (mainly of the *conspicuum* and *roxburghii* sub-race materials from Tanzania and Malawi) were clustered in Group 9b. Another large cluster of Guinea race accessions, mainly from western Africa (Mali, Ghana, Nigeria, Burkina Faso, etc.) and including sub-races *gambicum* and *guineense*, were found in Group 6. Accessions of the *margaritiferum* (Gma) sub-race from western Africa formed a separate Group 5b in close association with wild and weedy accessions.

Caudatum race accessions (577) were broadly dispersed. The vast majority originated from eastern Africa and grouped in and around Groups 7 and 9c. The others followed a geographic organization, with accessions from China in Group 1 and accessions from western Africa and southern Africa in Groups 6 and 10, respectively.

The Durra race was the most widely represented in the GCGC (656 accessions). Most were distributed across several major clusters, with a strong geographical organization. Most Durra accessions from the Indian subcontinent were in Group 2 along with related intermediate materials from that region. Accessions from eastern Asia (mostly from China) were found in Group 1 and accessions from the Middle East and eastern Africa fell in the components of Group 3, while smaller numbers of Durra accessions were in the periphery of Groups 6, 7 and 9c. Interestingly, five Durra accessions clustered with wild/weedy accessions in the vicinity of Group 5b.

The Kafir accessions (239) were mostly from southern Africa and fell in Group 10, together with Kafir-Caudatum and Kafir-Durra accessions from the same region.

The majority of intermediate race accessions were grouped according to their geographic origin. Guinea-Caudatum (GC) was the most common (361 accessions) and was scattered across all NJ sectors, with a majority in the vicinity of Group 7. Durra-Caudatum (DC) was the next most common intermediate race (330 accessions), and was geographically distributed around Group 6 (western Africa) and around Groups 1 and 2 (Mediterranean Basin and the Middle East). Caudatum-Bicolor (CB) accessions were predominantly from eastern Asia and fell in and around Group 1 whereas Durra-Bicolor (DB) accessions from the Indian subcontinent and eastern Africa fell in and around Groups 2 and 4, respectively. Ten intermediate race accessions grouped with wild/weedy accessions close to Group 5b.

A total 430 trait-specific accessions were included in the sorghum GCGC. Many of them were classified as race Caudatum, including accessions resistant to downy mildew, which were clustered according to their origins in Groups 2, 6, 7 and 9c. Stem borer resistant genotypes of race Durra from the Indian subcontinent and Africa were grouped together in Group 2. Genotypes with the capacity to germinate through crusted soil were found in various groups in accordance with their origin and race. Most midge resistant genotypes were found in Group 7. Most of the sweet stalk sorghums that are of increasing interest globally were observed to have Caudatum race background and fell into Group 7. Broom sorghum accessions of race Bicolor from USA formed a specific single Group 5a, whereas all pop sorghum accessions belonging to race Guinea from the Indian subcontinent grouped together in Group 9a. The latter two groups are both small in size and might actually exist because of an over-representation of specialty sorghums gathered for a targeted purpose and resting on a narrow genetic basis.

Global Differentiation Pattern

The differentiation between all the components derived from the confrontation of both classification methods was assessed using the F_{ST} estimate (Table S2 and Figure 5b). Pairwise F_{ST} estimates between the 13 groups identified were all significantly different from zero and varied from 0.130 to 0.531, with a mean value of 0.378.

The relationships based on the final groups, their mutual differentiations measured with F_{ST} estimate, the distribution of the various races and intermediates in the NJ dendrogram are summarized in Figure 5.

With the exception of Groups 5a and 9a, sorghum genetic diversity appears organized along a limited number of clearly differentiated groups in the West (Guinea-dominated, yet clearly different from one another, Groups 5b and 6), in the South (Kafir-dominated Group 10), in the East (multiracial Groups 1, 2 and 9a) and in the Center (Durra/Bicolor Group 4 and Durra-dominated Group 3), within a background that appears as a broad swarm in central and eastern Africa (weak structure between Groups 7, 8 and 9) with a frequent reference to the Caudatum race component.

Reference Set of Sorghum

A core reference set with 383 accessions was selected to capture the global genetic diversity of sorghum (Table 3). It includes 332 landraces, 28 breeding lines and 23 wild/weedy accessions, all five cultivated basic races, the 10 intermediate races and accessions of all different geographic origins except South America. It represents the global genetic diversity present in sorghum GCGC (Figure 6). This sorghum reference set captured 78.3% (613 alleles) of the SSR alleles detected in the GCGC, with an average of 14.9 alleles per SSR primer pair (Table 3), comparable to standardized allelic richness of the GCGC. For markers *mSbCIR*306 and *Xisep*0310, all alleles (5 and 10 alleles, respectively) detected in the GCGC were captured in the reference set. Average gene diversity (0.71) in the reference set is slightly larger than for the GCGC. Clustering of accessions in the reference set follows the pattern of race within geographic origin described above for the GCGC. In the case of Gma sub-race, 11 of 12 accessions included in the global composite germplasm collection (all from western Africa) were captured in the reference set.

Discussion

Maintenance and characterization of large germplasm collections is a huge task. Knowledge of the characteristics of the materials is essential for their efficient management. Both genetic and morpho-agronomical characterizations are required for breeders to better understand and use the available genetic resources. It increases the efficiency of selection of more diverse, adapted, germplasm parents in crop improvement programs. To serve as an entry point to large collections, representative subsets (often referred to as core or minicore collections) provide an economically and logistically attractive option for both gene banks and the breeding programs they serve. However, it is very important that such core collections represent the full range of diversity available at the time of the study. In this context, we used SSR markers to ascertain the population structure of a very large set of sorghum germplasm, in the framework of an international project (the Generation Challenge Programme), consisting of accessions assumed to be representative of global germplasm available for improvement of this crop. This set was used to fine-tune and complete previous knowledge on the evolutionary history and domestication pattern of sorghum. Using this information, a representative subset of this collection was chosen, of a more convenient size for detailed characterization of traits of economic importance to plant breeding programs and for the assessment of allelic diversity in genes associated with variation in such traits.

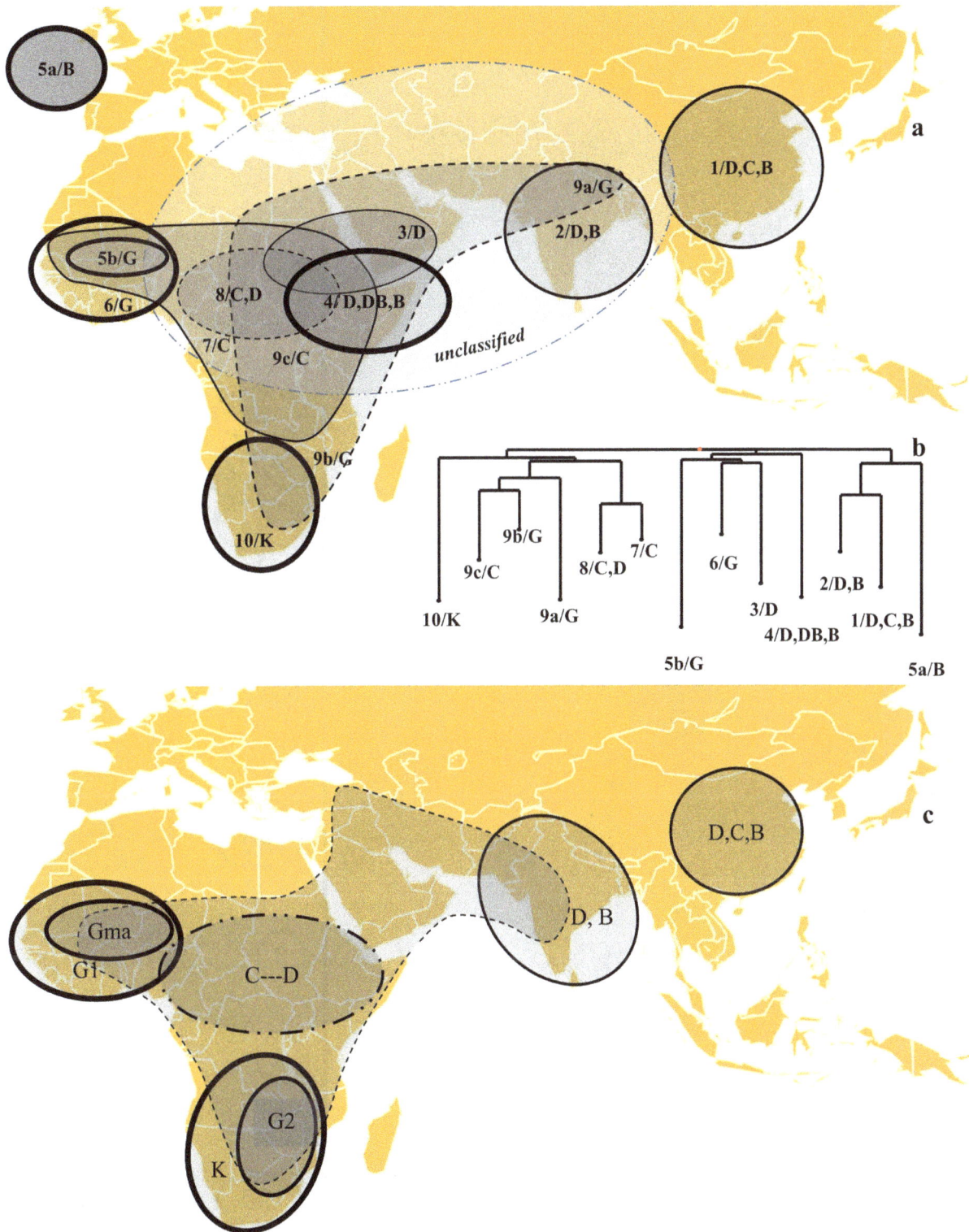

Figure 5. Schematic representation of the pattern of diversity in sorghum projected on a geographical map. a- The groups as identified in Figure 4 are drawn according to their geographical distribution and the predominant race(s) is (are) indicated. The groups are framed differently to reflect their higher (thicker frame) or lower (thin, dotted frame) levels of differentiation as estimated through the F_{ST} parameter and the distribution of intermediates. Group 5a actually originates from a collection in USA. b- NJ dendrogram of F_{ST} distances between groups identified as in Figure 4. c- Pure races and main regions are predominantly featured, but the intermediate types or regions fall in continuity with this landscape (dotted lines). Races are framed differently to reflect their higher (thicker frame) or lower (thin, dotted frame) levels of differentiation as estimated through the F_{ST} parameter as in Figure 5a.

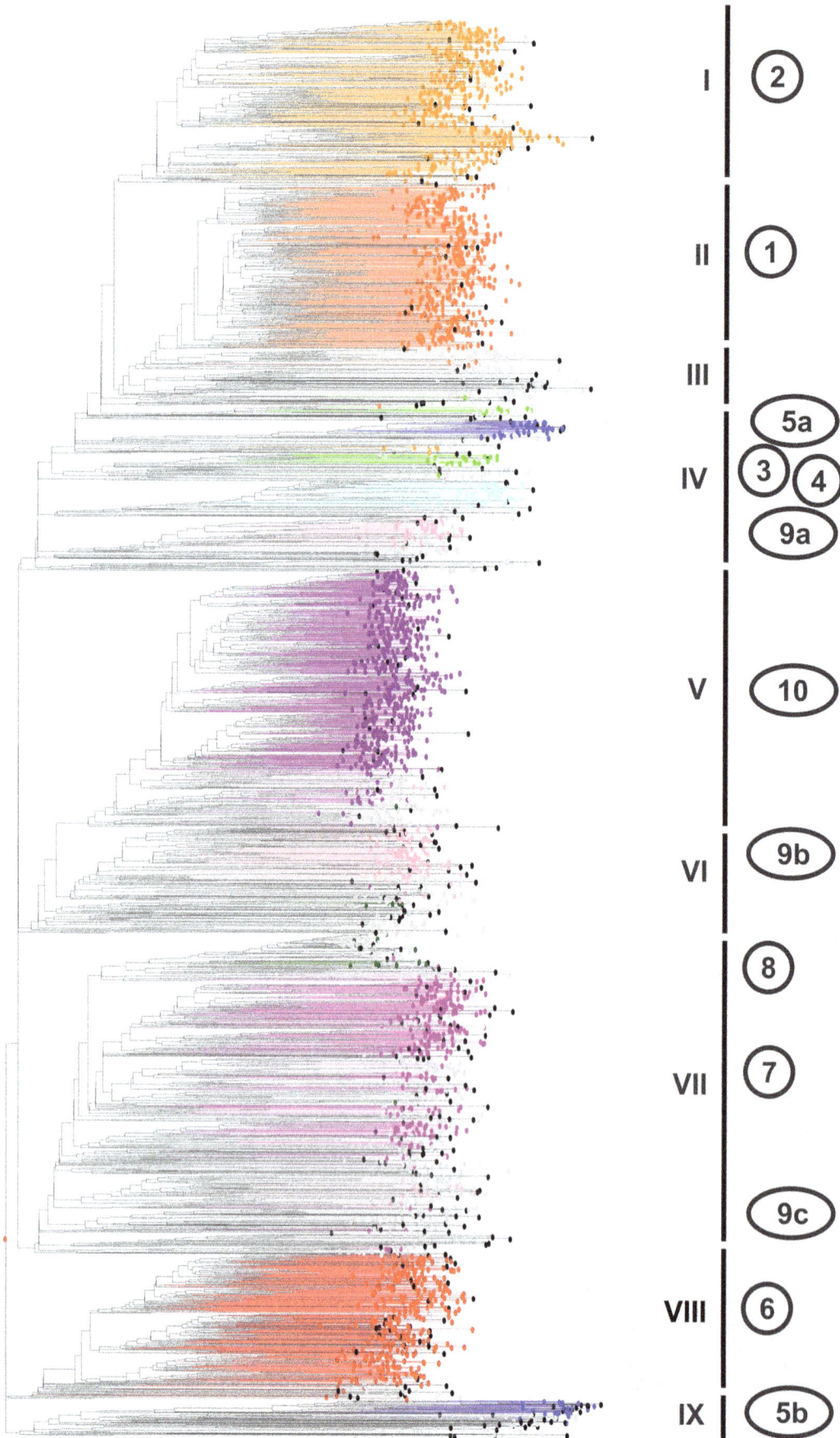

Figure 6. Selected sorghum Reference Set (383 accessions, in black) in relation with hierarchical NJ cluster analysis of 3367 sorghum accessions of a global composite germplasm collection based on allelic data from 41 SSR markers (simple matching distance). Accessions grouped by Bayesian analysis are represented in colors as in Figure 4: Group 1 in orange (C, CB and D from Eastern Asia), Group 2 in light orange (D and B from the Indian subcontinent) Group 3 in light green (D from Eastern Africa), Group 4 in light blue: B and DB from Eastern Africa), Group 5a in dark blue (B accessions assembled from North America) and Group 5b in dark blue (Gma), Group 6 in red (D, DC and G from Western Africa), Group 7 in light purple (C from Central and Eastern Africa), Group 8 in dark green (C and GC from Southern Africa), Group 9a in pink (G from the Indian subcontinent and East Asia), Group 9b in pink (G from Southern Africa), Group 9c in pink (C from Eastern Africa), and Group 10 in purple (GC, K and KC from Southern Africa). Unassigned accessions are presented in grey.

A Species-wide Scan Assessment of Neutral Genetic Diversity

Breadth of variation. To our knowledge, this is the largest study undertaken in a systematic way for exploring genetic diversity in global crop germplasm. The broad plant material coverage resulted in larger allele numbers (average of 19.2 alleles per locus) and higher diversity parameters than in most previous studies ([48], [49], [50], [51], [32]). It was comparable to the features reported in a focus study on Niger by [30], [52]. The same is true when considering each race separately.

The mean observed heterozygosity (Ho) was 0.037, indicating that most markers used detected only one allele per accession, and that the accessions are highly inbred, as expected for accessions of a largely self-pollinated species maintained in collections by enforced selfing. This comparison is notably different when using samples directly derived from landraces, e.g. 0.11 in a Cameroon village [29] or 0.09 in a mix of Guinea race accessions [35].

Relevance and distribution of taxonomic components. The high level of genetic diversity in sorghum is thought to be due to multiple origins for domesticated sorghum, intermating between products of these independent domestication events, and continued gene flow between wild and cultivated sorghums [53]. In this study we found substantial evidence of sorghum population structure based on geographic origin and race within geographic origin. This is congruent with previous studies with RFLP markers [11], SSRs [30], SSRs and SNPs [54], and also with recently developed DArT markers ([55], [22]). Yet the structure we observed led us to propose a schematic representation of population structure in sorghum (Figure 5). The periphery harbors types that are more clearly differentiated and more homogeneous. The center harbors more diverse types, with many more intermediates and a concentration of wild types that appear related to several cultivated forms. Among the cultivated accessions, there is hardly any coincidence between a race and a group based on markers, with the single exception of Kafir in Southern Africa. The "races" might better be referred to as morphotypes, or at least consider that races could encompass different morphotypes.

The 68 wild and weedy accessions presented the highest gene diversity and private allele numbers. The majority of wild/weedy sorghum felled in the periphery of Group 5b, but as previously discussed, they were not definitely assigned. The other wild and weedy accessions were distributed, yet on long branches (see Figure S3), in three other dendrogram sectors predominated by cultivated accessions. This contrasts with Aldrich and Doebley's (1992) results [12], who found a clear separation between the two compartments using RFLP markers, as well as Casa et al. 2005 [50] who confirmed this fact with SSR markers. Clearly, the exploration of diversity in a broader representation of wild sorghum is necessary. One can retain yet the broad distribution of wild and weedy accessions throughout the cultivated sorghum diversity patterns, which adds evidence to a corpus of results (including [9], [56], [57], [58], [31]) that suggests that there is considerable exchange of genetic material (gene-flow) between cultivated and wild accessions.

A global interpretation of sorghum genetic diversity. Altogether, the geographical pattern of differentiation, the limited congruence between marker-based classifications, the racial classification based on morpho-agronomic traits and the likely occurrence of profuse gene flow advocate for a diversity pattern largely determined by 1) geographical radiation in various directions from the center of origin, with both differential drift among lineages and possibly novel variation selected along the process, 2) common gene exchange among landraces and local wild types, ensuring population dynamics, and 3) selection for race-related trait associations responsible for phenotypic convergence between genetically differentiated sub-populations. Germplasm introduction explains the diversity of the materials contributed to the sorghum GCGC from North America, whereas loss of alleles due to drift appears to have contributed to the reduced diversity observed among samples from India and East Asia compared to those from Africa, with the latter contributing to the observed groupings.

In this scenario, it is likely that the genes that underlie the morphological differences between the most typical morphotypes are few in number and deploying visible polymorphism across geographically differentiated groups. This scenario will be testable when whole-genome genotyping is available in sorghum and may reveal footprints for natural and anthropogenic selection along the genome.

Community Resources

Data. Data generated in the present study was deposited in the GCP central registry (http://generationcp.org/research/research-themes/crop-information-systems, using Sorghum as a 'crop' filter, file G2005-01c_Sb_3393accX41SSR_V2.xlsx) and is accessible to the global community. They come in addition to passport data that are available in the germplasm banks and occasional evaluation data that may have been produced as part of searches for donors of specific traits to be used in breeding programs. The data can serve as a reference since it was obtained with an easily accessible kit of markers [24] that can be used on any new material for comparison.

Reference set of sorghum. We used the marker data and population structure of the sorghum GCGC from this study to identify a much smaller representative subset of accessions, called the 'Reference Set'. This Reference Set provides an entry point to sorghum germplasm globally, to identify geographic regions and racial subgroups from which sorghum accessions exhibiting interesting variability in a particular trait can be found. The general value of an internationally agreed set of representative germplasm to serve as a common reference for focussing characterization has been highlighted elsewhere [59].

This proposed Reference Set consists of 383 accessions, includes important germplasm lines used in crop breeding programs, wild accessions and a mini-core collection of genetically diverse accessions for which considerable phenotypic data is already available. Five basic morphological types, ten intermediate ones and wild/weedy accessions from nearly all geographic origins were captured in this sorghum Reference Set. This set represents most of the genetic diversity present in the GCP sorghum Global Composite Germplasm Collection, with all assignation groups and clusters represented. It has a population structure similar to that discussed above for the sorghum GCGC, yet with less redundancy in highly populated narrow clusters. Compared to previously described subsets ([5], used e.g. in [54]) which include converted lines with photoperiod-insensitivity and dwarfing genes, this reference set includes all types of material, enabling breeding choice in Africa. Besides, it also includes wild samples, is more balanced in terms of initial racial classification (more Guinea and less Caudatum in proportion), and represents all geographical origins (and correlatively to racial belonging, represents best West Africa). Seeds are maintained by ICRISAT and available upon request. All passport data published in the System Wide Information Network on Genetic Resources (SINGER), including Sorghum, are available in Genesys (http://www.genesys-pgr.org/), which aims at being the global information system on the germplasm held ex situ.

Perspectives. The core reference set is expected to stimulate links among sorghum scientists. The data have been analysed with several methods, which provide marker-based keys to germplasm classification and are meant to serve as a reference. Any new material can easily be compared to this reference; these markers are easily applicable for local studies with local questions in local laboratories, and yielding results that are comparable to other studies performed elsewhere, thanks to the use of a common kit of markers and standards.

This will be very useful for identifying germplasm action priorities, for enriching global collections if novel types are uncovered or for broadening the basis of a given breeding program.

Having the data available for the whole GCGC for 41 SSR loci provides a considerable backup for mining germplasm diversity. Molecular data can serve for complementing reference materials with additional germplasm targeted towards particular applications depending on the operational constraints, the biological constraints (e.g. phenology) and statistical power. Typically SSR data can be used to adjust a sample to a target size with the view to minimizing population structure in order to maximize resolution power in a given association analysis; the Maximum Length Subtree function of the DARwin software can help do this easily, quickly and rigorously [43]. This dynamics will also enable adjusting the reference set by making it inclusive of newly characterized diversity.

In the long term, helping a global community to focus on similar materials for all sorts of biological investigations will help accumulate and compile data in order to develop better biological understanding of sorghum, and of plant biology thanks to sorghum.

Supporting Information

Figure S1   Scree plot of the factorial analysis. Proportion of variance for each component, sorted in decreasing order of variance.

Figure S2   Graphical method (as in Evanno et al. 2005) allowing detection of the number of groups K (output of

CLUMPP process). L(K) for each K. Rate of change of the likelihood distribution (mean ± SD) calculated as $L'(K) = L(K) - L(K-1)$. Absolute values of the second order rate of change of the likelihood distribution (mean ± SD) calculated according to the formula: $|L''(K)| = |L'(K+1) - L'(K)|$. ΔK calculated as $\Delta K = m|L''(K)|/s[L(K)]$.

Figure S3   Hierarchical NJ cluster analysis of sorghum accessions of a global composite germplasm collection based on allelic data from 41 SSR markers (simple matching distance). All accessions except Gma are presented. Accessions grouped by Bayesian analysis are represented in colors as in figure 4: Group 1 in orange (C, CB and D from Eastern Asia), Group 2 in light orange (D and B from the Indian subcontinent) Group 3 in light green (D from Eastern Africa), Group 4 in light blue: B and DB from Eastern Africa), Group 5a in dark blue (B accessions assembled from North America), Group 6 in red (D, DC and G from Western Africa), Group 7 in light purple (C from Central and Eastern Africa), Group 8 in dark green (C and GC from Southern Africa), Group 9a in pink (G from the Indian subcontinent and East Asia), Group 9b in pink (G from Southern Africa), Group 9c in pink (C from Eastern Africa), and Group 10 in purple (GC, K and KC from Southern Africa). Unassigned accessions are presented in grey. Wild accessions are presented in black.

Table S1   List of accessions comprising the sorghum global composite germplasm (GCCG) collection and characterization. It includes each accession number, institution originally providing the material, biological status, species, subspecies, race, geographic origin (country), geographic origin (continent), subgroup assignation (K value and Cluster), included or not in the Reference Set. Countries are given in standard ISO 3166-1 alpha-3 codes. Continents are AUS (Australia), CAfrica (Central Africa), EAfrica (Eastern Africa), EAsia (Eastern Asia), Europe, IND (India), MedB (Mediterranean Basin), MidE (Middle East), NAmerica (Northern America), SAfrica (Southern Africa), SAmerica (Southern America), WAfrica (Western Africa). Subgroup assignations correspond to Structure K values combined with Cluster assignations. When an accession is included in the Reference Set, last column includes a "one", on the contrary a "zero".

Table S2   Pairwise F_{ST} differentiations between groups as identified in Figure 4.

Acknowledgments

We thank Laurence Dedieu (Cirad) and reviewers for fruitful reading.

Author Contributions

Conceived and designed the experiments: CB PR JFR JCG HU YL CTH. Performed the experiments: CB RR JFR PR LG TW PL. Analyzed the data: CB PR JFR JLN MD JCG CTH. Contributed reagents/materials/analysis tools: CB PR RR LG SB JFR HU EA YL JC RF. Wrote the paper: CB PR MD JCG JLN CTH.

References

1.   Paterson AH (2008) Genomics of sorghum. International Journal of Plant Genomics 2008: 6 pages, doi:10.1155/2008/362451.

2.   Paterson AH, Bowers JE, Bruggmann R, Dubchak I, Grimwood J, et al. (2009) The *Sorghum bicolor* genome and the diversification of grasses. Nature 457: 551–556.

3. Grenier C, Bramel-Cox PJ, Hamon P (2001) Core collection of sorghum. Crop Science 41: 234–240.

4. Upadhyaya HD, Pundir RPS, Dwivedi SL, Gowda CLL, Reddy VG, et al. (2009) Developing a mini core collection of sorghum for diversified utilization of germplasm. Crop Science 49: 1769–1780.

5. Casa AM, Pressoir G, Brown PJ, Mitchell SE, Rooney WL, et al. (2008) Community resources and strategies for association mapping in Sorghum. Crop Science 48: 30–40.

6. Wang M, Zhu C, Barkley N, Chen Z, Erpelding J, et al. (2009) Genetic diversity and population structure analysis of accessions in the US historic sweet sorghum collection. Theoretical and Applied Genetics 120: 13–23.

7. Xin Z, Wang M, Burow G, Burke J (2009) An induced sorghum mutant population suitable for bioenergy research. BioEnergy Research 2: 10–16.

8. Anas, Yoshida T (2004) Genetic diversity among Japanese cultivated sorghum assessed with simple sequence repeats markers. Plant Production Science 7: 217–223.

9. Harlan JR, de Wet JMJ (1972) A simplified classification of cultivated sorghum. Crop Science 12: 172–176.

10. Ollitrault P (1987) Evaluation génétique des sorghos cultivés (Sorghum bicolor Moench) par l'analyse conjointe des diversités enzymatiques et morpho-physiologique. Relation avec les sorghos sauvages: Université Paris XI, France.

11. Deu M, Rattunde H, Chantereau J (2006) A global view of genetic diversity in cultivated sorghums using a core collection. Genome 49: 168–180.

12. Aldrich PR, Doebley J (1992) Restriction fragment variation in the nuclear and chloroplast genomes of cultivated and wild Sorghum bicolor. Theoretical and Applied Genetics 85: 293–302.

13. Hulbert SH, Richter TE, Axtell JD, Bennetzen JL (1990) Genetic mapping and characterization of sorghum and related crops by means of maize DNA probes. Proceedings of the National Academy of Sciences of the United States of America 87: 4251–4255.

14. Pereira MG, Lee M, Bramel-Cox P, Woodman W, Doebley J, et al. (1994) Construction of an RFLP map in sorghum and comparative mapping in maize. Genome 37: 236–243.

15. Ragab RA, Dronavalli S, Maroof MAS, Yu YG (1994) Construction of a sorghum RFLP linkage map using sorghum and maize DNA probes. Genome 37: 590–594.

16. Xu GW, Magill CW, Schertz KF, Hart GE (1994) A RFLP linkage map of Sorghum bicolor (L.) Moench. Theoretical and Applied Genetics 89: 139–145.

17. Rami JF, Dufour P, Trouche G, Fliedel G, Mestres C, et al. (1998) Quantitative trait loci for grain quality, productivity, morphological and agronomical traits in sorghum (Sorghum bicolor L. Moench). Theoretical and Applied Genetics 97: 605–616.

18. Whitkus R, Doebley J, Lee M (1992) Comparative genome mapping of sorghum and maize. Genetics 132: 1119–1130.

19. Devos KM, Gale MD (1997) Comparative genetics in the grasses. Plant Molecular Biology 35: 3–15.

20. Tao YZ, Jordan DR, McIntyre CL, Henzell RG (1998) Construction of a genetic map in a sorghum recombinant inbred line using probes from different sources and its comparison with other sorghum maps. Australian Journal of Agricultural Research 49: 729–736.

21. Ventelon M, Deu M, Garsmeur O, Doligez A, Ghesquière A, et al. (2001) A direct comparison between the genetic maps of sorghum and rice. Theoretical and Applied Genetics 102: 379–386.

22. Bouchet S, Pot D, Deu M, Rami J-F, Billot C, et al. (2012) Genetic structure, linkage disequilibrium and signature of selection in sorghum: lessons from physically anchored DArT markers. PLoS ONE 7: e33470.

23. Nelson J, Wang S, Wu Y, Li X, Antony G, et al. (2011) Single-nucleotide polymorphism discovery by high-throughput sequencing in sorghum. BMC Genomics 12: 352.

24. Billot C, Rivallan R, Sall MN, Fonceka D, Deu M, et al. (2012) A reference microsatellite kit to assess for genetic diversity of Sorghum bicolor (Poaceae). American Journal of Botany 99: e245-e250.

25. Wang ML, Barkley NA, Yu J-K, Dean RE, Newman ML, et al. (2005) Transfer of simple sequence repeat (SSR) markers from major cereal crops to minor grass species for germplasm characterization and evaluation. Plant Genetic Resources 3: 45–57.

26. Burow G, Franks C, Acosta-Martinez V, Xin Z (2009) Molecular mapping and characterization of BLMC, a locus for profuse wax (bloom) and enhanced cuticular features of Sorghum (Sorghum bicolor (L.) Moench). Theoretical and Applied Genetics 118: 423–431.

27. Yonemaru J-i, Ando T, Mizobuyashi T, Kasuga S, Matsumoto T, et al. (2009) Development of genome-wide simple sequence repeat markers using whole-genome shotgun sequences of sorghum (Sorghum bicolor (L.) Moench). DNA Research 16: 187–193.

28. Ghebru BG, Schmidt RS, Bennetzen JB (2002) Genetic diversity of Eritrean sorghum landraces assessed with simple sequence repeat (SSR) markers. Theoretical and Applied Genetics 105: 229–236.

29. Barnaud A, Deu M, Garine E, McKey D, Joly HI (2007) Local genetic diversity of sorghum in a village in northern Cameroon: structure and dynamics of landraces. Theoretical and Applied Genetics 114: 237–248.

30. Deu M, Sagnard F, Chantereau J, Calatayud C, Hérault D, et al. (2008) Niger-wide assessment of in situ sorghum genetic diversity with microsatellite markers. Theoretical and Applied Genetics 116: 903–913.

31. Sagnard F, Deu M, Dembélé D, Leblois R, Touré L, et al. (2011) Genetic diversity, structure, gene flow and evolutionary relationships within the Sorghum bicolor wild/weedy crop complex in a western African region. Theoretical and Applied Genetics 123: 1231–1246.

32. Caniato F, Guimarães C, Schaffert R, Alves V, Kochian L, et al. (2007) Genetic diversity for aluminum tolerance in sorghum. Theoretical and Applied Genetics 114: 863–876.

33. Ali M, Rajewski J, Baenziger P, Gill K, Eskridge K, et al. (2008) Assessment of genetic diversity and relationship among a collection of US sweet sorghum germplasm by SSR markers. Molecular Breeding 21: 497–509.

34. Wang ML, Dean R, Erpelding J, Pederson G (2006) Molecular genetic evaluation of sorghum germplasm differing in response to fungal diseases: Rust (Puccinia purpurea) and anthracnose (Collectotrichum graminicola). Euphytica 148: 319–330.

35. Folkertsma RT, Rattunde H, Chandra S, Raju GS, Hash CT (2005) The pattern of genetic diversity of Guinea-race Sorghum bicolor (L.) Moench landraces as revealed with SSR markers. Theoretical and Applied Genetics 111: 399–409.

36. Mace E, Buhariwalla K, Buhariwalla H, Crouch J (2003) A high-throughput DNA extraction protocol for tropical molecular breeding programs. Plant Molecular Biology Reporter 21: 459–460.

37. Deu M, Hamon P, Dufour P, D'hont A, Lanaud C, et al. (1995) Mitochondrial DNA diversity in wild and cultivated sorghum. Genome 38: 635–645.

38. Kimura M, Ohta T (1978) Stepwise mutation model and distribution of allelic frequencies in a finite population. Proceedings of the National Academy of Sciences of the United States of America 75: 2868–2872.

39. Idury RM, Cardon LR (1997) A simple method for automated allele binning in microsatellite markers. Genome Research 7: 1104–1109.

40. Liu K, Muse SV (2005) PowerMarker: an integrated analysis environment for genetic marker analysis. Bioinformatics 21: 2128–2129.

41. Szpiech ZA, Jakobsson M, Rosenberg NA (2008) ADZE: a rarefaction approach for counting alleles private to combinations of populations. Bioinformatics 24: 2498–2504.

42. Goudet J (2005) Hierfstat, a package for R to compute and test hierarchical F-statistics. Molecular Ecology Notes 5: 184–186.

43. Perrier X, Jacquemoud-Collet JP (2006) DARwin software. http://darwin.cirad.fr/darwin.

44. Gao H, Williamson S, Bustamante CD (2007) A Markov chain Monte Carlo approach for joint inference of population structure and inbreeding rates from multilocus genotype data. Genetics 176: 1635–1651.

45. Pritchard J, Falush D, Stephens M (2002) Inference of population structure in recently admixed populations. American Journal of Human Genetics 71: 177–177.

46. Evanno G, Regnaut S, Goudet J (2005) Detecting the number of clusters of individuals using the software STRUCTURE: a simulation study. Molecular Ecology 14: 2611–2620.

47. Ehrich D (2006) AFLPDAT: a collection of R functions for convenient handling of AFLP data. Molecular Ecology Notes 6: 603–604.

48. Grenier C, Deu M, Kresovich S, Bramel-Cox PJ, Hamon P (2000) Assessment of genetic diversity in three subsets constituted from the ICRISAT sorghum collection using random vs non-random sampling procedures B. Using molecular markers. Theoretical and Applied Genetics 101: 197–202.

49. Shehzad T, Okuizumi H, Kawase M, Okuno K (2009) Development of SSR-based sorghum (Sorghum bicolor (L.) Moench) diversity research set of germplasm and its evaluation by morphological traits. Genetic Resources and Crop Evolution 56: 809–827.

50. Casa AM, Mitchell SE, Hamblin MT, Sun H, Bowers JE, et al. (2005) Diversity and selection in sorghum: simultaneous analyses using simple sequence repeats. Theoretical and Applied Genetics 111: 23–30.

51. Agrama HA, Tuinstra MR (2003) Phylogenetic diversity and relationships among sorghum accessions using SSRs and RAPDs. African Journal of Biotechnology 2: 334–340.

52. Deu M, Sagnard F, Chantereau J, Calatayud C, Vigouroux Y, et al. (2010) Spatio-temporal dynamics of genetic diversity in Sorghum bicolor in Niger. Theoretical and Applied Genetics 120: 1301–1313.

53. Doggett H (1988) Sorghum: Longman Scientific and Technical, Burnt Mill, Harlow, Essex, England; John Wiley and Sons, New York.

54. Brown PJ, Myles S, Kresovich S (2011) Genetic support for phenotype-based racial classification in Sorghum. Crop Science 51: 224–230.

55. Mace ES, Xia L, Jordan DR, Halloran K, Parh DK, et al. (2008) DArT markers: diversity analyses and mapping in Sorghum bicolor. BMC Genomics 9: 26.

56. Mann JA, Kimber CT, Miller FR (1983) The origin and early cultivation of sorghums in Africa. The Texas agricultural experimental station bulletin 1454: 1–21.

57. Tesso T, Kapran I, Grenier Cc, Snow A, Sweeney P, et al. (2008) The potential for crop-to-wild gene flow in Sorghum in Ethiopia and Niger: A geographic survey. Crop Science 48: 1425–1431.

58. Mutegi E, Sagnard F, Semagn K, Deu M, Muraya M, et al. (2011) Genetic structure and relationships within and between cultivated and wild sorghum (Sorghum bicolor (L.) Moench) in Kenya as revealed by microsatellite markers. Theoretical and Applied Genetics 122: 989–1004.

59. Glaszmann JC, Kilian B, Upadhyaya HD, Varshney RK (2010) Accessing genetic diversity for crop improvement. Current Opinion in Plant Biology 13: 167–173.

20

Managing Potato Biodiversity to Cope with Frost Risk in the High Andes: A Modeling Perspective

Bruno Condori[1], Robert J. Hijmans[2], Jean Francois Ledent[3], Roberto Quiroz[4]*

1 Liaison Office in Bolivia, International Potato Center, La Paz, La Paz, Bolivia, **2** Department of Environmental Science and Policy, University of California, Davis, Davis, California, United States of America, **3** Unité d'Ecophysiologie et d'Amélioration Végétale, Université Catholique de Louvain, Louvain-la-Neuve, Brabant, Belgium, **4** Integrated Crops and Systems Research Program, International Potato Center, La Molina, Lima, Peru

Abstract

Austral summer frosts in the Andean highlands are ubiquitous throughout the crop cycle, causing yield losses. In spite of the existing warming trend, climate change models forecast high variability, including freezing temperatures. As the potato center of origin, the region has a rich biodiversity which includes a set of frost resistant genotypes. Four contrasting potato genotypes –representing genetic variability- were considered in the present study: two species of frost resistant native potatoes (the bitter *Solanum juzepczukii*, var. Luki, and the non-bitter *Solanum ajanhuiri*, var. Ajanhuiri) and two commercial frost susceptible genotypes (*Solanum tuberosum* ssp. *tuberosum* var. Alpha and *Solanum tuberosum* ssp. *andigenum* var. Gendarme). The objective of the study was to conduct a comparative growth analysis of four genotypes and modeling their agronomic response under frost events. It included assessing their performance under Andean contrasting agroecological conditions. Independent subsets of data from four field experiments were used to parameterize, calibrate and validate a potato growth model. The validated model was used to ascertain the importance of biodiversity, represented by the four genotypes tested, as constituents of germplasm mixtures in single plots used by local farmers, a coping strategy in the face of climate variability. Also scenarios with a frost routine incorporated in the model were constructed. Luki and Ajanhuiri were the most frost resistant varieties whereas Alpha was the most susceptible. Luki and Ajanhuiri, as monoculture, outperformed the yield obtained with the mixtures under severe frosts. These results highlight the role played by local frost tolerant varieties, and featured the management importance –e.g. clean seed, strategic watering- to attain the yields reported in our experiments. The mixtures of local and introduced potatoes can thus not only provide the products demanded by the markets but also reduce the impact of frosts and thus the vulnerability of the system to abiotic stressors.

Editor: Ji-Hong Liu, Key Laboratory of Horticultural Plant Biology (MOE), China

Funding: This work has been made possible through the financial contribution of the following organizations: The Unit of Eco-physiology and Crop Breeding, Catholic University of Louvain in Belgium that granted the PhD scholarship of Bruno Condori under which the model was initially developed. The Canadian International Development Agency (CIDA) which funded the CIP conducted ALTAGRO project through which Bruno Condori's work was partially supported. The Swiss Development Cooperation (SDC) which funded the CIP conducted initiative PAPA ANDINA through which Bruno Condori's work was partially supported. The funders had no role in study design, data collection and analysis, decision to publish, or preparation of the manuscript.

Competing Interests: The author have declared that no competing interests exist.

* E-mail: r.quiroz@cgiar.org

Introduction

The Altiplano is a high tropical plateau located at 3600–4300 m above sea level in the Andes of Bolivia and Peru. Most of the cropland is located below 4000 masl; above that elevation land is mainly covered by natural grasslands and is only used for growing bitter potato landraces, which are adapted to cold conditions. Potato is by far the most important crop in the region, accounting for 44% of the gross value of crop production [1] from a cropping area of about 88,000 ha [2]. Potato production is limited by abiotic and biotic factors; Andean farmers manage these constraints mainly by the use of a high diversity of native species and cultivars that are often grown as mixtures in single plots [3,4,5]. As potato originated in the Andes [6], local genetic diversity in cultivated potato is large and includes several species, comprising both bitter -*Solanum juzepckzukii* (triploid), and 1 non-bitter frost resistant potatoes: Solanum *ajanhuiri* (diploid), but also the non-bitter frost susceptible conventional *Solanum tuberosum* subspecies *tuberosum* (tetraploid), and *Solanum tuberosum* ssp. *andigenum* (tetraploid), which are present in the Altiplano [7,8].

The principal role played by the diversity of potatoes grown in the Altiplano is related to smallholder's food security. Potato fresh yields in the area are low. In Peru and the northernmost part of the Bolivian Altiplano, yield average varies from 4 to 5.2 t/ha whilst in the southern Bolivian section [1] the average yield is 3.6 t/ha. The growing season in the Altiplano extends from October to March, when maximum annual temperature coincides with the rainy season. In the agricultural zones of the Altiplano, average maximum temperature is around 18°C whereas minimum temperature is around 4°C during the growing season. Precipitation is around 800 mm/year in the northeast of the Altiplano whereas in the southwestern Altiplano, it is about 200 mm/year, mostly occurring during the same growing season. Production risk for potato is high due to several recurrent factors, particularly drought, hail, and frost. Frost-free period averages 140 days in the northern Altiplano and 110 days in the Southern areas [9]. The high production risks presented by frost and other factors may also lead to reduced investment in agriculture, resulting in low production which in turn affects food availability.

The varieties of the species *Solanum tuberosum* ssp. *andigenum* are the most widely cultivated in the Andes. The *Solanum juzepczukii* stands out for its high frost and drought tolerance and its capacity to grow at 4000 masl and above [10]. However its tubers are bitter due to a high content of glycoalkaloids, requiring processing for direct human consumption [11]. This processing is an old Andean's strategy for conserving food – chuño: dehydrated potatoes - for several years [12]. It has been estimated that at least 25% of total area under potato in the Altiplano is planted with bitter varieties [13]. This assertion is supported by the estimate that bitter potatoes make up 15% of total potato area in Bolivia [11]. In fact, more bitter potatoes are found in the Altiplano than in most other zones. The varieties of *Solanum ajanhuiri* have characteristics of tolerance to frost and drought similar to those of the *Solanum juzepczukii* but they do not have high glycoalkaloid contents and are therefore non bitter. At intermediate and lower altitudes the *Solanum tuberosum* ssp. *tuberosum* varieties are the most widespread whereas the other species predominate at altitudes higher than 3500 masl. Notwithstanding, recent findings showed that the actual upper limit for all potato varieties has increased to around 4300 masl [14] in response to increased temperatures and disease pressure at lower altitudes brought about by climate change.

The Andes represent the largest and highest mountain range in the tropics, and thus a suitable ecosystem to study changes in climate and how they affect the natural resources and the livelihood they sustain. Recent studies based on local meteorological networks [15,16] have shown a significant warming trend after 1979 (0.32–0.34°C/decade). This warming trend shows some interesting features. Below 1,000 m there is a substantial difference between the eastern (Amazonian) and the western (Pacific watershed) slopes i.e. no warming trend on the eastern slope whereas on the western side there is a warming trend of 0.39°C/decade. On this flank, there is an almost linear decrease of the warming trend with altitude, reaching down to 0.16°C/decade above 4000 m. Thus there are differences between the Eastern and Western slopes and a definite vertical structure. This behavior differs from other mountain ranges in other latitudes e.g. European Alps and Tibet. A warming trend is also projected for the future [17], but differences as functions of position and altitudes are yet to be studied.

Summer frost events are caused by radiative cooling and are common throughout the Andean highlands and can occur at any time during the growing season [9,18]. Frost can cause partial or complete loss of foliage, leading to a reduction in photosynthate production and hence yield. In turn, crop failure caused by frost damage may lead to a decrease in the total area planted to potato in the subsequent season due to seed shortage [18]. The temperature at which leafage frost damage occurs depends on the species and the cultivar. For *Solanum tuberosum* subsp. *andigenum*, frost damage is likely to occur when the temperature drops to 2°C or lower [19]. Higher frost resistance exists in other cultivars and in wild potato species. For example, cultivated potato species such as *S. ajanhuiri* and *S. curtilobum* are damaged at −3 to −5°C, whereas *S. juzepczukii* generally resists temperatures down to −5°C and perhaps even lower [20].

Crop growth simulation models can be used to analyze constraints and opportunities for crop yields in complex production systems (e.g. [4,5]). One of the limitations for using simulation models in this context is that most available models are calibrated only for varieties of *S. tuberosum* subspecies. Almost all modeling work has been done for so-called 'modern' varieties from formal breeding programs; with a few exceptions [21]. This can be particularly problematic when working in areas of high genetic diversity, where landraces with rather distinct growth characteristics may coexist with modern cultivars. A more in-depth understanding of the most important crop growth limiting factors as affected by genetic diversity is needed to arrive at more robust recommendations to reduce the vulnerability and improve the productivity of potato in environmentally challenged complex cropping systems. Comparative growth analysis and its translation into potato growth models can constitute a useful tool for scientists, particularly under climate variability and global climate change conditions in areas of rich genetic diversity.

The objective of the present study was to conduct a comparative growth analysis of four contrasting genotypes from the species *S. juzepczukii*, *S. ajanhuiri* and *S. tuberosum* subspecies *andigenum* and *tuberosum* and translate it into a modeling of their agronomic response. It included assessing their performance under contrasting agroecological conditions in the high Andes as well as the impact of frost events.

Materials and Methods

This study did not involve neither human nor animal subjects. Field experiments were conducted in research stations and farmer fields and no law regulated permits were required because all materials used are commonly used by farmers with no environmental negative impacts. No protected species were sampled.

Permission to work on PROINPA's field research sites (Patacamaya 2, Patacamaya 3, Puchuni and Laurani) was granted by Mr. Enrique Carrasco, then Leader of PROINPA.

Permission to work on farmers' fields was granted by Mr. Vitaliano Mamani, Mrs. Maria Laura and Mr. Juan Carlos Huanca, under a collaboration agreement with PROINPA.

All authorizations were granted as verbal agreements on a *bona fide* basis.

Table 1. General description of the locations where the experiments were conducted.

Location	Latitude S	Longitude W	Altitude (masl)	Year	Rainfall (mm)	Minimum Temp (°C)	Maximum Temp (°C)	PAR (MJ m^{-2}d)	Soil texture
Patacamaya 2	17°14'	67°55'	3789	1998	437	4.6	18.7	10.4	SaCL
Patacamaya 3	17°16'	67°55'	3789	1998	350	4.6	18.7	10.4	SaCL
Puchuni	17°16'	68°13'	3950	1998	513	2.4	17.4	10.3	SiL
Laurani	17°14'	68°11'	3850	1999	430	3.3	16.8	10.3	CL
Wichukollu	17°01'	68°05'	3911	1999	445	3.0	16.5	10.1	CL

Where: C is clay, L is loam, Sa is sand and Si is silt. Temperatures and PAR are monthly values and rainfall is cumulated value during crop cycle.

Table 2. Abbreviations, list of variables and parameters used in the article.

Abbreviation	Description	Units
MCC	Maximum canopy cover	Fraction
LUE	Light use efficiency	g MJ^{-1}
DTY	Dry tuber yield	t ha^{-1}
HI	Harvest index	%
PAR	Photosynthetically active radiation	MJ m^{-2}
Fo	Initial fraction of light interception	Fraction
Ro	Relative rate of light interception increase	°C^{-1}d^{-1}
D	Twice the duration from $t_{0.5}$ to the end of the light interception	°Cd
$t_{0.5}$	Time when the fraction of light intercepted is reduced to 50% of its Maximum value attained at MCC	°Cd
TIO	Tuber initiation onset	°Cd
TM	Tuber growth cessation or maturation	°Cd
X_0	Point in thermal time when the maximum speed of translocation of assimilates is given	°Cd
TG_{max}	Maximum tuber growth or bulking rate	g m^{-2} °Cd^{-1}

Germplasm

Four contrasting germplasm (species, cultivars, varieties, and landraces) - three native landraces and one introduced cultivar - representing the genetic diversity cultivated in the high Andes were studied. The four genotypes were Bola luki, referred to as Luki in this paper, a bitter potato variety of *Solanum juzepczukii* (3× or triploid); Chiar ajanhuiri, referred to as Ajanhuiri, a diploid (2×) variety of *Solanum ajanhuiri*; Gendarme, a tetraploid (4×) variety of *Solanum tuberosum* ssp. *andigenum*; and Alpha, a European tetraploid variety of *Solanum tuberosum* ssp. *tuberosum* [7,22].

Experiments and Data Collection

Four field trials, comparing four potato genotypes, were carried out in farmer's fields or on experimental stations in different locations (Patacamaya 2, Patacamaya 3, Puchuni and Laurani) in high and semi-arid zones in Bolivia (Table 1), where potato is the dominant crop, grown as the first crop after a fallow period. Within each site, complete randomized block designs with 3 or 4 replications were implemented in plots of 25.2 m^2. The planting density was 4.76 plants/m^2, a density suitable for the potato plow used in the zone. The tuber seeds used were from certified quality

and were stored homogeneously during five months before planting. In addition, another experiment was conducted at Wichukollu locality, where the effect of frost (−2.51°C befall at 94 days after planting) on the four studied varieties was assessed.

Plots were homogeneously managed to assure non-limiting factors for achieving potential production. Based on previous fertilization trials in the region [23], N was applied at 60–80 kg/ha and P$_2$O$_5$ at 100–120 kg/ha, in addition to 5 t/ha of bovine manure. Potassium was not applied due to the high levels of this element in these soils. Crops were rain-fed, with supplemental irrigation to bring the soil to field capacity when soil moisture dropped below 75% of field capacity, as determined by weekly soil samplings. Ridomil MZ (metalaxyl-M and mancozeb) and Bravo 500 (chlorothalonil) were used against potato late blight (caused by *Phytophthora infestans* Mont. de Bary). Synthetic pyrethroids, Karate (lambda-cyhalothrine) and Lorsban (chlorpyrifos) were used to control Andean potato weevils (*Premnotrypes latithorax* Piercei, *P. solaniperda* Kuschel, and *Rigopsidius tucumanus* Heller) and potato tuber moths (*Phthorimaea operculella* Zeller, *Symmetrischema tangolias* Geyen and *Paraschema detectendum* Povolny).

Table 3. Mean and standard error of the mean of potato growth responses to changes in environment and genotypes.

Environment	LUE[ns]	SE	HI*	SE	MCC [ns]	SE	DTY [ns]	SE
Patacamaya 2	2.658	0.082	79.19	0.97	72.69	5.60	10.98	0.84
Patacamaya 3	2.800	0.085	81.58	0.61	64.83	5.36	10.28	0.69
Puchuni	2.788	0.066	85.33	1.35	66.69	3.79	11.93	0.72
Laurani	2.529	0.147	84.12	1.62	69.58	4.87	9.29	0.58
Genotype	ns	$S_{\bar{x}}$	***	$S_{\bar{x}}$	***	$S_{\bar{x}}$	***	$S_{\bar{x}}$
Alpha	2.687	0.091	86.55	1.55	40.05	1.18	7.21	0.38
Gendarme	2.661	0.104	83.13	1.17	84.21	2.76	13.07	0.78
Ajanhuiri	2.736	0.096	79.54	0.82	73.93	2.39	11.69	0.39
Luki	2.708	0.124	80.84	1.32	76.31	1.96	10.99	0.48

SE = Standard error of the mean; ns = non-significant P>0.05; ** = significant P<0.01; *** = significant P<0.001; Light Use Efficiency (LUE) in g DM MJ^{-1}; Harvest Index (HI) in %; Maximum Canopy Cover (MCC) in %; and Dry Tuber Yield (DTY) in t ha^{-1}.

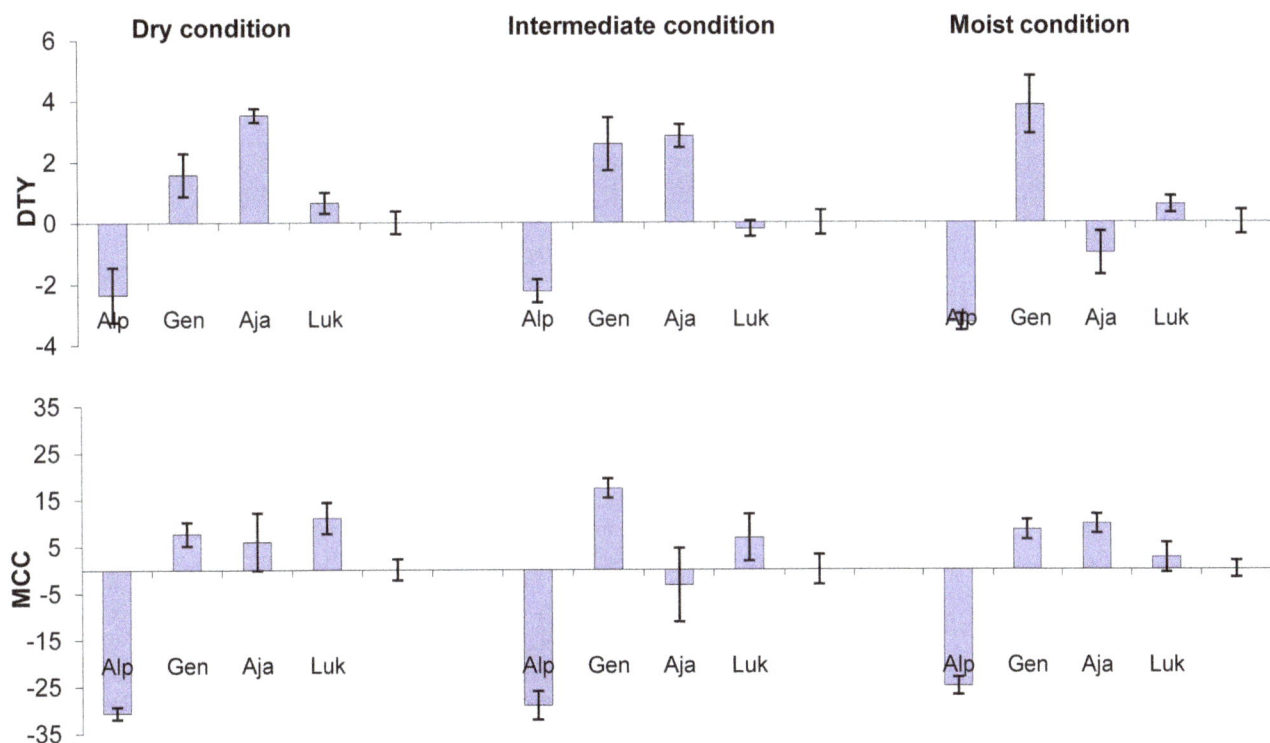

Figure 1. Genotype by Environment interaction for Dry Tuber Yield (DTY) and Maximum Canopy Cover (MCC) in three environment conditions in the high Andes. Residuals in Y axis, and standard error by variety.

Precipitation, minimum and maximum temperatures were measured in the field using Hobo data loggers (ONSET Pro Series H08-032-08, USA). Global solar radiation was measured at meteorological stations representative of the experimental sites. Photosynthetically active radiation (PAR) was calculated as half of the global solar radiation [24].

During the crop growing period, plants were sampled three to five times from the plots. Intermediate harvests consisted of 4 plants per plot, while the final harvest consisted of 20 plants. Biomass was divided into leaves, stems, and tubers. Roots were not measured since they are difficult to collect. For each biomass group, total fresh weight was determined and their dry matter fraction estimated using samples of 150 g fresh weight subsequently dried at 75°C until constant weight which occurred after 48 to 72 hours.

The fraction of green canopy cover was estimated in triplicate in each experimental unit, every first or second week, from

emergence to the final harvest. A wooden frame of 70 cm×90 cm divided into 100 cells of 7 cm×9 cm was used. The frame covered the area of three plants and cells were marked as covered if leaf surfaces occupied more than 50% of the cell area.

Data Processing

The analysis of variance for the comparative growth analysis of all four genotypes was conducted by pooling the results from the four locations, once the homogeneity of the variance across experiments was demonstrated with the Bartlett test. The crop growth responses tested included: Maximum Canopy Cover (MCC), Light Use Efficiency (LUE), Dry Tuber Yield (DTY) and Harvest Index (HI). A summary of the variables and parameters of this study is described in Table 2.

Table 4. Means of model parameters calibrated on four environments.

Genotype	F_0***	SE	R_0**	SE	d*	SE	$t_{0.5}$***	SE	TIO***	SE	TM***	SE
Alpha	0.0050	0.0005	0.0068	0.0004	398	35	1163	27	259	17	748	48
Gendarme	0.0023	0.0003	0.0073	0.0006	268	25	1327	34	440	14	889	29
Ajanhuiri	0.0034	0.0004	0.0063	0.0003	253	43	1360	39	469	30	944	21
Luki	0.0020	0.0001	0.0055	0.0005	321	38	1329	50	650	29	984	56

About 11 degree days (°Cd) is equivalent to 1 day. SE = Standard error of the mean; * = significant P<0.05; ** = significant P<0.01; *** = significant P<0.001; F_0: initial fraction of light interception; R_0: rate of relative increase of light interception in $°C^{-1}d^{-1}$; d: twice the duration from $t_{0.5}$ to the end of the light interception in °Cd; $t_{0.5}$: time when the fraction of light intercepted is reduced to 50% of its maximum value attained at MCC ($t_{0.5}$); TIO: tuber initiation onset in °Cd; and (TM): tuber growth cessation or maturation in °Cd.

Table 5. Coefficients of the logistic regression for dry matter translocation to tubers (HI, X_0 and R^2) and maximum tuber growth or bulking rate (TG_{max}).

Genotype	HI	SE	X_0	SE	R^2	T_{Gmax}	SE
Alpha	0.871	0.047	726.38	31.67	0.96**	1.50	1.18
Gendarme	0.814	0.029	874.92	20.50	0.99**	3.79	1.56
Ajanhuiri	0.808	0.043	959.44	25.60	0.99**	2.78	1.71
Luki	0.768	0.040	1030.25	25.27	0.97**	2.28	1.99

Where: HI is the maximum value that α reaches, X_0 is the point in thermal time (°Cd), where the maximum speed of translocation of assimilates is given, TG_{max} is the Maximum tuber growth or bulking rate (g m^{-2} °Cd^{-1}). SE is the Standard error; R^2 is the determination coefficient at P<0.01 (**).

The linear additive model used was:

$$Yijkl = \mu + Ei + Bj + Vk + Ei * Bj + Ei * Vk + Bj * Vk + \varepsilon ijkl$$

Where

$$Yijkl = \text{Crop growth responses}$$

$$\mu = \text{General mean}$$

$$Ei = \text{Environment effect (Fixed)}$$

$$Bj = \text{Block effect (Random)}$$

$$Vk = \text{Genotype effect (Fixed)}$$

$$\varepsilon ijkl = \text{General error term}$$

Growth analysis. Field measurements of plant growth components such as leaf area and the weight of plant parts provided for the parameters required for growth analysis. Maximum canopy cover MCC was calculated from the observed data through regression analysis. Growth parameters such as the rate of relative increase of light interception (Ro), the initial fraction of light interception at plant emergence (Fo), and the time when the fraction of light intercepted was reduced to 50% ($t_{0.5}$) were estimated using curve fitting. The tuber initiation onset (TIO), the maximum tuber growth or bulking rate (TG_{max}), and cessation of tuberization (TM) were calculated by fitting a logistic function (Sigma Plot V.9, USA) to the measured tuber weight as a function of thermal time (°Cd). The average LUE was determined as the slope of the curve between cumulative total dry matter and intercepted PAR [24].

Model description and calibration. Results from growth analysis can be integrated into mathematical models of the growth process, which constitute a robust tool for yield forecasting. This basic approach was followed in the present study to assess the response of the four potato genotypes and the results were used to

Table 6. Mean Bias Error (MBE), Root Mean Square Error (RMSE), Nash and R^2 validation coefficients for Dry Tuber Yield.

DTY	Alpha	Gendarme	Ajanhuiri	Luki
MBE	0.0696	0.4678	0.8964	0.3564
RMSE	0.7386	0.8638	1.4764	0.9737
Nash	0.96	0.98	0.95	0.89
R^2	0.89	0.96	0.88	0.81

calibrate the mathematical equations included in a slightly modified version of the LINTUL-potato model [24,25,4]. The model's routines were originally developed in Visual Basic of Microsoft Excel and, once validated, programmed in C++. The model operates on a daily time step and its equations and parameters are described in [21]. The results of two experiments (Patacamaya 2 and Puchuni) were selected to calibrate the model and the results of the remaining two experiments were used for validation.

For modeling purposes, single plot variety mixtures were constructed, based on common practice by farmers. Our simulated mixtures included 65% Gendarme, 25% Luki, 7% Ajanhuiri and 3% Alpha [11,13]. The yields from the mixtures in single-plots were compared against monoculture yield for each variety.

Model validation. Data sets from trials conducted in the other two independent sites (Patacamaya 3 and Laurani) were used to validate the calibrated model. Simulated data for MCC and DTY were compared with field results. Mean Bias Error (MBE), the Root Mean Square Error (RMSE), the Nash coefficient (Nash), and R^2 [26,27] were used to test the suitability of the calibrated model to simulate the growth of different genotypes under the experimental conditions.

Regression diagnostics of modeling. Regression diagnostics is the general class of techniques for detecting problems - with either the model or the data set - in regression analysis [28]. We adapted the residual analysis to evaluate simulated canopy cover data in time. This analysis was portrayed by plotting the phenological time series and residual values for detecting problem areas with outliers. The Durbin-Watson test was used to detect the presence of serial correlation in the residuals and influence statistics of points was measured through Cook's D, DFFITS, DFBETAS and COVRATIO statistics [28]. The time series plotting was made in Sigma Plot V.9 software and the statistical analysis in SAS V.8.

Scenario analysis. *What if* type scenarios were constructed based on the validated model into which a frost routine, based on the data coming from the Wichukollu experiment, was incorporated to estimate potato yield losses for every genotype [29,4]. Two issues were addressed with the scenario analysis: 1) How would different genotypes behave when exposed to progressive frosts (increasing by −1°C) during any of the four critical (30, 60, 90 and 120 DAE) phenological phases? 2) How would mixtures of varieties behave under extreme frosts, of frequent occurrence in the high Andes (−3°C at 60 and 90 DAE), compared to the monocultures? The model was set to simulate expected impact of progressive and extreme frosts on yield and it was compared to the production without frost stress.

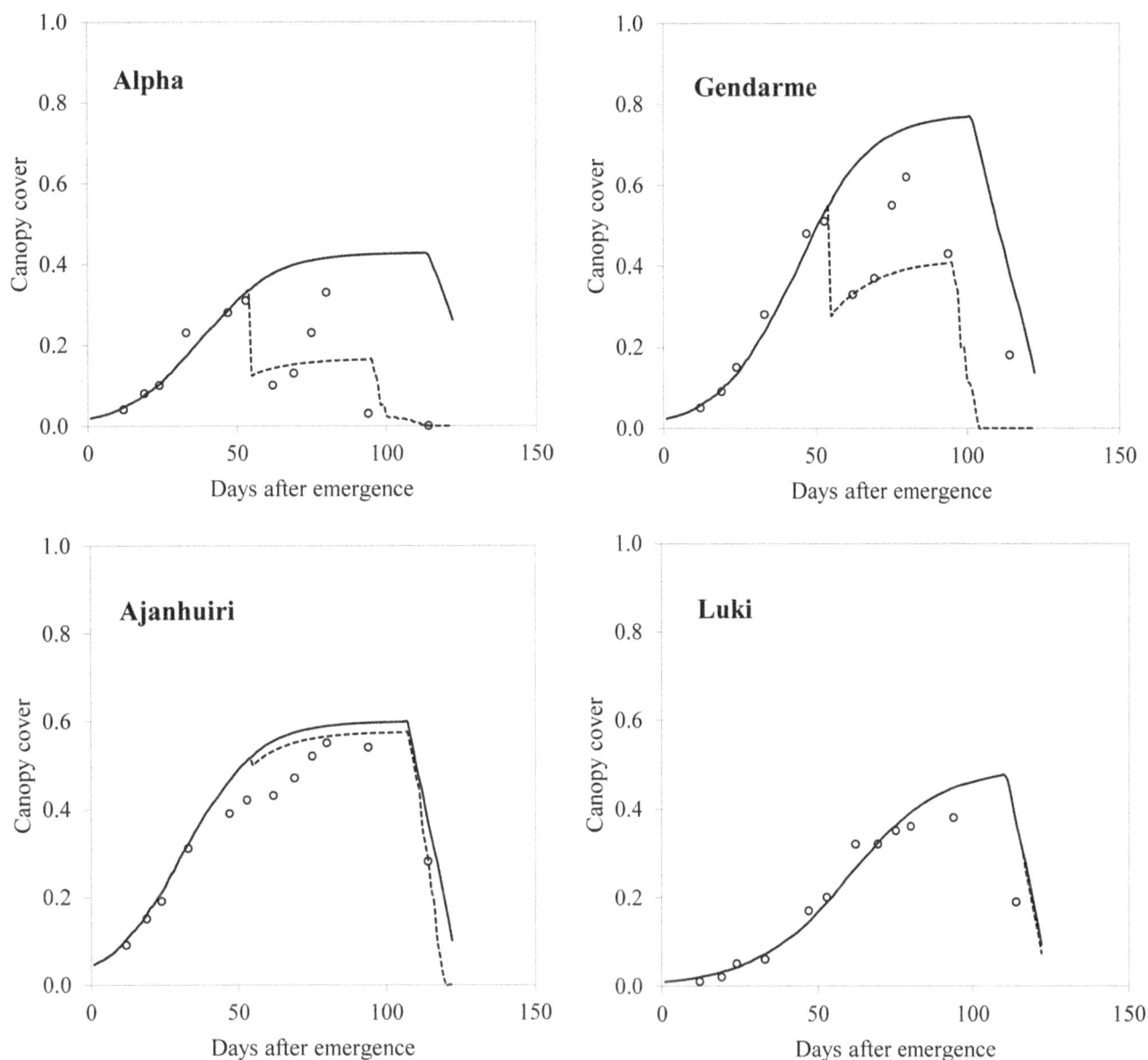

Figure 2. A frost (−2.51°C) impact on canopy cover at 94 days after planting in Wichukollu locality.

Results and Discussion

Growth Analysis

Harvest index was the only response variable affected (P<0.05) by location (Table 3). Average LUE, MCC and DTY did not differ (P>0.05) among locations. The highest HI was recorded in Puchuni whereas the lowest HI occurred in Patacamaya 2. The average yield across all locations was six to sevenfold higher than the national average.

MCC, DTY and HI differed (P<0.001) among genotypes (Table 3). Alpha showed the lowest MCC and the lowest DTY while Gendarme showed the highest MCC and the highest DTY. Ajanhuiri produced the lowest HI whereas the highest HI corresponded to Alpha. Luki gave intermediate values of MCC, DTY and HI, compared to the other genotypes evaluated. Changes in MCC explained 86% of the variation in DTY, whereas HI had no significant relationship with DTY (P>0.05). Average LUE did not differ (P>0.05) among genotypes.

The calculated values of the light use efficiency, (around 2.7 g DM MJ^{-1} as shown in Table 3) measured as the slope of the line describing the relationship between biomass production and absorbed PAR, were within the range of values reported in the literature for the *tuberosum* genotypes under temperate conditions [24,30,31].

Harvest index is generally regarded as the most important variable to estimate the productivity efficiency of the crops [32]. Generally the potato crop has higher HI values (about 0.80) than the other important crops (e.g. oil seeds 0.3 and cereals 0.5). In our case, HI was ≥0.8. However, in spite of the genetic diversity included in the study, the correlation with tuber yield was low, contrasting with literature findings [10]. This was probably due to MCC values that did not reach the level of a full coverage (MCC<1), which is likely due to the low F_0 and slow R_0 values in Andean genotypes, that might be explained by the low temperatures to which they are usually exposed during their growth in the

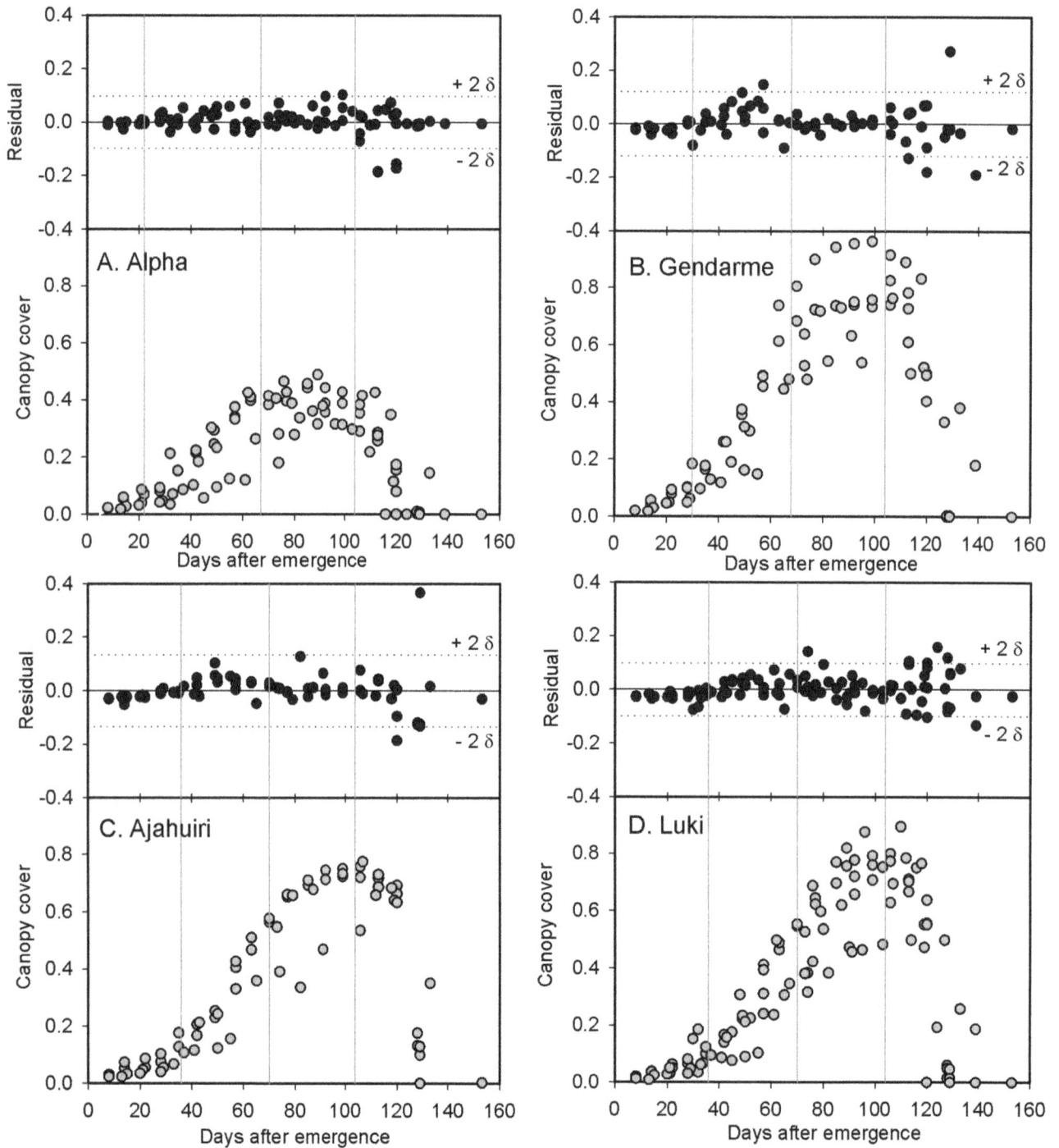

Figure 3. Residual dynamics analysis for Canopy Cover in Alpha, Gendarme, Luki and Ajanhuiri varieties.

Andean highlands. This poor value of the coverage level in the high Andes associated to its short duration contributes to a low biomass production [30,33].

Genotype by Environment Interaction

DTY (P<0.01) and MCC (P<0.05) were significant for genotype x environment interactions. Figure 1, shows the performance of each genotype in the four experiments in which frost was not considered, against the average performance of all

genotypes in each contrasting environment defined by cumulated rainfall as dry, intermediate and moist condition. Under all conditions, Gendarme outperformed the average yield whereas Ajanhuiri outperformed the average yield in dry and intermediate conditions only but was negatively affected by moist conditions. Alpha yielded much less than the average in all environments, a behavior explained by the overall low MCC. Luki was not influenced by environments and its yield was similar to the average.

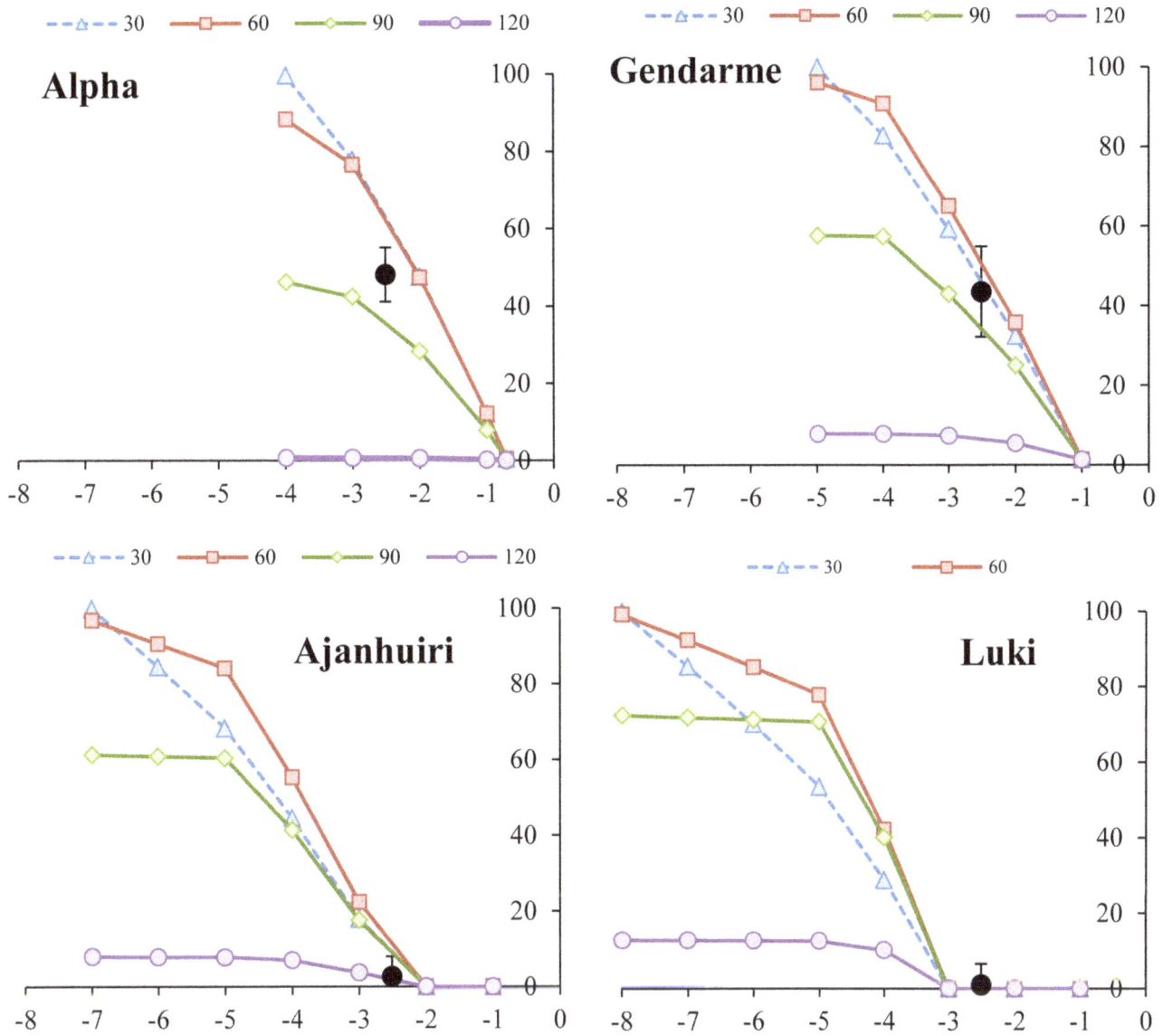

Figure 4. Percentage of tuber yield loss affected by frost (°C) at 30, 60, 90 and 120 days after planting on four potato varieties. In circles real losses on yield caused by frost (−2.51°C) occurred at 94 DAP.

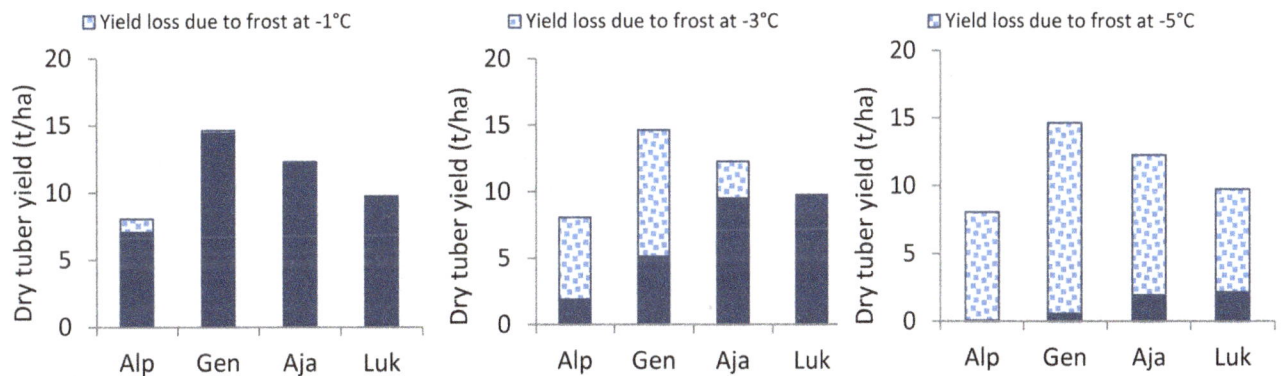

Figure 5. Impact on dry tuber yield loss due to frost at −1°C, −3°C and −5°C simulated at 60 days after emergence.

Table 7. Cumulative total dry yield (tons) under monoculture and mixture (Mix) conditions: without frost; with −3°C frost at 60 and 90 DAE.

Frost free

Genotype	Monoculture				Mixture
Alpha	8.1	–	–	–	0.2
Gendarme	–	14.6	–	–	9.5
Ajanhuiri	–	–	12.3	–	0.9
Luki	–	–	–	9.8	2.4
Total	8.1	14.6	12.3	9.8	13.0

Frost (−3°C at 60 and 90 DAE)

Genotype	Monoculture				Mixture
Alpha	1.5	–	–	–	0.0
Gendarme	–	3.5	–	–	2.3
Ajanhuiri	–	–	8.1	–	0.6
Luki	–	–	–	9.8	2.5
Total	1.5	3.5	8.1	9.8	5.3

Mixtures were composed of Alpha at 0.03, Gendarme at 0.65, Ajanhuiri at 0.07, and Luki at 0.25 of the land.

Modeling

The adapted LINTUL model adequately ($R^2 > 0.88$) reproduced observed field responses in the calibration experiments (Table 4). Alpha showed the highest (5×10^{-3}) light interception at emergence F_0 value whereas the lowest value was shown by Luki (2×10^{-3}). Rate of increase of light interception R_0 was highest for Gendarme (7.3×10^{-3} °Cd^{-1}), and lowest for Luki (5.5×10^{-3} °Cd^{-1}). The duration of senescence (d) was longest for Alpha (approximately 38 days after the attainment of MCC) and shortest for Ajanhuiri (23 days following maximum canopy cover MCC). In the Bolivian Andes, Alpha is clearly an early maturing type [time of reduction to 50% of light interception $t_{0.5} = 106$ Days after Planting (DAE) and tuber initiation onset TIO = 24 DAE] while Luki is a very late maturing with a $t_{0.5}$ of 120 DAE and TIO of 60 DAE (Table 4).

The growth and developmental differences between Alpha and Luki genotypes is further evidenced by the time when maximum bulking rate TG_{max} is attained; 60 DAE and 90 DAE for Alpha and Luki, respectively. The intercept and the slope were equal (P>0.05) to 0 and 1, respectively. The tuberization dynamics and parameter values for all genotypes are shown in Table 5.

The R^2 values (Table 6) show that the model explained more than 81% of the variations in DTY obtained experimentally for all varieties. For Luki and Ajanhuiri, the model explained more than 81 and 88% of the variation in the measured variable, respectively. For Alpha it explained more that 89% of the variation whereas for Gendarme, the model performed quite well, with a R^2 value of 96% for DTY. All other statistical metrics used to test the validity of the model (MB, RMSE and Nash) to simulate potato yield and yield components under the conditions encountered in the high Andes also showed the model's robustness (Table 6).

The *S. tuberosum* ssp. *andigenum*, *S. ajanhuiri* and *S. juzepczukii* seem to be well fitted for the harsh and variable environmental conditions prevalent in the high Andes, as suggested by all varieties. Concerning frost, differences in frost tolerance among

the tested germplasm were evidenced by the differences in canopy cover retention following a frost episode. Figure 2 shows that Luki was not affected by frost; the effect on Ajanjuiri was minimal. On the other hand, Alpha was severely affected by the same frost episode. Nonetheless, genotypes from the subspecies *tuberosum* might play a role for specific objectives and climatic conditions, as suggested by Alpha performance under less stressful conditions. The genetic diversity of potatoes in the high Andes and their different responses to stresses suggests that potato production there can be highly increased with an adequate choice of germplasm, crop management and seed quality [34,33].

Tuberization dynamics provides a clear differentiation among the genotypes studied [21]. Earliness defined as tuber initiation TIO is one of the growth parameters that express differences among genotypes. The earliness of Alpha was probably prompted by environmental conditions in the experimental sites where mean temperature was around 12°C with around 12 h of photoperiod as it has been reported that short photoperiods and low mean temperatures trigger an early onset of tuber growth and bulking in *S. tuberosum* ssp. *Tuberosum* [35].

It seems that the simultaneous occurrence of early tuber bulking and fast bulking rate traits is of high importance in the environmental conditions found in the high Andes. In spite of its earliness, the slow bulking rate of Alpha was associated with its comparative lower yields. On the contrary, Gendarme with its intermediate bulking initiation but fast-bulking rates produced the highest yield. Intermediate tuber initiation TIO combined with intermediate bulking rates produced intermediate yields.

Alpha accumulates assimilates earlier thus relying on a mechanism for escaping frost damage provided it happens late in the growing season. On the other hand, Luki presents a delayed TIO and a slow tuberization rate and thus extreme frost events (temp<−5°C) can disrupts the accumulation of assimilates and might cause losses.

The *juzepczukii* (var. Luki) and *ajanhuiri* genotypes outperformed the yield obtained with the mixtures under severe frosts. These results highlight the importance of the role played by local varieties tolerant to frost and drought stresses. Since the simulations were based on models parameterized with experimental results, yields obtained are feasible if good quality seeds and adequate management are implemented.

Modeling Diagnostics

The residual analysis showed a good spread of the values throughout the growth cycle. Luki presented a few points out of the accepted confidence envelop during the MCC and senescent phases. This was probably due to the rosette-type morphology which makes this variety susceptible to flattening (Figure 3).

For the emergence, maximum growth rate and MCC phases, the residual values were distributed within two standard deviations from the mean. Nonetheless, the regression diagnostics (Cook's D, DFFITS, DFBETAS and COVRATIO), showed flagging points during senescence. This might be due to the lack of accuracy of the visual estimation of canopy cover, as the human eye cannot differentiate photosynthetically active leaves from senescent ones, particularly at the early stages of senescence. The use of digital cameras and appropriate segmentation techniques can reduce these problems (unpublished results from Production Systems and Environment subprogram at CIP).

Scenario Analysis

The validated model was used to run *what if* type scenarios, to assess the likely impact of frost events - at specific phenologic stages – on yields. A threshold for absolute growth cessation with no regrowth was established in the model for each genotype (Alpha −4°C; Gendarme −5°C; Ajanhuiri −7°C; and Luki −8°C) occurring at different DAE. Overall, highest losses were evidenced for early frost events with average yield reductions ranging from 70 to 100% when occurring at 30 and 60 DAE. When the frosts were simulated at 90 DAE, the average reduction was around 50%. This late frost affected Luki the most with yield reductions of around 70%. In contrast, the early variety Alpha was the least affected by late frost with only about 40% tuber yield reduction. The latest simulated frost at 120 DAE caused minimal damages on tuber yields (<10%) in all genotypes (Figure 4).

It is noteworthy that when the frost event occurs at 60 DAE, there is a drastic yield reduction in all genotypes. This seems to be associated to the fact that this is the time when canopy maximum growth rate and tuber initiation are also occurring. Alpha was the most affected with losses increasing exponentially as the severity of the frost augmented. Total loss in Alpha was caused by a −5°C frost. Gendarme presents a minimal damage at −1°C (0.2 t/ha), important losses at −3°C (9.5 t/ha) and a severe loss of tuber production at −5°C (14 t/ha). Ajanhuiri, in turn, was not affected at −1°C, presented a minor reduction at −3°C (2.7 t/ha) and an important reduction of tuber production at −5°C (10.3 t/ha). Luki was not affected at −1 and −3°C, but at −5°C was almost devastated (Figure 5).

We also simulated total yield comparing monoculture and single plot mixtures containing fractions of total acreage with four varieties distributed in the field (Alpha at 0.03, Gendarme at 0.65, Ajanhuiri at 0.07, and Luki at 0.25) the proportions mimicking the actual genotype distributions encountered in farmer fields. Scenarios without and with frost occurrence (−3°C at 60 and 90 DAE) were compared (Table 7). Frost free seasons resulted in average yields of 13 t/ha for the mixture. Gendarme in monoculture out yielded the simulated mixture. On the other hand, the mixture yielded 5.3 t/ha when affected by frost simulations. This yield was higher than the resulting from the commercial varieties Alpha and Gendarme but lower than those produced by Ajanhuiri and Luki as monocultures. If the tendency to a reduced planting of frost tolerant genotypes (Ajanhuiri and Luki) is confirmed, the production based on commercial potatoes can attain only 3.5 t/ha (Gendarme) when frosts occur (Table 7), thus local frost tolerant varieties provide an insurance under uncertain climate.

The varietal single plot mixtures used by the high Andes farmers seem to be a robust strategy to cope with climatic risks. This is particularly true under uncertain climate when a portfolio of options is warranted. The inclusion of native potatoes especially bitter varieties (*S. juzepczukii*) in the mix is a must under highly variable climatic conditions. The scenarios tested were purposely limited to fixed frosts at specific times during the growing period, just as examples of the modeling possibilities. The reality is that frosts are likely to occur randomly at different times thus highlighting the possible role of mixing varieties – as practiced for millennia by local farmers – to cope with those extreme events. Although the projection of future climate lean towards warmer nights, the increased variability of both night temperature and precipitation together with increased radiation cooling [17] warrant the inclusion of the drought and frost tolerance traits of *juzepczukii* and *ajanhuiri* genotypes. Therefore, the genetic diversity must be maintained since temperature and rainfall variability is expected to increase.

Conclusions

The correspondence between the simulations and the results of the independent set of experiments conducted under various agroecological conditions demonstrated the adequacy of the model. This robustness applies not only to yield results under non-limiting factors but also to the forecast of the effects of frost events on canopy cover and correlated yield, which is a plus of the model. All statistical metrics used to test the validity of the model to simulate potato yield and yield components under the conditions encountered in the high Andes showed the model's robustness.

The results indicate that Ajanhuiri and Gendarme varieties seem to be well suited for conditions with intermediate moisture levels, whereas, Ajanhuiri showed a better adaptation to drier areas. Ajanhuiri is a non-bitter variety tolerant to frost. Reintroducing good quality seeds into high Andean farming systems could enhance yields given its hardiness to cope with frequent frosts. For moister – rainfed or with strategic irrigation - areas the clear recommendation seems to be Gendarme and the like genotypes. Luki, on the contrary seems to be insensitive to environmental variations within the Altiplano.

Current conditions in the high Andes include a prolonged growing period of around 180 days. Small farmers diversify their risk through an assortment of genotypes in single plots. The observed diversity of tuberization dynamics of different genotypes supports this risk management strategy given the climatic and altitudinal variability and the low opportunity cost of the land. As the access to land decreases and direct as well as indirect effects of climate change impinge on the crop, new strategies must be sought. One such a strategy could be to develop genotypes with an early TIO and a very fast early bulking rate, able to escape from early biotic or abiotic shocks while sustaining a sizable yield.

Acknowledgments

The authors thank all those who made this study possible, especially to field researchers of PROINPA Foundation in Bolivia; to the Unit of Ecophysiology and Crop Breeding, Catholic University of Louvain in Belgium; to the International Potato Center through PAPA ANDINA and ALTAGRO Projects. This research was conducted under the CGIAR Research Program on Climate Change, Agriculture and Food Security (CCAFS). Special thanks to Dr. Victor Mares for his support in the technical editing several versions of the final manuscript.

Author Contributions

Conceived and designed the experiments: BC RJH JFL RQ. Performed the experiments: BC. Analyzed the data: BC JFL RQ. Contributed reagents/materials/analysis tools: BC RJH JFL RQ. Wrote the paper: BC JFL RQ.

References

1. PNUMA - Programa de las Naciones Unidas para el Medio Ambiente (2011) Perspectivas del Medio Ambiente en el Sistema Hídrico Titicaca-Desaguadero-Poopó-Salar De Coipasa (TDPS). GEO Titicaca. Oficina Regional para Latinoamérica y el Caribe, Panamá.

2. INE - Instituto Nacional de Estadísticas (2009) Encuesta Nacional Agropecuaria 2008. Ministerio de Planificación del Desarrollo de Bolivia. La Paz.

3. Terrazas F, Valdivia G (1998) Space dynamics of in situ preservation: handling of the genetic diversity of Andean tubers in mosaic systems Candelaria, Cochabamba (Bolivia). Plant Genet Resour Newsl 114, 9–15.

4. Hijmans RJ, Condori B, Carrillo R, Kropff MJ (2003) A quantitative and constraint-specific method to assess the potential impact of new agricultural technology: the case of frost resistant potato for the Altiplano (Peru and Bolivia). Agric Syst 76, 895–911.

5. Hijmans RJ (2003) The effect of climate change on global potato production. Am J Potato Res 80, 271–280.

6. Spooner DM, McLean K, Ramsay G, Waugh R, Bryan GJ (2005) A single domestication for potato based on multilocus amplified fragment length polymorphism genotyping. PNAS 102, 14694–14699.

7. Ochoa CM (1990) The Potatoes of South America: Bolivia. Cambridge University Press, United Kingdom. 535 p.

8. Estrada N (1999) La Biodiversidad en el Mejoramiento Genético de la Papa. Programa de Investigación de la Papa (PROINPA). Centro de Información para el Desarrollo (CID) y Centro Internacional de la Papa (CIP), La Paz, Bolivia, 372 p.

9. Le Tacon Ph (1989) Manifestation des risques climatiques à l'échelle de l'exploitation agricole, conséquences sur les pratiques paysannes. Cas de l'Altiplano Bolivien. Mémoire de stage. ORSTOM, ENSSAA, CNEARC, ENSAM, Montpellier.

10. Tourneux C, Devaux A, Camacho MR, Mamani P, Ledent JF (2003) Effects of water shortage on six potato genotypes in the highlands of Bolivia (I): morphological parameters, growth and yield. Agronomie 23, 169–179.

11. Rea J (1992) Vigencia de las papas nativas en Bolivia. In: Rea J, Vacher JJ, editors. La Papa Amarga. Primera Mesa Redonda: Perú, Bolivia. ORSTOM. La Paz, 15–23.

12. Woolfe JA, Poats SB (1987) The Potato in the Human Diet. Cambridge University Press. 231 p.

13. Hijmans RJ (1999) Estimating frost risk in potato production on the Altiplano using interpolated climate data. In: International Potato Center, Impact on a Changing World, Program Report 1997–1998. International Potato Center, Lima, 373–380.

14. De Haan S, Juarez H (2010) Land use and potato genetic resources in Huancavelica, central Peru. Journal of Land Use Change 5(3): 179–195.

15. Vuille M, Bradley R (2000) Mean annual temperature trends and their vertical structure in the tropical Andes. Geophysical Research Letters 27(23): 3885–3888.

16. Vuille M, Bradley RS, Werner M, Keimig F (2003) 20th century climate change in the tropical Andes: observations and model results, Climatic Change, 59, 75–99.

17. Thibeault JM, Seth A, Garcia M (2010) Changing climate in the Bolivian Altiplano: CMIP3 projections for temperature and precipitation extremes. J Geophys Res Vol. 115.

18. Francois C, Bosseno R, Vacher JJ, Seguin B (1999) Frost risk mapping derived from satellite and surface data over Bolivian Altiplano. Agriculture and Forest Meteorology 95 (2), 113–137.

19. Carrasco E, Devaux A, García W, Esprella R (1997) Frost tolerant potato varieties for the Andean Highlands. In: International Potato Center, Program Report 1995–1996. International Potato Center, Lima, 227–232.

20. Tapia N, Saravia G (1997) Biodiversidad en papas amargas. Provincia Tapacarí, Departamento de Cochabamba. AGRUCO, Cochabamba.

21. Condori B, Hijmans RJ, Quiroz R, Ledent JF (2010) Quantifying the expression of potato genetic diversity in the high Andes: Growth analysis and modeling. Field Crops Res 119 (2010): 135–144.

22. Gabriel JL, Carrasco E, García W, Equise H, Navia O, et al. (2001) Experiencias y logros sobre mejoramiento convencional y selección participativa de cultivares de papa en Bolivia. Revista Latinoamericana de la Papa 169–192.

23. Devaux A, Vallejos J, Hijmans R, Ramos J (1997) Respuesta agronómica de dos variedades de papa (spp. *tuberosum* y *andigena*) a diferentes niveles de fertilización mineral. Revista Latinoamericana de la Papa 9/10: 123–139.

24. Spitters CJT (1988) An analysis of variation in yield among potato cultivars in terms of light absorption, light utilization, and dry matter partitioning. Acta Hortic 214, 71–84.

25. Spitters CJT (1990) Crop growth models: their usefulness and limitation. Acta Hortic 267, 349–368.

26. Willmott CJ (1982) Some comments on the evaluation of model performance. Bull Am Met Soc 63, 1309–1313.

27. Willmott CJ, Robeson SM, Matsuura K (2011) A refined index of model performance. Int. J. Climatol. (2011). Published online in Wiley Online Library DOI: 10.1002/joc.2419.

28. Rawlings JO, Pantula SG, Dickey DA (1998) Applied Regression Analysis: A Research Tool. Second Edition. Springer Texts in Statistics. Department of Statistics North Carolina State University, Raleigh. 657 p.

29. Condori B (2002) Validation du modèle LINTUL-POTATO (*Solanum* spp.) pour estimer le rendement potentiel et les pertes provoquées par le gel dans la région andine bolivienne. Dissertation de la Maîtrise en Sciences Biologiques, Agronomiques et Environnementales. Université Catholique de Louvain, Belgique.

30. Stol W, De Koning GHJ, Kooman PL, Haverkort AJ, Van Keulen H, et al. (1991) Agro-ecological Characterization for Potato Production. A Simulation Study at the Request of the International Potato Center (CIP), Lima, Peru (CABO-DLO, Report 155). CABO-DLO, Wageningen.

31. Kooman PL, Haverkort AJ (1995) Modeling development and growth of the potato crop influenced by temperature and daylength: LINTUL-POTATO. In: Haverkort AJ, Mackerron DKL, editors. Potato Ecology and Modelling of Crops Under Conditions Limiting Growth. Current Issues in Production Ecology, vol. 3. Kluwer Academic Publishers, Dordrecht, Netherlands, 41–59.

32. Moriondo M, Bindi M, Sinclair T (2005) Analysis of *Solanaceae* species harvest organ growth by linear increase in harvest index and harvest-organ growth rate. J Am Soc Hortic Sci 130, 799–805.

33. Condori B, Mamani P, Botello R, Patiño F, Devaux A, et al. (2008) Agrophysiological characterisation and parametrisation of Andean tubers: Potato (*Solanum* sp.), oca (*Oxalis tuberosa*), isaño (*Tropaeolum tuberosum*) and papalisa (*Ullucus tuberosus*). Eur J Agron 28, 526–540.

34. Iriarte V, Badani AG, Villarroel CL, Aguirre G, Fernández-Northcote EN (2001) Priorización, Limpieza Viral, Producción de Semilla de Calidad Básica y Devolución de Cultivares Nativos Libres de Virus. Revista Latinoamericana de la Papa 12: 72–95.

35. Van Dam J, Kooman PL, Struik PC (1996) Effects of temperature and photoperiod on early growth and final number of tubers in potatoes (*Solanum tuberosum* L.). Potato Res 39, 51–62.

21

Crop Diversity for Yield Increase

Chengyun Li[1], Xiahong He[1], Shusheng Zhu[1], Huiping Zhou[1], Yunyue Wang[1], Yan Li[1], Jing Yang[1], Jinxiang Fan[2], Jincheng Yang[3], Guibin Wang[4], Yunfu Long[5], Jiayou Xu[6], Yongsheng Tang[7], Gaohui Zhao[8], Jianrong Yang[9], Lin Liu[1], Yan Sun[1], Yong Xie[1], Haining Wang[1], Youyong Zhu[1]*

1 Key Laboratory of Agro-Biodiversity and Pest Management of Education Ministry of China, Yunnan Agricultural University, Kunming, Yunnan, China, **2** Plant Protection Station of Honghe Prefecture, Mengzi, Yunnan, China, **3** Agroscience Research Institute of Yuxi City, Yuxi, Yunnan, China, **4** Plant Protection Station of Chuxiong Prefecture, Chuxiong, Yunnan, China, **5** Plant Protection Station of Shiping County, Shiping, Yunnan, China, **6** Agroscience Research Station of Hongxi Town, Mile, Yunnan, China, **7** Agricultural Technology Extension Centre of Qujing City, Qujing, Yunnan, China, **8** Agricultural Technology Extension Centre of Zhaotong City, Zhaotong, Yunnan, China, **9** Agricultural Technology Extension Centre of Lincang City, Lincang, Yunnan, China

Abstract

Traditional farming practices suggest that cultivation of a mixture of crop species in the same field through temporal and spatial management may be advantageous in boosting yields and preventing disease, but evidence from large-scale field testing is limited. Increasing crop diversity through intercropping addresses the problem of increasing land utilization and crop productivity. In collaboration with farmers and extension personnel, we tested intercropping of tobacco, maize, sugarcane, potato, wheat and broad bean – either by relay cropping or by mixing crop species based on differences in their heights, and practiced these patterns on 15,302 hectares in ten counties in Yunnan Province, China. The results of observation plots within these areas showed that some combinations increased crop yields for the same season between 33.2 and 84.7% and reached a land equivalent ratio (LER) of between 1.31 and 1.84. This approach can be easily applied in developing countries, which is crucial in face of dwindling arable land and increasing food demand.

Editor: Dorian Q. Fuller, University College London, United Kingdom

Funding: This work was supported by the National Basic Research Program (No. 2006CB100200) and The Ministry of Science and Technology of China. The funders had no role in study design, data collection and analysis, decision to publish or preparation of the manuscript.

Competing Interests: The authors have declared that no competing interests exist.

* E-mail: yyzhu@ynau.edu.cn

Introduction

It has been recognized that biodiversity is key to securing global food supply[1]. Zhu *et al.* reported that planting a mixture of rice varieties was effective in boosting yields[2] and decreasing diseases[3]. The approach had been extended to 1.57million ha from 2000 to 2004 in 11 provinces of China. It increased 675 kg/ha yield in average and 259 million US$ of income and cost-saving. Rice blast in mixtures was 67% less severe than that in monoculture[4]. In natural ecological systems, it has been shown that biomass production can be elevated with increasing biodiversity[5,6]. For example, Tilman *et al.* showed that biomass production from experimental fields in which 16 grass species were grown in a mixture was increased by 2.7 times compared with those in which single species were grown alone[7]. They also demonstrated that the more plant species a field contained the more stable the ecological system was from year to year[8]. This accords with observations by Li *et al.*[9], Morgado and Willey[10] and Dybzinski *et al.*[11] that biodiversity could increase soil fertility and LER. In crop systems, there is great potential for the use of mixed cropping to enhance productivity, but this must be tested at a scale relevant to agricultural production[12–13].

Results

In collaboration with farmers and extension personnel in 10 counties in Yunnan Province, we tested intercropping of tobacco-maize, sugarcane-maize, potato-maize, and wheat-broad bean,

either by overlapping growing seasons or by mixing crop species based on differences in their heights. The four crop combinations were compared with their respective monocrops in adjacent plots. A schematic illustration of the planting arrangements is shown in Figure 1 with details of the planting and harvesting dates.

Yunnan is the key plantation region for tobacco in China, with a cultivation area of over 400,000 ha. Local farmers grow tobacco in summer and wheat or barley in winter; tobacco is normally harvested in mid-August and planting of wheat or barley does not begin until November, leaving the fields unutilized for three months. By planting maize in the tobacco field in mid-July and harvesting in November, an additional crop can be grown in this period.

We tested intercropping of tobacco-maize in Mile, Yao'an and Chuxiong counties, and this pattern was adopted by local farmers in 325 ha and 4,162 ha of farmland in 2006 and 2007, respectively. The results show that the yields of tobacco were comparable in both systems. Intercropping resulted in additional maize production of 5.88 and 5.91 t/ha in 2006 and 2007, respectively (Table 1), constituting 84.7 and 84.5% of the production from the monocrops, with LERs of 1.84 and 1.83. Severity of tobacco brown leaf spot disease in the two systems was comparable, but northern maize leaf blight in intercropped plots was decreased by 17.0 and 19.7% in 2006 and 2007, respectively, compared with the monocropped controls (Fig. 2).

This approach was also applied to crops with long cultivation seasons. Sugarcane, with a cultivation area of about 300,000 ha in Yunnan, has a year-long cultivation season. As the plants are short

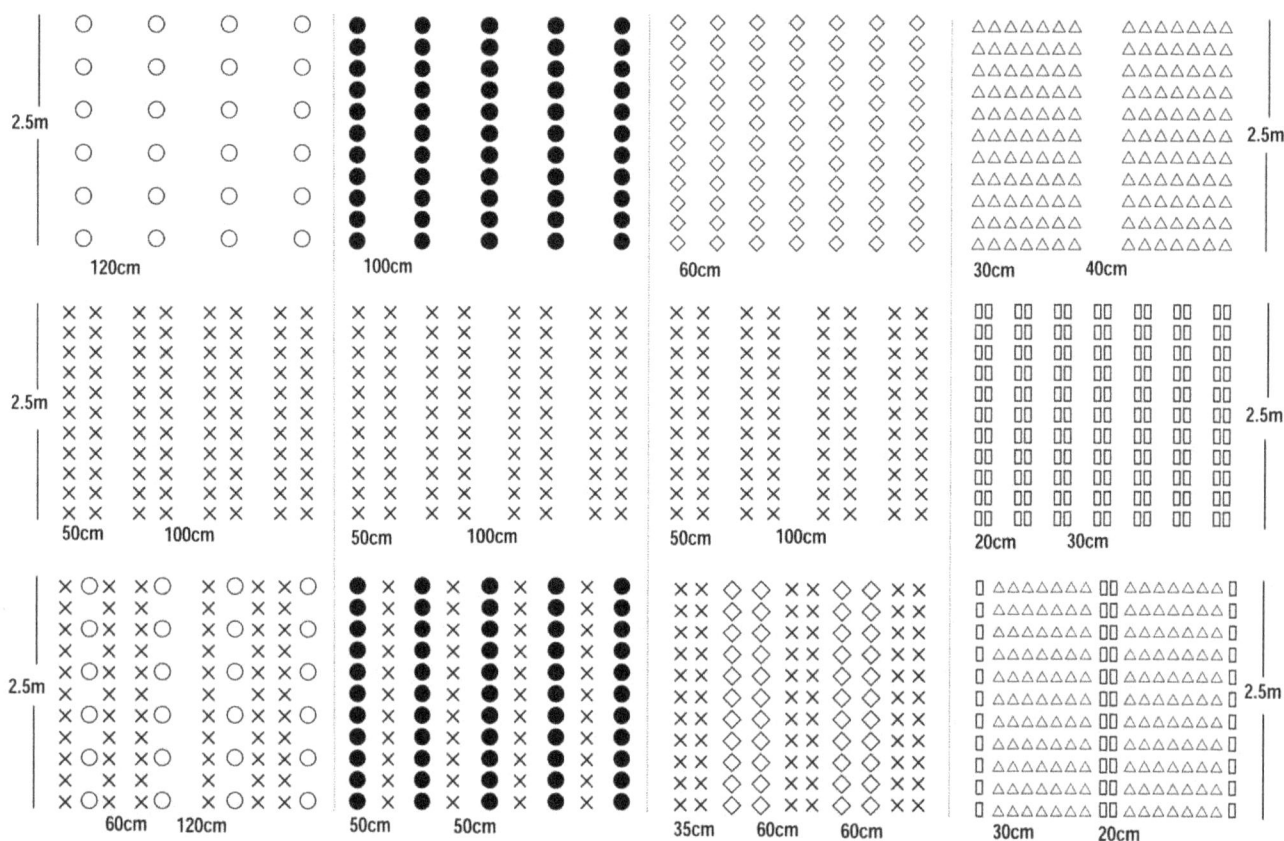

Figure 1. Crop patterns in intercropping and monoculture experiment plots. Each symbol represents a plant (hill) of a different crop species: tobacco (○); maize (×); sugarcane (●); potato (◇); wheat (△); broad bean (□).

in the first half of the season, there is sufficient light available to grow an additional crop of maize in this period. Sugarcane-maize was tested in Mile, Shiping and Yongde counties, and adopted by local farmers in 80 ha and 1,582 ha of farmland in 2006 and 2007 respectively.

The results show that the yields of sugarcane were comparable between monocropped and intercropped plots. The intercropped maize produced an additional 4.77 and 4.72 t/ha in 2006 and 2007, respectively (Table 1), constituting 64.0 and 63.2% of the production from the respective monocrops, with LERs of 1.63 and 1.64. Severity of sugarcane eye spot disease in the two systems was comparable, but northern maize leaf blight in intercropped plots decreased by 55.9 and 49.6% in 2006 and 2007, respectively, compared with the monocrops (Fig. 2). This reduction in maize disease may be a result of less rainfall during the earlier growing period of the intercropped plots.

We also intercropped short and tall crops in the same field, increasing the spatial utilization of farmland. By working with farmers in Xuanwei, Huize and Zhaotong counties, we inter-cropped potato-maize in 1,685 and 5,658 ha of farmland in 2004 and 2005, respectively. The maize yields from intercropping were 147% in both years compared with equal areas of the monocrops. The intercropped potato yields in these two years were 115 and 120% compared with equal areas of monocrops (Table 1), resulting in LERs of 1.31 and 1.33. Severity of potato late blight in the intercropping system was decreased by 32.9 and 39.4% in 2004 and 2005, respectively, compared to the monocrops, while northern maize leaf blight in intercropped plots was decreased by 30.4 and 23.1% (Fig. 2).

Similar experiments were also carried out with crops of similar cultivation seasons. In 2004 and 2005, we intercropped wheat and broad bean in 358 ha and 1,452 ha of farmland, respectively, in Hongta and Yimeng counties (Fig. 1). The results show that wheat yields from intercropping were comparable with the monocrops in both years. Intercropping resulted in additional broad bean production of 0.98 and 0.97 t/ha, in 2004 and 2005, respectively (Table 1), constituting 34.2 and 33.2% of the production from the monocrops, giving LERs of 1.34 and 1.33 in 2004 and 2005 (Table 2). Severity of broad bean chocolate spot disease in the intercrops decreased by 33.8 and 31.7% in 2004 and 2005, respectively, compared with the monocrops (Fig. 2). This could result from reduction in disease spread among broad bean plants because they were separated by rows of wheat plants.

Discussion

These large-scale experiments demonstrate the advantages of cultivating a mixture of crop species in the same field through temporal and spatial management. Intercropping maize in tobacco and sugarcane enhanced utilization of land space and physical resources during the late growing period of tobacco and early cultivation phase of sugarcane, adding a season of maize production. In the conventional practice, only one crop was cultivated in the field during growing season. For the tested patterns in the study, all the practice for tobacco and sugarcane was exactly adopted as same as conventional practice. Maize was extra crop for tobacco and sugarcane fields. The combination of potatoes with maize as well as wheat with broad bean took

Table 1. Yield and monetary value for different crops.

Crop	Variety	Plants m^{-2}	Yield ± s. e. m (t/ha)		Crop value (US$ per ha)	
			1st year	2nd year	1st year	2nd year
Tobacco	Yunyan-87	1.67	2.82±0.003	2.86±0.007	5829	5912
Maize	Huidan-4	5.35	6.94±0.003	6.99±0.017	1972	1986
Intercropping		6.67	8.69	8.75	7477	7477
Tobacco	Yunyan-87	1.67	2.81±0.006	2.84±0.017	5808	5870
Maize	Huidan-4	5.00	5.88±0.004	5.91±0.017	1671	1679
Sugarcane	Xintaitan-2	9.62	105.87±0.851	105.23±0.256	2529	2514
Maize	Xundan-7	5.35	7.54±0.006	7.47±0.030	2142	2123
Intercropping		13.45	110.35	111.67	3878	3878
Sugarcane	Xintaitan-2	9.45	105.58±0.575	106.95±0.409	2522	2555
Maize	Xundan-7	4.00	4.77±0.005	4.72±0.020	1355	1341
Potato	Hui-2	6.67	31.86±0.105	31.27±0.380	2058	2020
Maize	Huidan-4	5.35	7.17±0.022	7.13±0.026	2037	2026
Intercropping		7.42	23.71	23.99	2687	2687
Potato	Hui-2	3.71	18.45**(115)**	18.75**(120)**	1192	1211
Maize	Huidan-4	3.71	5.26**(147)**	5.24**(147)**	1495	1489
Wheat	Yumai-3	277.36	5.31±0.013	5.32±0.016	1577	1580
Broad bean	Dabaidou	13.65	2.87±0.011	2.92±0.011	1389	1413
Intercropping		280.05	6.27	6.28	2045	2045
Wheat	Yumai-3	277.36	5.29±0.020	5.31±0.017	1571	1577
Broad bean	Dabaidou	2.69	0.98±0.012	0.97±0.007	474	469

Crop yield determined by grain weight for rice, wheat and broad bean, dry leaf weight for tobacco, fresh stem and tuber weight for sugarcane and potato. Crop values based on market prices of 2067.02 US$ per ton for tobacco, 284.15 US$ per ton for maize, 23.89 US$ per ton for sugarcane, 64.59 US$ per ton for potato, 296.98 US$ per ton for wheat, 483.97 US$ per ton for broad bean. Crop yield and value were for individual species within intercropping. Yields of tobacco-maize, sugarcane-maize and wheat-broad bean patterns were additional production compared with monocrops. Yields of potato intercropped with maize and maize intercropped with potato, compared with equal areas of monocrops are shown in **(bold)**. Statistical analyses: each survey plot was considered to be an experimental unit, and analyses were based on actual mean plot yields. Statistical analyses were conducted by software SPSS 13.0. One-tailed t-tests were used to determine if the yield differed significantly (p≤0.05).

advantage of the differences in their heights. Such intercropping resulted in the formation of three-dimensional crop assemblies in the fields, possibly improving growth through a more favorable microclimate. These systems boosted yields and reduced disease, produced high LERs and increased farmers' incomes (Table 1; Table 2; Fig. 2), although they required higher labors inputs, more seeds and fertilizer.

Intercropping short and tall plants may benefit crop growth by increasing light and air diffusion. The reduction in potato late blight disease in intercropped plots may be a result of less rainfall during the growing period between April and July compared with the monocrops between June and August, when the disease normally peaks[14]. After the potato crop was harvested, the ambient humidity and leaf wetness of the maize decreased because of the distance between the rows of plants, which may limit the spread of the northern leaf blight. Because of these beneficial effects, this intercropping design has been adopted by most of the local farmers.

Taken together, our large-scale intercropping experiments in 15,302 ha of farmland have provided an unprecedented amount of data that demonstrate intercropping's clear advantages of boosting yields and preventing disease. Our results support the findings on the relationship between biodiversity and biomass production based on perennial plant populations in experimental models[15]. Our studies involved farmers from the beginning, an approach that helped them to understand the rationale behind the technique and enabled rapid assimilation of research results among the local communities.

In addition to pointing to the importance of crop diversity, our findings have wide implications for food security. Increasing food production by intercropping is very simple and can be easily applied in developing countries, which is crucial in the face of dwindling arable land and increasing food demand. It has been recognized that reduction in arable land is one of the key factors in causing the current food crisis[16]. In China, the area of arable land was reduced by 4.7 Mha, or about 4.5% between 1978 and 1996[17]. A report by FAO projects that the area of arable land per person in the world might decrease below the critical level of 0.1 hectare by 2050 due to increasing desertification and urbanization[16]. As the reduction in arable land is unavoidable, increasing LER and food production per unit area is crucial for securing food supply. The crop diversity techniques described in this paper have been listed by the Yunnan Provincial Government as a key strategy to boost food production and is applied to 1.5 Mha per year, according to Provincial Government statistics. It is crucial that such a simple, effective approach to boosting crop yields and increasing LERs is widely adopted in the global challenge of securing the food supply.

Materials and Methods

Field Study

The field experiment sites of different crops combination were located in different areas which were suitable for these crops

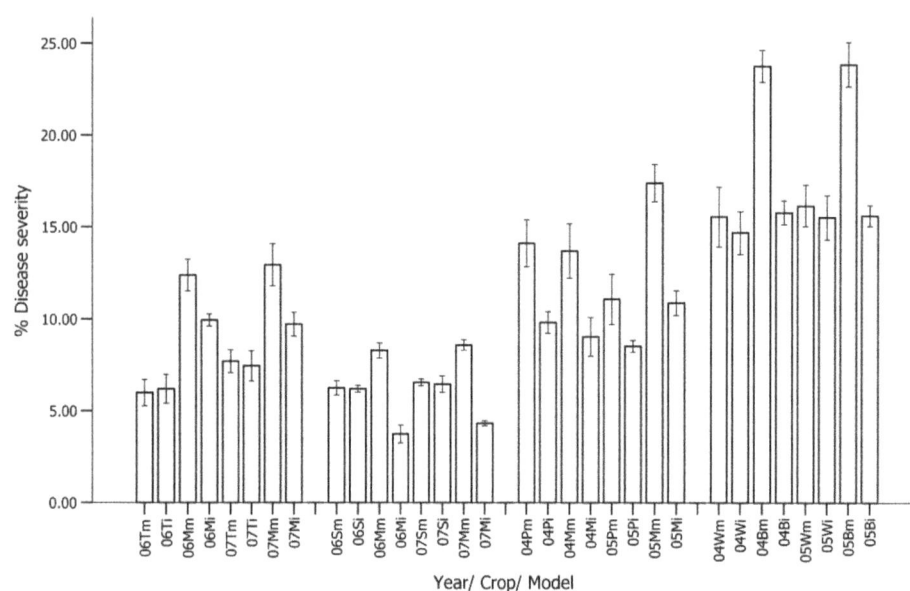

Figure 2. Severity of main diseases of the crops in monocropping and intercropping systems. T = Tobacco brown leaf spot (*Alternaria alternate* Keissler); M = Maize northern leaf blight (*Setosphaeria turcica* Leonard); S = Sugarcane eye spot (*Bipolaris sacchari* (Butl) Shoemaker); P = Potato late blight (*Phytophthora infestans* (Mont.) de Bary); W = Wheat Stripe Rust (*Puccinia striiformis* West); B = Broad bean *chocolate* spot (*Botrytis fabae* Sard). m = disease severity for crop species grown in monoculture control plots; i = disease severity for the same crop species grown in intercropping plots in the same fields. Error bars are one s. e. m; n = 3. Statistical analyses were conducted by software SPSS 13.0. All differences between pairs are significant at P≤0.05 based on one-tailed t-test.

growth. The experiment sites were chosen in the most suitable cultivation area for different crop combinations. Tobacco-maize sites were located in Mile, Yao'an and Chuxiong Counties ranging from 1000 m to 1600 m (a.s.l.), Sugarcane-maize sites in Shiping, Mile and Yongde Counties under 1200 m (a.s.l), Maize-potato sites in Xuanwei, Huize and Zhaotong Counties ranging from 1600 m to 2100 m (a.s.l.), Wheat-broad bean sites in Hongta and Yimen Counties ranging from 1500 m to 1900 m (a.s.l.). Each crop combination was tested in three different experiment sites. Each site included three treatments, i.e. one treatment for intercropping, and the other two treatments for monoculture of the two crops (Fig. 1). Each treatment with three replicates (3×3), nine plots for each site were located in the same field by randomized blocks design.

Field surveys were carried out in 2004–2007, with experimental crop patterns adopted by farmers in Yunnan Province. There were three experimental plots (each about 200 m^2) for each crop combination each year. Planting and harvest time for each crop combination were the same for each year and field management

was conducted by farmers according to local practice. For tobacco-maize, tobacco seedlings (Yunyan-87) were planted on 22 April and harvested progressively between 15 June and 18 August. To intercrop, maize seeds (Huidan-2) were planted in the tobacco fields on 10 July and harvested on 30 October. For monocropping, tobacco was grown in the same way as the intercrop, whereas maize cultivation followed the usual timeframe – sown 25 May and harvested 18 September. For sugarcane-maize, sugarcanes (Xintaitang-2) were planted on 5 January and harvested on 25 December; maize seeds (Xundan-7) were planted in the sugarcane fields on 20 February and harvested on 30 June. For monocropping, sugarcane was grown in the same way as the intercrop, maize in the usual timeframe, as above. For potato-maize, the potato cultivation season was shifted seven weeks earlier: the Hui-2 variety was planted 30 March and harvested 15 July. Maize (Huidan-4) was grown in the usual timeframe, as above. For monocropping, cultivation of both potato and maize followed their usual timeframe, with potatoes planted 20 May and harvested 2 September, and with maize planted and harvested as described above. With wheat-broad bean, for both intercropping and monoculture, wheat (Yunmai-3) was sown 28 October and harvested 25 April; broad bean (Dabaidou) was planted 10 October and harvested 5 April.

Yield and Monetary Value Surveys

The yield data in Table 1 were based on whole plots harvest. Crop yield was determined by grain weight for Maize, wheat and broad bean, dry leaf weight for tobacco, and fresh stem and tuber weight for sugarcane and potato. Crop values were based on market prices of 2067.02 US$ per ton for tobacco, 284.15 US$ per ton for maize, 23.89 US$ per ton for sugarcane, 64.59 US$ per ton for potato, 296.98 US$ per ton for wheat, and 483.97 US$ per ton for broad bean. Crop yield and value were for individual species within intercropping. Yields of tobacco-maize, sugarcane-maize

Table 2. Land equivalent ratios for crop yields produced by intercropping.

Intercropping	First year	Second year
Tobacco/Maize	1.84	1.83
Sugarcane/Maize	1.63	1.65
Potato/Maize	1.31	1.33
Wheat/Broad bean	1.34	1.33

Land equivalent ratios (LERs) were calculated as (yield ha^{-1} of crop A in intercropping/yield ha^{-1} of crop A in monoculture)+(yield ha^{-1} of crop B in intercropping/yield ha^{-1} of crop B in monoculture).

and wheat-broad bean patterns were additional production compared with monocrops. Each survey plot was considered to be an experimental unit, and analyses were based on actual mean plot yields. Statistical analyses were conducted by software SPSS 13.0. One-tailed t-tests were used to determine if the yield differed significantly (p≤0.05).

Severity of Crop Diseases

One of the most serious diseases for each crop was surveyed. Survey standard of tobacco brown leaf spot (*Alternaria alternate* Keissler) is based on the YC/T40-1996, P.R. China[18]; Maize northern leaf blight (*Setosphaeria turcica* Leonard) on the NY/T 1248.1-2006, P.R. China[19]; Wheat Stripe Rust (*Puccinia striiformis* West) on the NY/T 1443.1-2007, P.R. China[20]; Disease survey of sugarcane eye spot (*Bipolaris sacchari* (Butl) Shoemaker) followed Dai's report (1993) [21]; Potato late blight (*Phytophthora infestans* (Mont.) de Bary) followed Handbook of Crop Pest Forecasting (2006) [22]; Broad bean *chocolate* spot (*Botrytis fabae* Sard) followed Wallen's report (1957) [23].

All investigated diseases were assessed at five sampling points in each plot, distributed in a uniform pattern. Sampling number, sampling plant part and disease categories were different described as below.

Tobacco brown leaf spot (*Alternaria alternate* Keissler) disease: whole leaves of twenty plants were evaluated at each sampling point. Five disease scales were rated in terms of the percentage of symptomatic leaf area: 0, no disease; 0.5, less than 1%; 1, 1–5%; 2, 5–10%; 3, 10–20%, 4, >20% leaf area affected.

Maize northern leaf blight [*Setosphaeria turcica* (Pass.) Leonard & Suggs] disease was investigated at the wax ripeness stage. Twenty plants were evaluated at each sampling point. Six disease scales were rated based on the percentage of the total leaf area affected: 0, no disease; 1, few lesion below the ear leaf covering less than 5% of leaf surface; 3, lesion below the ear leaf cover between 6 and 10% of leaf surface, few lesion above the ear leaf; 5, lesion below the ear leaf cover 11–30% of leaf surface, a few lesion above ear leaf; 7, lesion area below the ear leaf cover 31–70% of leaf surface, lots of lesion above the ear leaf; 9, large coalesced lesions, covering more than 70% of the leaf surface, and foliage completely destroyed.

Wheat stripe rust (*Puccinia striiformis* West. f. sp. *tritici* Erikss) disease severity was surveyed at milk-ripe stage. 100 flag leaves were evaluated at each sampling point. Disease severity were rated according to a linear scale of percentage of symptomatic leaf area from 0, 1%, 5%, 10%, 20%, 40%, 60%, 80% and 100%.

Potato late blight [*Phytophthora infestans* (Mont.) de Bary] disease was assessed at twenty days after potato flowering. Twenty plants were randomly selected to evaluate at each sampling point. Six disease scales were rated according to the percentage of leaflets area affected: 0, none lesions; 1, less than 5%; 3, more than 6%

but less than 10%; 5, more than 11% but less than 20%; 7, more than 20% but less 50%, 9, more than 50% leaf area affected.

Sugarcane eye spot [*Bipolaris sacchari* (Butler) Shoemaker] disease was surveyed at the late stage of stalk elongation. Thirty plants were surveyed at each sampling point, Six disease scales was scored: 0, no disease; 1, few small brown spot on leaf; 2, lesion area about 3 mm×1 mm and the number less than 10; 3, lesion area about 4 mm×1.5 mm and the number more than 10; 4, lesion area about 5~8 mm×1.5~2.0 mm and the number more than 20, some lesions coalesced; 5, lesion area about 5~8 mm×1.5~2.0 mm and the number more than 30, some lesions coalesced and leaf destroyed.

Broad bean chocolate spot (*Botrytis fabae* Sard.) disease was surveyed on pod-setting stage. Thirty plants were randomly selected to evaluate at each sampling point. Five disease scales was scored: 0, no lesions or few small brown, non-sporulating specks, covering up to 1% of leaf surface; 1, few small, discrete, brown, circular, nonsporulating lesions (2–3 mm in diameter) covering between 1.1 and 2% of leaf surface.; 2 is lesions common (3–5 mm in diameter) some coalesced, covering 2.1–5% of leaf surface, with some defoliation and very poor sporulating.; 3 is large coalesced irregular lesions which are blackish, sporulating, and cover 5.1–10% of leaf surface, average defoliation, flower drop, and some dead plants.; 4 is extensive large coalesced irregular lesions which are blackish, heavily sporulating, and cover more than 10% of the leaf surface, severe defoliation, stem girdling, and death of great majority of plants.

Disease severity was summarized within each plot as $\{[(n_1 \times 1) + (n_2 \times 2) + (n_3 \times 3) + ... + (n_N \times N)]/N \times (n_1 + n_2 + n_3 + ... + n_N)\} \times 100$, where $n_1... n_N$ is the number of leaves in each of the respective disease categories, N is the highest scoring of the disease; m = disease severity for crop species grown in monoculture control plots; i = disease severity for the same crop species grown in intercropping plots in the same fields. Error bars are one s.e.m; n = 3. Statistical analyses were conducted by software SPSS 13.0. All differences between pairs are significant at P≤0.05 based on one-tailed t-test.

Acknowledgments

We thank the personnel of the provincial and county Plant Protection Stations and participating farmers for their contributions to this project, Dr. J. Qiu, Professors T. J. Hocking and M. A. Fullen for manuscript modification.

Author Contributions

Conceived and designed the experiments: CL YZ. Performed the experiments: CL XH SZ HZ YW YL JY JF JY GW YL JX YT GZ JY LL YS YX HW YZ. Analyzed the data: XH YZ. Wrote the paper: CL XH YZ.

References

1. Thrupp LA (2000) Linking agricultural biodiversity and food security: The valuable role of agrobiodiversity for sustainable agriculture. Int Affairs 76: 265–281.

2. Zhu Y, Chen H, Fan J, Wang Y, Li Y, et al. (2000) Genetic diversity and disease control in rice. Nature 406: 707–716.

3. Leung H, Zhu Y, Revilla-Molina I, Fan JX, Chen H, et al. (2003) Using genetic diversity to achieve sustainable rice disease management. Plant Dis 87: 1156–1169.

4. Zhu YY, Fang H, Wang YY, Fan JX, Yang SS, et al. (2005) Panicle blast and canopy moisture in rice cultivar mixtures. Phytopathology 95: 433–438.

5. Flombaum P, Sala OE (2008) Higher effect of plant species diversity on productivity in natural than artificial ecosystems. Proc Natl Acad Sci U S A 105: 6087–6090.

6. Fridley JD (2002) Resource availability dominates and alters the relationship between species diversity and ecosystem productivity in experimental plant communities. Oecologia 132: 271–277.

7. Tilman D, Reich PB, Knops J, Wedin D, Mielke T, et al. (2001) Diversity and productivity in a long-term grassland experiment. Science 294: 843–845.

8. Tilman D, Reich PB, Knops JMH (2006) Biodiversity and ecosystem stability in a decade-long grassland experiment. Nature 441: 629–632.

9. Li L, Li SM, Sun JH, Zhou LL, Bao XG, et al. (2007) Diversity enhances agricultural productivity via rhizosphere phosphorus facilitation on phosphorus-deficient soils. Proc Natl Acad Sci U S A 104: 1192–1196.

10. Morgado LB, Willey RW (2008) Optimum plant population for maize-bean intercropping system in the Brazilian semi-arid region. Sci Agric 65: 474–480.

11. Dybzinski R, Fargione JE, Zak DR, Fornara D, Tilman D (2008) Soil fertility increases with plant species diversity in a long-term biodiversity experiment. Oecologia 158: 85–93.

12. Altieri MA (1999) The ecological role of biodiversity in agro-ecosystems. Agr Ecosyst Environ 74: 19–31.

13. Willey RW (1979) Intercropping- its importance and research needs. Part 1. Competition and yield advantages. Field Crop Abstracts 32: 1–10.

14. Wang L, Sun ML, Yang YL, Ma YC, Qian CX, et al. (2005) Studies on regional epidemiology of potato late blight in Yunnan. Southwest China Jour of Agric Sci (*Chinese version*) 18 (2): 157–162.

15. van Ruijven J, Berendse F (2005) Diversity-productivity relationships: initial effects, long-term patterns, and underlying mechanisms. Proc Natl Acad Sci U S A 102: 695–700.

16. Fischer G, van Velthuizen H, Shah M, Nachtergaele F (2002) Global agro-ecological assessment for agriculture in the 21st century. Report of International Institute for Applied Systems Analysis (IIASA) and Food and Agriculture Organization (FAO). Available: http://www.iiasa.ac.at/.

17. Yang H, Li X (2000) Cultivated land and food supply in China. Land Use Policy 17: 73–88.

18. State Tobacco Monopoly Administration (1996) The principle of the agricultural standard in the People's Republic of China YC/T40-1996. The method of testing agrochemicals control to tobacco disease. (China Standard Press, Beijing).

19. Ministry of Agriculture of the People's Republic of China. (2007) The principle of the agricultural standard in the People's Republic of China NY/T 1248.1-2006. Rules for evaluation of maize for resistance to pests Part 1: Rules for evaluation of maize for resistance to northern corn leaf blight. Beijing: China Agricultural Press.

20. Ministry of Agriculture of the People's Republic of China. (2007) The principle of the agricultural standard in the people's republic of China NY/T 1443.1-2007. Rules for Resistance Evaluation of Wheat to Diseases and Insect Pests Part 1: Rule for Resistance Evaluation of Wheat to Yellow Rust (*Puccinia striiformis* West.f.sp. *tritici* Eriks.et Henn.). Beijing: China Agricultural Press.

21. Dai XY (1993) Studies on resistance of *Saccharum arundinaceum* Retz to *Drechslere sacchari* (Butler) Subran and Jain. Journal of Yunnan Agricultural University 8(2): 143–145.

22. The National Agro-Tech Extension and Service Center (NATESC) (2006) Handbook of Crop Pest Forecasting. The forecasting and survey method of potato late blight. Beijing: China Agricultural Press. pp 225–227.

23. Wallen VR (1957) The identification and distribution of physiologic races of Ascochyta pisi Lib. in Canada. Can J Plant Sci 37: 337–341.

22

Nucleotide Polymorphisms and Haplotype Diversity of *RTCS* Gene in China Elite Maize Inbred Lines

Enying Zhang[1,2◉], **Zefeng Yang**[1*◉], **Yifan Wang**[1], **Yunyun Hu**[1], **Xiyun Song**[2], **Chenwu Xu**[1*]

1 Key Laboratory of Crop Genetics and Physiology of Jiangsu Province, Key Laboratory of Plant Functional Genomics of the Ministry of Education, College of Agriculture, Yangzhou University, Yangzhou, China, **2** College of Agronomy and Plant Protection, Qingdao Agricultural University, Qingdao, China

Abstract

The maize *RTCS* gene, encoding a LOB domain transcription factor, plays important roles in the initiation of embryonic seminal and postembryonic shoot-borne root. In this study, the genomic sequences of this gene in 73 China elite inbred lines, including 63 lines from 5 temperate heteroric groups and 10 tropic germplasms, were obtained, and the nucleotide polymorphisms and haplotype diversity were detected. A total of 63 sequence variants, including 44 SNPs and 19 indels, were identified at this locus, and most of them were found to be located in the regions of UTR and intron. The coding region of this gene in all tested inbred lines carried 14 haplotypes, which encoding 7 deferring RTCS proteins. Analysis of the polymorphism sites revealed that at least 6 recombination events have occurred. Among all 6 groups tested, only the P heterotic group had a much lower nucleotide diversity than the whole set, and selection analysis also revealed that only this group was under strong negative selection. However, the set of Huangzaosi and its derived lines possessed a higher nucleotide diversity than the whole set, and no selection signal were identified.

Editor: Gregory Tranah, San Francisco Coordinating Center, United States of America

Funding: This work was supported by grants from the National Program on the Development of Basic Research (2011CB100100), the Priority Academic Program Development of Jiangsu Higher Education Institutions, the National Natural Science Foundations (31200943 and 31171187) and the Natural Science Foundations of Jiangsu Province (BK2012261). The funders had no role in study design, data collection and analysis, decision to publish, or preparation of the manuscript.

Competing Interests: The authors have declared that no competing interests exist.

* E-mail: zfyang@yzu.edu.cn (ZY); qtls@yzu.edu.cn (CX)

◉ These authors contributed equally to this work.

Introduction

In the past, fundamental researches on increasing shoot biomass and seed yield attracted most attentions of the crop scientists, and the relevance of the root system for food production has often been overlooked [1,2]. However, a healthy and well-developed root stock architecture is especially important for the developing of plant, because it is the organ absorbing water and inorganic nutrients, in addition to anchoring of the plant body to the ground [1,3]. Maize (*Zea mays* L.), one of the most widely grown grain crop in the world, possesses a unique and complex root stock architecture composed of embryonic and postembryonic roots [4,5]. The embryonic roots, defined by the primary root and a variable number of seminal roots, play important roles for early vigor of the maize seedlings. However, at the postembryonic stage, shoot-borne system forms the major backbone of the adult stock [6].

Recently, several genes controlling the development of maize shoot-borne roots, lateral roots, and root hairs have been isolated [7,8,9]. Among them, the gene *RTCS* (rootless concerning crown and seminal roots) was demonstrated to play a central role in the auxin-mediated initiation of seminal and shoot-borne roots in maize [5,9] and the mutant of this gene was impaired in the formation of these roots. Map-based cloning revealed that this gene was located in the short arm of chromosome 1, and encoded a LOB domain protein. Sequence analysis illustrated the maize *RTCS* gene was composed of 2 exons, separated by a 96-bp intron, and its protein product contained 244 amino acid residues. The

maize *RTCS* gene is preferentially expressed in root tissues [5] and its protein product showed typical features of a transcription factor including nuclear localization, DNA-binding and downstream gene activation [6].

Although the favorable root architecture plays critically important roles for the development of plant, root architecture was rarely considered as a selection criterion or traits for maize improvement, mainly because of the practical difficulties with their evaluation under field conditions [3]. Recent researches in maize revealed that changes in root architecture can strongly affect the yield [10]. Because increasing crop yield through improvement of plant type and growing use of fertilizer has reached a maximum, much attention should be focused on improving the root system [1]. Researches on the sequence polymorphisms of key genes are important not only for crop improvement but also for efficient management and conservation of plant genetic resources [11,12,13]. However, rare researches in genetic variants in the DNA sequence have focused on the genes controlling the development of plant roots. In addition, the genetic diversity at the DNA level of maize *RTCS* gene is not known at present. Therefore, we detected nucleotide polymorphisms, haplotype diversity and evolutionary factors of the gene *RTCS* by direct sequencing 73 China elite inbred lines, including the lines from 5 temperate heterotic groups and some tropic germplasms.

Materials and Methods

Plant Materials

A total of 73 China maize elite inbred lines were used in this study (Table 1). Among these inbred lines, 63 temperate and 10 tropic germplasms were used. The 63 temperate inbred lines were from 5 heterotic groups, including 15 from Tangsipingtou, 9 from Lvdahonggu, 11 from Lancaster, 13 from Reid, and 14 from P group.

DNA-extraction and Sequencing RTCS Gene

Genomic DNA of was extracted from young leaves of the tested inbred lines at the seedling stage using CTAB (cetyl trimethyl ammonium bromide) method based on the modified protocol [14]. The sequences of the RTCS gene in 73 inbred lines was sequenced by BGI Life Tech Co., Ltd. using the target sequence capture sequencing technology on the NimbleGen platform [15].

Sequence Analysis

Multiple sequence alignment was performed using Clustal X [16] and was further edited manually. The software DNASP 5.0 [17,18] was used to analyze sequence nucleotide polymorphism and allelic diversities. Two parameters of nucleotide diversity, π and θ were estimated. Where π is the average number of nucleotide differences per site between any two DNA sequences, and θ is derived from the total number of segregating sites and corrected for sampling size. Tajima's D [19] and Fu and Li's [20] statistical tests were used to test the evidence of neutral evolution within each group and each defined region. The minimum number of recombination events [21] was estimated in the period of evolution of RTCS gene among these inbred lines.

Results

Nucleotide Diversity and Selection of RTCS Gene in China Elite Inbred Lines

Sequence polymorphisms were detected among 73 maize inbred lines across 1279 bp of sequence, which covers a 167 bp 5′ untranslated region (UTR), a 735 bp coding region, a 104 bp intron region, and a 273 bp 3′ UTR. Nucleotide substitutions and indels at the RTCS locus were identified, and the results were summarized in Table 2. From the putative genomic sequences of the 73 maize inbred lines, a total of 44 SNP sites were identified, and among them, 16 and 28 sites belonged to singleton variable sites and parsimony informative sites, respectively. In addition, a total of 19 indel events covering 90 sites were identified in the genomic sequences (Table S1). For all the 73 inbred lines, the overall nucleotide diversity (π) of RTCS locus was 0.00666. Among 4 regions of the gene RTCS, the coding region showed much lower nucleotide polymorphism than others, while the intron region had the highest frequency of all sequence variants. This might be caused by the variant of indels, because this region had the highest

Table 1. List of the 73 inbred lines included in this study.

No.	Inbred line	Heterotic group	No.	Inbred line	Heterotic group	No.	Inbred line	Heterotic group
1	QH19612[a]	Tangsipingtou	26	4CV	Lancaster	50	178	P group
2	Chang7-2[a]	Tangsipingtou	27	Qi232	Lancaster	51	QP1721	P group
3	LX9801[a]	Tangsipingtou	28	OH43	Lancaster	52	Exhan	P group
4	107	Tangsipingtou	29	MO17	Lancaster	53	xy35	P group
5	Huang518[a]	Tangsipingtou	30	BJ-4	Lancaster	54	P138	P group
6	k12[a]	Tangsipingtou	31	BEM	Lancaster	55	6819	P group
7	H21[a]	Tangsipingtou	32	BJ-1	Lancaster	56	Dan988	P group
8	Ji853[a]	Tangsipingtou	33	BJ-3	Lancaster	57	319B	P group
9	Za107	Tangsipingtou	34	BJ-5	Lancaster	58	Qi319	P group
10	Huangzaosi[a]	Tangsipingtou	35	412	Lancaster	59	Qi318	P group
11	502[a]	Tangsipingtou	36	8112	Reid	60	Shen137	P group
12	Luyuan92	Tangsipingtou	37	K8112	Reid	61	91158	P group
13	10168	Tangsipingtou	38	Wu314[a]	Reid	62	s80	P group
14	QZ01[a]	Tangsipingtou	39	4866	Reid	63	Danhuang25	P group
15	Y53	Tangsipingtou	40	3189	Reid	64	11099	Tropic
16	Dan598	Lvdahonggu	41	Tie9206	Reid	65	suwan	Tropic
17	Zong3	Lvdahonggu	42	Benyu15	Reid	66	11118	Tropic
18	E28	Lvdahonggu	43	Chun2433	Reid	67	11200	Tropic
19	Dan340	Lvdahonggu	44	478Xuan	Reid	68	10533-1	Tropic
20	Zi330	Lvdahonggu	45	Zheng58	Reid	69	GB28	Tropic
21	S122	Lvdahonggu	46	7922	Reid	70	RCML15	Tropic
22	340Gai	Lvdahonggu	47	8605-2	Reid	71	DK3110	Tropic
23	JH3372	Lvdahonggu	48	JB	Reid	72	RBS11	Tropic
24	Dan99	Lvdahonggu	49	B73	Reid	73	FLB01	Tropic
25	nx335	Lancaster						

[a]The inbred lines of Huangzaosi and its derived lines.

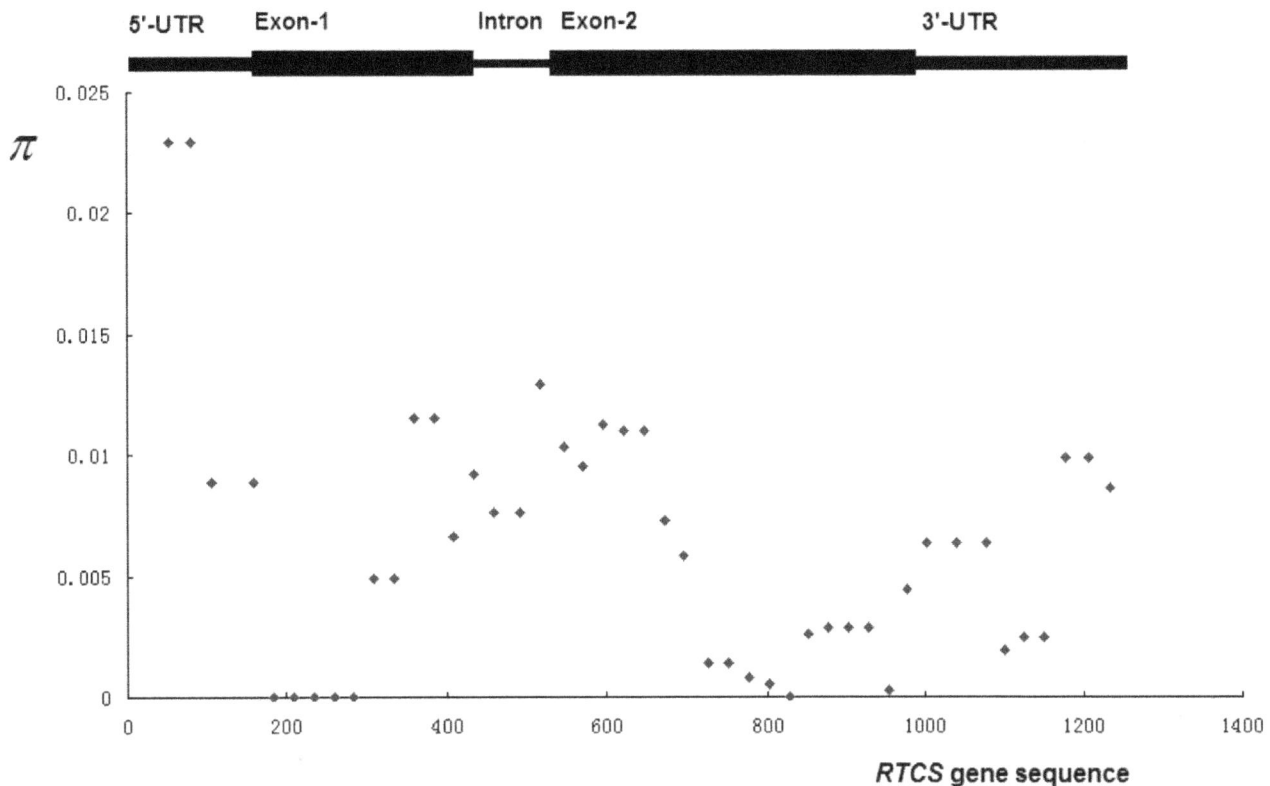

Figure 1. The nucleotide diversity (π) estimated along the *RTCS* **gene sequences.** π is calculated in sliding windows of 100 bp using a step size of 25 bp. 5 regions of *RTCS* gene, including 5′ UTR, Exon-1, Intron, Exon-2, and 3′ UTR, were indexed on the top of the coordinate.

frequency of indels per bp. However, the frequency of nucleotide substitutions in 5′UTR was higher than other regions. When we used the sliding window of 100 bp under a step size of 25 bp, the result revealed each region of the *RTCS* sequence possessed high frequency of polymorphic sites (Figure 1). The highest nucleotide diversity was within 1–159 bp in 5′-UTR with $\pi = 0.02296$, while the lowest value $(\pi = 0)$ was found in regions of exon-1 and exon-2, respectively. The observed distribution of SNP sites and indel sites was found to be significantly different (for SNP, $\chi^2 = 11.147$, $df = 3$, $P < 0.05$; for indel, $\chi^2 = 19.923$, $P < 0.01$) from an expected even distribution across the four defined regions

(Table 3). The uneven distribution of polymorphisms might be particularly due to the low frequency of variants in coding region.

The Tajima's D test is a widely used test to identify sequences which do not fit the neutral theory model at equilibrium between mutation and genetic drift [19]. All the values of Tajima's D in the present study were not statistically significant, illustrating no significant selection existed in the entire *RTCS* sequences. In addition, Fu and Li's D* and F* were also not significant in almost all regions except for intron. Although these results could not reject the hypothesis of mutation drift equilibrium, a lack of footprint of positive selection in most regions of *RTCS* was suggested.

Table 2. Summary of the frequency of polymorphisms.

Parameters	5′UTR	Coding region	Intron	3′UTR	Entire region
Total length of amplicons (bp)	167	735	104	273	1279
Number of all sequence variants (SNPs and indels)	16	16	11	20	63
Frequency of all sequence variants	0.0958	0.0218	0.1058	0.0733	0.0493
Number of nucleotide substitutions (bp)	10	14	6	14	44
Frequency of polymorphic sites per bp	0.0599	0.0190	0.0577	0.0513	0.0344
Number of indels	6	2	5	6	19
Number of indel sites	34	6	12	38	90
Average indel length	5.6667	3	2.4	6.3333	4.7368
Frequency of indels per bp	0.0359	0.0027	0.0481	0.0220	0.0146

Table 3. Nucleotide and allele diversities of *RTCS* gene by analyzing 73 maize inbred lines.

Parameters	5′UTR	Coding region	Intron	3′UTR	Entire region
π	0.01726	0.00436	0.00860	0.00702	0.00666
θ	0.01547	0.00395	0.01342	0.01226	0.00761
Tajima′ D	0.31120	0.29566	−0.85026	−1.22021	−0.40940
Fu and Li's D*	1.38655	−1.63477	−2.51473*	−1.63477	−1.52039
Fu and Li's F*	1.20720	−1.13533	−2.32168*	−1.76828	−1.30730

*indicates the significance at *P*<0.05 level.

Nucleotide Diversity and Selection in Each Heterotic Group

The inbred lines used in this study can be classified into 6 groups, including 5 temperate heterotic groups and the group of tropic germplasms. We also tested the nucleotide diversity of both entire region and coding region of *RTCS* sequences for each group, and the result revealed that the nucleotide diversities of 5 groups were higher than or very near to the whole set (Table 4). The tropic group possessed the highest value nucleotide diversity, and its haplotype diversity (*Hd*) is 1 for the entire region of *RTCS*, suggesting each inbred line carried a haplotype. Only the P heterotic group had much lower nucleotide and haplotype diversities than the whole set both for the entire region and coding region. This result suggested that the P group was more conserved in *RTCS* locus than other groups. In addition, we also noticed that the statistics for Tajima's D, Fu and Li's D* and F* were all statistically significant in P group. This result illustrated that the *RTCS* gene in P group were not evolved neutrally, and also suggesting that selection might only acted the evolution of *RTCS* gene in this group.

Huangzaosi is believed to be the representative line of the Tangsipingtou heterotic group and was used as a key maize inbred line in China [22]. Among all inbred lines used in this study, at least 11 lines were Huangzaosi and its derived lines. We also tested the sequence polymorphisms of *RTCS* gene in Huangzaosi and its derived lines. The result revealed that the nucleotide diversity (π) is higher than the whole set, illustrating that there were abundant nucleotide variations in Huangzaosi and its derived lines. In addition, none of the statistics for Tajima's D, Fu and Li's D* and F* were statistically significant for Huangzaosi and its derived lines, suggesting that selection was not included in the *RTCS* locus of this population.

Haplotype Diversity

Based on the whole length of the *RTCS* gene sequenced in 73 maize inbred lines, a total of 34 haplotypes were detected with a *Hd* equal to 0.8992 (Table S2). The inbred lines were unbalancedly distributed in these haplotypes. Among the haplotypes identified in this analysis, 26 contained only one inbred line. The most frequent haplotype was Hap_8, which contained 21 inbred lines. It should be mentioned that nearly all the inbred lines in P group belonged to this haplotype except for 91158, which was assigned to the haplotype Hap_27.

In the coding region of the gene *RTCS*, 16 sequence variants, including 2 indels and 14 SNPs, were detected. Both of the 2 indels contained 3 nucleotide acids, respectively, and this can not result in frameshift of the codons. When we used the coding sequences to identify the hapotype diversity, a total of 14 haplotypes were

Table 4. Nucleotide and allele diversities of *RTCS* gene in each heterotic group.

Parameters	Tangsipingtou		Lvdahonggu		Lancaster		Reid		P group		Tropic lines		Huangzaosi and its derived lines	
	Entire region	CDS	Entire region	CDS	Entire region	CDS	Entire region	CDS	Entire region	CDS	Entire region	CDS	Entire region	CDS
π	0.00941	0.00484	0.0064	0.00402	0.00746	0.00392	0.00706	0.00413	0.00081	0.00078	0.01155	0.00661	0.00737	0.00447
θ	0.00828	0.00336	0.00659	0.00302	0.00858	0.00373	0.00768	0.00352	0.00178	0.00172	0.01219	0.00679	0.00806	0.00373
Hd	0.9714	0.8476	0.8889	0.7500	0.8727	0.7091	0.978	0.8718	0.2747	0.1429	1	0.9333	0.9818	0.8182
Tajima′ D	0.57302	1.62023	−0.14261	1.45615	−0.60687	0.21418	−0.34215	0.67257	−2.01359*	−1.79759*	−0.25538	−0.11714	−0.3956	0.82009
Fu and Li's D*	0.5794	1.36102*	−0.32781	1.37224	−1.28402	−0.46234	−0.7663	0.88217	−2.60025*	−2.2738*	−0.14763	−0.41245	−0.74657	0.46649
Fu and Li's F*	0.66505	1.63999*	−0.31725	1.55087	−1.25975	−0.32956	−0.7467	0.94243	−2.79240*	−2.44883*	−0.19709	−0.38199	−0.74484	0.62774

*indicates the significance at *P*<0.05 level.

```
CDS_Hap_1/2          : MTGFGSPCGACKFLRRKCVRGCVFAPYFCHEQGAAHFAAIHKVFGASNVSKLLAHLPLADRAEAAVTISYEAQARLRDPIYGC :  83
CDS_Hap_3            : MTGFGSPCGACKFLRRKCVRGCVFAPYFCHEQGAAHFAAIHKVFGASNVSKLLAHLPLADRAEAAVTISYEAQARLRDPIYGC :  83
CDS_Hap_4            : MTGFGSPCGACKFLRRKCVRGCVFAPYFCHEQGAAHFAAIHKVFGASNVSKLLAHLPLADRAEAAVTISYEAQARLRDPIYGC :  83
CDS_Hap_5/6/7/14     : MTGFGSPCGACKFLRRKCVRGCVFAPYFCHEQGAAHFAAIHKVFGASNVSKLLAHLPLADRAEAAVTISYEAQARLRDPIYGC :  83
CDS_Hap_8/10/12/13   : MTGFGSPCGACKFLRRKCVRGCVFAPYFCHEQGAAHFAAIHKVFGASNVSKLLAHLPLADRAEAAVTISYEAQARLRDPIYGC :  83
CDS_Hap_9            : MTGFGSPCGACKFLRRKCVRGCVFAPYFCHEQGAAHFAAIHKVFGASNVSKLLAHLPLADRAEAAVTISYEAQARLRDPIYGC :  83
CDS_Hap_11           : MTGFGSPCGACKFLRRKCVRGCVFAPYFCHEQGAAHFAAIHKVFGASNVSKLLAHLPLADRAEAAVTISYEAQARLRDPIYGC :  83
                               LOB domain

CDS_Hap_1/2          : VAHIFALQQQVMTLQAQLASLKAQAAQGQQGVHEDAKGYVGSAAAEQLGYGYPWCSGNGGAAAAAGAVGAPAAQPGAYGNGA : 166
CDS_Hap_3            : VAHIFALQQQVMTLQAQLASLKAQAAQGQQGVHEDAKGYVGSAAAEQLGYGYPWCSGNGGAAAAAGAVGAPAAQPGAYGNGA : 166
CDS_Hap_4            : VAHIFALQQQVMTLQAQLASLKAQAAQGQQGVHEDAKGYVGSAAAEQLGYGYPWCSGNGG-AAAAGAVGAPAAQPDAYGNGA : 165
CDS_Hap_5/6/7/14     : VAHIFALQQQVMTLQAQLASLKAQAAQGQQGVHEDAKGYVGSAAAEQLGYGYPWCSGNGG-AAAAGAVGAPAAQPGAYGNGA : 165
CDS_Hap_8/10/12/13   : VAHIFALQQQVMTLQAQLASLKAQAGQGQQGVHEDAKGYVGSAAAEQLGYGYPWCSGNGG-AAAAGAVGAPAAQPGAYGNGA : 165
CDS_Hap_9            : VAHIFALQQQVMTLQAQLASLKAQAGQGQQGVHEDAKGYVGSAAAEQLGYGYPWCSGNGG-AAAAGAVGAPAAQPGAYGNGA : 165
CDS_Hap_11           : VAHIFALQQQVMTLQAQLASLKAQAGQGQQGVHEDAKGYVGSAAAEQLGYGYPWCSGNGG--AAATGAVGAPAAQPGAYGYGA : 164

CDS_Hap_1/2          : HESLTALLGSSDYMQQSLYHAFEQAGADDDDGRQGYAFEAAAESSSLGAEESGWRSSSGYQDCEDLQSVAYAYLNHRS : 244
CDS_Hap_3            : HESLTALLGSSDYMQQSLYHAFEQAGADDDDGRQGYAFEAAAESSSLGAEESGWRSSSGYQDCEDLQSVAYAYLNHRS : 244
CDS_Hap_4            : HESLTALLGSSDYMQQSLYHAFEQAGADDDDGRQGYAFEAAAESSSLGAEESGWRSSSGYQDCEDLQSVAYAYLNHRS : 243
CDS_Hap_5/6/7/14     : HESLTALLGSSDYMQQSLYHAFEQAGADDDDGRQGYGFEAAAESSSLGAEESGWRSSSGYQDCEDLQSVAYAYLNHRS : 243
CDS_Hap_8/10/12/13   : HESLTALLGSSDYMQQSLYHAFEQAGADDDDGRQGYGFEAAAESSSLGAEESGWRSSSGYQDCEDLQSVAYAYLNHRS : 243
CDS_Hap_9            : HESLTALLGSSDYMQQSLYHAFEQAGADDDDGRQGYAFEAAAESSSLGAEESGWRSSSGYQDCEDLQSVAYAYLNHRS : 243
CDS_Hap_11           : HESLTALLGSSDYMQQSLYHAFEQAGADDDDGRQGYAFEAAAESSSLGADESGWRSSSGYQDCEDLQSVAYAYLNHRS : 242
```

Figure 2. Sequence alignment of RTCS proteins encoded by different CDS haplotypes. The hyplotypes identified by coding sequences of *RTCS* gene were used as sequence name. Polymorphisms from inferred amino acid were indicated by shade.

identified for these 73 inbred lines (Table S3), and the hapotype diversity was 0.7705. Among the haplotypes identified according to CDS, 9 contained only one inbred line. The most frequent CDS haplotype was CDS_Hap_5, which contained 29 inbred lines from all 6 groups. In addition, CDS_Hap_7 and CDS_Hap_8 were also haplotypes with high frequency, and only no inbred lines in P group and tropic lines carried them, respectively.

Among 14 SNPs in the coding region, 8 were synonymous sites, and the other 6 were nonsynonymous sites. The nonsynonymous sites and the indels will lead to the changes of protein sequences. When we translated the CDS into amino acid sequences, 7 types of RTCS protein sequences were found to be encoded by these inbred lines (Figure 2). Haplotypes CDS_Hap_5/6/7/14 encoded the most frequent type of RTCS protein, and contributed to more than half of all the inbred lines (42 out of 73). The variation of RTCS protein sequences was the result of combinations of 6 nonsynonymous mutations and 2 indels in the coding region. All of the variants at the protein level were found to be outside the LOB domain region (Figure 2), and in other words, the region of LOB domain of RTCS protein showed 100% identify in all the tested inbred lines.

Evidence of Recombination

The polymorphic sites in the entire *RTCS* sequence were used to detect the evidence of recombination. The patterns of the polymorphisms identified in inbred lines surveyed in this study indicated the history of recombination at *RTCS* gene, which contributed to the haplotype diversity. Under the algorithm of Hudson and Kaplan [21], at least 6 recombination events were found to be responsible to the polymorphism of *RTCS* gene. The recombination events were detected in the informative sites of every region, and they were found in the positions between 5′-UTR and exon-1 (82–353), between exon-1 and intron (398–468), between intron and exon-2 (499–553), the exon-2 (597–667), between exon-2 and 3′-UTR (879–1038), and the 3′UTR (1038–1154), respectively. The consequences of recombination events are evident in the pattern of polymorphisms when compared the sequence of one haplotype with others. For example, the 5′ UTR sequence of the Hap_1 was the same as that of Hap_2. However, across the coding region and intron region, there were 4 variants between them, including 3 SNP and 1 indel covering 3 sites. The 3′UTR region of Hap_1 was found to be again virtually identical to Hap_2. This result suggested that the *RTCS* sequence in Hap_2 has resulted from at least two recombination events in the past relative to Hap_1.

Discussion

The abundant genetic variations are the foundation for crop improvement. The analysis of the genetic diversity of plant variants is critical for understanding the genetic background of phenotypic variation, and in turn will provide great help for crop improvement [23]. In this study, we detected the nucleotide polymorphisms and the haplotype diversity of the gene *RTCS*, an important regulator for the developing of roots, in 73 China elite maize inbred lines. The identification of nucleotide variations exerting functional effects, especially those causing changes of amino acid composition, is the primary focus of association mapping [13]. Although most variants were found to be located in the non-coding region, the SNP sites and indels in the coding region also classified the tested inbred lines into 14 haplotypes. In addition, a total 7 deferring RTCS proteins were encoded by this gene in all the tested inbred lines. The nucleotide polymorphisms of *RTCS* gene in this study would be helpful in identifying alleles for further genetic analysis, and might also provide foundation for maize improvement.

Heterotic groups are of primary importance in hybrid breeding. Crosses between inbred lines from different heterotic groups generally result in vigorous F1 hybrids with significantly more heterosis than F1 hybrids from inbred lines within the same heterotic group [24]. Heterotic groups are created by plant breeders to classify inbred lines, and can be progressively improved by reciprocal recurrent selection [25]. Although the classification of inbred lines into heterotic groups was based on their general combining ability (GCA) and specific combining ability (SCA) effects, the inbred lines within one heterotic group were generally believed to possess lower genetic divergence than those between different groups. Thus, molecular data, especially

SSR molecular markers, was thought to be the efficient method in assigning inbred lines to specific heterotic groups [24,26,27,28,29]. The nucleotide polymorphisms of the *RTCS* locus were investigated in 73 elite inbred lines from different heterotic groups. The results revealed that sequence variants within each group were higher or very near to those of the whole set except for P group for both the entire region and the coding sequences. Because breeders mainly focused on increasing shoot biomass and seed yield in maize improvement in the past, the relevance of the root system for crop improvement has often overlooked [1,2]. The abundant variants within one heterotic group might be the result of overlook in the selection by breeders, although this gene plays important roles in formation of seminal and shoot-borne roots.

The purpose of the selection test is to distinguish between a DNA sequence evolving randomly (neutrally) and one evolving under a non-random process, including directional selection or balancing selection, demographic expansion or contraction, genetic hitchhiking, or introgression [19]. The randomly evolving mutations are called "neutral", while mutations under selection are "non-neutral". In this study, we performed selective analysis for each heterotic group, and the results revealed that only P group was influenced by strong negative selection. Other groups have not influenced by selection, suggesting that a bottleneck for the usage of this locus in breeding in these heterotic groups. In addition, the haplotype detection also found that P group has a lower value of haplotype diversity than others. This might be the result of that this group was used in China for a short period after 1980s, and most of the inbred lines of this group in China were selected from the pioneer hybrid P78599 [26]. The consistency of the genetic

background for the inbred lines in P group resulted in the low frequency of nucleotide variants. Huangzaosi is the most used maize inbred line in China, and more than 42 hybrids and 70 derived lines used this inbred line since it was first bred in 1971 [22]. 11 inbred lines of Huangzaosi and its derived lines were used to test the nucleotide polymorphisms. The results revealed that this set has a higher nucleotide diversity than the whole set, and no selection was identified in this set. These result suggested that the *RTCS* locus was not adopted when the breeders used Huangzaosi as a key inbred line.

Author Contributions

Conceived and designed the experiments: ZY CX. Performed the experiments: EZ YW YH. Analyzed the data: ZY EZ YW. Contributed reagents/materials/analysis tools: EZ ZY XS. Wrote the paper: ZY EZ.

References

1. Den Herder G, Van Isterdael G, Beeckman T, De Smet I (2010) The roots of a new green revolution. Trends Plant Sci 15: 600–607.
2. Smith S, De Smet I (2012) Root system architecture: insights from Arabidopsis and cereal crops. Philos Trans R Soc Lond B Biol Sci 367: 1441–1452.
3. Cai H, Chen F, Mi G, Zhang F, Maurer HP, et al. (2012) Mapping QTLs for root system architecture of maize (Zea mays L.) in the field at different developmental stages. Theor Appl Genet 125: 1313–1324.
4. Hochholdinger F, Woll K, Sauer M, Dembinsky D (2004) Genetic dissection of root formation in maize (Zea mays) reveals root-type specific developmental programmes. Ann Bot 93: 359–368.
5. Taramino G, Sauer M, Stauffer JL, Multani D, Niu X, et al. (2007) The maize (Zea mays L.) RTCS gene encodes a LOB domain protein that is a key regulator of embryonic seminal and post-embryonic shoot-borne root initiation. Plant J 50: 649–659.
6. Majer C, Xu C, Berendzen KW, Hochholdinger F (2012) Molecular interactions of ROOTLESS CONCERNING CROWN AND SEMINAL ROOTS, a LOB domain protein regulating shoot-borne root initiation in maize (Zea mays L.). Philos Trans R Soc Lond B Biol Sci 367: 1542–1551.
7. Wen TJ, Hochholdinger F, Sauer M, Bruce W, Schnable PS (2005) The roothairless1 gene of maize encodes a homolog of sec3, which is involved in polar exocytosis. Plant Physiol 138: 1637–1643.
8. Hochholdinger F, Wen TJ, Zimmermann R, Chimot-Marolle P, da Costa e Silva O, et al. (2008) The maize (Zea mays L.) roothairless 3 gene encodes a putative GPI-anchored, monocot-specific, COBRA-like protein that significantly affects grain yield. Plant J 54: 888–898.
9. Hochholdinger F, Tuberosa R (2009) Genetic and genomic dissection of maize root development and architecture. Curr Opin Plant Biol 12: 172–177.
10. Hammer GL, Dong Z, McLean G, Doherty A, Messina C, et al. (2009) Can Changes in Canopy and/or Root System Architecture Explain Historical Maize Yield Trends in the US Corn Belt? Crop Science 49: 299–312.
11. Achon MA, Larranaga A, Alonso-Duenas N (2012) The population genetics of maize dwarf mosaic virus in Spain. Arch Virol.
12. Li L, Hao Z, Li X, Xie C, Li M, et al. (2011) An analysis of the polymorphisms in a gene for being involved in drought tolerance in maize. Genetica 139: 479–487.
13. Lestari P, Lee G, Ham TH, Reflinur, Woo MO, et al. (2011) Single nucleotide polymorphisms and haplotype diversity in rice sucrose synthase 3. J Hered 102: 735–746.
14. Fulton TM, Chunwongse J, Tanksley SD (1995) Microprep protocol for extraction of DNA from tomato and other herbaceous plants. Plant Molecular Biology Reporter 13: 207–209.
15. Nuwaysir EF, Huang W, Albert TJ, Singh J, Nuwaysir K, et al. (2002) Gene expression analysis using oligonucleotide arrays produced by maskless photolithography. Genome Res 12: 1749–1755.
16. Larkin MA, Blackshields G, Brown NP, Chenna R, McGettigan PA, et al. (2007) Clustal W and Clustal X version 2.0. Bioinformatics 23: 2947–2948.
17. Rozas J (2009) DNA sequence polymorphism analysis using DnaSP. Methods Mol Biol 537: 337–350.
18. Librado P, Rozas J (2009) DnaSP v5: a software for comprehensive analysis of DNA polymorphism data. Bioinformatics 25: 1451–1452.
19. Tajima F (1989) Statistical method for testing the neutral mutation hypothesis by DNA polymorphism. Genetics 123: 585–595.
20. Fu YX, Li WH (1993) Statistical tests of neutrality of mutations. Genetics 133: 693–709.
21. Hudson RR, Kaplan NL (1985) Statistical properties of the number of recombination events in the history of a sample of DNA sequences. Genetics 111: 147–164.
22. Zhang J, He M, Liu Y, Liu H, Wei B, et al. (2012) Sequence polymorphism characteristics in the See2beta gene from maize key inbred lines and derived lines in China. Biochem Genet 50: 508–519.
23. Garcia-Arenal F, Fraile A, Malpica JM (2003) Variation and evolution of plant virus populations. Int Microbiol 6: 225–232.
24. Reif JC, Melchinger AE, Xia XC, Warburton ML, Hoisington DA, et al. (2003) Genetic distance based on simple sequence repeats and heterosis in tropical maize populations. Crop Science 43: 1275–1282.
25. Berilli AP, Pereira MG, Goncalves LS, da Cunha KS, Ramos HC, et al. (2011) Use of molecular markers in reciprocal recurrent selection of maize increases heterosis effects. Genet Mol Res 10: 2589–2596.
26. Yu Y, Wang R, Shi Y, Song Y, Wang T, et al. (2007) Genetic diversity and structure of the core collection for maize inbred lines in china. Maydica 52: 181–194.
27. Aguiar CG, Schuster I, Amaral AT, Scapim CA, Vieira ES (2008) Heterotic groups in tropical maize germplasm by test crosses and simple sequence repeat markers. Genet Mol Res 7: 1233–1244.
28. Longin CF, Utz HF, Melchinger AE, Reif JC (2007) Hybrid maize breeding with doubled haploids: II. Optimum type and number of testers in two-stage selection for general combining ability. Theor Appl Genet 114: 393–402.
29. Lu H, Li JS, Liu JL, Bernardo R (2002) Allozyme polymorphisms of maize populations from southwestern China. Theor Appl Genet 104: 119–126.

23

Selection Strategies for the Development of Maize Introgression Populations

Eva Herzog[1], Karen Christin Falke[2], Thomas Presterl[3], Daniela Scheuermann[3], Milena Ouzunova[3], Matthias Frisch[1]*

1 Institute of Agronomy and Plant Breeding II, Justus Liebig University, Giessen, Germany, 2 Institute for Evolution and Biodiversity, University of Münster, Münster, Germany, 3 KWS Saat AG, Einbeck, Germany

Abstract

Introgression libraries are valuable resources for QTL detection and breeding, but their development is costly and time-consuming. Selection strategies for the development of introgression populations with a limited number of individuals and high-throughput (HT) marker assays are required. The objectives of our simulation study were to design and compare selection strategies for the development of maize introgression populations of 100 lines with population sizes of 360–720 individuals per generation for different DH and S_2 crossing schemes. Pre-selection for complete donor chromosomes or donor chromosome halves reduced the number of simultaneous backcross programs. The investigated crossing and selection schemes differed considerably with respect to their suitability to create introgression populations with clearly separated, evenly distributed target donor chromosome segments. DH crossing schemes were superior to S_2 crossing schemes, mainly due to complete homozygosity, which greatly reduced the total number of disjunct genome segments in the introgression populations. The S_2 crossing schemes were more flexible with respect to selection and provided economic alternatives to DH crossing schemes. For the DH crossing schemes, increasing population sizes gradually over backcross generations was advantageous as it reduced the total number of required HT assays compared to constant population sizes. For the S_2 crossing schemes, large population sizes in the final backcross generation facilitated selection for the target segments in the final backcross generation and reduced fixation of large donor chromosome segments. The suggested crossing and selection schemes can help to make the genetic diversity of exotic germplasm available for enhancing the genetic variation of narrow-based breeding populations of crops.

Editor: Lewis Lukens, University of Guelph, Canada

Funding: Funding from the German Federal Ministry of Education and Research (BMBF Grant 0315951) is gratefully acknowledged. [http://www.bmbf.de/en/index.php] The funders had no role in study design, data collection and analysis, decision to publish, or preparation of the manuscript.

Competing Interests: The authors have the following interests: Thomas Presterl, Daniela Scheuermann and Milena Ouzunova are employed by KWS Saat AG. There are no patents, products in development or marketed products to declare related to KWS SAAT AG and the subject matter of the publication.

* E-mail: matthias.frisch@uni-giessen.de

Introduction

Introgression libraries are valuable resources for the identification of alleles of agricultural interest in exotic germplasm. They facilitate the introduction of new genetic variation into elite breeding germplasm by providing favorable chromosome segments from wild or exotic species in an adapted genetic background [1,2]. Ideally, an introgression library consists of a set of homozygous introgression lines (ILs) which carry short marker-defined chromosome segments from an exotic donor in a common genetic background. The concept was first described in tomato [3]. In the mean time, introgression libraries have been developed for the model species *Arabidopsis thaliana* [4,5], and in many agriculturally important crops, such as rice [6,7], barley [8,9], wheat [10,11], maize [12,13] and rye [14].

Introgression libraries are usually developed by marker-assisted backcrossing followed by selfing or production of double haploid (DH) lines. The backcross process for their development is costly and labor-intensive if complete coverage of the donor genome by short evenly distributed target chromosome segments is to be achieved. Often additional backcross programs have to be run for

the developed ILs in order to close gaps in donor genome coverage, or to shorten donor chromosome segments by additional recombination events [3,9]. In spite of the high resource requirements, only incomplete donor genome coverage has been achieved for most of the reported introgression libraries [9,14].

In previous simulation studies on introgression libraries, two generations of selfing were investigated for line development [15,16]. Recent genetic studies in maize were based on ILs that underwent two to five generations of selfing [17–19]. The use of DH technology has to our knowledge not yet been investigated in simulation studies on the development of introgression libraries. However, *in vivo* induction of maternal haploids is currently a routine method of DH production in commercial maize breeding programs. The main advantage of the DH technology is that complete homozygosity can be obtained after only two generations. Inspite of this time-saving, the production of DH lines is still considerably more costly than conventional selfing [20]. Moreover, a current drawback of *in vivo* induction of maternal haploids in maize is that on average only one viable DH line can be derived from one backcross individual. It is therefore of economic interest to compare this method with S_2 crossing schemes which require the same number of generations to evaluate the benefits of DH lines.

A possible approach to tackle the high costs required for the development of ideal introgression libraries would be to resort to introgression populations which are not perfect in appearance, but carry some additional donor segments outside the actual target segments. Such introgression populations could be developed with fewer individuals and marker assays. Complete coverage of the donor genome is desirable in order to capture the whole wealth of alleles of agricultural interest in the exotic donor. It is therefore one component of a minimum standard which introgression populations should meet. A second component are short, evenly distributed target donor chromosome segments in a clean adapted background, as they facilitate the use of the ILs in the following breeding process.

The design of the crossing scheme and the selection strategy are the most important factors that influence the distribution of donor chromosome segments in the introgression population. Falke et al. [16] suggested for the development of ideal introgression libraries that a chromosome-based selection strategy which pre-selects individuals carrying the donor alleles on complete chromosomes in generation BC_1 saves resources. Adapting and advancing this concept to crossing schemes with small population sizes might be an efficient approach to develop introgression populations with a limited number of marker assays.

The objectives of our simulation study were (1) to design selection strategies and crossing schemes for the development of maize introgression populations with limited resources, (2) to compare these selection strategies with respect to the distribution and length of donor chromosome segments and the required investments in terms of time, individuals and marker assays, (3) to give guidelines for the optimal experimental design for constructing introgression populations.

Materials and Methods

Software

All simulations were conducted in R version 3.0.0 [21] with the software package SelectionTools, which is available from http://www.uni-giessen.de/population-genetics/downloads.

Genetic Model

A genetic model of maize with 10 equally sized chromosomes of 200 cM length was used for the simulations. Genetic markers for selection were equally spaced. The distance between two adjacent marker loci was 1 cM. All markers were polymorphic between donor and recipient. It was assumed that markers were analyzed with high-throughput (HT) assays. One HT assay comprised genotyping one individual at all marker loci in the linkage map. Recombination was modelled assuming no interference in crossover formation [22]. Each simulation of an introgression population of 100 ILs was replicated 1,000 times in order to reduce sampling effects and to obtain results with high numerical accuracy and a small standard error.

Crossing Schemes

Four crossing schemes were investigated: BC_2DH, BC_3DH, BC_2S_2, BC_3S_2. Each crossing scheme started with the cross of a homozygous donor and a homozygous recipient to create one F_1 individual. The F_1 individual was backcrossed to the recipient to create a BC_1 population of size n_{BC1}. From the BC_1 population, the best individuals with the highest values of selection indices for the respective selection strategy were selected. Each of the selected BC_1 individuals was backcrossed to the recipient to create BC_2

sub-populations of size n_{BC2}. From these BC_2 sub-populations, the best individuals with the highest values of the respective selection indices were selected. For the DH crossing schemes, in vivo induction of maternal haploids was assumed with a success rate of one viable DH line per backcross individual. For the BC_2DH schemes, one DH line was thus created from each of the selected BC_2 individuals. For the BC_2S_2 crossing schemes, the selected BC_2 individuals were selfed to create a fixed number of S_1 individuals. Each of the S_1 individuals was selfed again and one S_2 individual was created. For the BC_3 crossing schemes, each of the selected BC_2 individuals was backcrossed to the recipient to create BC_3 sub-populations of size n_{BC3}. From these BC_3 sub-populations, the best individuals with the highest values of the respective selection indices were selected. The generations S_1, S_2 or DH of the BC_3 crossing schemes were carried out as described for the BC_2 crossing schemes.

Evaluation of Selection Candidates

The final introgression populations should consist of 100 ILs which guarantee an acceptable resolution of QTL detection in maize, and which can be immediately used in further breeding steps. Each IL should ideally carry a 20 cM chromosome segment from the donor to provide a complete and even coverage of the donor genome without overlap. The 20 cM chromosome segments are hereafter simply referred to as "target segments". To determine the selection index for an individual with respect to a given target segment, we denote with t_c the donor genome proportion of the chromosome on which the target segment is located, with t_h the donor genome proportion of the chromosome half on which the target segment is located and with t_s the donor genome proportion of the target segment itself. The values for the genetic background b_c, b_h, b_s correspond to t_c, t_h, t_s and denote the recipient genome proportion outside the respective chromosome region. Depending on the selection strategy, t and b are used to define selection indices.

Selection Strategies

We considered generations $g = \{BC_1, BC_2, BC_3, DH, S_1, S_2\}$ for selection. Generation DH was the generation in which homozygous diploid DH lines were available for selection. In each generation g, the genome was divided into selection regions that could either be 10 complete chromosomes, 20 chromosome halves or 100 target segments. For selection for complete donor chromosomes, a fixed number n_{sel} of best individuals for each of the chromosomes $c = 1, 2, ..., 10$ with the highest values for selection index $i = t_c + b_c$ were selected. For selection for donor chromosome halves, a fixed number n_{sel} of best individuals for each of the chromosome halves $h = 1, 2, ..., 20$ with the highest values for selection index $i = t_h + b_h$ were selected. For selection for donor target segments, a fixed number n_{sel} of best individuals for each of the target segments $s = 1, 2, ..., 100$ with the highest values for selection index $i = t_s + b_s$ were selected.

Selection for complete donor chromosomes, donor chromosome halves and donor target segments were combined to form different selection strategies. Selection for complete donor chromosomes in a backcross generation is denoted by a C in the strategy name, selection for donor chromosome halves is denoted by an H, and selection for donor target segments is denoted by an S. For example, for strategy CH, selection for complete donor chromosomes was conducted in generation BC_1 while selection for donor chromosome halves was conducted in generation BC_2. An overview of the investigated selection strategies is presented in Table 1. The investigated combinations of crossing scheme and selection strategy are listed in the first column of Table 2. For all

Table 1. Definition of the selection index i in generations BC_1, BC_2, BC_3, DH, S_1, S_2 for different selection strategies for developing introgression populations.

	Generation				
Strategy	**BC_1**	**BC_2**	**BC_3**	**S_1**	**DH/S_2**
C	$t_c + b_c$	–	–	–	$t_s + b_s$
H	$t_h + b_h$	–	–	–	$t_s + b_s$
CC	$t_c + b_c$	$t_c + b_c$	–	–	$t_s + b_s$
HH	$t_h + b_h$	$t_h + b_h$	–	–	$t_s + b_s$
CH	$t_c + b_c$	$t_h + b_h$	–	–	$t_s + b_s$
CCC	$t_c + b_c$	$t_c + b_c$	$t_c + b_c$	–	$t_s + b_s$
HHH	$t_h + b_h$	$t_h + b_h$	$t_h + b_h$	–	$t_s + b_s$
CHH	$t_c + b_c$	$t_h + b_h$	$t_h + b_h$	–	$t_s + b_s$
HHS	$t_h + b_h$	$t_h + b_h$	$t_s + b_s$	–	$t_s + b_s$

Selection for complete donor chromosomes (C), selection for donor chromosome halves (H) and selection for donor target segments (S) were combined to form different selection strategies (left column). t_c, t_h and t_s denote the donor genome proportions of the chromosome on which the target segment is located, of the chromosome half on which the target segment is located and of the target segment itself. b_c, b_h and b_s correspond to t_c, t_h, t_s and denote the recipient genome proportion outside the respective chromosome region.

selection strategies, the best 100 ILs for selection index $i = t_s + b_s$ were selected in generation DH or S_2, depending on the crossing scheme.

Population Sizes and Simulation Series

We investigated population sizes of $n_{tot} = 360 - 720$ individuals per backcross generation. This should be within a range which can be realized in practical maize breeding programs. Variations in population size were investigated to determine both the effect on preserving the target segments up to line development as well as on recovering the genotype of the recipient outside the target segments.

In the first series of simulations, basic crossing schemes were investigated. Selection was carried out in generation BC_1 for basic crossing schemes with two backcross generations, and in generations BC_1 and BC_2 for basic crossing schemes with three backcross generations. The total population size per generation was kept constant at $n_{tot} = 360$ individuals in every generation g.

In the second series of simulation, crossing schemes with high selection intensity were investigated. Population size was doubled compared to the basic crossing schemes ($n_{tot} = 720$) in every generation g, while the number of selected individuals was the same as for the basic crossing schemes. The crossing schemes with high selection intensity are denoted by $BC_3 - CC'$, $BC_3 - HH'$ and $BC_3 - CH'$ (Table 2). In the first and second series of simulations, all backcross individuals generated in the final backcross generation were used for line development for both DH and S_2 crossing schemes. One IL was derived from one backcross individual.

In the third series of simulations, crossing schemes with selection in the final backcross generation were investigated. n_{tot} was doubled to 720 individuals in the final backcross generation for the DH crossing schemes $BC_2DH - CC$, $BC_2DH - HH$, $BC_2DH - CH$, $BC_3DH - CCC$, $BC_3DH - HHH$, $BC_3DH - CHH$. This increase in population size was necessary to enable selection and to keep n_{tot} at 360 individuals in generation DH. For the corresponding S_2 schemes, n_{tot} was kept at 360 individuals also in the final backcross generation.

In the fourth series of simulations, crossing schemes with increasing population sizes were investigated. Selection was conducted in the final backcross generation. The crossing schemes with increasing population sizes are denoted by $BC_3 - HHH^*$ and $BC_3 - HHS^*$. The details concerning the total population size n_{tot} and population sizes in the sub-populations n_g for all investigated combinations of crossing scheme and selection strategy are summarized in Table 2. Schematic representations of the crossing schemes $BC_3DH - HHH^*$ and $BC_3S_2 - HHS^*$ are given in Figure 1 and Figure 2 for illustration.

Measures

To evaluate and compare introgression populations originating from different crossing and selection schemes, the following measures were determined: (a) the genome coverage of the donor O in percent, which is defined as the proportion of the donor genome which is covered by the introgression population, irrespective of whether by the target segments or other donor segments in the genetic background, (b) the depth of donor genome coverage T, which is defined as the average number of ILs in which each donor allele appears in the introgression population, (c) the number of disjunct genome segments in the introgression population S, (d) the resolution of the introgression population R in cM, which is defined as the total genome length of the genetic model in cM divided by S, (e) the average number of donor segments per IL N, (f) the average length of donor segments per IL L in cM, (g) the average total donor genome proportion of the introgression population D_t in percent, (h) the average donor genome proportion of the chromosomes carrying the respective target segments D_c in percent, (i) the average donor genome proportion of the target segments D_s in percent.

Results

High values for the donor genome coverage O around 99% were observed for all crossing schemes (Table 3). However, the resulting introgression populations differed substantially in the values for the number of disjunct genome segments S, the total donor genome proportion D_t, the donor genome proportion of the carrier chromosomes D_c and the donor genome proportion of the target segments D_s. BC_3 crossing schemes resulted in 2–3% lower values for D_t than BC_2 crossing schemes, even if the number of

Table 2. Subdivision of the total population sizes n_{tot} into sub-population sizes n_g in generations $g = BC_1, BC_2, BC_3, S_1, DH, S_2$ for different crossing and selection schemes for developing introgression populations.

Scheme	Generation				
	BC$_1$	BC$_2$	BC$_3$	S$_1$	DH/S$_2$
Basic crossing schemes					
BC$_2$DH$-$C	$1 \times 1 \times 360$	$1 \times 10 \times 36$	$-$	$-$	$10 \times 36 \times 1$
BC$_2$DH$-$H	$1 \times 1 \times 360$	$1 \times 20 \times 18$	$-$	$-$	$20 \times 18 \times 1$
BC$_3$DH$-$CC	$1 \times 1 \times 360$	$1 \times 10 \times 36$	$10 \times 1 \times 36$	$-$	$10 \times 36 \times 1$
BC$_3$DH$-$HH	$1 \times 1 \times 360$	$1 \times 20 \times 18$	$20 \times 1 \times 18$	$-$	$20 \times 18 \times 1$
BC$_3$DH$-$CH	$1 \times 1 \times 360$	$1 \times 10 \times 36$	$10 \times 2 \times 18$	$-$	$20 \times 18 \times 1$
Crossing schemes with high selection intensity					
BC$_3$DH$-$CC$'$	$1 \times 1 \times 720$	$1 \times 10 \times 72$	$10 \times 1 \times 72$	$-$	$10 \times 72 \times 1$
BC$_3$DH$-$HH$'$	$1 \times 1 \times 720$	$1 \times 20 \times 36$	$20 \times 1 \times 36$	$-$	$20 \times 36 \times 1$
BC$_3$DH$-$CH$'$	$1 \times 1 \times 720$	$1 \times 10 \times 72$	$10 \times 2 \times 36$	$-$	$20 \times 36 \times 1$
Crossing schemes with selection in the final BC generation					
BC$_2$DH$-$CC	$1 \times 1 \times 360$	$1 \times 10 \times 72$	$-$	$-$	$10 \times 36 \times 1$
BC$_2$DH$-$HH	$1 \times 1 \times 360$	$1 \times 20 \times 36$	$-$	$-$	$20 \times 18 \times 1$
BC$_2$DH$-$CH	$1 \times 1 \times 360$	$1 \times 10 \times 72$	$-$	$-$	$10 \times (2 \times 18) \times 1$
BC$_3$DH$-$CCC	$1 \times 1 \times 360$	$1 \times 10 \times 36$	$10 \times 1 \times 72$	$-$	$10 \times 36 \times 1$
BC$_3$DH$-$HHH	$1 \times 1 \times 360$	$1 \times 20 \times 18$	$20 \times 1 \times 36$	$-$	$20 \times 18 \times 1$
BC$_3$DH$-$CHH	$1 \times 1 \times 360$	$1 \times 10 \times 36$	$10 \times 2 \times 36$	$-$	$20 \times 18 \times 1$
Crossing schemes with increasing population sizes					
BC$_3$DH$-$HHH*	$1 \times 1 \times 180$	$1 \times 20 \times 18$	$20 \times 1 \times 27$	$-$	$20 \times 18 \times 1$
BC$_3$DH$-$HHS*	$1 \times 1 \times 180$	$1 \times 20 \times 18$	$20 \times 1 \times 30$	$-$	$20 \times (5 \times 3) \times 1$
Basic crossing schemes					
BC$_2$S$_2$$-$C	$1 \times 1 \times 360$	$1 \times 10 \times 36$	$-$	$10 \times 36 \times 1$	$10 \times 36 \times 1$
BC$_2$S$_2$$-$H	$1 \times 1 \times 360$	$1 \times 20 \times 18$	$-$	$20 \times 18 \times 1$	$20 \times 18 \times 1$
BC$_3$S$_2$$-$CC	$1 \times 1 \times 360$	$1 \times 10 \times 36$	$10 \times 1 \times 36$	$10 \times 36 \times 1$	$10 \times 36 \times 1$
BC$_3$S$_2$$-$HH	$1 \times 1 \times 360$	$1 \times 20 \times 18$	$20 \times 1 \times 18$	$20 \times 18 \times 1$	$20 \times 18 \times 1$
BC$_3$S$_2$$-$CH	$1 \times 1 \times 360$	$1 \times 10 \times 36$	$10 \times 2 \times 18$	$20 \times 18 \times 1$	$20 \times 18 \times 1$
Crossing schemes with high selection intensity					
BC$_3$S$_2$$-CC'$	$1 \times 1 \times 720$	$1 \times 10 \times 72$	$10 \times 1 \times 72$	$10 \times 72 \times 1$	$10 \times 72 \times 1$
BC$_3$S$_2$$-HH'$	$1 \times 1 \times 720$	$1 \times 20 \times 36$	$20 \times 1 \times 36$	$20 \times 36 \times 1$	$20 \times 36 \times 1$
BC$_3$S$_2$$-CH'$	$1 \times 1 \times 720$	$1 \times 10 \times 72$	$10 \times 2 \times 36$	$20 \times 36 \times 1$	$20 \times 36 \times 1$
Crossing schemes with selection in the final BC generation					
BC$_2$S$_2$$-$CC	$1 \times 1 \times 360$	$1 \times 10 \times 36$	$-$	$10 \times 1 \times 36$	$10 \times 36 \times 1$
BC$_2$S$_2$$-$HH	$1 \times 1 \times 360$	$1 \times 20 \times 18$	$-$	$20 \times 1 \times 18$	$20 \times 18 \times 1$
BC$_2$S$_2$$-$CH	$1 \times 1 \times 360$	$1 \times 10 \times 36$	$-$	$10 \times 2 \times 18$	$20 \times 18 \times 1$
BC$_3$S$_2$$-$CCC	$1 \times 1 \times 360$	$1 \times 10 \times 36$	$10 \times 1 \times 36$	$10 \times 1 \times 36$	$10 \times 36 \times 1$
BC$_3$S$_2$$-$HHH	$1 \times 1 \times 360$	$1 \times 20 \times 18$	$20 \times 1 \times 18$	$20 \times 1 \times 18$	$20 \times 18 \times 1$
BC$_3$S$_2$$-$CHH	$1 \times 1 \times 360$	$1 \times 10 \times 36$	$10 \times 2 \times 18$	$20 \times 1 \times 18$	$20 \times 18 \times 1$
Crossing schemes with increasing population sizes					
BC$_3$S$_2$$-$HHH*	$1 \times 1 \times 180$	$1 \times 20 \times 18$	$20 \times 1 \times 27$	$20 \times 1 \times 18$	$20 \times 18 \times 1$
BC$_3$S$_2$$-$HHS*	$1 \times 1 \times 180$	$1 \times 20 \times 18$	$20 \times 1 \times 23$	$20 \times 5 \times 4$	$100 \times 1 \times 4$

The total population size in generation g is defined as $n_{tot} = n_{pop} \times n_{sel} \times n_g$. n_{pop}: number of sub-populations in generation $g-1$; n_{sel}: number of individuals selected from the sub-populations in generation $g-1$; n_g: population size per sub-population in generation g.

generations of selection was the same. For example, the basic crossing scheme **BC$_3$DH$-$CC** resulted in a D_t of only 5.0%, while crossing scheme **BC$_2$DH$-$CC** with selection in the final backcross generation resulted in a D_t of 7.8%. An additional generation of selection in **BC$_2$** schemes only resulted in minor improvements of D_t of 0.4–1.4% compared to the basic crossing schemes without selection. For example, scheme **BC$_2$DH$-$CC** improved D_t only by 0.5% compared to scheme **BC$_2$DH$-$C**.

The DH crossing schemes had in most cases better values for T, D_t, D_c, D_s and especially S than the **S$_2$** crossing schemes (Table 3).

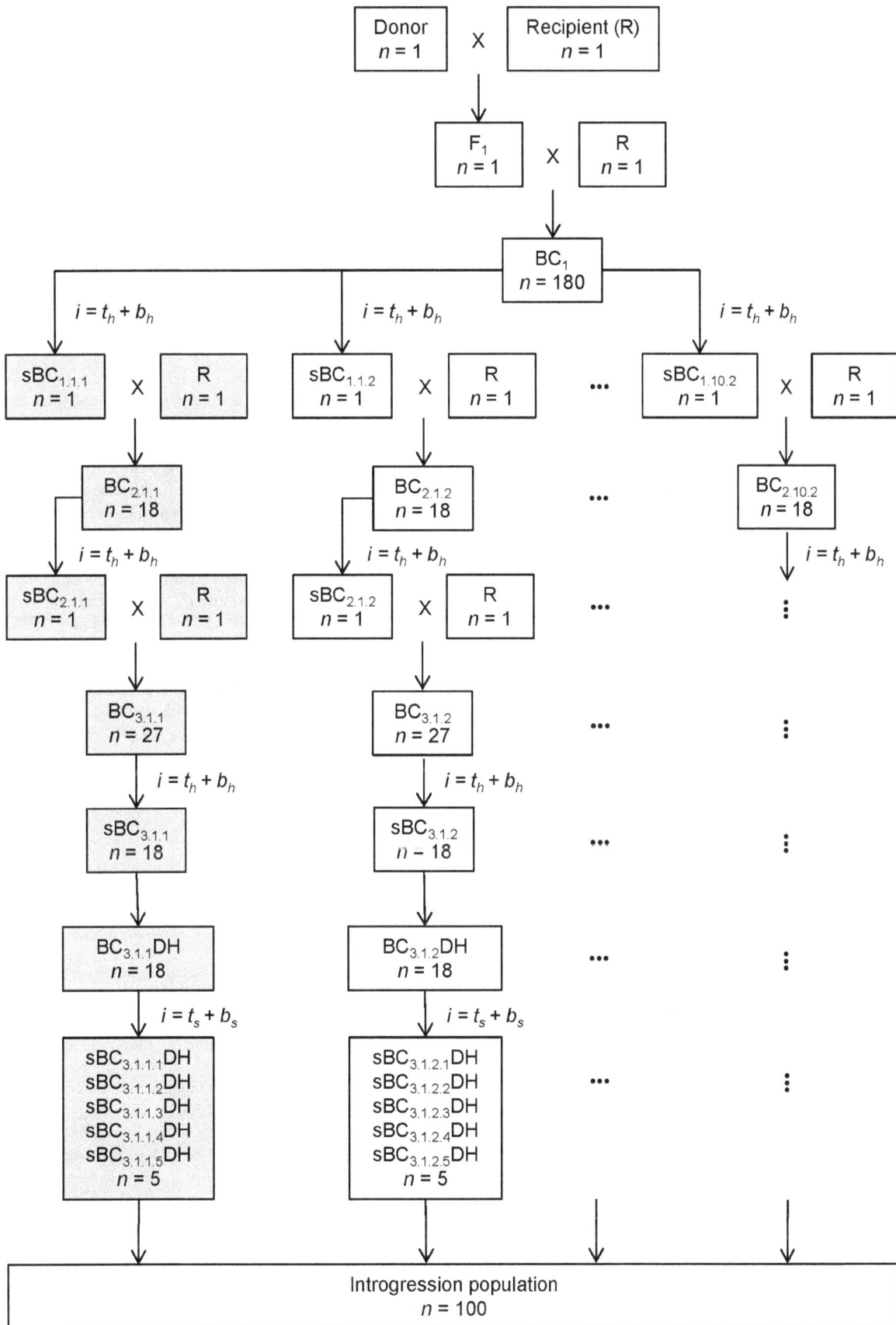

Figure 1. Schematic representation of crossing scheme $BC_3DH - HHH^*$. Crossing scheme $BC_3DH - HHH^*$ is characterized by increasing population sizes in the backcross generations and selection for donor chromosome halves in the final backcross generation. The parts highlighted in gray represent one branch of the crossing scheme. Sub-populations are indexed by BC_g, $BC_{g.c.h}$ and $BC_{g.c.h.s}$, where g is the respective backcross generation, c is the respective chromosome, h is the respective chromosome half, s is the respective target segment; $sBC_{g.c.h}$ and $sBC_{g.c.h.s}$ denote individuals selected for the respective selection regions.

Very high values of $S > 1000$ segments were observed for the basic crossing schemes $BC_2S_2 - C$ and $BC_2S_2 - H$. These crossing schemes had on average $N = 1$ additional donor segment per IL compared to the corresponding DH crossing schemes. However, they were also characterized by incomplete homozygosity (Figure 3B). The S_2 crossing schemes with selection in the final backcross generation required 360 individuals and HT assays less then the corresponding DH crossing schemes (Tables 2 and 3). Nevertheless, the differences between DH and S_2 crossing schemes then diminished. For example, scheme $BC_3S_2 - HHH$ resulted in similar values for most measures as the corresponding scheme $BC_3DH - HHH$ (Table 3).

The differences in the total donor genome proportion D_t between selection for complete donor chromosomes and selection for donor chromosome halves ranged only between 0.1–0.7% for the same number of backcross generations and generations of selection. However, substantial differences were observed for the donor genome proportion of the carrier chromosomes D_c and the donor genome proportion of the target segments D_s. For selection for complete donor chromosomes, high values for D_c of up to 48% were observed. They were clearly visible in the graphical genotypes for schemes $BC_2S_2 - CC$ and $BC_3DH - CC$ (Figure 3B and C). For selection for donor chromosome halves, the values for D_c were much lower and did not exceed 42% (Table 3). Without selection in the final backcross generation, selection for donor chromosome halves resulted in substantially reduced values for D_s. For example, the basic crossing schemes $BC_3DH - HH$ and $BC_3S_2 - HH$ resulted in values for D_s of only 94% and 90%. Moreover, the ranges for D_s for these crossing schemes were substantially greater (Figure 4 for S_2 crossing schemes, for DH data not shown).

The basic crossing schemes $BC_3DH - CH$ and $BC_3S_2 - CH$ which combined selection for complete donor chromosomes and selection for donor chromosome halves resulted in similarly low values for D_s of 93.7% and 89.9% as selection for donor chromosome halves only (Table 3). In addition, the combined strategies CH and CHH resulted in high values for D_c of up to 45.9%. The low values for D_s and the high values for D_c were reflected in the graphical genotype of scheme $BC_3DH - CH$, e.g. in ILs 47 and 54 (Figure 3A).

Doubling population sizes n_{tot} from 360 to 720 individuals in the crossing schemes with high selection intensity reduced the total donor genome proportion D_t from 5.0–5.1% to 3.6–3.8% compared to the basic DH crossing schemes, and from 5.3–5.7% to 4.3–4.4% compared to the basic S_2 crossing schemes (Table 3). The donor genome proportion of the carrier chromosomes D_c was reduced by about 4.2–7.5% for the DH crossing schemes, and by about 0.9–6.9% for the S_2 crossing schemes. The reduction of the donor genome proportion of the target segments D_s in combination with increased ranges that was observed with selection for donor chromosome halves in the basic crossing schemes was not observed in the crossing schemes with high selection intensity (Table 3 and Figure 4). D_s was increased by 5.2% for crossing scheme $BC_3DH - HH'$ and by 8.6% for crossing scheme $BC_3S_2 - HH'$ compared to the basic crossing schemes $BC_3DH - HH$ and $BC_3S_2 - HH$. However, these

improvements were only achieved with 2160 HT assays compared to 1080 HT asssays in the basic crossing schemes (Table 3).

The crossing schemes with selection in the final backcross generation resulted in values for D_t that were 1.1–1.2% higher for the DH crossing schemes und 0.6–1.4% higher for the S_2 crossing schemes compared to the crossing schemes with high selection intensity. The ranges of D_s for selection for donor chromosome halves were about the same size as for the crossing schemes with high selection intensity (Figure 4). The average values for D_s were 0.5% lower for scheme $BC_3DH - HHH$ and 0.3% lower for scheme $BC_3S_2 - HHH$ (Table 3). The number of required HT assays was reduced by 360 for the DH crossing schemes and by 720 for S_2 crossing schemes compared to the crossing schemes with high selection intensity. For the crossing schemes with selection in the final backcross generation, selection for donor chromosome halves was the most advantageous selection strategy with respect to the genetic background and to the target segments. Most notably, the crossing schemes $BC_3DH - HHH$ and $BC_3S_2 - HHH$ resulted in the lowest values for the donor genome proportion of the carrier chromosomes D_c. Compared to the most efficient basic crossing schemes $BC_2DH - C$ and $BC_3DH - CC$, the crossing schemes $BC_2DH - HH$ and $BC_3DH - HHH$ resulted in small improvements of both the genetic background and D_s. However, in both cases 720 additional HT assays had to be invested. For the S_2 crossing schemes with selection in the final backcross generation, high values of D_c of 38.1–48.3% were observed. Large donor chromosome segments on the carrier chromosomes were also visible in the graphical genotypes for schemes $BC_2S_2 - CC$ and $BC_3S_2 - HHH$ (Figure 3B and D). The high values for D_c were associated with a considerable reduction of the number of disjunct genome segments S of > 200 segments for the BC_2 crossing schemes and of 100–200 segments for the BC_3 crossing schemes compared to the basic S_2 crossing schemes (Table 3).

The crossing schemes with increasing population sizes reduced the number of required HT assays for DH crossing schemes by 360 in comparison to the crossing schemes with selection in the final backcross generation and constant population sizes. The crossing schemes $BC_3DH - HHH^*$ and $BC_3DH - HHS^*$ resulted in similar values for most measures as the crossing scheme $BC_3DH - HHH$. However, D_c and D_s were slightly reduced for crossing scheme $BC_3DH - HHS^*$. Compared to the most efficient basic crossing scheme $BC_3DH - CC$, crossing scheme $BC_3DH - HHH^*$ required 360 additional HT assays, but reduced D_c by 1.9% and increased D_s by 0.6%. The crossing scheme $BC_3S_2 - HHH^*$ resulted with 38.0% in a much higher D_c than the crossing scheme $BC_3S_2 - HHS^*$ with 30.4%. For crossing scheme $BC_3S_2 - HHS^*$, the average D_s was only 96.2% and the range for D_s was higher than for the crossing schemes $BC_3S_2 - HHH^*$ and $BC_3S_2 - HHH$ (Figure 4). However, D_t and D_c were the lowest for all investigated crossing schemes, with the exception of the crossing schemes with high selection intensity and $n_{tot} = 720$ (Table 3). The clear-cut separation of the target segments is also visible in the graphical genotype (Figure 3F).

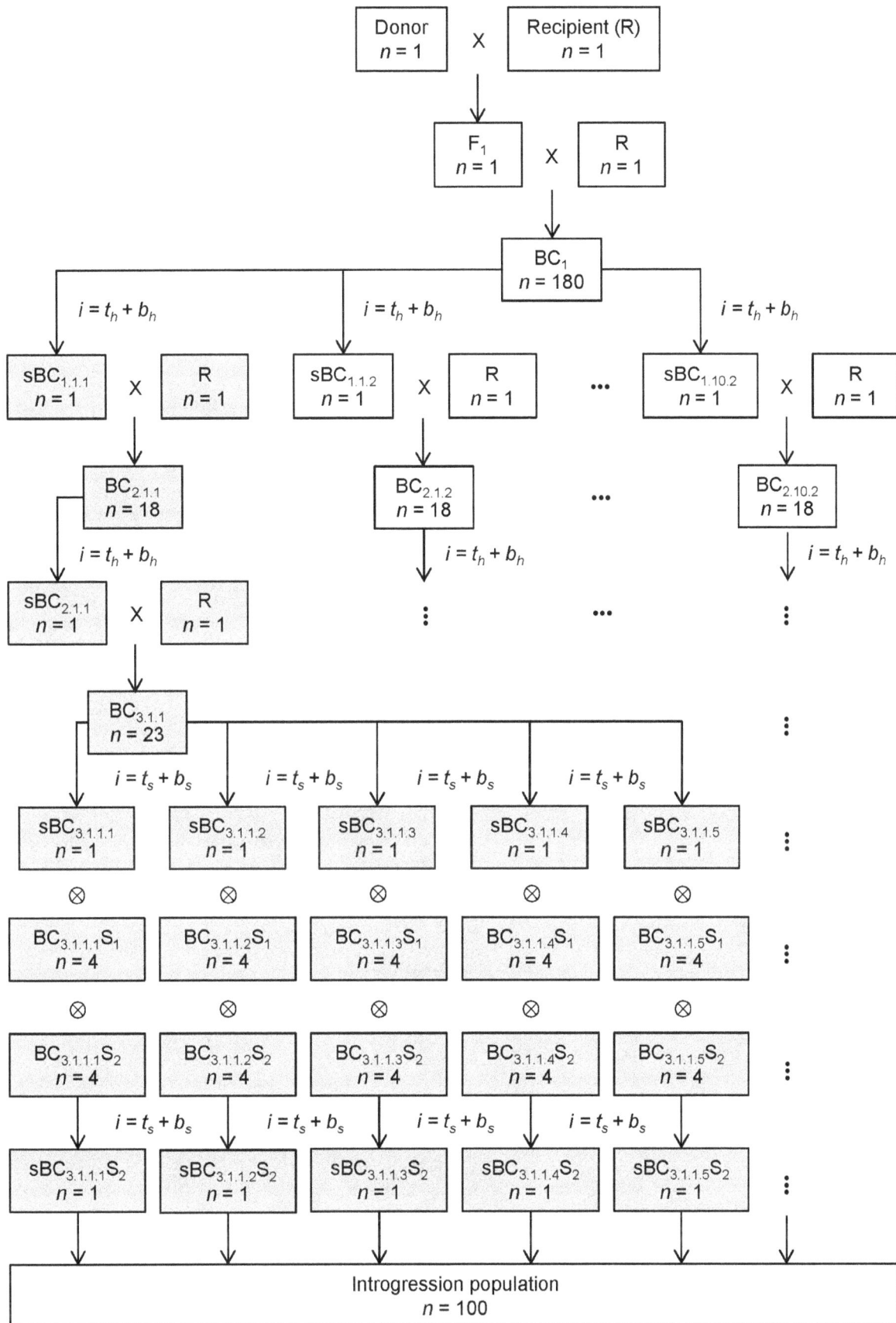

Figure 2. Schematic representation of crossing scheme $BC_3S_2-HHS^*$. Crossing scheme $BC_3S_2-HHS^*$ is characterized by increasing population sizes in the backcross generations and selection for target segments in the final backcross generation. The parts highlighted in gray represent one branch of the crossing scheme. Sub-populations are indexed by BC_g, $BC_{g.c.h}$ and $BC_{g.c.h.s}$, where g is the respective backcross generation, c is the respective chromosome, h is the respective chromosome half, s is the respective target segment; $sBC_{g.c.h}$ and $sBC_{g.c.h.s}$ denote individuals selected for the respective selection regions.

Discussion

Measures for Characterizing Introgression Populations

Measures for the description of introgression populations should allow to distinguish between introgression populations of different structure. Complete donor genome coverage O is desirable in order to make the complete genetic variation of the donor available for the breeding process. However, high values for O can also be caused by donor segments outside the target segments which could not be removed from the genetic background. O is therefore only informative if interpreted in relation to measures which reflect the distribution of the donor genome in the introgression population. A distinctive description of introgression populations is possible with the total donor genome proportion D_t, the donor genome proportion on the carrier chromosomes D_c and the donor genome proportion of the actual target segments D_s.

A high total donor genome proportion D_t is often associated with a high number of disjunct genome segments S. S determines the resolution R, which is an important parameter for the accuracy of QTL detection. However, if S is greater than the number of ILs, the problem of overparameterization arises with classical linear model approaches. This issue has only in part been resolved by using statistical methods which pre-select a reduced number of ILs for the linear model [23].

High values for the donor genome proportion on the carrier chromosomes D_c and the depth of donor genome coverage T reflect undesired donor segments attached to the actual target segments. Such large donor segments which overlap between ILs have been reported to increase the risk of false-positive effects in QTL detection and reduce the power of QTL detection [24]. This is mainly a problem if linkage maps with large distances between adjacent markers of 10 cM or more are employed, because QTLs located between the last marker of the target segment and the next marker outside the target segment are incorrectly assigned to the target segments. With dense marker maps which are now available this problem should be overcome. However, large donor segments also increase the risk of linkage drag in the breeding process and often require further steps of separation [24].

Low values for the donor genome proportion of the target segments D_s indicate a loss of target segments and potentially useful alleles. This is a problem that arises with small population sizes as were investigated in the present study [16]. Even if the missing target segments are present in the genetic background of other ILs, this might impair QTL detection and the further use of the ILs for the breeding progress.

We therefore argue that short non-overlapping target segments in a clean recipient background are advantageous also with dense marker maps. For 20 cM target segments and a genomic model of 10 equally sized chromosomes of 200 cM length, this corresponds to $D_t=1\%$, $D_c=10\%$ and $D_s=100\%$ in the ideal case. The effort and time required for developing introgression populations with such characteristics is beyond the scope of most breeding programs. With the limited population sizes and number of HT assays investigated in this study, these ideal values could not be achieved with two or three backcross generations (Table 3). We therefore considered those crossing and selection schemes as efficient which with a given limited resource input resulted in the highest coverage of target segments D_s in combination with low overlap of target segments reflected in D_c and T and a low total donor genome proportion D_t.

With respect to QTL detection, it can be expected that the optimal values for the suggested measures will depend on the statistical method and the genetic architecture of the trait. They could be determined for a given statistical method by including QTLs of different number and effect in future simulation studies. We plan further investigations in this area of research.

Crossing Schemes

BC_3 crossing schemes had 2–3% lower values for the total donor genome proportion D_t than BC_2 crossing schemes (Table 3), even if no selection for the genetic background was conducted in generation BC_3. Selection in generation BC_2, as was investigated with the crossing schemes BC_2-CC, BC_2-HH and BC_2-CH, only resulted in a reduction of D_t of 0.4–1.4% compared to the basic crossing schemes BC_2-C and BC_2-H (Table 3). An explanation for this comparatively small reduction is that the limiting factor for the reduction of D_t is the number of recombinations during meiosis. Hence, even though BC_2 crossing schemes have a time advantage, the effect of a third backcross generation cannot be compensated by investing in additional marker analyses. We therefore conclude that BC_3 crossing schemes result in introgression populations with an improved structure, and that the time investment in the additional backcross generation is worthwhile.

DH crossing schemes were for most measures superior to the corresponding S_2 crossing schemes. The differences were most pronounced in the number of disjunct genome segments S. Even though the S_2 schemes on average had a slightly higher number of donor segments per IL N, it seems that the very high values for S that were observed especially in the BC_2-S_2 crossing schemes mainly had to be attributed to incomplete homozygosity (Figure 3B). It can be expected that introgression populations with $S>1000$ segments in 100 ILs (Table 3) are not suitable for effective QTL detection. We therefore conclude that the DH method is essential for short crossing schemes with only two backcross generations.

A drawback of the DH method is that with current protocols of in vivo DH induction of maternal haploids, only a very limited number of viable DH lines can be derived from one backcross individual. We expect that our assumption of one DH line per backcross individual is a conservative, but realistic estimate. In contrast, with selfing, many progenies can be derived from one selected backcross individual. In the S_2 crossing schemes, it is consequently comparatively cheap and easy to conduct selection in the final backcross generation. For the DH schemes, selection in the final backcross generation could only be conducted if population size in this generation was higher than the desired number of final DH lines. As a result, the S_2 crossing schemes with selection in the final backcross generation required 360 HT assays less than the corresponding DH schemes (Table 3). Moreover, the selected fractions of best backcross individuals were much greater for the DH than for the S_2 crossing schemes (Table 2). This resulted in a lower selection intensity for both the selection region of the final backcross generation and the genetic background in

Table 3. Measures evaluated for introgression populations resulting from different crossing and selection schemes.

Scheme	O	T	S	R	N	L	D_t	D_c	D_s	HT
Basic crossing schemes										
BC_2DH-C	99.9	8.3	691	2.9	6.2	28.7	8.3	38.9	98.7	720
BC_2DH-H	100.0	8.9	751	2.7	6.4	29.2	8.9	37.3	94.1	720
BC_3DH-CC	99.2	5.1	457	4.4	3.8	29.8	5.0	35.2	97.8	1080
BC_3DH-HH	99.8	5.2	487	4.1	3.9	29.8	5.1	33.3	94.2	1080
BC_3DH-CH	99.6	5.2	469	4.3	3.8	30.5	5.1	35.1	93.7	1080
Crossing schemes with high selection intensity										
BC_3DH-CC'	99.3	3.6	389	5.1	3.0	27.8	3.6	27.7	98.8	2160
BC_3DH-HH'	99.9	3.8	406	4.9	3.0	29.7	3.8	29.1	99.4	2160
BC_3DH-CH'	99.6	3.9	399	5.0	3.0	30.0	3.8	29.9	99.0	2160
Crossing schemes with selection in the final BC generation										
BC_2DH-CC	99.9	7.8	676	3.0	5.9	28.7	7.8	41.0	98.9	1440
BC_2DH-HH	100.0	8.1	716	2.8	6.0	29.0	8.1	38.9	99.3	1440
BC_2DH-CH	99.9	8.5	679	2.9	6.0	30.4	8.5	43.7	98.8	1440
$BC_3DH-CCC$	99.1	4.9	457	4.4	3.7	30.2	4.8	35.8	98.0	1800
$BC_3DH-HHH$	99.9	4.8	464	4.3	3.6	31.0	4.7	33.9	98.9	1800
$BC_3DH-CHH$	99.6	4.8	450	4.4	3.5	31.4	4.7	35.0	98.5	1800
Crossing schemes with increasing population sizes										
$BC_3DH-HHH^*$	99.9	5.0	492	4.1	3.8	29.7	5.0	33.3	98.4	1440
$BC_3DH-HHS^*$	99.8	4.9	484	4.1	3.8	29.4	4.8	32.5	97.5	1440
Basic crossing schemes										
BC_2S_2-C	100.0	11.4	1021	2.0	7.3	26.4	9.3	41.4	97.6	720
BC_2S_2-H	100.0	11.4	1073	1.9	7.4	25.9	9.3	36.3	90.4	720
BC_3S_2-CC	99.3	6.9	684	2.9	4.5	27.7	5.7	39.2	97.0	1080
BC_3S_2-HH	99.9	6.3	702	2.8	4.4	26.3	5.3	33.0	90.3	1080
BC_3S_2-CH	99.7	6.4	681	2.9	4.3	26.9	5.3	35.0	89.9	1080
Crossing schemes with high selection intensity										
BC_3S_2-CC'	99.5	5.1	585	3.4	3.6	26.6	4.3	32.3	98.8	2160
BC_3S_2-HH'	99.9	5.1	591	3.4	3.5	27.8	4.4	32.1	98.9	2160
BC_3S_2-CH'	99.7	5.2	581	3.4	3.5	28.2	4.4	33.4	98.4	2160
Crossing schemes with selection in the final BC generation										
BC_2S_2-CC	99.3	10.5	795	2.5	6.5	27.2	8.3	48.3	98.0	1080
BC_2S_2-HH	99.9	9.7	785	2.5	6.2	27.0	7.9	42.8	98.8	1080
BC_2S_2-CH	99.8	9.7	761	2.6	6.1	27.6	7.9	45.9	98.4	1080
BC_3S_2-CCC	98.3	7.1	588	3.4	4.2	30.5	5.7	44.6	97.5	1440
BC_3S_2-HHH	99.7	5.9	510	3.9	3.7	31.0	5.0	38.1	98.6	1440
BC_3S_2-CHH	99.3	6.0	509	3.9	3.7	31.3	5.0	39.5	98.2	1440
Crossing schemes with increasing population sizes										
$BC_3S_2-HHH^*$	99.7	5.8	508	3.9	3.8	30.7	4.9	38.0	98.7	1440
$BC_3S_2-HHS^*$	99.8	5.1	596	3.4	3.7	26.1	4.3	30.4	96.2	1400

O: donor genome coverage in percent; T: depth of donor genome coverage; S: number of disjunct genome segments; R: resolution; N: number of donor segments per IL; L: length of donor segments per IL in cM; D_t: total donor genome proportion in percent; D_c: donor genome proportion of carrier chromosomes in percent; D_s: donor genome proportion of target segments in percent; HT: the required number of HT assays. Measures are arithmetic means over 1,000 replications.

the DH crossing schemes. We therefore suggest that a comparison of DH and S_2 crossing schemes should take the distinctive features of both methods into account. The evaluation of efficiency should also be based on the number of required HT assays. Considering this, S_2 crossing schemes which exploit their selection advantages represent economic and easy-to-handle alternatives to DH crossing schemes.

Selection Strategies for Small and Constant Population Sizes

For a given genetic model and crossing scheme, the selection strategy is the most important factor that influences the structure of the resulting introgression population. In the following paragraphs, different aspects such as the length of the selection

A

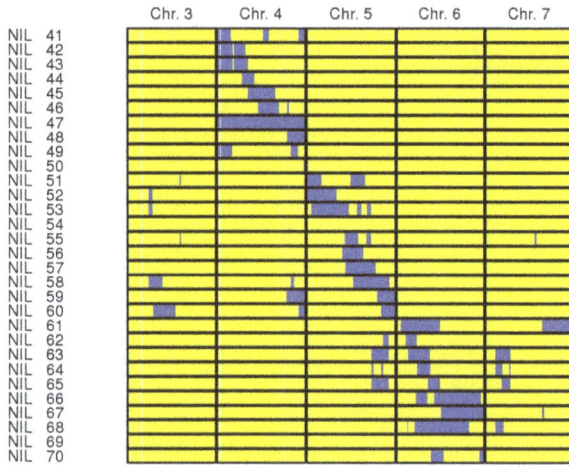

B

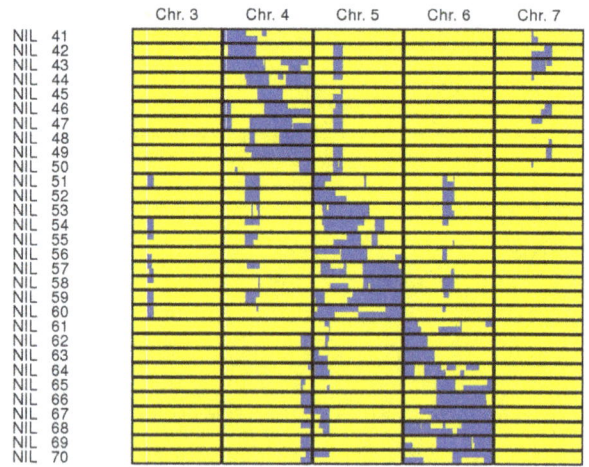

C

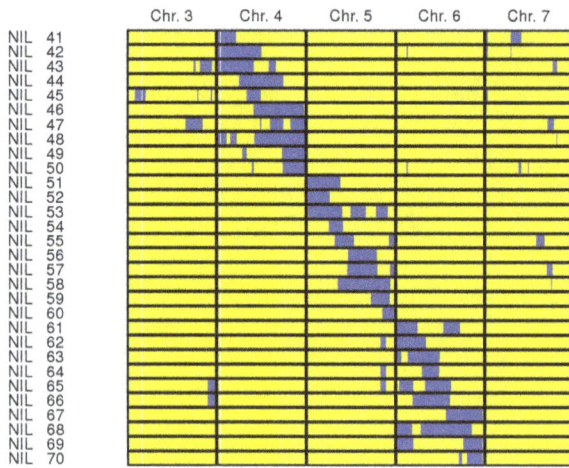

D

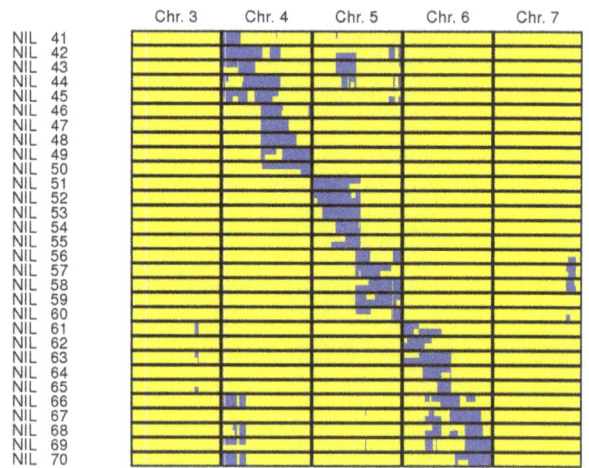

E

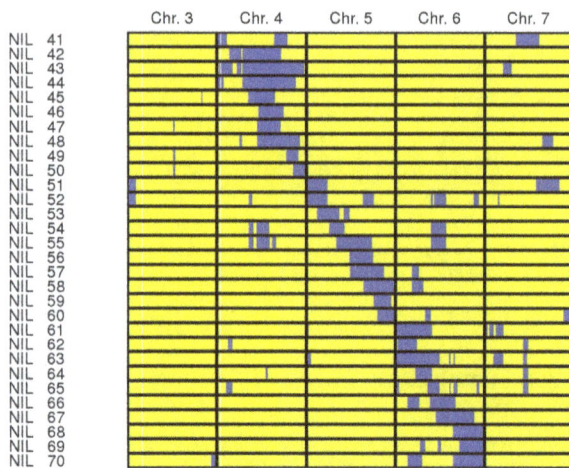

F

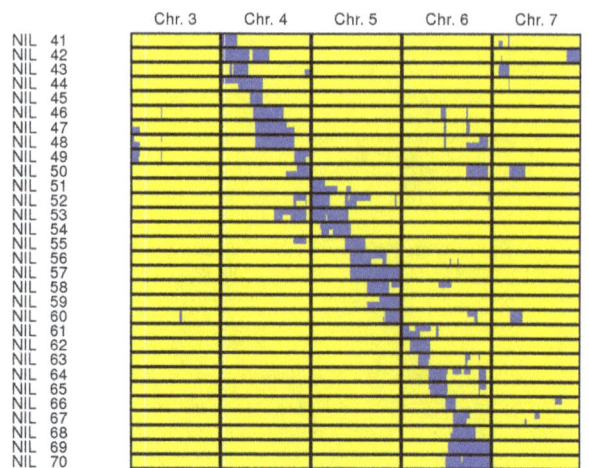

Figure 3. Graphical genotypes of introgression populations resulting from six different crossing schemes. A: BC_3DH-CH; B: BC_2S_2-CC; C: BC_3DH-CC; D: BC_3S_2-HHH; E: $BC_3DH-HHH^*$; F: $BC_3S_2-HHS^*$. The graphical genotypes display the chromosomes 3 to 7 of ILs 41–70 and are examples from one simulation run. Chromosome segments which stem from the donor are displayed in blue, whereas chromosome segments which stem from the recipient are displayed in yellow. The graphical genotypes illustrate the differences between the alternative crossing schemes with respect to their suitability to create introgression populations with complete donor genome coverage and clearly separated, evenly distributed target donor chromosome segments.

regions, the number of generations of selection and the required population sizes for effective selection are discussed.

Selection strategies which pre-select individuals carrying complete donor chromosomes reduce the number of simultaneous backcross programs to the number of chromosomes [16]. They are therefore suitable for breeding programs with limited resources. However, for long chromosomes of 200 cM length, selection for complete donor chromosomes preserved large donor chromosome segments on the carrier chromosomes up to line development (Figure 3C and B). This was reflected in high values for the proportion of donor genome on the carrier chromosomes D_c of up to 48% (Table 3). The selection regions for selection in the backcross generations were therefore reduced to donor chromosome halves for selection strategies H, HH and HHH. In all four series of simulations, selection for donor chromosome halves resulted in the desired reduction of D_c compared to selection for complete donor chromosomes (Table 3). Other measures for the genetic background were approximately equivalent. We therefore conclude that for crop species with long chromosomes such as maize, wheat or rapeseed, selection for donor chromosome halves reduces the length of the donor segments attached to the actual target segments and the risk of linkage drag.

However, for crossing schemes without selection in the final backcross generation and constant population sizes of $n_{tot} = 360$ individuals, selection for donor chromosome halves resulted in a considerable reduction of the donor genome proportion of the target segments D_s of up to 7%. Moreover, the estimated values for D_s were less reliable for these crossing schemes, *e.g.*, in schemes BC_2S_2-H and BC_3S_2-HH (Figure 4). These findings have to be attributed to the small population sizes n_g in the sub-populations and the structure of the selection index i. In generation DH or S_2, population sizes were reduced to $n_g = 18$ individuals with selection for donor chromosome halves (Table 2). Without selection in the final backcross generation, around 50% of the ILs developed from the backcross individuals are expected to carry no donor allele at a given locus within the respective target segment. The probability to find five ILs with complete donor target segments for the introgression population was therefore even further reduced. As the selection index $i = t_s + b_s$ weighed the target segments and the genetic background equally, a clean genetic background sometimes outweighed a reduced D_s and led to the observed loss of target segments in these small sub-populations. We therefore conclude that a sufficiently large population size is the crucial

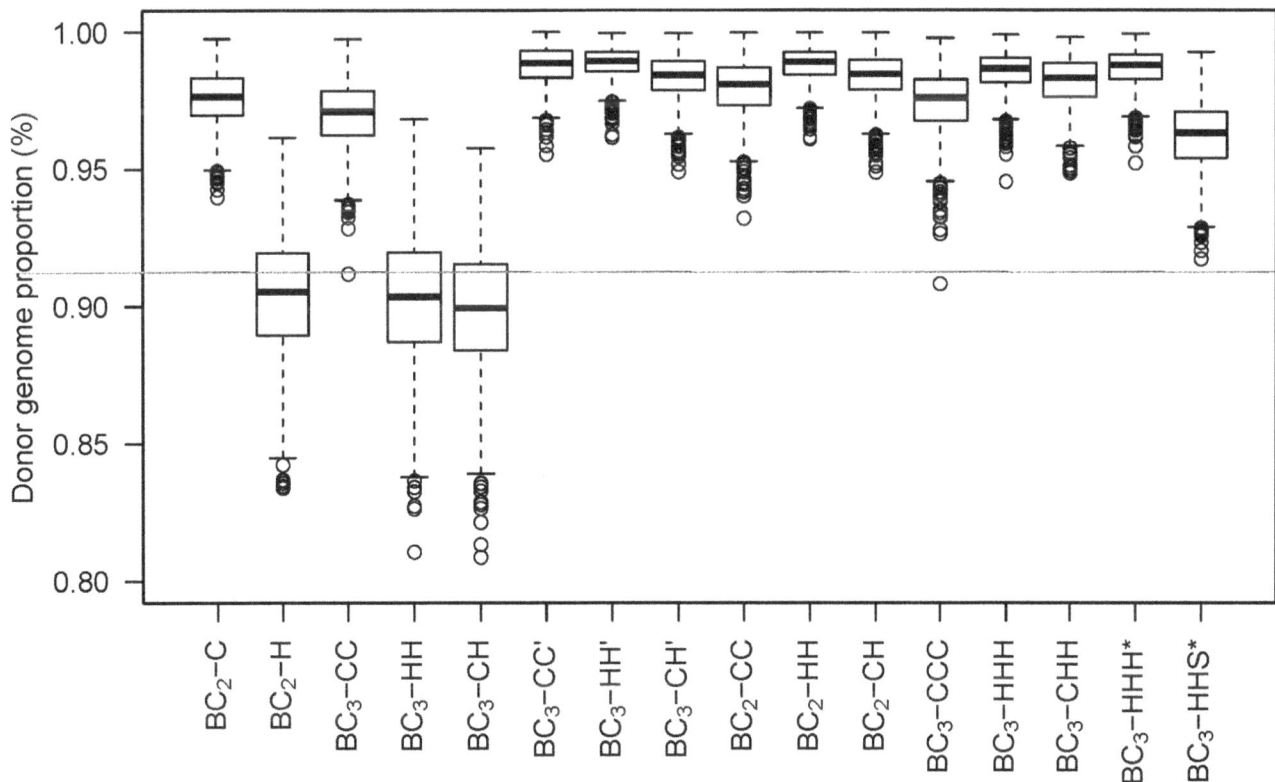

Figure 4. Donor genome proportion of target segments D_s for all investigated S_2 crossing schemes. The boxplots represent the distribution over 1,000 replications of the simulations. The basic crossing schemes BC_2S_2-H, BC_3S_2-HH and BC_3S_2-CH which select for donor chromosome halves are characterized by higher ranges for D_s.

factor for the successful application of selection for donor chromosome halves.

A loss of target segments caused by small population sizes was also observed for the basic combined selection strategy CH which selected for complete donor chromosomes in generation BC_1 and for donor chromosome halves in generation BC_2. In addition, the combined strategies CH and CHH resulted in high values for the donor genome proportion on the carrier chromosomes D_c of up to 45% (Table 3). This can be explained by the efficient selection for complete donor chromosomes from the comparatively large BC_1 population of $n_g = n_{tot} = 360$ individuals (Table 2). The pre-selected complete donor chromosomes are in large part preserved up to line development. The combination of missing target segments with large donor chromosome segments on carrier chromosomes was also reflected in the graphical genotype for scheme $BC_3DH - CH$, $e.g.$, in IL 47 and 54 (Figure 3A). The selection strategies CH and CHH therefore combine the drawbacks of both selection for complete donor chromosomes and selection for chromosome halves. They are not suitable for crossing schemes with small and constant population sizes, in which the population sizes n_g in the sub-populations are subsequently reduced over the backcross generations. We conclude that for for small breeding programs with a constant population size of $n_{tot} = 360$ and a limited number of HT assays for selection, selection strategies which only select for complete donor chromosomes in the backcross generations should be employed in both DH and S_2 crossing schemes to avoid the loss of target segments.

Finding more Carriers of Donor Target Segments for Line Development

To employ selection for donor chromosome halves effectively for reducing the donor genome proportion of the carrier chromosomes D_c without losing the target segments, it is necessary to increase the frequency of carriers of donor target segments for line development. Using larger population sizes is a straightforward solution for this problem, which in addition can improve the overall structure of introgression populations. The crossing schemes with high selection intensity and double population sizes of $n_{tot} = 720$ individuals resulted in small improvements of the total donor genome proportion D_t of about 1–1.5% compared to the basic crossing schemes (Table 3). The desired increase in the donor proportion of the target segments D_s was achieved. For selection for donor chromosome halves, D_s was increased by 5.2–8.6%. Selection for donor chromosome halves was then even superior to selection for complete donor chromosomes. Moreover, the donor genome proportion on the carrier chromosomes D_c was reduced by up to 7.5%, indicating an improved separation of target segments. The observed improvements were greater for the DH than for the S_2 crossing schemes. Nevertheless, the comparatively small improvements of the introgression populations required 1080 additional HT assays. We therefore conclude that such large population sizes are only suitable for breeding programs with access to DH technology, less stringent resource restrictions and high requirements with respect to the genetic background. If the requirements concerning the structure of the introgression population are not that high, it might be more economic to increase population size only in the final backcross generation and/or to invest in additional HT assays only in this generation.

For crossing schemes with selection in the final backcross generation, the total donor genome proportion D_t was similar to the values of the basic crossing schemes, and about 1% higher than for the crossing schemes with higher selection intensity

(Table 3). However, the average values for the donor genome proportion of the target segments D_s were similar to the crossing schemes with higher selection intensity (Table 3) and the ranges were effectively reduced (Figure 4). Moreover, the number of required HT assays was reduced by 360 for the DH crossing schemes and by 720 for S_2 crossing schemes compared to the crossing schemes with higher selection intensity (Table 3). The decision for doubling population sizes requires the same resources as would be required for generating an additional introgression population. This large effort seems not to be justified by the relatively small improvements compared to the basic crossing schemes. We therefore conclude that selection in the final backcross generation is the more efficient solution for both DH and S_2 crossing schemes.

Selection for donor chromosome halves was the best strategy with selection in the final backcross generation for both DH and S_2 crossing schemes (Table 3). However, for the DH schemes, only small improvements for schemes $BC_2DH - HH$ and $BC_3DH - HHH$ were observed compared to the most efficient basic crossing schemes $BC_2DH - C$ and $BC_3DH - CC$ (Table 3). For these small improvements, 720 additional individuals and HT assays had to be invested. For the S_2 schemes, considerable reductions in S of 174 and 236 segments were observed for schemes $BC_2S_2 - HH$ and $BC_3S_2 - HHH$ with selection in the final backcross generation compared to the basic crossing schemes $BC_2S_2 - C$ and $BC_2S_2 - CC$. D_t was only slightly reduced. However, the donor genome proportion on carrier chromosomes D_c was in general very high for the crossing schemes with selection in the final backcross generation with 38–48%. This indicates a fixation of the selection regions of the final backcross generation (Figure 3B and D). In schemes $BC_2S_2 - CC$ and $BC_3S_2 - HH$, complete donor chromosomes and donor chromosome halves still appear as blocks around the target segments. These blocks lead to an overlap of donor segments between ILs that reduces the effective resolution of the introgression population for QTL detection. The overlap also hampers the further use of the ILs in the breeding process, as further steps of separation of the target segments by backcrossing are required. We therefore conclude that the crossing schemes with selection in the final backcross generation have the potential to improve the resulting introgression populations at moderate cost. However, for the DH crossing schemes, the number of required HT assays and individuals has to be reduced. For the S_2 crossing schemes, the fixation of large donor chromosome segments has to be avoided. Optimizations of the respective crossing schemes are presented in the following.

Increasing Population Sizes Over Backcross Generations

With constant population sizes of $n_{tot} = 360$ individuals, the population size in generation BC_1 was large in relation to the genetic gains that could be achieved by selecting a comparatively small fraction of 10 or 20 individuals (Table 2). Starting with smaller population sizes in generation BC_1 and gradually increasing population sizes in the following backcross generations was therefore an efficient option to reduce the overall number of required individuals and HT assays for selection in the final backcross generation. Larger population sizes in generation BC_3 also enabled selection for target segments, which was investigated as an option to avoid the fixation of large donor chromosome segments especially for the S_2 crossing schemes.

The schemes $BC_3DH - HHH^*$ and $BC_3DH - HHS^*$ resulted in similar values for all measures (Table 3). However, D_c and D_s were slightly lower for scheme $BC_3DH - HHS^*$. We therefore conclude that selection for target segments already in the final backcross generation is not efficient for DH crossing schemes. In comparison to the best but also very expensive scheme $BC_3DH-HHH$ with selection in the final backcross generation, scheme $BC_3DH-HHH^*$ can be considered equivalent, but required 360 individuals and HT assays less. In comparison to the more economic basic crossing scheme BC_3DH-CC, scheme $BC_3DH-HHH^*$ improved D_c and D_s and thus the separation of target segments. This is also visible in the graphical genotype (Figure 3). The investment in the additional 360 HT assays seems therefore worthwhile (Table 3).

The scheme $BC_3S_2-HHS^*$ resulted in better values than the schemes BC_3S_2-HHH and $BC_3S_2-HHH^*$. Most notably, it resulted in a much lower D_c of 30% compared to 38%. Scheme $BC_3S_2-HHS^*$ resulted in a D_s that was 2.4% lower compared to schemes BC_3S_2-HHH and $BC_3S_2-HHH^*$ and the ranges for D_s were higher (Figure 4). Nevertheless, it resulted in the lowest values of D_t and D_c and the best separation of target genes of all investigated DH and S_2 crossing schemes with comparable population sizes. The comparatively high value of S of 596 segments in combination with reduced values for D_t can in this case be explained by a greatly improved separation of target segments compared to the other S_2 schemes with selection in the final backcross generation. The improved separation of target segments is also visible in the graphical genotype (Figure 3F). This was achieved with 40 HT assays less (Table 3). We therefore expect that scheme $BC_3S_2-HHS^*$ will result in an improved power of QTL detection, and recommend selection for target segments in the final backcross generation for S_2 crossing schemes.

Compared to the best but expensive comparable DH crossing scheme $BC_3DH-HHH$, the S_2 crossing scheme $BC_3S_2-HHS^*$ resulted in similar values and required 400 HT assays less. Overall, we conclude that increasing population sizes over backcross are advantageous and economic for both DH and S_2 crossing schemes. Moreover, crossing scheme $BC_3S_2-HHS^*$ can provide a cheap alternative to comparable DH crossing schemes.

Conclusions

Our study has shown that introgression populations with complete coverage of the donor genome and reasonably clean recipient background can be developed with a limited number of backcross individuals and HT assays. It has provided further insight on how different crossing and selection schemes influence the structure of the resulting introgression populations. The guidelines which have been derived for maize are transferable to other crop species with similar number and length of chromosomes. For crops with different genome size, some considerations are discussed in the following.

Rapeseed is a crop with a large genome of 19 chromosomes, for which efficient protocols of microspore culture are available for DH production. For the large genome of rapeseed, it can be expected that the values for the total donor genome proportion D_t will be lower than those observed for the smaller genome of maize. With the investigated selection index i, the selection pressure on the carrier chromosomes will be reduced with increasing genome size and number of chromosomes. It might therefore be an interesting option for rapeseed to put more weight on the background markers on the carrier chromosomes to achieve an efficient reduction of D_c. As with microspore culture many DH lines can usually be derived from one backcross individual, the advantages of DH production should be more pronounced than for maize. However, the optimal selection strategies for DH crossing schemes in rapeseed should then be similar to those for selfing in maize.

Sugar beet is a crop with a small genome of 9 chromosomes, for which the guidelines for selfing should be most relevant. In smaller genomes, equivalent values of D_t can usually be reached with smaller population sizes and with fewer backcrosses. However, the average length of the chromosomes in cM is also much shorter than in maize. This implies that fewer crossovers occur per meiosis, and that it might require more individuals and backcross generations to effectively separate the target segments. The combined effects of genome size and chromosome length will also depend on the desired number and length of the target segments.

Simulations can considerably facilitate the planning process for the development of introgression populations in different crop species. The derived guidelines can help breeders and geneticists to enhance the genetic variation of narrow based breeding populations of crops.

Author Contributions

Conceived and designed the experiments: EH KCF TP DS MO MF. Performed the experiments: EH KCF. Analyzed the data: EH. Wrote the paper: EH MF.

References

1. Zamir D (2001) Improving plant breeding with exotic genetic libraries. Nat Rev Genet 2: 983–989.
2. McCouch S (2004) Diversifying selection in plant breeding. PLoS Biol 2: e347.
3. Eshed Y, Zamir D (1994) A genomic library of *Lycopersicon pennellii* in *L. esculentum*: a tool for fine mapping of genes. Euphytica 79: 175–179.
4. Keurentjes JJ, Bentsink L, Alonso-Blanco C, Hanhart CJ, Blankestijn-De Vries H, et al. (2007) Development of a near-isogenic line population of *Arabidopsis thaliana* and comparison of mapping power with a recombinant inbred line population. Genetics 175: 891–905.
5. Törjék O, Meyer RC, Zehnsdorf M, Teltow M, Strompen G, et al. (2008) Construction and analysis of 2 reciprocal Arabidopsis introgression line populations. J Hered 99: 396–406.
6. Lin S, Sasaki T, Yano M (1998) Mapping quantitative trait loci controlling seed dormancy and heading date in rice, *Oryza sativa* L., using backcross inbred lines. Theor Appl Genet 96: 997–1003.
7. Cheema KK, Bains NS, Mangat GS, Das A, Vikal Y, et al. (2008) Development of high yielding IR64× *Oryza rufipogon* (Griff.) introgression lines and identification of introgressed alien chromosome segments using SSR markers. Euphytica 160: 401–409.
8. Matus I, Corey A, Filichkin T, Hayes P, Vales M, et al. (2003) Development and characterization of recombinant chromosome substitution lines (RCSLs) using *Hordeum vulgare* subsp. *spontaneum* as a source of donor alleles in a *Hordeum vulgare* subsp. *vulgare* background. Genome 46: 1010–1023.
9. Schmalenbach I, Körber N, Pillen K (2008) Selecting a set of wild barley introgression lines and verification of QTL effects for resistance to powdery mildew and leaf rust. Theor Appl Genet 117: 1093–1106.
10. Liu S, Zhou R, Dong Y, Li P, Jia J (2006) Development, utilization of introgression lines using a synthetic wheat as donor. Theor Appl Genet 112: 1360–1373.
11. Pumphrey MO, Bernardo R, Anderson JA (2007) Validating the *Fhb1* QTL for Fusarium head blight resistance in near-isogenic wheat lines developed from breeding populations. Crop Sci 47: 200–206.
12. Ribaut JM, Ragot M (2007) Marker-assisted selection to improve drought adaptation in maize: the backcross approach, perspectives, limitations, and alternatives. J Exper Bot 58: 351–360.
13. Szalma S, Hostert B, LeDeaux J, Stuber C, Holland J (2007) QTL mapping with near-isogenic lines in maize. Theor Appl Genet 114: 1211–1228.
14. Falke K, Sušić Z, Hackauf B, Korzun V, Schondelmaier J, et al. (2008) Establishment of introgression libraries in hybrid rye (*Secale cereale* L.) from an Iranian primitive accession as a new tool for rye breeding and genomics. Theor Appl Genet 117: 641–652.
15. Syed N, Pooni H, Mei M, Chen Z, Kearsey M (2004) Optimising the construction of a substitution library in *Arabidopsis thaliana* using computer simulations. Mol Breeding 13: 59–68.
16. Falke KC, Miedaner T, Frisch M (2009) Selection strategies for the development of rye introgression libraries. Theor Appl Genet 119: 595–603.
17. Welcker C, Sadok W, Dignat G, Renault M, Salvi S, et al. (2011) A common genetic determinism for sensitivities to soil water deficit and evaporative

demand: meta-analysis of quantitative trait loci and introgression lines of maize. Plant Physiol 157: 718–729.

18. Belcher AR, Zwonitzer JC, Santa Cruz J, Krakowsky MD, Chung CL, et al. (2012) Analysis of quantitative disease resistance to southern leaf blight and of multiple disease resistance in maize, using near-isogenic lines. Theor Appl Genet 124: 433–445.

19. Mano Y, Omori F (2013) Flooding tolerance in interspecific introgression lines containing chromosome segments from teosinte (*Zea nicaraguensis*) in maize (*Zea mays subsp. mays*). Ann Bot-London 112: 1125–1139.

20. Lübberstedt T, Frei UK (2012) Application of doubled haploids for target gene fixation in backcross programmes of maize. Plant Breeding 131: 449–452.

21. R Core Team (2013) R: A Language and Environment for Statistical Computing. R Foundation for Statistical Computing, Vienna, Austria. URL http://www.R-project.org/.

22. Stam P (1979) Interference in genetic crossing over and chromosome mapping. Genetics 92: 573–594.

23. Mahone GS, Borchardt D, Presterl T, Frisch M (2012) A comparison of tests for QTL mapping with introgression libraries containing overlapping and nonoverlapping donor segments. Crop Sci 52: 2198–2205.

24. Falke K, Frisch M (2011) Power and false-positive rate in QTL detection with near-isogenic line libraries. Heredity 106: 576–584.

Permissions

The contributors of this book come from diverse backgrounds, making this book a truly international effort. This book will bring forth new frontiers with its revolutionizing research information and detailed analysis of the nascent developments around the world.

We would like to thank all the contributing authors for lending their expertise to make the book truly unique. They have played a crucial role in the development of this book. Without their invaluable contributions this book wouldn't have been possible. They have made vital efforts to compile up to date information on the varied aspects of this subject to make this book a valuable addition to the collection of many professionals and students.

This book was conceptualized with the vision of imparting up-to-date information and advanced data in this field. To ensure the same, a matchless editorial board was set up. Every individual on the board went through rigorous rounds of assessment to prove their worth. After which they invested a large part of their time researching and compiling the most relevant data for our readers.

The editorial board has been involved in producing this book since its inception. They have spent rigorous hours researching and exploring the diverse topics which have resulted in the successful publishing of this book. They have passed on their knowledge of decades through this book. To expedite this challenging task, the publisher supported the team at every step. A small team of assistant editors was also appointed to further simplify the editing procedure and attain best results for the readers.

Apart from the editorial board, the designing team has also invested a significant amount of their time in understanding the subject and creating the most relevant covers. They scrutinized every image to scout for the most suitable representation of the subject and create an appropriate cover for the book.

The publishing team has been an ardent support to the editorial, designing and production team. Their endless efforts to recruit the best for this project, has resulted in the accomplishment of this book. They are a veteran in the field of academics and their pool of knowledge is as vast as their experience in printing. Their expertise and guidance has proved useful at every step. Their uncompromising quality standards have made this book an exceptional effort. Their encouragement from time to time has been an inspiration for everyone.

The publisher and the editorial board hope that this book will prove to be a valuable piece of knowledge for researchers, students, practitioners and scholars across the globe.

List of Contributors

Chiara Marchi and Volker Loeschcke
Department of Bioscience, Aarhus University, Aarhus, Denmark

Liselotte Wesley Andersen
Department of Bioscience, Aarhus University, Rønde, Denmark

Rachit K. Saxena, Hari D. Upadhyaya, Kulbhushan Saxena and Rajeev K. Varshney
International Crops Research Institute for the Semi-Arid Tropics (ICRISAT), Hyderabad, Andhra Pradesh, India

Eric von Wettberg
Department of Biological Sciences, Florida International University, Miami, Florida, United States of America
Fairchild Tropical Botanic Garden, Kushlan Institute for Tropical Science, Miami, Florida, United States of America

Serah Songok
International Crops Research Institute for the Semi-Arid Tropics (ICRISAT), Hyderabad, Andhra Pradesh, India
Egerton University, Egerton, Kenya

Vanessa Sanchez
Florida International University, Department of Earth and Environment, Miami, Florida, United States of America

Paul Kimurto
Egerton University, Egerton, Kenya

Joanne Russell, Ian K. Dawson, Allan Booth and Robbie Waugh
Cell and Molecular Sciences, The James Hutton Institute, Invergowrie, Scotland, United Kingdom

Maarten van Zonneveld
Regional Office for the Americas, Bioversity International, Cali, Colombi

Brian Steffenson
Department of Plant Pathology, University of Minnesota, Saint Paul, Minnesota, United States of America

Manish Roorkiwal, Abhishek Rathore, Rajeev K. Varshney and Hari D. Upadhyaya
International Crops Research Institute for the Semi-Arid Tropics (ICRISAT), Hyderabad, Andhra Pradesh, India

Eric J. von Wettberg and Emily Warschefsky
Department of Biological Sciences, Florida International University, Miami, Florida, United States of America
Center for Tropical Plant Conservation, Fairchild Tropical Botanic Garden, Miami, Florida, United States of America

Maarten van Zonneveld
Bioversity International, Regional Office for the Americas, Cali, Colombia
Ghent University, Faculty of Bioscience Engineering, Gent, Belgium

Xavier Scheldeman
Bioversity International, Regional Office for the Americas, Cali, Colombia

Pilar Escribano and María A. Viruel
Instituto de Hortofruticultura Subtropical y Mediterránea, (IHSM-UMA-CSIC), Estación Experimental La Mayora, Algarrobo-Costa, Málaga, Spain

Patrick Van Damme
Ghent University, Faculty of Bioscience Engineering, Gent, Belgium
World Agroforestry Centre (ICRAF), GRP1 - Domestication, Nairobi, Kenya

Willman Garcia
PROINPA, Oficina Regional Valle Norte, Cochabamba, Bolivia

César Tapia
Instituto Nacional Autónomo de Investigaciones Agropecuarias (INIAP) Panamericana sur km1, Quito, Ecuador

José Romero
Naturaleza y Cultura Internacional (NCI), Loja, Ecuador

Manuel Sigueñas
Instituto Nacional de Innovación Agrícola (INIA), La Molina, Lima, Peru

JoséI. Hormaza
Instituto de Hortofruticultura Subtropical y Mediterránea, (IHSM-UMA-CSIC), Estación Experimental La Mayora, Algarrobo-Costa, Málaga, Spain

Jirong Wu, Mingzheng Yu, Jianhong Xu, Juan Du, Fang Ji, Fei Dong and Jianrong Shi
Institute of Food Safety and Detection, Jiangsu Academy of Agricultural Sciences, Nanjing, China
Key Lab of Food Quality and Safety of Jiangsu Province—State Key Laboratory Breeding Base, Nanjing, China
Jiangsu Center for GMO evaluation and detection, Nanjing, China

Xinhai Li
Institute of Crop Sciences, Chinese Academy of Agricultural Sciences, Beijing, China
Ron Ophir, Amir Sherman, Mor Rubinstein, Ravit Eshed and Michal Sharabi Schwager
Department of Fruit Tree Sciences, Agricultural Research Organization, Volcani Center, Bet Dagan, Israel

Rotem Harel-Beja, Irit Bar-Yáakov and Doron Holland
Department of Fruit Tree Sciences, Agricultural Research Organization, Newe Yáar Center, Ramat Yishai, Israel

Xiaobai Li
State Key Lab of Rice Biology, International Atomic Energy Agency Collaborating Center, Zhejiang University, Hangzhou, People's Republic of China
Institue of Horticulture, Zhejiang Academy of Agricultural Sciences, Hangzhou, People's Republic of China

Wengui Yan, Aaron Jackson and Anna McClung
Agricultural Research Service, United States Department of Agriculture, Dale Bumpers National Rice Research Center, Stuttgart, Arkansas, United States of America

Hesham Agrama and Karen Moldenhauer
University of Arkansas, Rice Research and Extension Center, Stuttgart, Arkansas, United States of America

Limeng Jia
Agricultural Research Service, United States Department of Agriculture, Dale Bumpers National Rice Research Center, Stuttgart, Arkansas, United States of America
Agricultural Research Service, United States Department of Agriculture, Dale Bumpers National Rice Research Center, Stuttgart, Arkansas, United States of America
University of Arkansas, Rice Research and Extension Center, Stuttgart, Arkansas, United States of America

Kathleen Yeater
Agricultural Research Service, United States Department of Agriculture, Southern Plains Area, College Station, Texas, United States of America

Dianxing Wu
State Key Lab of Rice Biology, International Atomic Energy Agency Collaborating Center, Zhejiang University, Hangzhou, People's Republic of China

Eva Konečná and Pavel Hanáček
Department of Plant Biology, Mendel University in Brno, Brno, Czech Republic
CEITEC MENDELU, Mendel University in Brno, Brno, Czech Republic

Dana Šafářová and Milan Navrátil
Department of Cell Biology and Genetics, Palacky University in Olomouc, Olomouc, Czech Republic

Clarice Coyne
Western Regional Plant Introduction Station - USDA, Pullman, Washington, United States of America

Andrew Flavell
Division of Plant Sciences, University of Dundee at James Hutton Institute, Invergowrie, United Kingdom

Margarita Vishnyakova
Vavilov Institute of Plant Industries, SaintPetersburg, Russian Federation

Mike Ambrose
John Innes Centre, Norwich, United Kingdom

Robert Redden
Australian Grains Genebank, Horsham, Victoria, Australia

Petr Smýkal
Department of Botany, Palacky University in Olomouc, Olomouc, Czech Republic

Stephanie M. Craig, Lauren E. Resnick and Ana L. Caicedo
Biology Department, University of Massachusetts, Amherst, Massachusetts, United States of America

Michael Reagon
Department of Evolution, Ecology, and Organismal Biology, Ohio State University, Lima, Ohio, United States of America

Erica N. C. Renaud and Edith T. Lammerts van Bueren
Wageningen UR Plant Breeding, Plant Sciences Group, Wageningen University, Wageningen, The Netherlands

James R. Myers
Department of Horticulture, Oregon State University, Corvallis, Oregon, United States of America

Maria João Paulo and Fred A. van Eeuwijk
Biometris, Plant Sciences Group, Wageningen University, Wageningen, The Netherlands

Ning Zhu and John A. Juvik
Department of Crop Sciences, University of Illinois, Urbana, Illinois, United States of America

Qiusheng Kong, Jingxian Yuan, Lingyun Gao, Shuang Zhao, Wei Jiang, Yuan Huang and Zhilong Bie
Key Laboratory of Horticultural Plant Biology, Ministry of Education/College of Horticulture and Forestry, Huazhong Agricultural University, Wuhan, China

Anhui Huang and Xiaodong Cai
Department of Electrical and Computer Engineering, University of Miami, Coral Gables, Florida, United State of America

Shizhong Xu
Department of Botany and Plant Sciences, University of California Riverside, Riverside, California, United State of America

Gopala Krishnan S.
Southern Cross Plant Science, Southern Cross University, Lismore, New South Wales, Australia
Division of Genetics, Indian Agricultural Research Institute, New Delhi, India

Daniel L. E. Waters
Southern Cross Plant Science, Southern Cross University, Lismore, New South Wales, Australia

Robert J. Henry
Queensland Alliance for Agriculture and Food Innovation, The University of Queensland, Brisbane, Queensland, Australia

Gajanan Tryambak Behere
Department of Genetics, Bio21 Molecular Science and Biotechnology Institute, The University of Melbourne, Parkville, Melbourne, Victoria, Australia

Division of Entomology, Indian Council of Agricultural Research, Research Complex for North Eastern Hill Region, Shilong, Meghalaya, India

Wee Tek Tay
CSIRO Ecosystem Sciences, Canberra, Australian Capital Territory, Australia

Derek Alan Russell
Department of Agriculture and Food Systems, The University of Melbourne, Parkville, Melbourne, Victoria, Australia

Keshav Raj Kranthi
Central Institute for Cotton Research, Nagpur, Maharashtra, India

Philip Batterham
Department of Genetics, Bio21 Molecular Science and Biotechnology Institute, The University of Melbourne, Parkville, Melbourne, Victoria, Australia

Frances M. Shapter, Graham J. King, Michael Cross Gary Ablett and Sylvia Malory
Southern Cross Plant Science, Southern Cross University, Lismore, New South Wales, Australia

Ian H. Chivers
Southern Cross Plant Science, Southern Cross University, Lismore, New South Wales, Australia
Native Seeds Pty Ltd, Sandringham, Victoria, Australia

Robert J. Henry
Queensland Alliance for Agriculture and Food Innovation, University of Queensland, Brisbane, Queensland, Australia

Hao Cheng, Jiao Wang, Shanshan Chu, Hong-Lang Yan and Deyue Yu
National Center for Soybean Improvement, National Key Laboratory of Crop Genetics and Germplasm Enhancement, Nanjing Agricultural University, Nanjing, China

Vanesse Labeyrie, Monique Deu, Caroline Calatayud, Marylène Buiron, Jean-Christophe Glaszmann and Christian Leclerc
UMR AGAP, CIRAD, Montpellier, France

Adeline Barnaud
UMR AGAP, CIRAD, Montpellier, France

Peterson Wambugu
National Genebank of Kenya, KARI, Nairobi, Kenya

Stéphanie Manel
UMR LPED, UniversitéAix- Marseille/IRD, Marseille, France
UMR AMAP, CIRAD, Montpellier, France

Claire Billot, Sophie Bouchet, Jacques Chantereau, Monique Deu, Laetitia Gardes, Jean-Louis Noyer, Jean-François Rami, Ronan Rivallan and Jean-Christophe Glaszmann
Cirad, UMR AGAP, Montpellier, France

Punna Ramu, Rolf T. Folkertsma, Hari D. Upadhyaya and C. Thomas Hash
International Crops Research Institute for the Semi-Arid Tropics (ICRISAT), Hyderabad, Andhra Pradesh, India

Yu Li, Ping Lu and Tianyu Wang
Institute of Crop Science, Chinese Academy of Agricultural Sciences (CAAS), Beijing, China

Elizabeth Arnaud
Bioversity International, Montpellier, France

Bruno Condori
Liaison Office in Bolivia, International Potato Center, La Paz, La Paz, Bolivia

Robert J. Hijmans
Department of Environmental Science and Policy, University of California, Davis, Davis, California, United States of America

Jean Francois Ledent
Unité d'Ecophysiologie et d'Amélioration Végé tale, Université Catholique de Louvain, Louvain-la-Neuve, Brabant, Belgium

Roberto Quiroz
Integrated Crops and Systems Research Program, International Potato Center, La Molina, Lima, Peru

Jennifer A. Moore
Biology Department, Grand Valley State University, Allendale, Michigan, United States of America
Department of Fisheries and Wildlife, Michigan State University, East Lansing, Michigan, United States of America

Hope M. Draheim
Department of Zoology, Michigan State University, East Lansing, Michigan, United States of America

Dwayne Etter
Wildlife Division, Michigan Department of Natural Resources, East Lansing, Michigan, United States of America

Scott Winterstein and Kim T. Scribner
Department of Fisheries and Wildlife, Michigan State University, East Lansing, Michigan, United States of America

Enying Zhang
Key Laboratory of Crop Genetics and Physiology of Jiangsu Province, Key Laboratory of Plant Functional Genomics of the Ministry of Education, College of Agriculture, Yangzhou University, Yangzhou, China
College of Agronomy and Plant Protection, Qingdao Agricultural University, Qingdao, China

Zefeng Yang, Yifan Wang, Yunyun Hu and Chenwu Xu
Key Laboratory of Crop Genetics and Physiology of Jiangsu Province, Key Laboratory of Plant Functional Genomics of the Ministry of Education, College of Agriculture, Yangzhou University, Yangzhou, China

Xiyun Song
College of Agronomy and Plant Protection, Qingdao Agricultural University, Qingdao, China

Eva Herzog and Matthias Frisch
Institute of Agronomy and Plant Breeding II, Justus Liebig University, Giessen, Germany

Karen Christin Falke
Institute for Evolution and Biodiversity, University of Münster, Mü nster, Germany

Thomas Presterl, Daniela Scheuermann and Milena Ouzunova
KWS Saat AG, Einbeck, Germany

Index

A

A-diversity Parameters, 40

Abiotic Stressors, 212

Allele Distributions, 95, 97, 99, 101, 103, 105

Association Mapping, 15, 74-75, 77-83, 148, 211

Austral Summer Frosts, 212

B

Beneficial Arthropod, 1, 3, 5, 7

Breeding Programs, 8, 10, 27, 38, 107, 133, 170, 197, 206, 209, 211, 213, 237, 245-246

Broccoli Cultivar, 106-107, 109, 111-113, 115, 117, 119, 121

C

Cajanus Spp., 8-9, 11, 13, 15, 36

Cereal Production, 164

Cherimoya (annona Cherimola Mill.), 25, 37-39, 41, 43, 45, 47, 49-50

Cicer Species, 27, 29, 35-36

Climate Threats, 17

Complex Trait, 74-75, 77, 79, 81, 83, 139

Crop Diversity, 50, 182-183, 190, 193, 195, 223, 225, 227

Crop Domestication, 15, 26, 36, 49, 82, 165, 171-172, 196

Crop Evolution, 16, 50, 93, 182, 193

Cucurbitaceae Crops, 122

Cultivated Soybean, 173-175, 177, 179-181

D

Demographic History, 8-9, 11, 13, 15-16, 36, 102

Dispersal Pattern, 1, 3, 5, 7

Diversifying Selection, 173-175, 177, 179-181, 247

Domestication Process, 27, 29, 31, 33, 35, 164, 169, 173, 178

E

Eblasso Model, 134, 140

Enzyme Activity, 51-52, 54, 58-61, 71

Epic-pcr Dna Markers, 150

Ethnolinguistic Diversity, 182-183, 185, 187, 189, 191, 193-195

Eukaryotic Translation, 84, 93-94

F

Flavanone 3-hydroxylase, 173, 175, 177, 179, 181

Food Security, 17-19, 25, 50, 85, 144, 212, 221, 225, 227

Frost Risk, 212-213, 215, 217, 219, 221-222

G

Gene Expression Normalization, 122-123, 125, 127, 129, 131-132

Genetic Diversity, 2-4, 7-9, 11-28, 31-34, 36-39, 41-45, 47, 49-50, 60-62, 69, 72-75, 79, 83-84, 86, 91-93, 96, 101-102, 107, 144, 148, 151, 153, 155, 164, 173, 181-183, 185, 188- 190, 192-195, 197-201, 206, 209-211, 213-214, 217, 220-222, 227, 229, 233-235

Genetic Interrelations, 62, 71

Genetic Structure, 2-4, 6, 8, 18, 20-21, 25, 46, 50, 62, 67- 72, 75, 79, 83, 150-151, 153, 155, 157, 159, 161-163, 175, 180, 182, 185-188, 192-195, 211

Genetic Variation, 8-9, 17-21, 24, 28, 34-35, 38, 42-43, 45, 63, 68, 75, 114, 116, 148, 151, 156, 162-163, 174, 197, 235, 242, 247

Genome-wide Snps, 8-9, 11, 13, 15, 36

Geographical Gradient, 84-85, 87, 89, 91, 93

Germplasm Collections, 10, 27, 35, 49, 72, 85, 93, 196-197

Germplasm Diversity, 8, 15, 27, 29, 31, 33, 35, 67, 84-85, 91, 210

Germplasm Genetic Diversity, 62

Glucobrassicin, 106-119

H

Haplotype Diversity, 91, 93, 174-175, 180, 229, 231-234

Harvest Index, 74-83, 214-215, 217, 222

Helicoverpa Armigera, 150-151, 153, 155-159, 161-163

Hierarchical Clustering, 64, 72

High-throughput Sequencing, 63, 164-165, 167, 169, 171- 172, 211

Hybrid Incompatibility Loci, 95, 97, 99, 101-103, 105

Hybrid Sterility, 95-96, 100, 102-105

I

Introgression Libraries, 235-236, 247-248

Isoflavone Synthase Genes, 173, 175, 177, 179, 181

L

Land Equivalent Ratio (ler), 223

Land Management Strategies, 1, 3-5, 7

Lineage Sorting, 27

Lob Domain Transcription Factor, 229

M

Maize Rtcs Gene, 229

Microlaena Stipoides, 164-165, 167-169, 171

Mutagenesis, 164-165, 167, 169-171

N

Neoglucobrassicin, 106-107, 109-119

Niche Modelling, 17-21, 23-25

Nucleotide Polymorphisms, 17, 25, 62-63, 72, 144-146, 148, 173-174, 177, 229, 231, 233-234

Null Allele Effect, 150

O

Organic Agriculture, 106, 118-119, 121

Oryza Rufipogon, 82-83, 105, 144, 148, 247

P

Pea (pisum) Germplasm, 84-85, 87, 89, 91, 93

Pea Seed Borne Mosaic Virus (psbmv), 85

Phytochemical Content, 106-107, 109, 111-115, 117-121

Plant Genetic Resources, 25, 36-38, 45-47, 49-50, 93, 183, 195, 211, 229

Polymorphism Deserts, 144-145, 147-149

Polymyxa Graminis, 51-52

Polyphagous Nature, 150

Polyphagus Arthropod, 1

Polyploid Grass, 164

Pomegranate Transcriptome, 62-65, 67, 69-71, 73

Potato Biodiversity, 212-213, 215, 217, 219, 221

Potyviruses, 84-85, 87, 89-91, 93

Pre-domestication Origin, 144-145, 147, 149

Prezygotic Factors, 95

Principal Components Analysis (pca), 56, 74-75, 81

Q

Qrt-pcr Analysis, 122-123, 125, 127, 129, 131

Quantitative Trait Loci (qtl), 8, 95

Quantitative Trait Locus (qtl), 133

R

Range Expansion, 17, 19-21, 23-25

Reference Genes, 122-132, 167, 173-175, 178-181

Rhizosphere Soil, 51-53, 55-56, 58-61

Rice Genome, 81-82, 100, 143-149, 171

S

Single Nucleotide Polymorphism (snps), 8

Sorghum Collection, 73, 83, 195-197, 199, 201, 203, 205, 207, 209, 211

Sorghum Genetic Patterns, 182-183, 185, 187, 189, 191, 193, 195

Spatial Analysis, 17, 25-26, 37-39, 41, 43, 45, 47-50

Subsistence Farming Systems, 182-183, 185, 187, 189, 191, 193, 195

Subtropical Tree Species, 37

T

Transgenic Wheat, 51-53, 55-61

Tropical Japonica (trj), 75

W

Watermelon (citrullus Lanatus), 122, 132

Weedy Rice, 95-96, 99, 104-105

Wheat Yellow Mosaic Virus (wymv), 51

Whole-genome Markers, 133

Wild Barley, 17-26, 170, 172, 247

Wild Barley Diversity Collection (wbdc), 18

Wild Chickpea Taxa, 27

Wild Progenitors, 173

Y

Yield Increase, 133, 223, 225, 227